I0823028

Oliver Lenzen

SAND UP CLOSE

As objects are lifted out of nothingness only by a human point of view, so when the points of view vanish, the objects, too, return to nothingness.

—*J. W. v. Goethe,* ***Maxims and Reflections***

My parents, my family

SAND UP CLOSE

Aesthetics, Philosophy, and Science

Oliver Lenzen

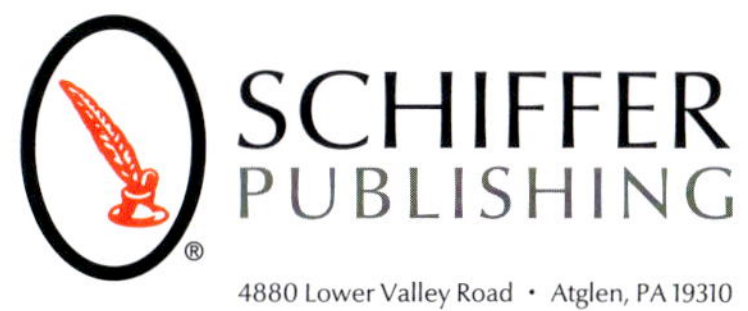

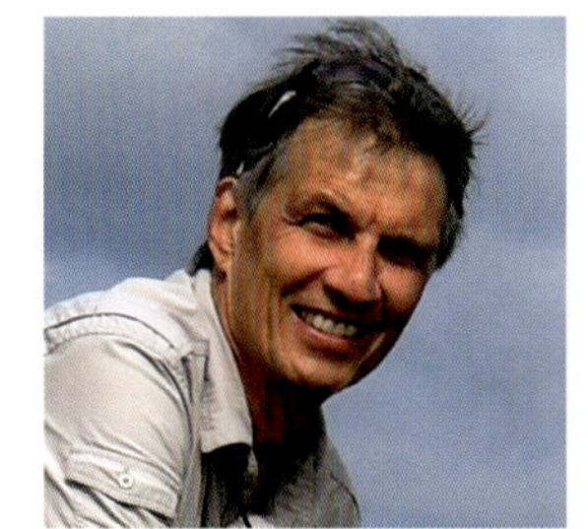

Oliver Lenzen, professor and doctor of engineering, was born in Berlin in 1960. Since his childhood, he has been interested in the natural sciences, especially geology, mineralogy, and astronomy, and later microscopy and the relationships among the natural sciences, philosophy, and technology. He also has a special interest in Goethe as a natural scientist. After receiving his doctorate and working in industry for many years, he was appointed professor at the University of Heilbronn in 2005, which he has led as rector since 2017.

Translated from the German by Rachel Joerges. Originally published as *Das große Buch vom Sand*, Oliver Lenzen, © 2022, Haupt, Bern, Switzerland
Designed by pooldesign.ch, CH-Zurich

English-language edition published by

Library of Congress Control Number: 2024941765

Type set in: Roboto
Cover design by: Molly Shields
ISBN: 978-0-7643-6914-8
ePub: 978-1-5073-0539-3
Printed in China
10 9 8 7 6 5 4 3 2 1

Published by Schiffer Publishing, Ltd.
4880 Lower Valley Road
Atglen, PA 19310
Phone: (610) 593-1777; Fax: (610) 593-2002
Email: info@schifferbooks.com
Web: *www.schifferbooks.com*

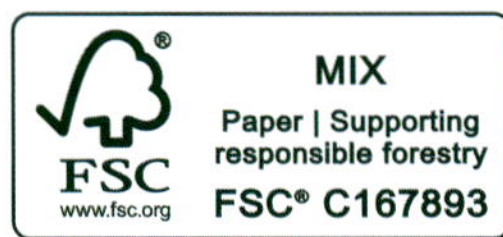

Note: This book uses the metric unit system, including metric tons, in scientific contexts.

Table of Contents

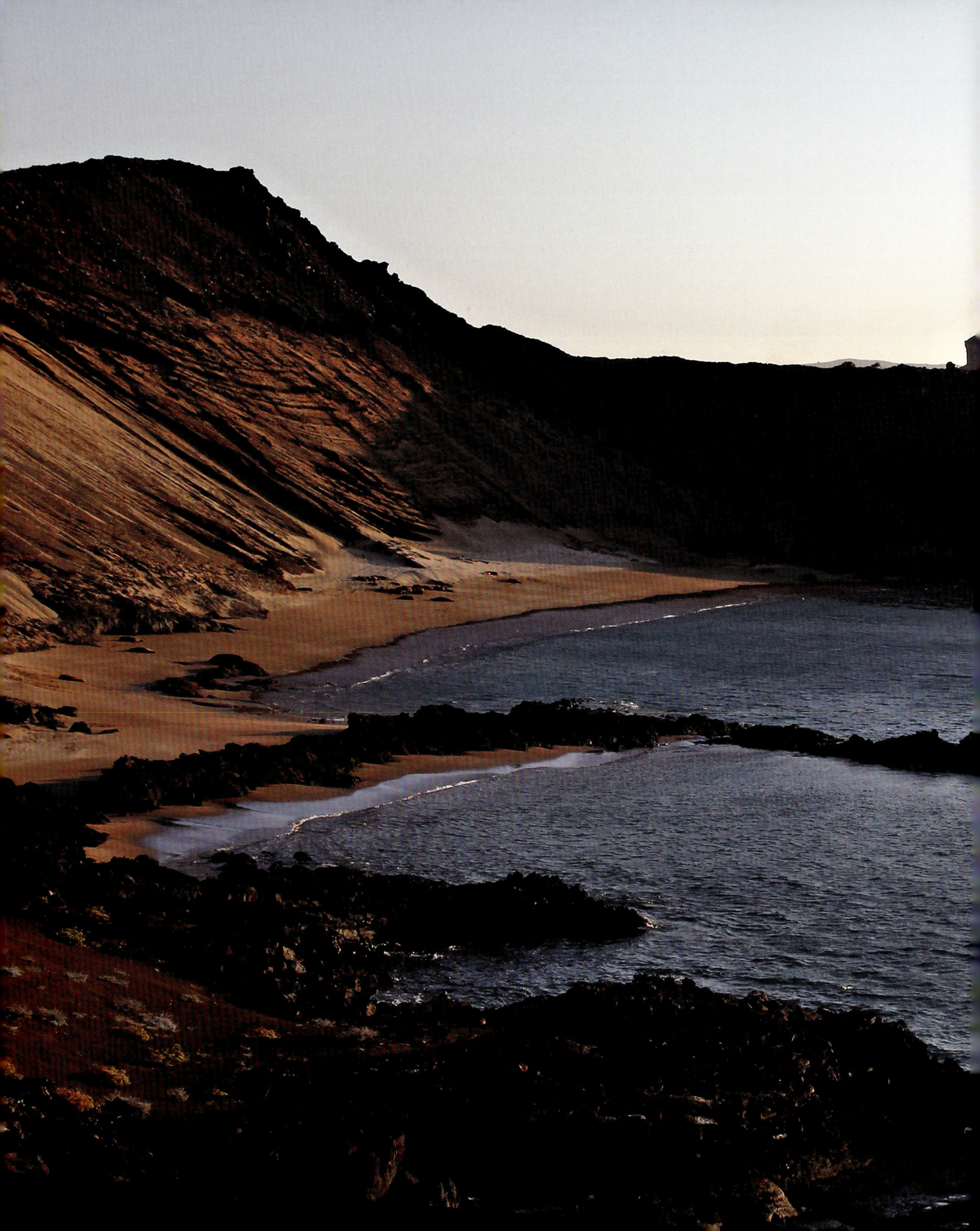

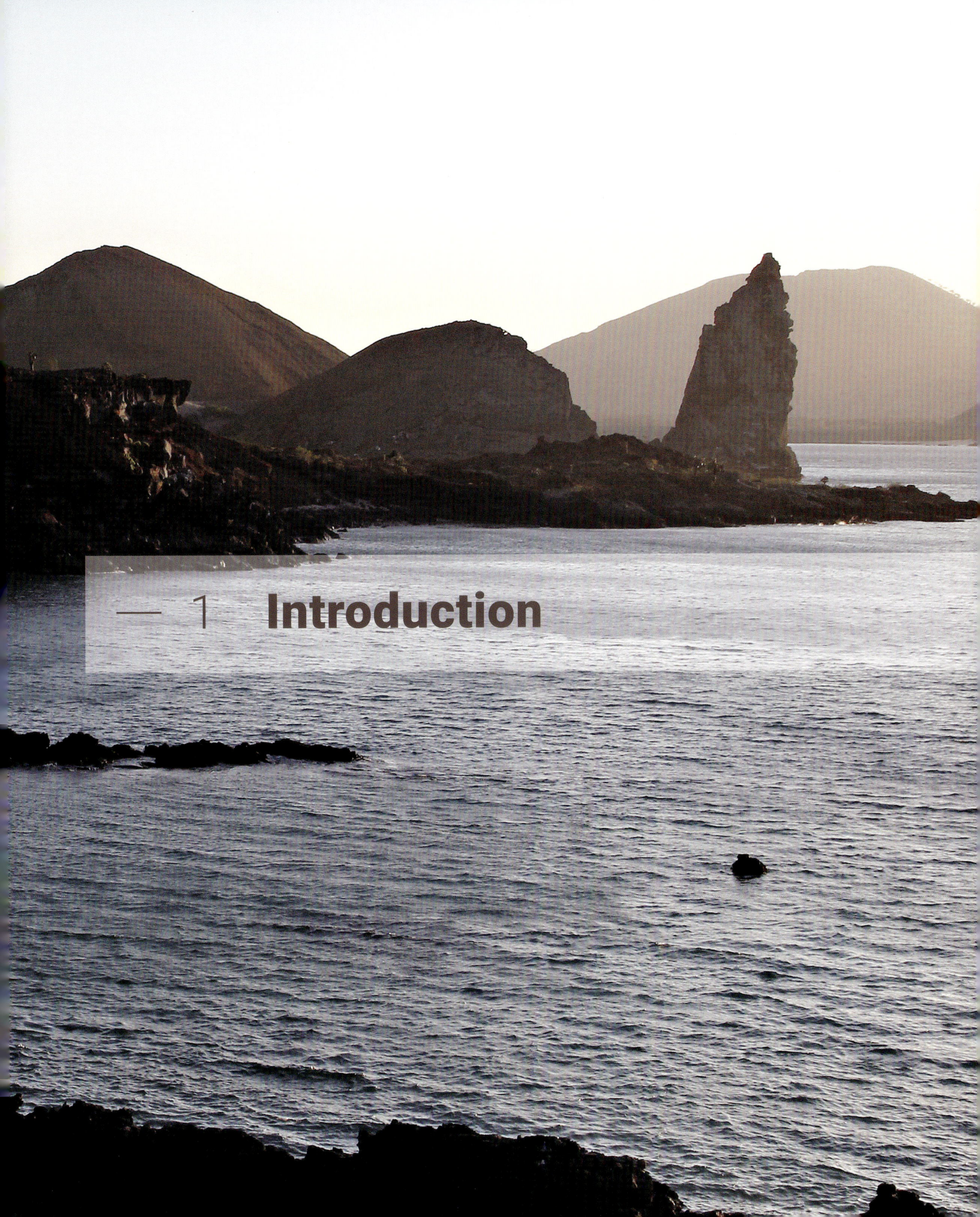

— 1 Introduction

Philosophy begins with wonder.
Wonder over the trivial.

—*C. F. v. Weizsäcker*, ***Zeit und Wissen***

Time was short, and the ship would certainly not wait for me. While my traveling companions were planning to visit the old fortress of Vardøs, I stole away and searched for the quickest way across the country down to the coast. I hoped to find the sand there that I had been interested in for quite a while and was still missing from my small collection. Only small, because I took and cataloged only samples that I either collected myself or that friends and acquaintances brought me from their own travels. But even in this manner, about seven hundred kinds of sand had come together. It wasn't actually even a collection anymore, more like a compilation, a stock. I wasn't seeking a certain number or completeness—I was looking to present something representative of the diversity of sand. The samples were to be used for direct examination, curiosity about this strange substance: sand.

After a while, I found a small bay with a quite natural shoreline and the beach I had been hoping for. Admittedly, not the kind of beach we know from vacation brochures: water, sand, and sun, accompanied by omnipresent heat and mostly restrained, subdued nature. No one seemed to care about this place, although the traces of civilization were clearly visible here. The bay opened to the west to a sea passage between the mainland, the Varanger Peninsula, and the offshore island where the small town of Vardø (fig. 1.1) is located. The opposite shore was clearly visible, with the sparse and low vegetation of the Nordic tundra covering a flat and treeless plateau reaching to the horizon.

But the interesting aspect was the geology of this lonely and inhospitable area. The Trollfjord-Komagelv fault zone stretches a little south of my location. It runs roughly northwest–southeast and separates two continental plate parts that push past each other laterally. Here, north of the fault, I was on the sediments of the Barents Sea region, of the so-called Lokvikfjell Group, a sedimentary sequence whose deposition began over 850 million years ago. So, I was standing on Neoproterozoic and thus ancient sediments. Over the course of millions of years, plate tectonic processes folded the sediments with unimaginably large forces and tilted the sand layers, which were originally deposited in the deep sea and were several miles thick.

When I reached the bay, I could see the sloping stratigraphic sequences, the large mass of which, of course, have long since eroded and worn away. Scattered everywhere were fist-sized, well-rounded boulders that did not resemble the sloping dark-layered rocks at all. It looked as if someone had dumped them on top (fig. 1.2).

Previous double page:
Isla Bartholome, Galápagos, Ecuador

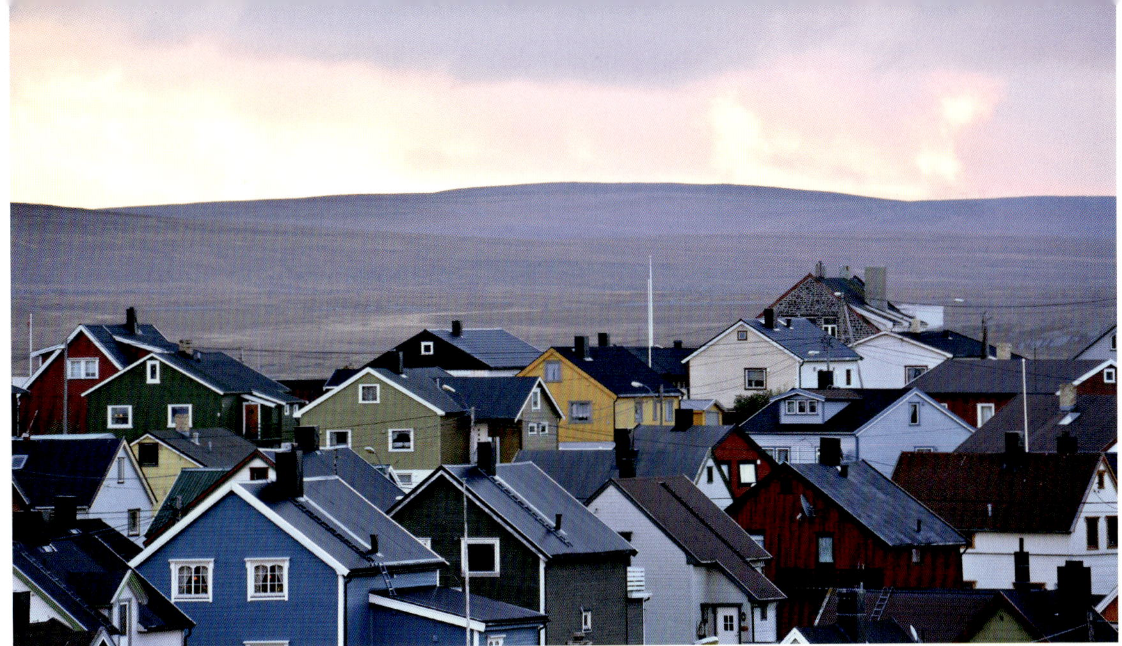

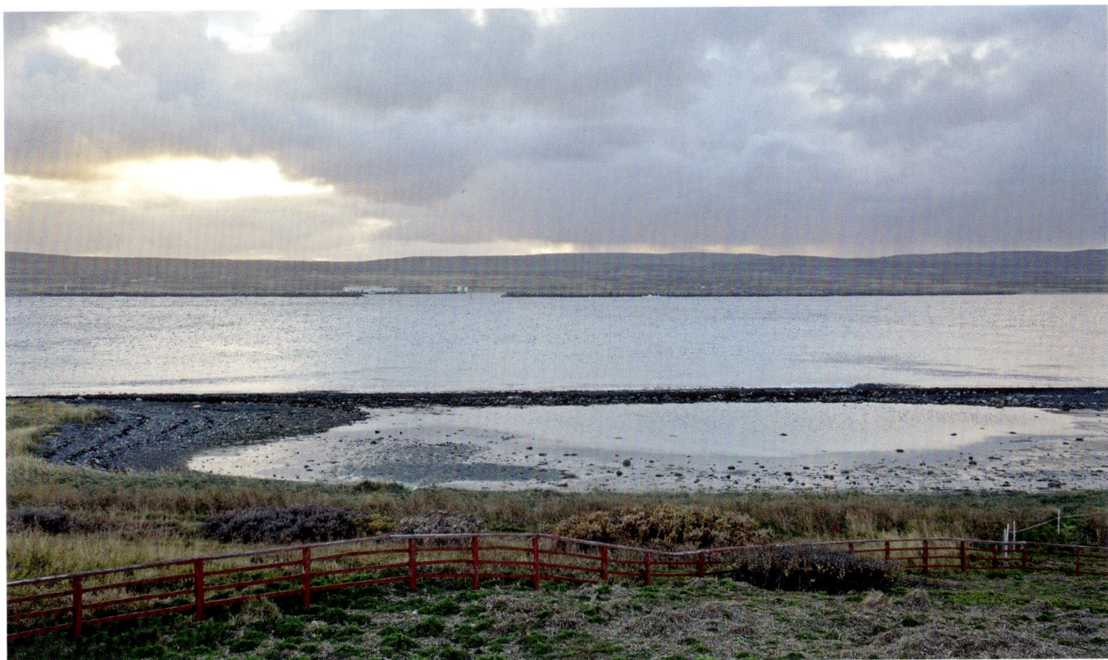

These are debris from one of the great glaciations that once covered these regions. After millions of years of tropical heat, a sudden climate change occurred here about 650 million years ago. As a result of the Varanger glaciation, the Baltic Shield was covered with mighty sheets of ice that helped shape the land in its current form. I didn't have much time left to imagine all that had once happened in this place. So, I quickly packed up a few stones lying around and tried to obtain a small sample of the stratigraphic sequences. Then I went on to the shore. The sand there looked quite ordinary, sand colored, perhaps a little darker.

Figure 1.1 The small town of Vardø against the backdrop of the glacially dominated Nordic plateaus. Below, the small bay at Vardø with a view to the west across the strait to the mainland.

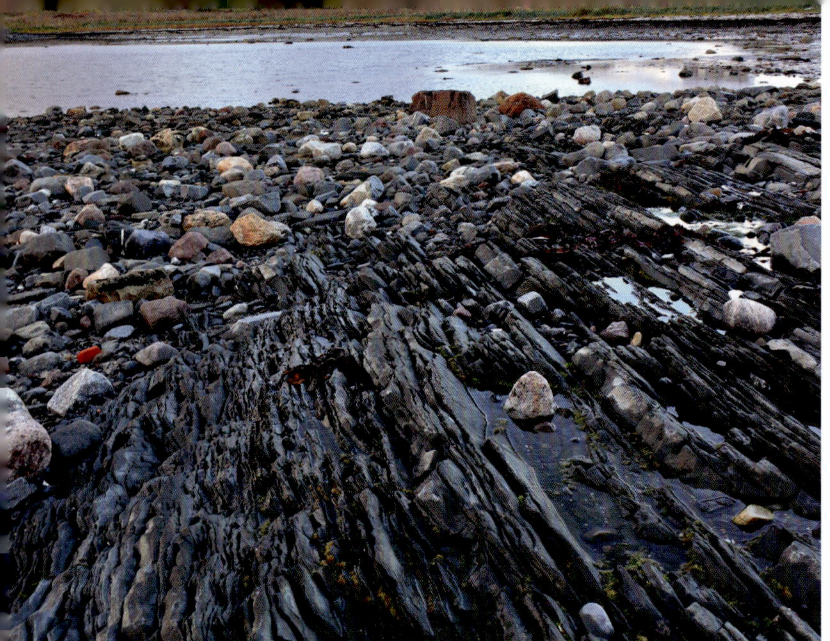

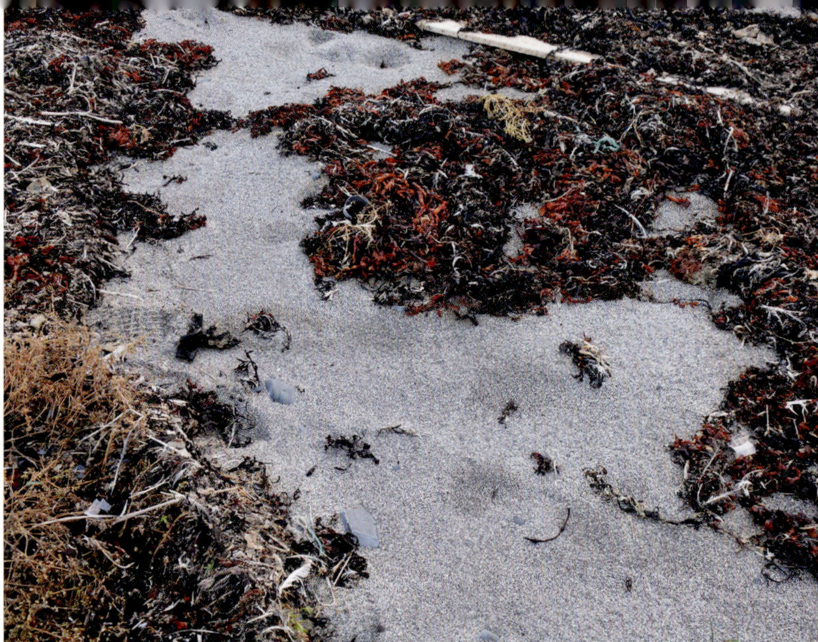

But as so often occurs in nature, a second, closer look is certainly worthwhile. What binoculars are to birdwatchers, magnifying glasses and microscopes are to sand observers. Only these optical aids enable us to immerse ourselves in a world that would otherwise remain closed. With these instruments made of glass, brass, and steel, one can get closer and closer to the sand. While the magnifying glass still allows us to see many thousands of grains, the microscope makes it possible to portray a single grain, to walk on its surface, and, even further, to penetrate its interior.

I quickly packed everything up and said goodbye to this hardly spectacular yet so incredibly interesting place, which would, I hoped, entrust me with its further secrets when I looked through the microscope in the laboratory at home. I reached the ship in time.

The two photographs (fig. 1.4 and fig. 1.5) are typical of those to come. Sand photos that show the composition of the sand moderately enlarged with image widths of 2–8 mm, as well as portrait-like representations of individual sand grains with high enlargements and image widths of around 1 mm. At a minimum for the last group, a microscope is necessary. There are, of course, many options to gain access to a microscope. You can simply buy one. Relatively inexpensive USB microscopes, which cost less than a dinner, now perform amazingly well and are completely sufficient for first explorations. There are hardly any limits upward.

Figure 1.2 The bay, with a view to the north. The sloped Proterozoic sediment layers with boulders on top. Components that make up the sand of the small bay. *Right*: Sand, seaweed, bird feathers, and washed-up debris in the bay. The footprint is from the photographer.

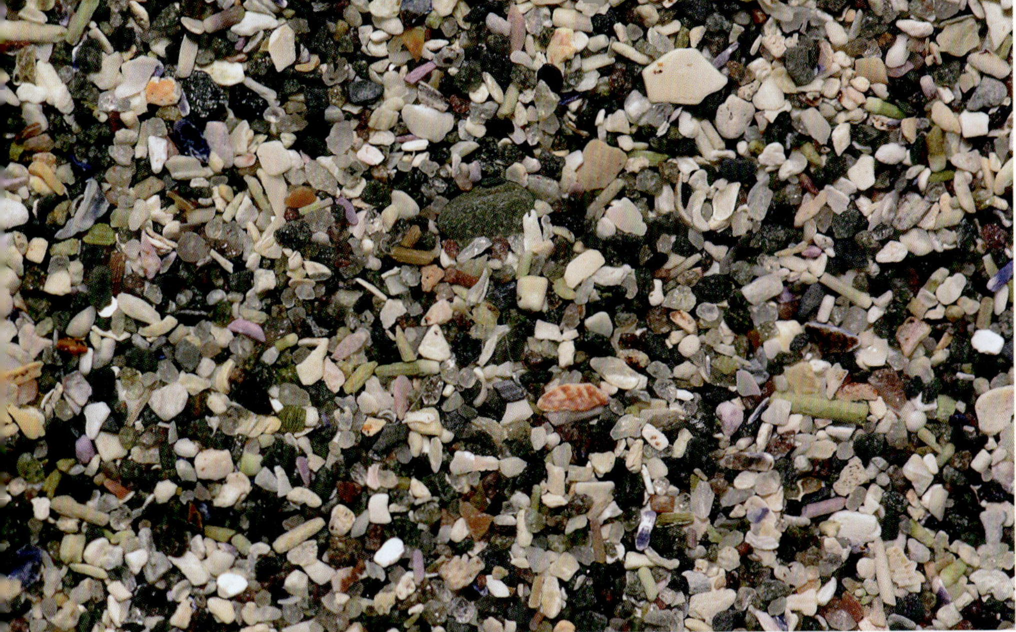

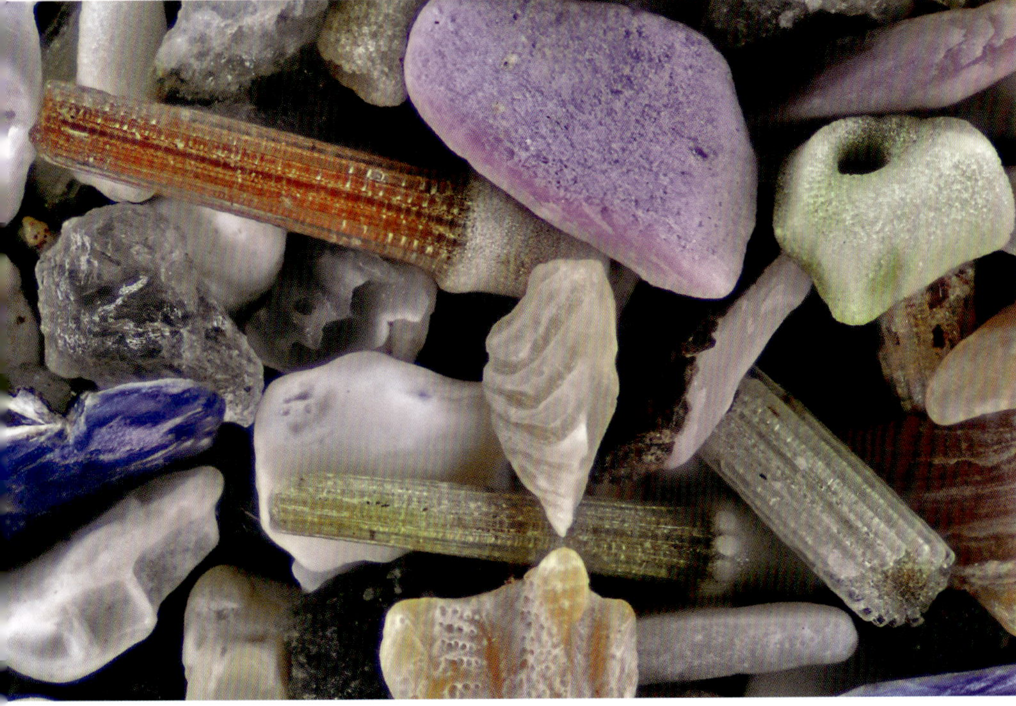

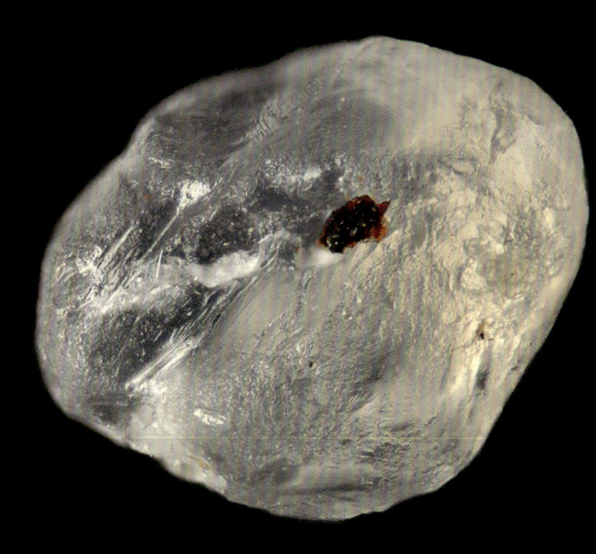

Figure 1.3 The amazingly colorful sand from the small bay near Vardø, seen in a magnifying glass perspective. Image width approximately 30 mm.

Figure 1.4 Colorful fragments of marine life such as mussel shells and sea urchins. *On the left*, a clear quartz grain with only moderate traces of erosion. Microscope image of the sand from Vardø, using reflected light procedure. Image width approximately 2.9 mm.

Figure 1.5 Portrait of a single grain of sand from the Vardø sample. The grain is made of quartz and shows physical and chemical traces of erosion as well as a colorful inclusion. Microscopic image in reflected light. Grain size 750 x 820 μm (1000 μm = 1 mm).

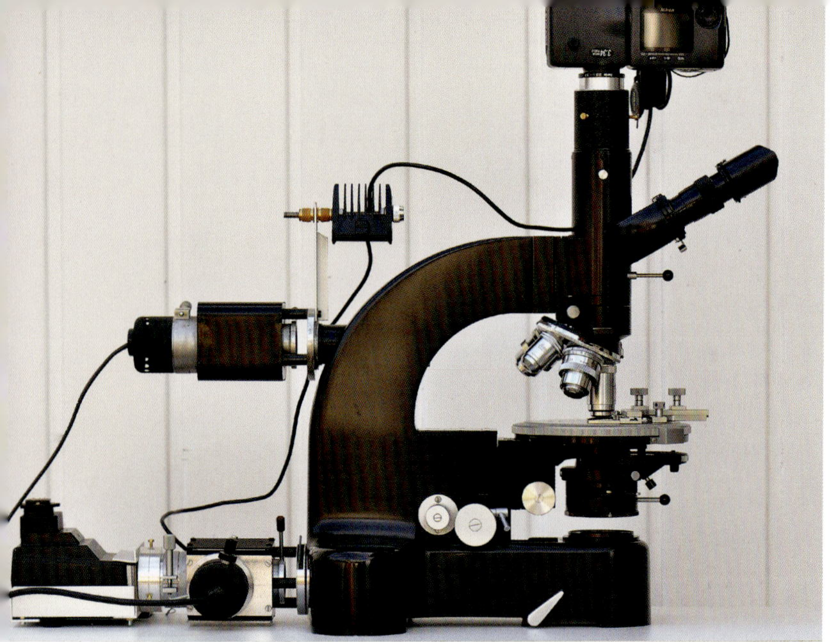

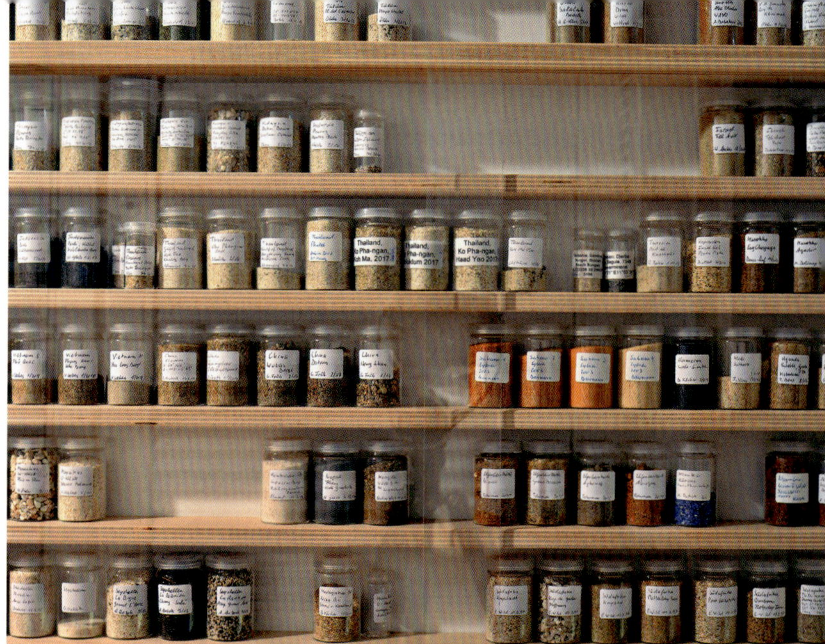

Every book holds a little promise. If the book and the reader are a good match, the reader should be satisfied to some extent regarding his or her expectations after reading it, or, perhaps even better, surprised. What can be expected from a book about sand? Something scientific, cultural, aesthetic, astonishing, or lexical?

A hobby is freeing from compulsory usefulness. However, it also does not free one from being precise, forming links, bringing concepts together, and theorizing. In this respect, this book is a thoroughly personal approach, a collection of topics on a very special natural material. In my opinion, it is almost unavoidable for the sensitive observer to be interested in one way or another in almost everything that the world has to offer. The book reflects this approach. The texts meander in some places between factual topics and associative. They flow, but the reference to the basic theme of "sand" remains. Just as sand seemingly becomes comprehensible out of its elusive totality, it nevertheless flows through one's fingers. It must be grasped and contemplated, and only then does it attain an unexpected beauty, which becomes fully apparent only in detail, in the individual grain. The task is to transform the sand into images and words.

Each of the chapters begins with synoptic sentences, a motto, and usually a quoted text passage. Some of it comes from Goethe's works. Why Goethe, and why in a book about sand? In Goethe's collections there are tin sands from the Fichtelgebirge, which he treasured very much. He probably didn't deal with sand as a general theme, but with natural sciences throughout his life, especially with rock science and mineralogy, both of which form the basis of the formation of sand. Thus, wherever he traveled, his gaze was also directed earthward to take samples and form a detailed image of the natural environment around him. Without being part of the classical scientific environment, he created a very special approach committed to the pure observation of things; he was able to marvel.

Figure 1.6 The Leitz Ortholux 1 microscope used for the pictures in the book, here with transmitted-light equipment. Only the camera has since been replaced with a higher-performance one. *Right*: Part of the collection of about seven hundred sand samples from all around the world.

Why are we still interested in this today, when many of his findings from that time are long outdated? Because, as lovers of the scattered subject area around sand, we may feel very much the same. And because Goethe always lets themes interact: the whole and the detail, nature and culture, spiritual and sensual. The synopsis as a method and the awareness of the possibility of a synopsis is as cognitive as it is instructive, and especially recommended as a supplement to the modern, predominantly and inevitably dissecting way of looking at things in the natural sciences. "We must go slowly and gently to work with nature, if we would get anything out of her," he expressed to Eckermann on October 1, 1828, a few years before his death. In the preface to his *Theory of Colors*, what he writes could also be taken as a motto for this book of sand:

> "Surely the mere inspection of a subject can profit us but little. Every act of seeing leads to consideration, consideration to reflection, reflection to combination, and thus it may be said that in every attentive look on nature we already theorise."

With unagitated matter-of-factness, Goethe brought together the natural sciences and the humanities as well as practical action and theoretical contemplation. His writings are an expression of a brotherliness in dealing with nature.[1] And we can profit from it.

The pictorial representation of sand and individual grains of sand is per se pleasantly free of purpose. The aim here is not to conduct scientific research, but to show connections. Many of the images are intended to be simply beautiful in addition to their informational content. Their subject matter is transient, showing what has become, not what has been made. Where not preserved as a permanent sample, the protagonists, the grains of sand, have long since been submerged again in the nameless mass of their contemporaries and will remain there forever. The traces of their existences are images that give them longevity through the moment of observation. The book will be useful to all those who want to enjoy the subtle beauty and diversity of sand but, at the same time, are eager to learn more about all the amazing backgrounds of this natural material and its composition.

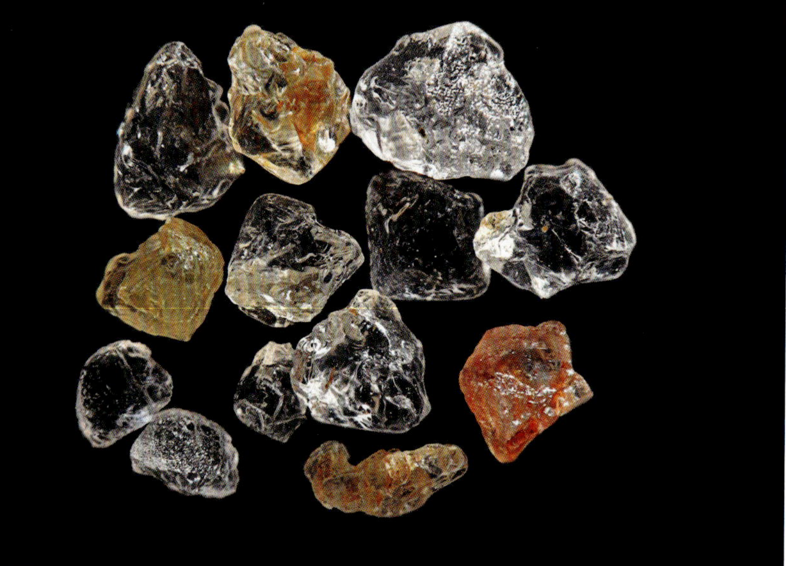

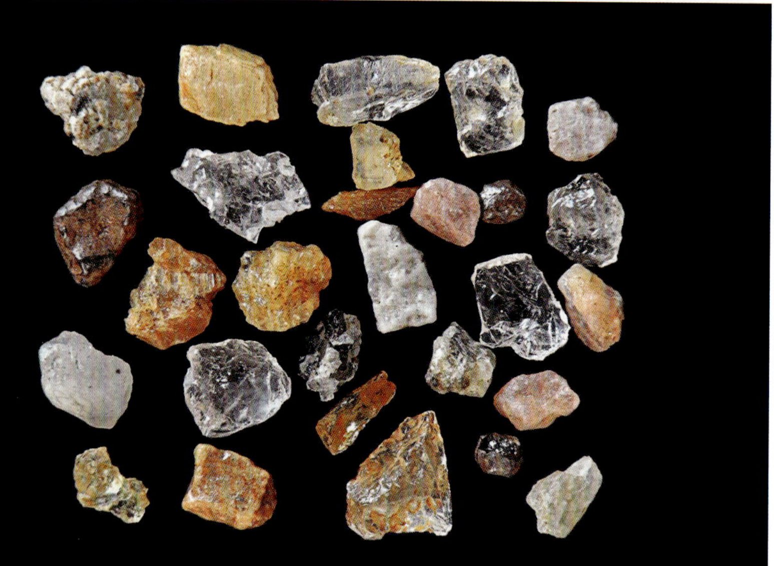

Figure 1.7 Ice-age sand from the largest European inland migratory dune near Klein Schmölen in the UNESCO biosphere reserve of the Elbe River in Mecklenburg–Western Pomerania. During the last ice age, these extensive dune areas formed along the glacial valley of the Elbe River. In the picture, quartz grains from presumably northern European bedrock predominate and are well rounded due to the glaciers, wind, and water. Image width 3.8 mm.

Figure 1.8 Quartz grains from Playa Santa Clara, Panama, Pacific coast. The extraordinarily shiny, almost-polished-looking surface of these grains is striking—they are probably of volcanic origin. Image width approximately 5 mm.

Figure 1.9 Fietjetjörn, Valle, Norway. Quartz grains on black glass. The only slightly rounded grains transported by glacier are colored by marshy, ferrous soils. Grain size each about 0.6–1.0 mm. Reflected light.

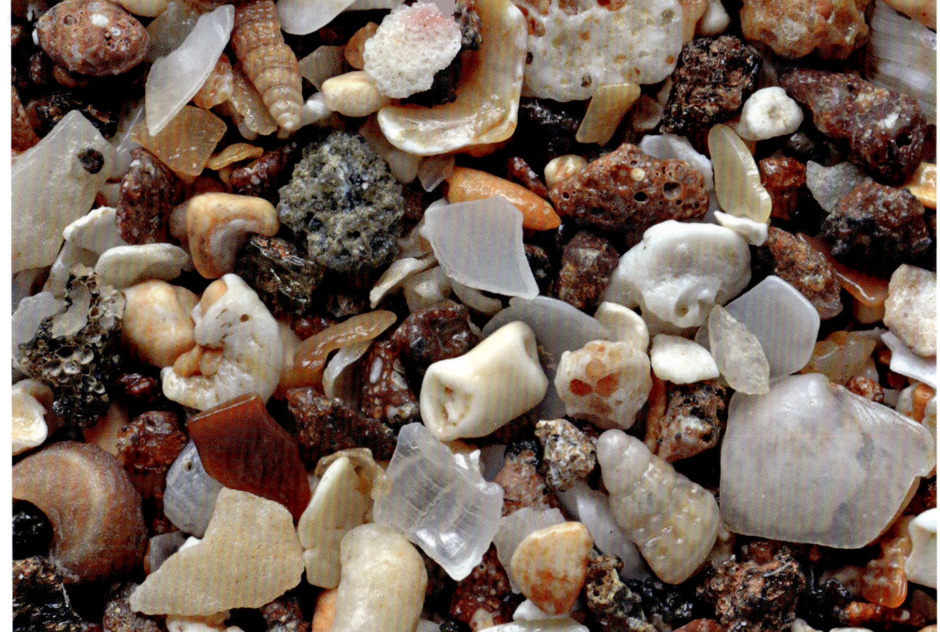

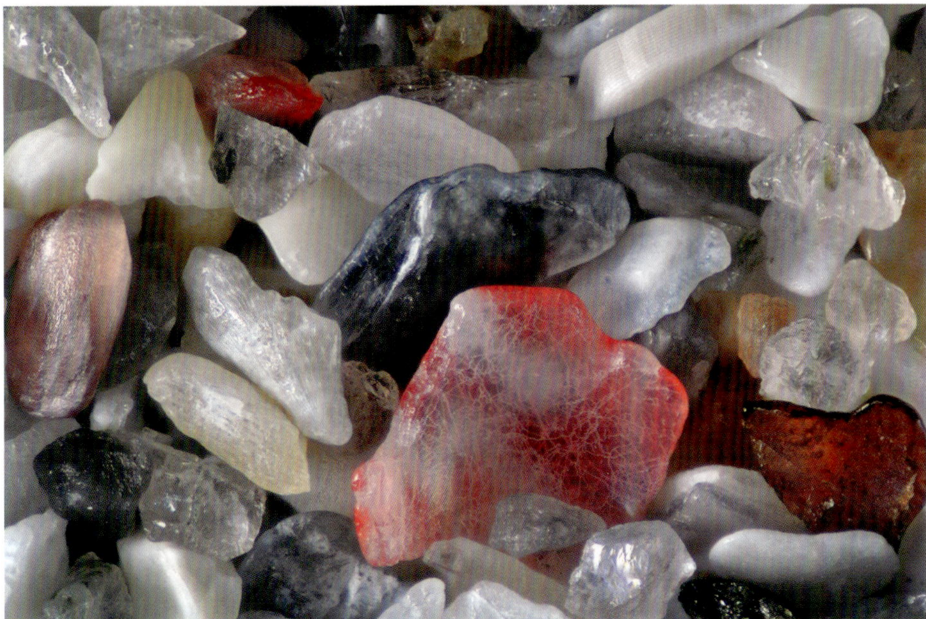

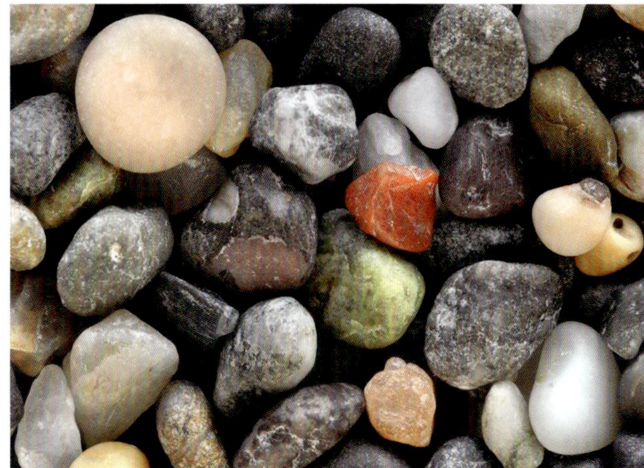

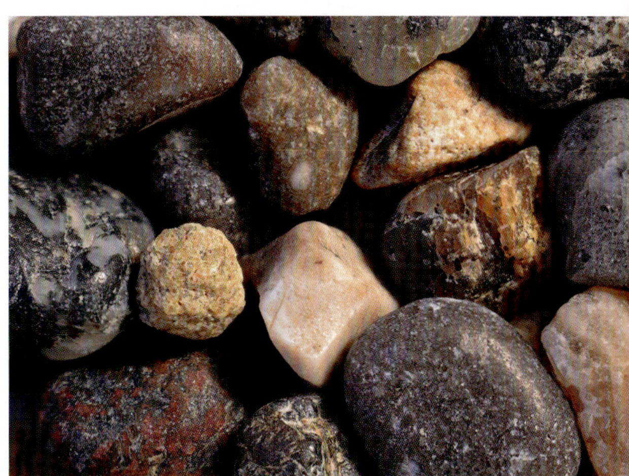

Figure 1.10 Isla Bartholome, Galápagos, Ecuador. A mixture of rounded and polished grains from the sea, composed of shell debris, sea urchins, snail fragments, and sand grains from bubbly lava. Image width 8.7 mm.

Figure 1.11 Sand from the Outer Hebrides, Vatersay Beach, Scotland. *In the middle*, a shell fragment with microboring traces of algae. *To the right*, a brown piece of biotite mica. Quartz-rich, continental sand with many biogenic fragments. Image width 2.2 mm.

Figure 1.12 Sand from the ocean coast at Jerapetra, Crete. Many of the grains of sand are still recognizable rock fragments, therefore polymineralic. Image width 12 mm.

Figure 1.13 Sand from polymineralic grains (rock fragments) from Point Pelee, Lake Erie, Canada. The high image resolution makes the sand seem as if it is a collection of pebbles. Image width 7.7 mm.

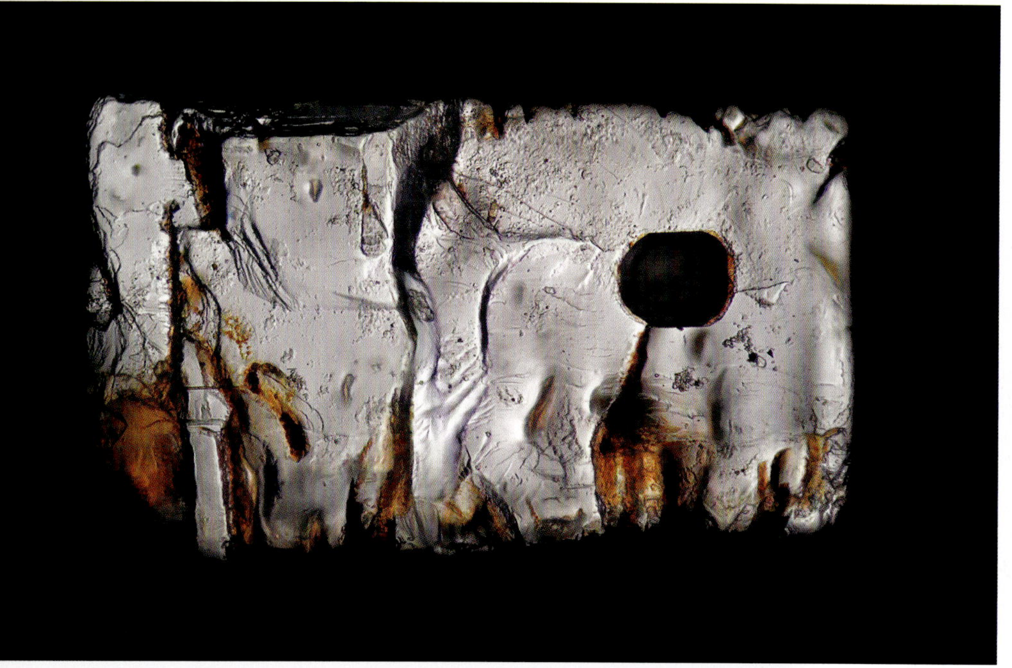

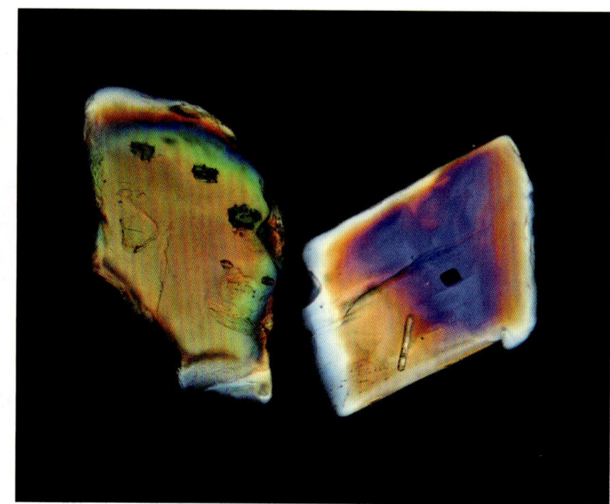

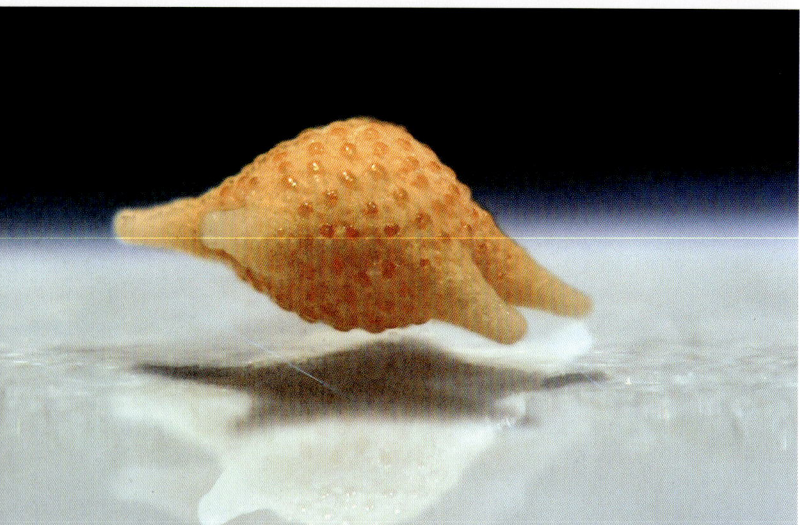

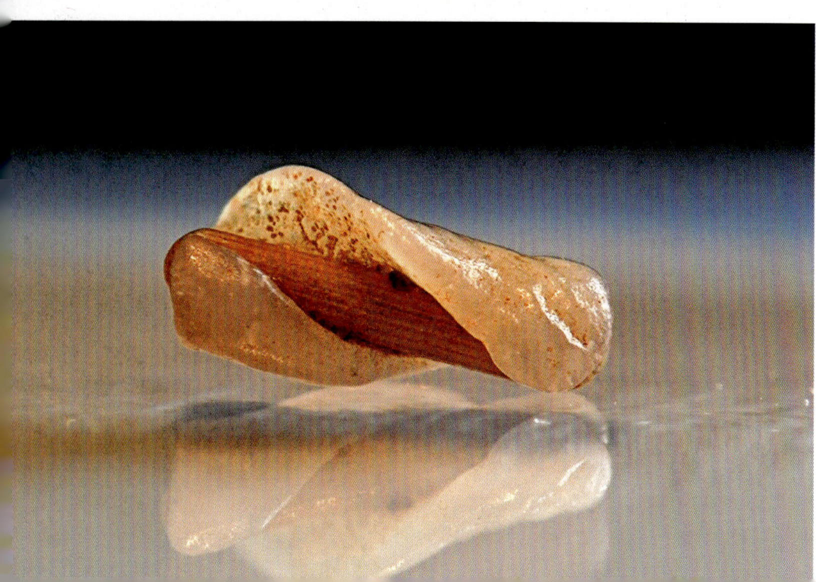

Figure 1.14 *Top left*: Grain of sand from Abad Turtle Beach, Kerala, India. The high resolution reveals all the details of the plastic-looking surface of the grain of sand. Transmitted light using a polarizer. Image width 670 µm.

Figure 1.15 *Top right*: Grains of sand with inclusions from Lovina, Bali, Indonesia. By using crossed polarizers, brightly colored interference colors become visible. Image width 670 µm. Transmitted light.

Figure 1.16 *Middle*: Michelmas Cay, Australia. Foraminifer *Baculogypsina sphaerulata*. Diameter without extensions around 1,000 µm. The grain of sand is sitting on a slide made of glass, which is why it seems to be floating. Reflected light.

Figure 1.17 *Below*: Isla Bartholome, Galápagos, Ecuador. A sea urchin spine apparently got lost in a snail spindle. The grain of sand is sitting on a slide made of glass. Grain length 1.9 mm. Reflected light.

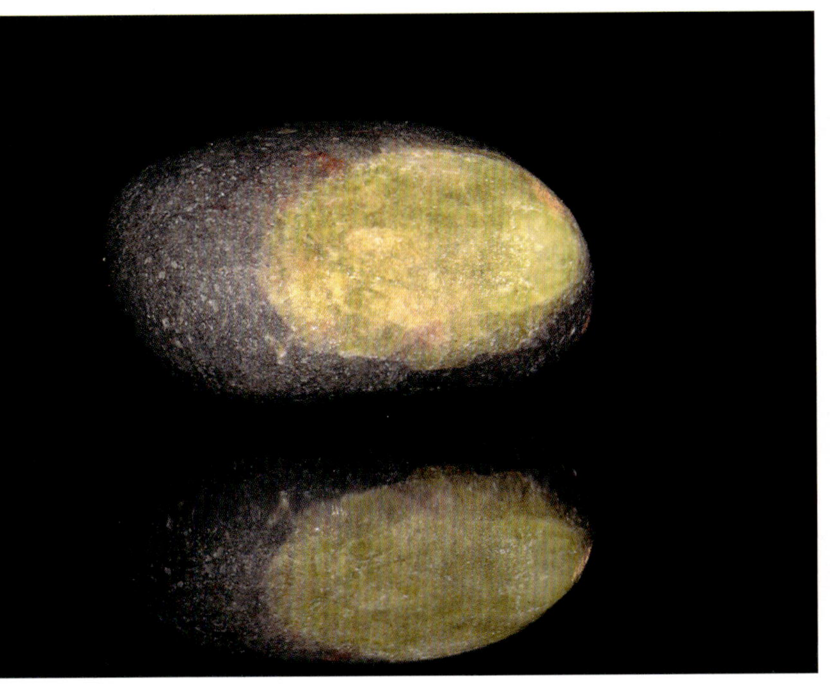

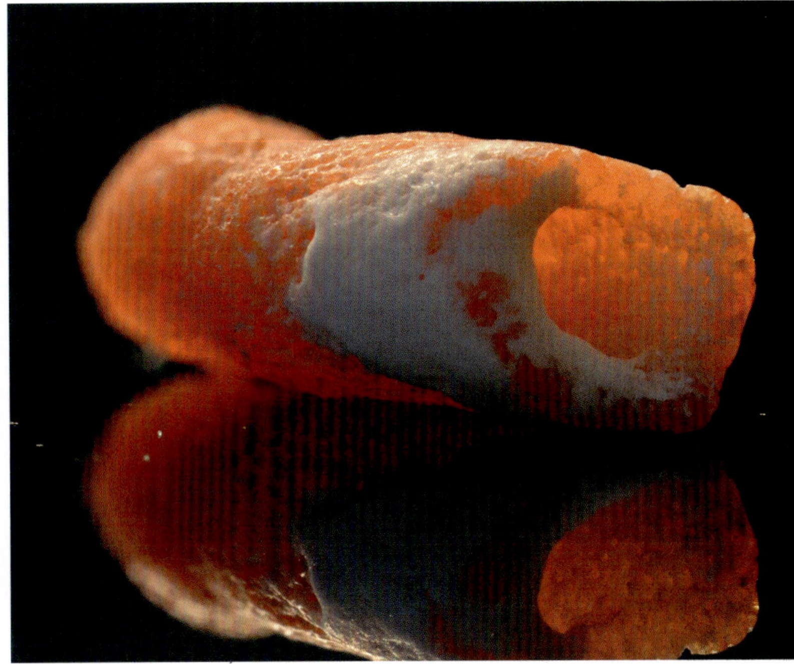

Figure 1.18 Jerapetra, Crete, Greece. Grain from basaltic rock with an olivine inclusion. Reflected light, grain on black glass. Grain size 2.6 mm, and therefore by definition categorized as fine gravel.

Figure 1.19 Michelmas Cay, Australia. Grain of sand, composed of a small coral branch. Grain on black glass lit by using reflected light from behind. Diameter around 1.6 mm.

Figure 1.20. Grain of sand from the central Norwegian highlands. The grain is quite fresh and shows hardly any weathering and only minor signs of abrasive wear on the protruding edges. The sand grain consists of several minerals (white quartz and black hornblende, as well as the golden-weathering mica muscovite) and thus represents a rock fragment. The enlarged image and the rather-high resolution hide the fact that the grain has a width of only 900 µm. The image was taken under the microscope and is composed of eighty single images, with the help of a panorama stack. The further you go with this effort, the more detailed the result will be. However, a high resolution is not proportional to the aesthetic quality of a photo, which depends on other, quite subjective, characteristics such as depth effect, color scheme, shape, lighting, blur, and the like.

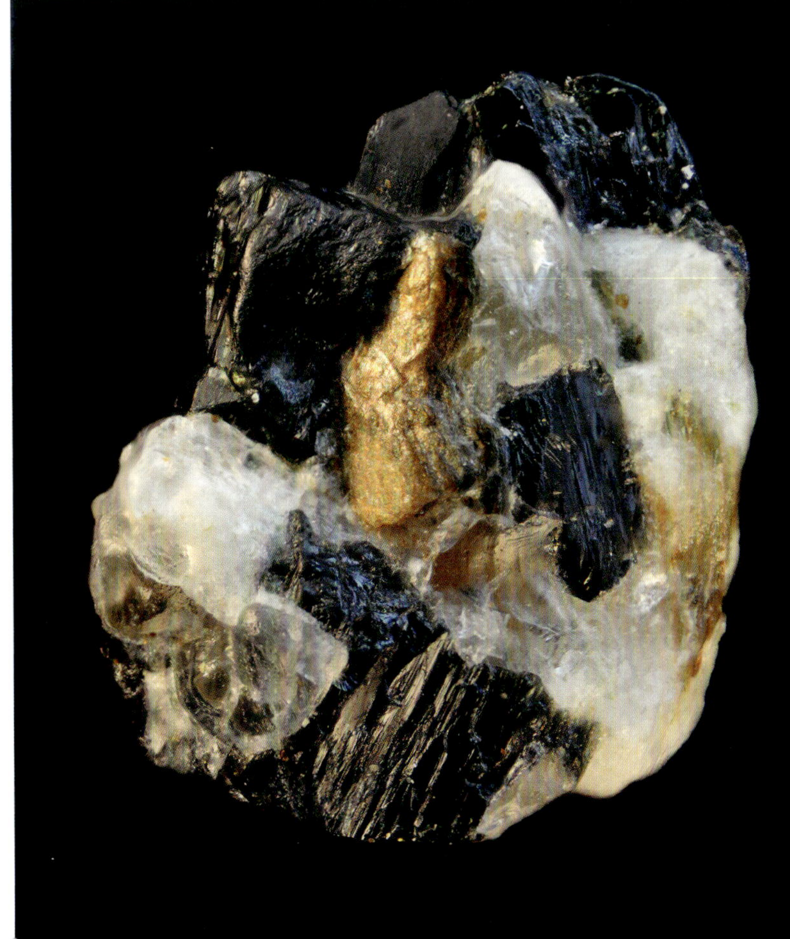

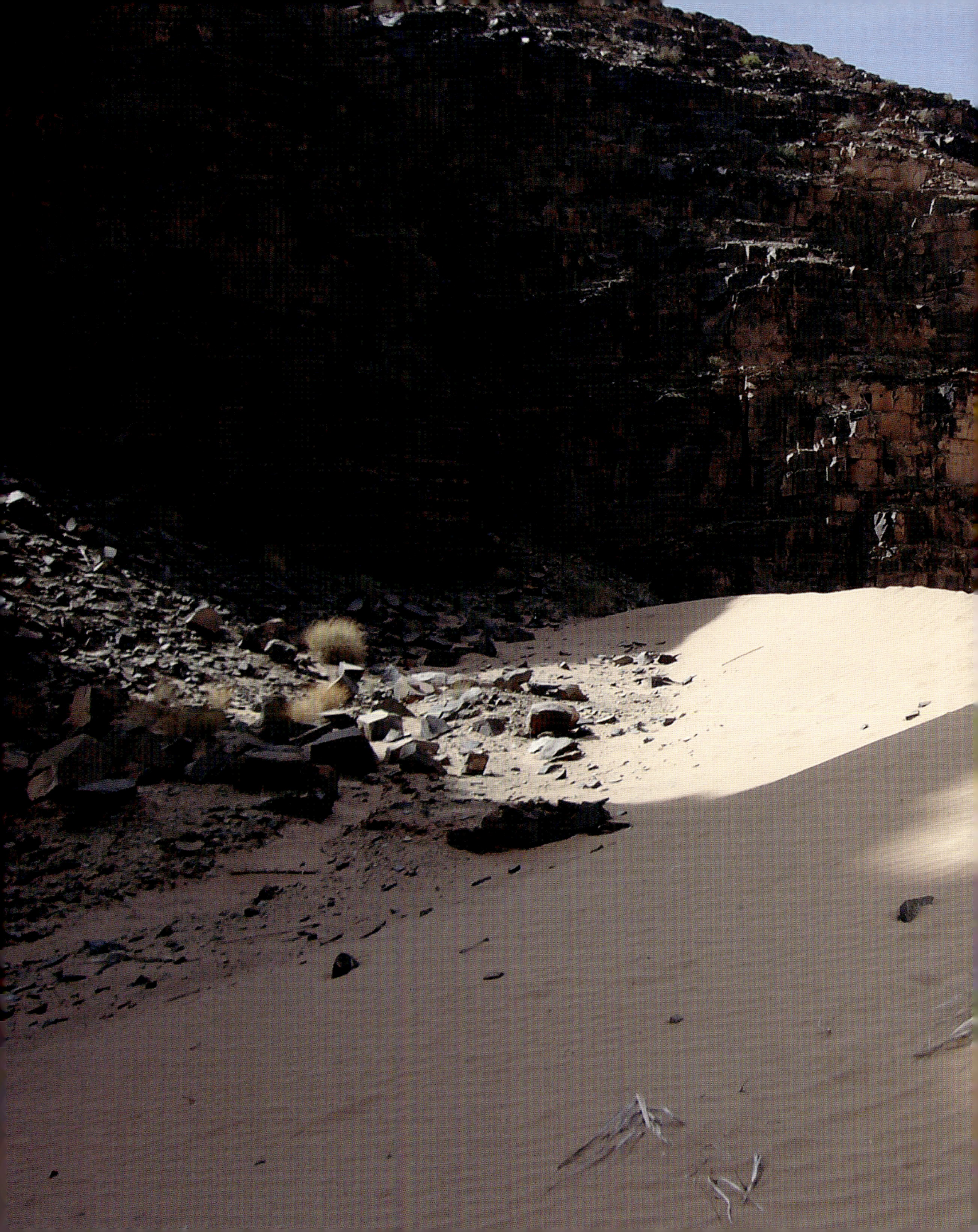

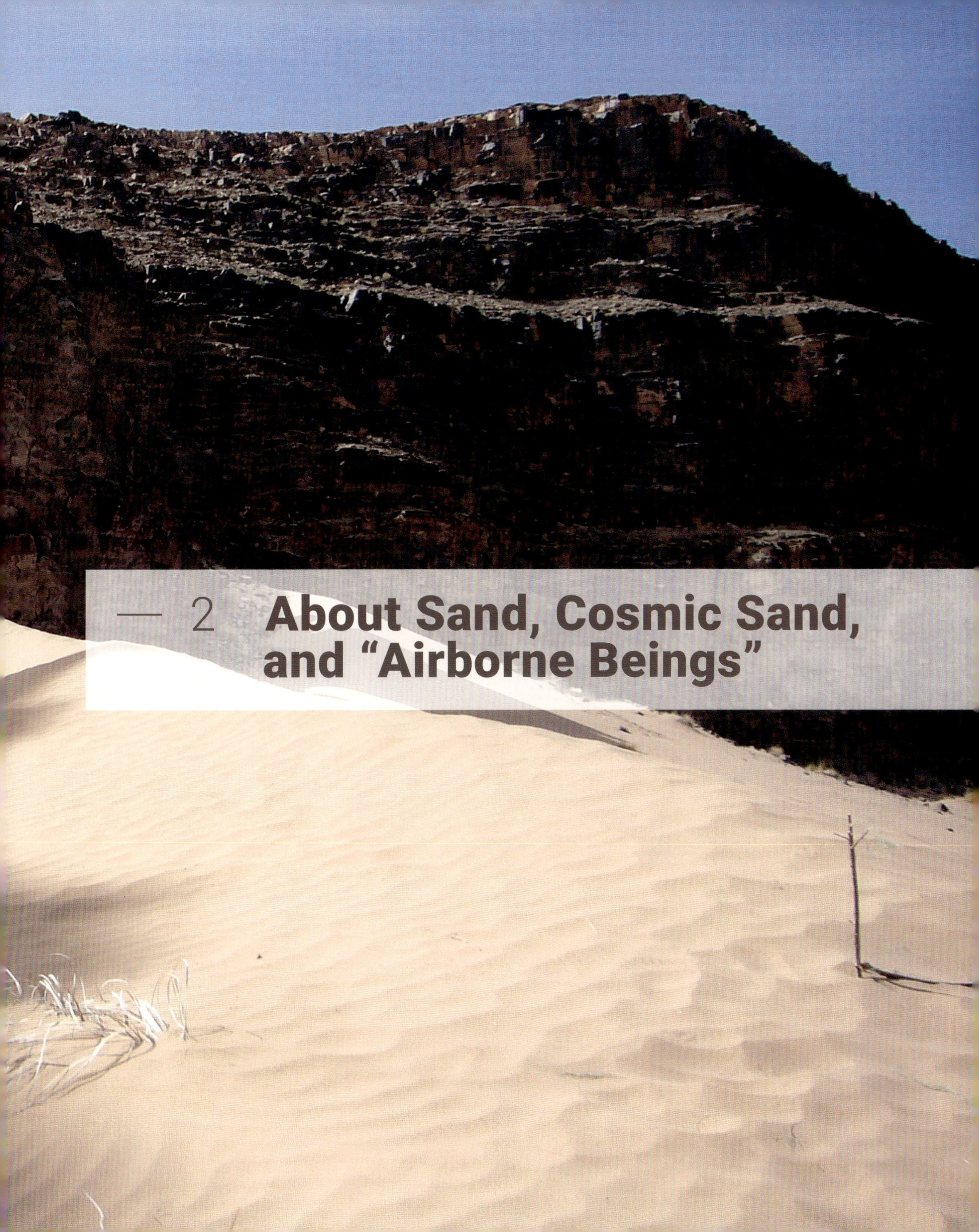

— 2 About Sand, Cosmic Sand, and “Airborne Beings”

I bring my microscope with me . . .

—*J. W. v. Goethe to Charlotte v. Stein*[2]

Before we turn to the core of sand—namely, the grain of sand itself, we let our thoughts roam quite far in this chapter. Along the way, we meet a country of sand, old Mesopotamia, the cradle of civilization. And we get to know small silicate spheres, the first "sand grains" in the solar system, which can still be discovered in some meteorites today. A journey to the origins of sand. Again and again, we will refer to exactly these origins along the further course of the book. Humboldt and Goethe will also be mentioned. Sand has many connections.

Our first care was to leave the dry sandy beach, which could afford us no discoveries in our science, and to examine the plantations, which from the ships had an enchanting appearance, notwithstanding the brownish cast which the time of the year had given. We found them indeed to answer to the expectations we had formed of a country described as an Elysium by M. de Bougainville."[3]

A grain of sand. There is nothing more ordinary that you can think of, and not much may occur to one spontaneously. Apparently, the scholar and natural scientist Georg Forster, a distant acquaintance of Goethe, felt the same way when, in the wake of the famous Captain Cook, he set foot on the "sandy" beach of the island of O-Tahiti in the South Seas in August 1773. There were probably no discoveries to be expected here in the way of science, he remarked. He was wrong—although the research voyage of the time certainly pursued a completely different goal.

The subject of sand can be approached in very different ways. Although sand is almost omnipresent, its nature remains quite anonymous and hidden. The variety of ways of looking at it seems to contradict the aforementioned ordinariness of the object. While there is no separate science dedicated to sand, there are a multitude of disciplines to be applied to understand the phenomena behind the simple image. Among others, sedimentology, mineralogy, geology, tribology, fluid mechanics, physics (especially optics), and microscopy should be mentioned. Chapter 8 extends this list by a further aspect, in that the concept of sand is grasped less physically than symbolically through time and quantity.

Previous double page:
Draa-Tal, Morocco

But what is sand, what is it about, and what connections does it have? To clarify these questions, we turn our gaze far back. In early Mesopotamia, the Mesopotamia of the Bible, sequences of characters were pressed with triangular styluses into clay tablets or other basic bodies; this writing is called cuneiform after the shape of its characters. In the Sumerian language, "sand" was assigned the following cuneiform characters:

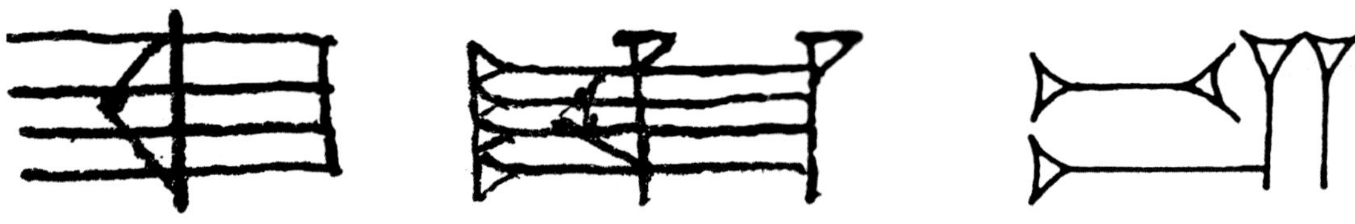

In the top sequence, the oldest character can be seen on the left,[4] which then developed further to the sign in the middle illustration,[5] which originates from the year of the ruler Gudea of Lagash, who lived around 2100 BCE and ordered the building of the Eninnu Temple. In the sign on the far right, a somewhat more modern form is shown.[6] The phonetic value, the pronunciation thus, can be rendered in each case with "sahar," whose meaning equates to dust, sand, earth. It is composed of the sound "sa_5" which means "red, brown," and the sound "hara," which means "crushed, pulverized." Sand, "sahar," can thus be literally translated as pulverized, reddish material. Quite remarkable here is the homophony to the largest modern sand desert in the world, the Sahara. Also, the land between the large Euphrates and Tigris Rivers had and has characteristics of a desert. In the time of the Sumerians, the inhabitants didn't have much more than calcareous clays and loams at their disposal next to the omnipresent sand. However, they were skillful at processing these. For example, to improve the properties, they used additives such as quartz sand and chopped straw, which can be seen today in microscopic examinations. Metals and stones were also not available and must be obtained at great expense by using the trade routes. In special cases, people resorted to the rare and coveted iron of meteorites to produce ritualistic objects. Thus, the two words "sky" (AN) and "iron" (BAR) have a common Sumerian language root. The syllabic symbols for iron can be rendered as "sky," "to shine," and "to cut." Cutting was made possible by using the iron from a meteorite falling from the sky.

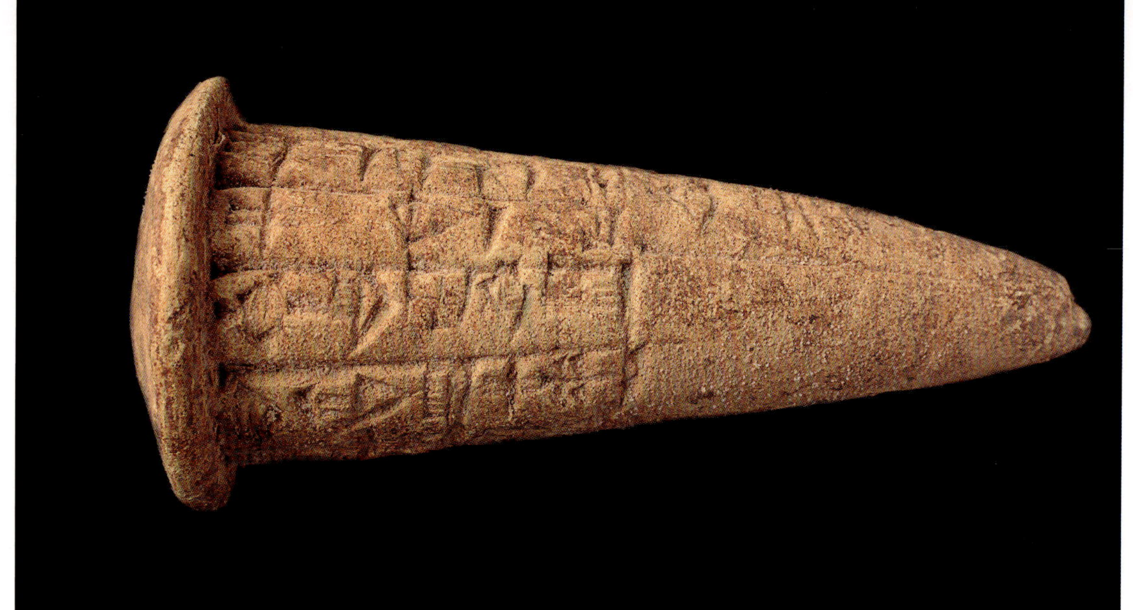

Even today, significant meteorite discoveries occur in desert sand and in the glaciers of Antarctica. In desert sand, "stones that fall from the sky" are easy to discover and remain on the surface of the sand for a long time before they disintegrate through erosion and are mixed undetectably with the desert sand. The epic of Gilgamesh even mentions the fall of a meteorite, the "piece of Anum," which is said to have been so heavy that King Gilgamesh could not lift it.

Mesopotamia was regularly hit by floods from the Tigris and Euphrates Rivers. The famous archeologist Leonard Woolley thought that he had proved one of the floods to be the biblical flood, the story of which goes back almost word for word to the Sumerian epic of Gilgamesh, which is more than 1,000 years older.[8] Over the course of time, the salts that were carried along with the floods accumulated in the clay and sand of the dry soil, which remained behind when the water evaporated. In many desert areas, the salts, which dissolve during precipitation and their subsequent crystallization in cracks or pores, lead to permanent disintegration of the rocks and of the art treasures hidden in the soil. During the crystallization of the salts, their volume increases by up to 100% in relation to the dissolved state, with the resulting bursting forces gradually disintegrating the mineral bond (salt weathering).[9]

The original meaning of the term "sahar" has not changed much. However, sand refers more to a property than a particular substance. According to its definition, sand is an amount of small, mineral grains of the same or different composition with a diameter between 0.063 and 2.0 mm. Sand is thus determined exclusively by the property of quantity and membership in the size class of the so-called psammites (from the Greek ποάμμοζ, for "sand"). Anything bigger is referred to as gravel (size class of psephite, from the Greek ποέφοζ, for "pieces"), and anything smaller as silt or clay (size class of pelite, from the Greek πέλοζ, for "mud").

Figure 2.1 The dedication foundation cone covered in cuneiform from the time of the ruler Gudea of Lagash, around 2100 BCE, fired clay. Dedication foundation cones had cultural purpose and are comparable to cornerstones. Most frequently, they were put in the sand with the point up and ultimately were covered with construction material. The dedication of the Enninu Temple by the ruler of Lagash in ancient South Babylonia is described. The city name can be read in the middle line. The characters were pressed into the soft clay with a wedge-shaped reed stylus before the cone was fired. Length approximately 104 mm.

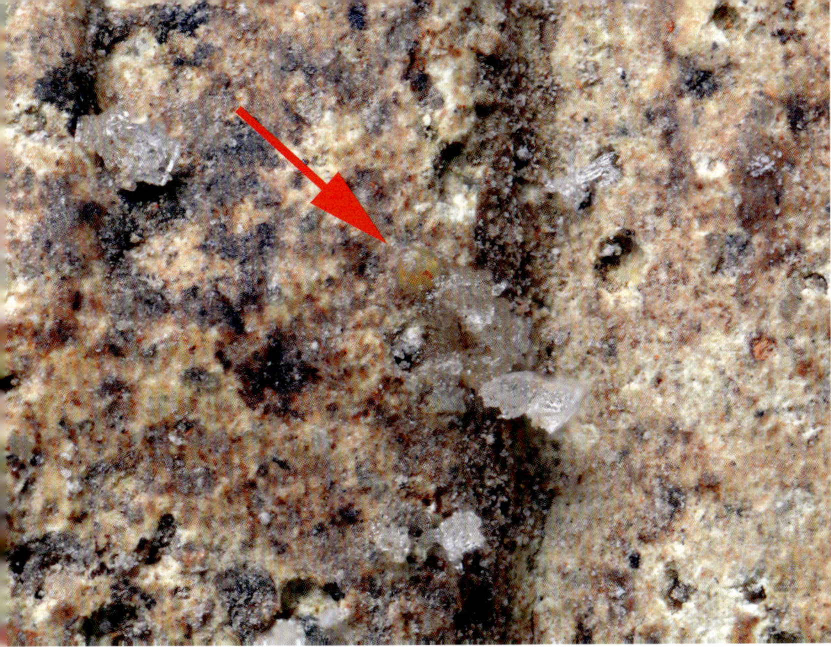

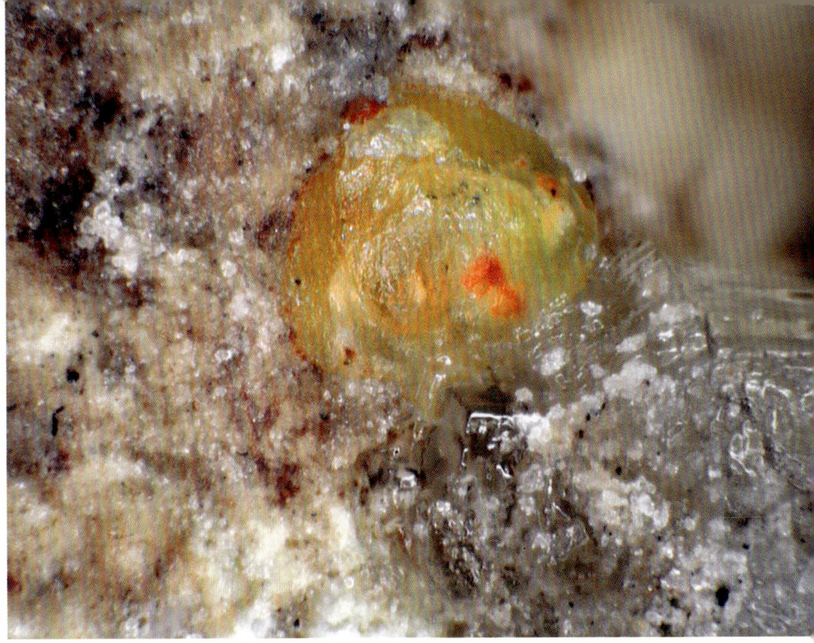

This classification has since been established for describing so-called clastic sediments, meaning sediments consisting of the debris of other parts, and has also been standardized. It is important to note that the physical properties of large particles are fundamentally and nonlinearly different from those of smaller particles. A grain with a diameter of 2 mm has a volume 10,000 times larger than one at the lower end of the scale with 0.063 mm. Accordingly, the behavior of these sand grains in their respective environments is also different. The structural characteristic of shape is not considered in the definition, but we will deal with it in detail elsewhere.

Other than the purely descriptive sphere, a clear allure emanates from the term "sand." Sand usually has positive connotations: from the sandbox at a young age or the sandman, who brings a beautiful dream with every grain of sand, to the white sand of a beach vacation of a tropical paradise, such as was described by Forster; from industrial raw material to the construction sand for our houses, to the protective sandbag in case of flooding. As far as sayings go, sand often doesn't fare as well. You bury your head in the sand, put sand in the gears, and, at worst, build on sand. As we continue, however, we would rather pursue the positive connotations.

Sand characterizes the border between the two great habitats on Earth: water and land. A significant portion of all the sand in the world can be found on the coasts.

Figure 2.2 *Left*: Detailed shot of the surface of the dedication foundation cone from fig. 2.1 In the center of the image, a greenish, round grain of sand embedded in the mass of fired clay can be seen in addition to the bleached salt crystals adhering to the surface. Image width 4.8 mm. *Right*: Enlargement of the grain of sand (microscope picture) from the figure to the left. The greenish grain, with a diameter of 210 µm, is well rounded, is clearly not out of quartz, and does not have any inclusions. *To the right*, under the grain is a salt crystal with a cubic crystal form. Just imagine: When Alexander the Great crossed the Mesopotamian plain from Babylon to Susa on his Persian campaign, the pictured dedication foundation cone was already buried in the desert sand not far from the route for about two thousand years. The small grain of sand has a history that began many millions of years ago. Supposedly it was swept onto the plain by one of the two great currents from the surrounding but distant hill country and was rounded by the wind until it was finally mixed into the clay of the foundation cone by a craftsman near the town of Lagash.

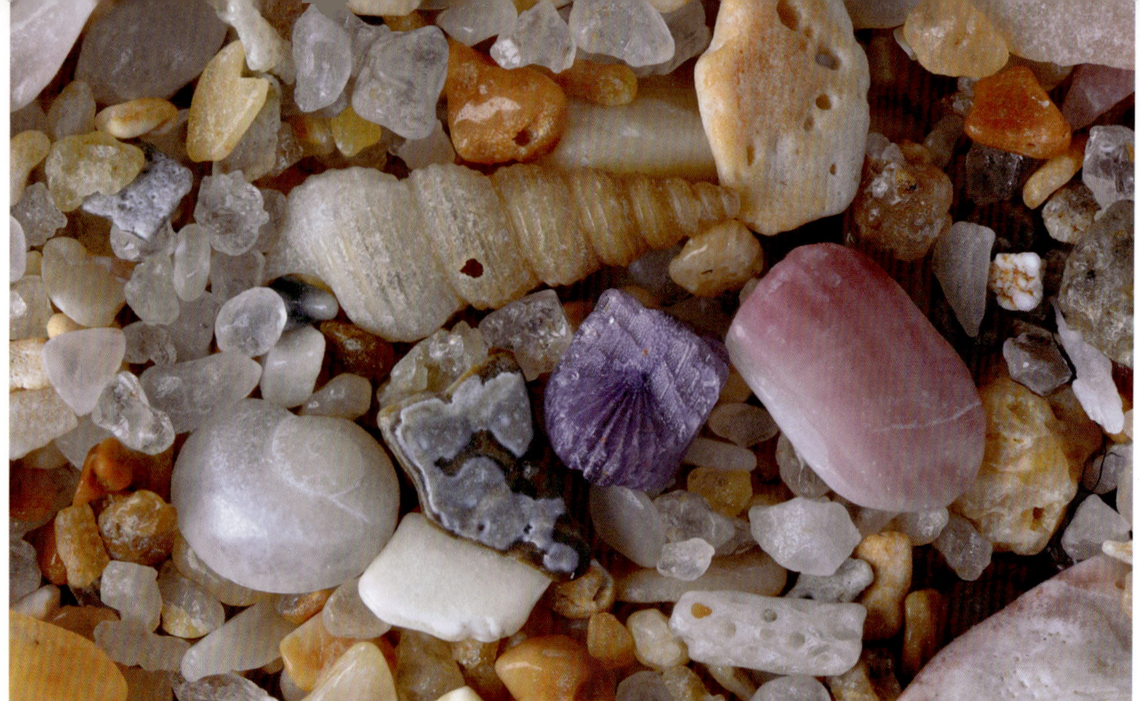

About five hundred million years ago, when the first marine organisms began to undergo the first decisive step toward land and, with it, to push into a new and completely foreign habitation, sand was probably the first material that they encountered. Settlement on land had its beginnings in sand. Perhaps this has become so deeply embedded in our collective memory (after all, we are descended from these creatures) that we can explain why many people perceive sand as something pleasantly primeval, almost archaic. Sand has a calming effect on the psyche but also evokes creative potential. Almost automatically we begin to enter a tactile relationship with sand: digging, piling up, or letting the sand trickle through our fingers. Both characteristics make sand useful for several therapeutic applications. Last but not least, the mentally centering Zen gardens should be mentioned, which—depending on the size—have sand or gravel as a design element and try to imitate the structures of moving water by introducing, for example, meandering patterns.

In our perception, sand oscillates between singular and multiple, solid and flowing, irrelevance and longing. Emerging from the symbols of durability and power, the rocks and mountain massifs, sand never retains its form: it is fluid and always only a potential or existing shape, taxis and cosmos in equal parts. While the Greek term "taxis" describes a state of order that is due to an aligning impulse or gradient, "cosmos" can be understood as a state of order that aims at the whole and cannot be attributed to any specific law. In this sense, the cosmos is a spontaneous and evolved order beyond a temporal assignment. Alexander von Humboldt, the well-traveled, universal scholar and contemporary of Goethe, notes the following regarding the inquiring human mind's understanding of nature:

Figure 2.3 Sand from Cape Trafalgar, Gibraltar, Spain. A colorful mix of mineral grains and shell remains of many different marine creatures (snails, clams, urchins, bryozoans). Reflected light. Image width 7.9 mm.

"In interrogating the history of the past, we trace the mysterious course of ideas yielding the first glimmering perception of the same image of a Cosmos, or harmoniously ordered whole, which dimly shadowed forth to the human mind in the primitive ages of the world, is now fully revealed to the mature intellect of mankind as the result of long and laborious observation."[10]

Even though the collection of many trillions of grains of sand to a state of order seems contradictory, it is exactly this collection, whose structure one views on a large scale, that can be found as an emergent property with the most-astonishing manifestations. It should be mentioned—without going into more detail as we continue—that above all, the forms of sand dunes in deserts, which have meanwhile been well scientifically investigated, can be connected to the fluid formation of ripples. The term "emergence" can be used to understand and summarize phenomena or properties that cannot be explained or derived directly from the properties of their constituent parts. For example, the thoughts that are evoked in the reader of these sentences cannot be derived by any means in the world from the chemical properties of the printing ink used, but are nevertheless ultimately caused by them. Our thoughts are, seen in this way, in relation to the partial properties of the book, emergent properties. The same applies to the relationship between grains of sand and sand dunes or even other system properties. The consistently taxic, organizing element, which is responsible for many observable sand-related phenomena and essentially for the sand grains themselves, is the universally acting gravity as a driving force for erosion.

During our expedition into the world of sand, the single, indispensable tool next to the magnifying glass will be the microscope. Humboldt used instruments, including the microscope, intensively and extensively and understood them as additional organs of the human being, enabling him to enjoy nature and interpret it scientifically. On these interpretations:

"While in the latter, this sense of enjoyment springs from a definite knowledge of the phenomena of nature. When man began to interrogate nature and, not content with observing, learned to evoke phenomena under definite conditions; when once he sought to collect and record facts, in order that the fruit of his labors might aid investigation after his own brief existence had passed away, the philosophy of nature cast aside the vague and poetic garb in which she had been enveloped from her origin, and, having assumed a severer aspect, she now weights the value of observations and substitutes induction and reasoning for conjecture and assumption."[11]

Here, Humboldt embodies the type of modern scientist who, far from any mystification or obscuration, approaches phenomena with the help of suitable tools and devises experiments but clearly distances himself from unreflective, inductive, and purely empirical knowledge gain.

His contemporary Goethe, on the other hand, had a very divided relationship with the microscope, which he believed, as an optical aid—as an aid for the senses—obscures the direct view of what is to be investigated more than it illuminates.

> "Insofar as he makes use of his healthy senses, man himself is the best and most exact scientific instrument possible. The greatest misfortune of modern physics is that its experiments have been set apart from man, as it were, [and] physics refuses to recognize nature in anything not shown by artificial instruments, and even uses this as a measure of its accomplishments. It is the same with calculating. There is much that is true that cannot be calculated, just like there is a lot that cannot be formed into a decisive experiment. But for this, man stands so high that what is otherwise unrepresentable is represented in him."[12]

The separation of the observer and the observed phenomenon began to become the greatest endeavor in the physics of that time. It was the search for objectivity, interpersonal observation, arbitrary reproducibility under identical and, by everyone, comprehensible conditions. An orientation, therefore, that ran counter to Goethe's very holistic view of the world, as well as his respect for the simply observed event, which presents itself to the senses and can be grasped by them directly and without a third party.

Newer physics brings the observer more and more into calculation. Observer and observed are connected by the system, and both are objects of investigation. Senses have since long been replaced by instruments, but it is still the judgment that transfers the pure facts of measurements to the evaluable and subsequently interpretable information. The chain from the instrument to the measured value to the interpretation and to the meaning is not closed.

Despite his reservations, Goethe could not escape the fascination of optical magnification and rarely failed to look through a microscope when the opportunity arose. For experiments with the appearance of light and color, he had a number of instruments specially made to his specifications. Goethe knew how to make good use of the advantages of the microscope, especially for his rather-little-known biology studies. In 1785, for example, he intensively and systematically studied infusoria, mostly unicellular organisms that can be observed when water or other liquids are applied to organic matter.

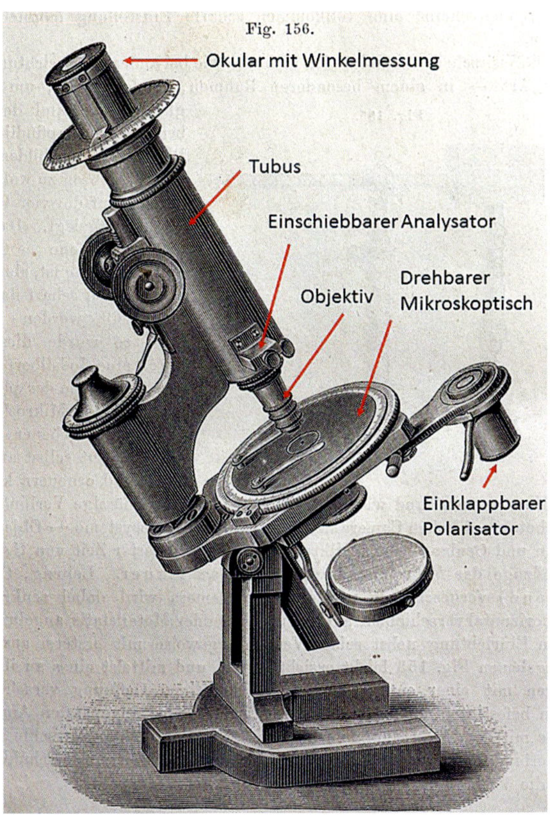

He varied the ingredients, alternating among hay, pepper, and bananas, and found a diverse world in miniature. Later, it was discovered after reconstructing the experiments on the basis of his notes that he found and described bacteria as well as vorticella, paramecia, and amoebae, although he did not identify them as such. In his letter to Charlotte v. Stein, whom he admired, he wrote on June 25, 1785:

> "I will bring my microscope with me. It is the best time to see the dances of the infusion animals. They have brought me great pleasure."

Unfortunately, it was to be some forty years before Goethe resumed microscopic investigations. Some of the instruments he used, including microscopes, are still in his legacy collection. Despite his critical attitude, he had apparently often used it profitably for his numerous studies. Goethe probably would have found our use of it in context with sand interesting, considering that the conveying instrument, the glass lens in the microscope, whose basic material is quartz, is as a rule most closely connected in material terms to the object being observed: the grain of sand.

Figure 2.4 Historical polarized light microscope from 1885, which includes all the important components of modern instruments. Figure modified from L. Dippel, *Grundzüge der Allgemeinen Mikroskopie* (Braunschweig, 1885).

Let us return to sand. Although grains of sand can be perceived directly by humans, all the properties that will be dealt with in the following chapters are still concealed. An optical journey through four decimal orders of magnitude—1, 10, 100, 1,000—reveals this vividly. Without additional magnification, a grain of sand can be easily identified as such under suitable lighting. Even at 10× magnification, its shape is quite perceptible, and the color stands out. This can be done with a good magnifying glass. At 100× magnification, surface structures become visible: to see them, a microscope is already needed. Only with correspondingly sophisticated equipment and the associated procedures is closer observation of the surface of the sand grain finally possible at 1,000× magnification. Mountains, valleys, furrows, and ridges become visible, and scientific interpretation becomes possible, or you can simply marvel at the aesthetics of the tiny mountain ranges, which owe their creation to the same basic forces as the actual mountains from which they originate. Furthermore, the imposing comparison with the fissured surface of a celestial body at the other end of the size scale is fascinating, whose surface the observer approaches bit by bit.

Every year, the earth collects between 35,000 and 100,000 tons of extraterrestrial matter on its orbit around the sun, the predominant amount of which has the particle size of sand grains at the most, but mostly significantly smaller (interplanetary dust). Sand-grain-sized parts produce typical shooting-star appearances (meteors) when entering the earth's atmosphere. Also in interplanetary space, the size of the particles plays a decisive, physically relevant role. Interplanetary dust with dimensions under 1 µm is gradually pushed out of our solar system by the radiation pressure of the permanent particle stream coming from the sun (solar wind). In contrast, larger particles up to about 10 µm tend to be slowed down in their motion and move in wide, spiral-shaped orbits toward the sun, where they are finally swallowed up like a big vacuum cleaner. Their lifetime is estimated[16] to be 10,000 years at most, a vanishingly short period by cosmic standards. Even particles with even-larger dimensions survive hardly longer than 10,000 years, since they break up through collisions among themselves again into fragments of the described size. If a piece of matter actually reaches the ground after flying through the atmosphere, it is called a meteorite. Their flight speed can reach over 25,000 mph (40,000 km/h).

Along with a range of minerals that are exclusively extraterrestrial, meteorites consist predominantly of minerals already known on Earth, whereby only about 3% of all meteorites are metallic and are composed of the iron or iron alloys so sought after in prehistoric times. However, the structure of the minerals and the iron alloys is distinguishable from the earthly variants, depending on the type. This is due to the very extreme crystallization conditions that prevail in space.

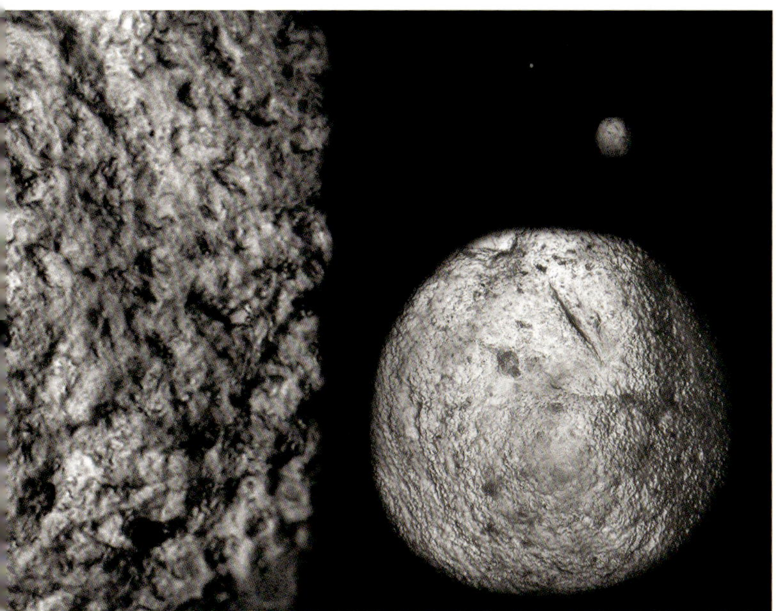

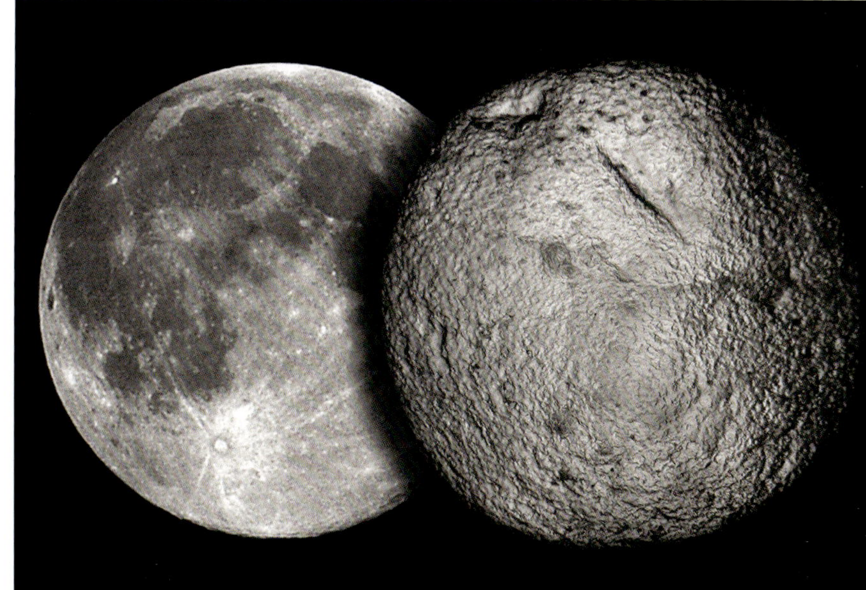

Studies suggest that iron meteorites in particular cool very slowly, only about 50 Kelvin in a million years, and travel in space for a very long time, up to six hundred million years, thus exposing them to radiation. The distribution of isotopes is also different from that on Earth. However, it is possible to distinguish these particles reliably only from their earthly counterparts, the sand grains and dust particles, with the help of complex analysis technology. Each meteorite has a very specific fingerprint, determined by mineral status, structure, and isotopic status, indicating a defined origin from an extraterrestrial "mother body." Some of these mother bodies, mostly asteroids but also Mars and the moon, can be identified beyond any doubt.

Figure 2.5 A trip through the dimensions. Quartz sand grain from the Sahara (original size 1 mm) enlarged 1, 10, 100, and 1,000 times.

Figure 2.6 Earth's moon and a grain of sand in a photomontage. The ratio of the diameters is about 1:4.6 billion. The bright highland areas of the lunar surface consist of crustal material and bright and light anorthosite rock. The dark zones, called "maria" (from *mare* = sea), consist of former huge basalt oceans. These were formed from the impacts of large meteorites.

Figure 2.7 Sikhote-Alin meteorite. The 77-ton (70,000 kg) Sikhote-Alin meteorite fell in eastern Sibiria on February 12, 1947. Upon entering the earth's atmosphere, it broke into thousands of smaller and larger fragments. One of them is shown here and shows the original melting crust with regmaglypts (flow structures), which were found during the entry into the earth's atmosphere. A melted rim can also be seen as a raised bulge that also formed during entry. The meteorite is composed of 93% iron, 5.9% nickel, 0.42% cobalt, and 0.06% phosphorus. Width of the fragment is 35 mm.

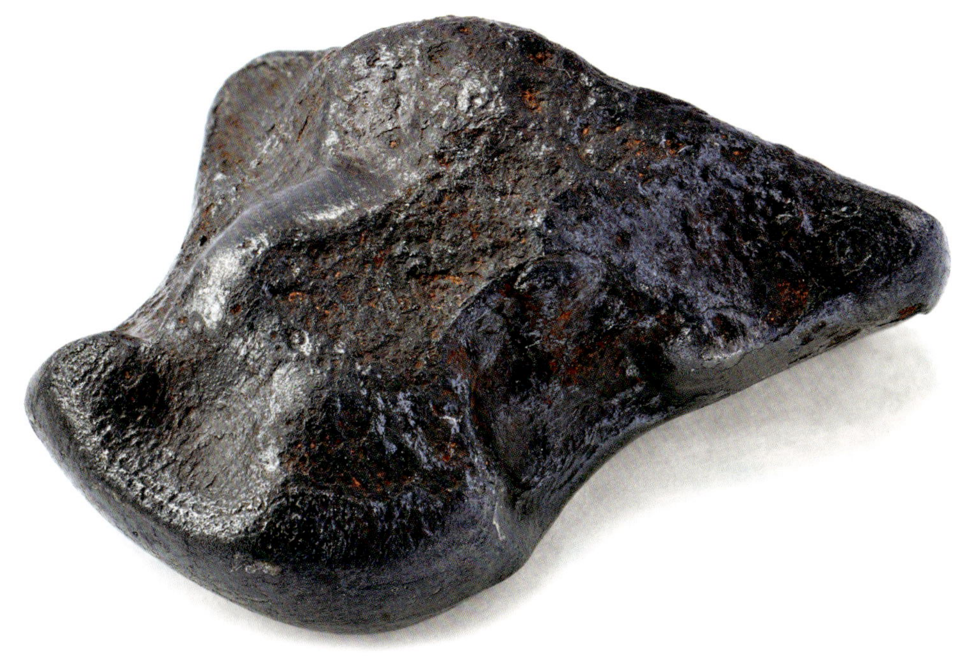

There is another connection to our sand theme that is subtle, but remarkable. Somewhat simplified, it can be said, In the beginning, there was sand! The predominant number of the meteorites that fall to Earth do not consist of the nickel iron, such as in fig. 2.7, but of rock, which can be seen, upon closer examination, to be composed of a characteristic mixture of small, silicate spheres (so-called chondrules), metallic particles, and further components embedded in a fine-grained matrix. The older the meteorites are, the more clearly and distinctly the chondrules are formed. Their diameter is in the range of less than 2 mm, thus falling pretty much within the definition of sand. The chondrules represent condensate and melt droplets of the early solar primeval nebula. At about 4.566 billion years old, they belong to the oldest form of matter formed in the solar system. They are, if you will, the oldest sand we know. The exact history of their formation is still a subject of research and hasn't been completely clarified. It is known that hydrogen, helium, heavier elements, and dust particles of interstellar space condensed to a primeval nebula that, with increasing rotational speed, produced the sun as a central star over the course of time. The process was probably caused by a supernova (star explosion) near the primeval nebula 4.568 billion years ago. The sun as a central body continuously attracted further matter, especially hydrogen and helium, and was in this early phase just before the start of an atomic reaction. Only about a million years later, high-melting particles rich in calcium and aluminum, the so-called CAI, formed and spread to the outer regions of the cloud. The remaining very thin particulate dust collected in a disk-shaped ring cloud, condensed, and melted into billions upon billions of silicate spheres, the chondrules.[17] The early solar system was thus filled with cosmic sand, which gradually, beginning 4.565 billion years ago, agglomerated together with CAI particles to form planetesimals and eventually planets. The remains of the solar disk of dust and its agglomerates are still found in the asteroid belt between Mars and Jupiter. The chondrule-containing meteorites originate precisely from this location, which can be examined today by scientists as relics from the origins of our solar system.[18] This is how the oldest cosmic sand comes into our hands: condensates from the matter of the solar primeval nebula, which show impressively beautiful structures in the microscope beyond the information content, and iridescent colorful sand grains in the polarized light, some of them even with relics from interstellar space.

Science owes a significant contribution to the understanding of the early solar system and thus also of our Earth to February 8, 1969. On that day, a meteor shower with a very large scattering field fell over the Mexican town of Pueblito de Allende in the province of Chihuahua, Mexico. The total mass of this meteorite, named after the village of Allende, is estimated at 4,500 pounds (2,000 kg).

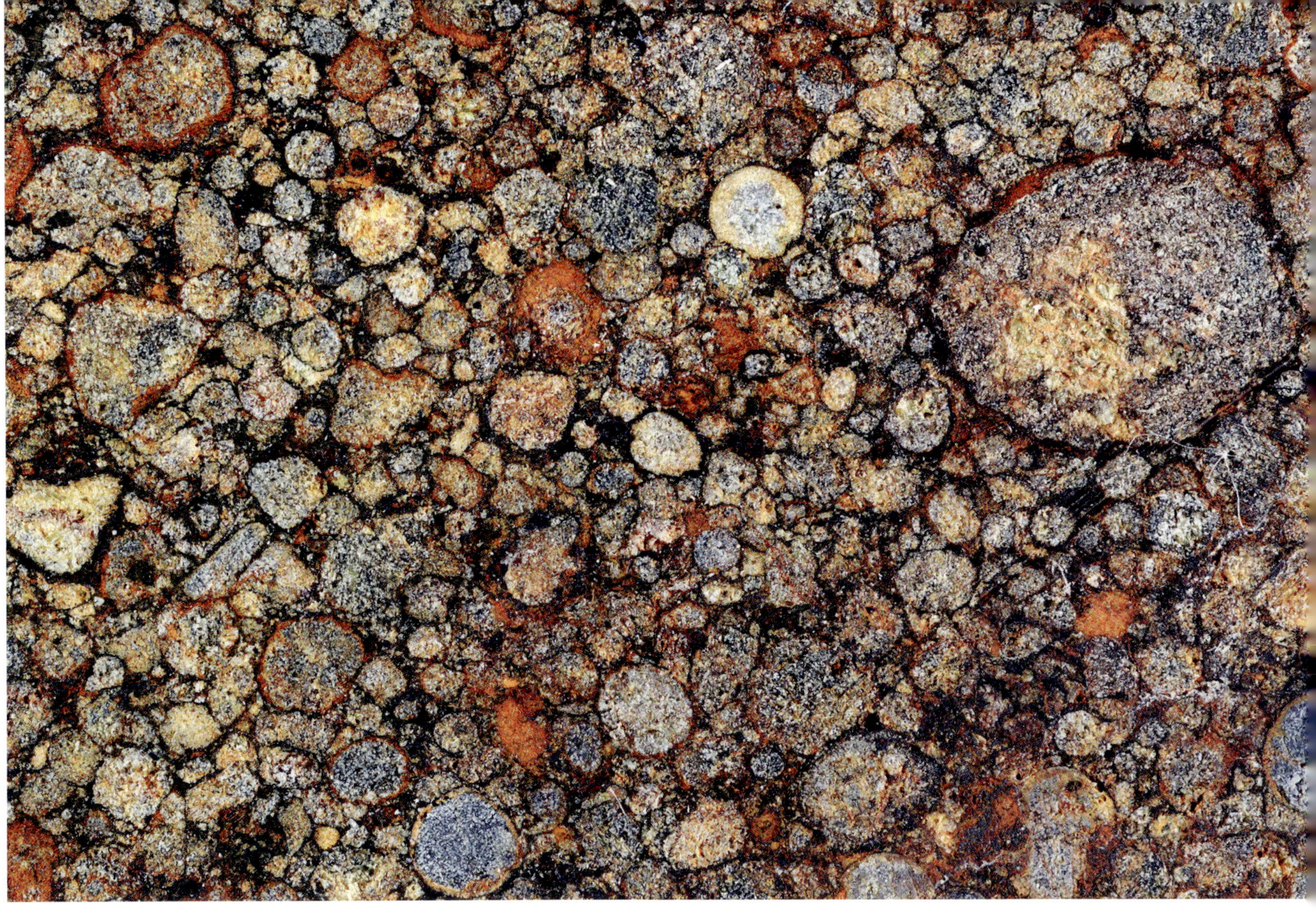

It belongs to the group of the until then very rare carbonaceous chondrites (CV-3 chondrites). In this type, the chondrules are well preserved (i.e., neither strongly altered by the high temperatures nor secondarily changed through high pressures). The meteorite contains grains of metallic nickel iron along with minerals such as olivine, plagioclase, and pyroxene (augite).[19] Also, this circumstance indicates that the material was probably not subjected to any further melting process; otherwise the metal grains scattered in the matrix would not have been preserved. Furthermore, minerals were found in the meteorite, which so far have been only extraterrestrial (e.g., kangite, davisite), as well as a new mineral called panguite, which was found exclusively in the Allende meteorite. The irregularly formed, fragment-shaped, white specks are particularly noticeable in the gray matrix. These are the CAI particles already mentioned above, which compose about 10% of the meteorite. In these CAIs, the aluminum isotope was found[26] in a ratio that does not occur otherwise on Earth, and that can be formed only in supernova explosions. Also, the isotopes of krypton, xenon, nitrogen, and other elements indicate an origin not only of extraterrestrial but of extrasolar origin (i.e., matter originating outside the solar system). The age of the CAIs is 4,567.2±0.26 million years, on the basis of the newest dating.

Figure 2.8 Who can escape the impression of having sand or sandstone in front of them here? This impression, however, is deceptive: this is "cosmic sand" in the form of a meteorite, Dar al Gani 785 LL3, which fell into the Libyan desert in January 2000. The recognizable grains are the chondrules referred to in the text, extraterrestrial melted droplets from the solar primeval nebula, over 4.5 billion years old, and baked to stone. Partial view of the surface of the meteorite. Image width 22 mm.

This makes CAIs the oldest and most-alien solid objects that a human being has ever touched, provided he finds himself in the enviable situation of being allowed to hold such a fragment in his hand.

The extraterrestrial origin of meteorites has by no means always been universally known. Goethe was also fond of meteorites and possessed among other things a piece of the Ensisheim meteorite in his Weimar collection of minerals and rocks, which included over 18,000 specimens. Regarding their origin, he accepted a common theory of the time. According to this theory, meteorites were formed as "airborne beings" or "aeroliths" from atmospheric vapors and gases mainly in connection with thunderstorms or volcanic activities. In this context, Goethe maintained correspondence with the founder of acoustics and modern meteorite research, Ernst Friedrich Florens Chladni (1756–1827). In 1791, Chladni published his groundbreaking work, *Über Feuermeteore und über die mit denselben herabstürzenden Massen* ("On fiery meteors and the masses falling with them"), and, within it, propagated the first convincing theory on the extraterrestrial origin of meteorites. Nevertheless, a little can be done to save the integrity of the "airborne beings" theory. Perhaps they do exist, even if in another form. On top of that, there is an unsuspected connection to our topic—sand. Along with the meteorites we just described, whose origin is without a doubt extraterrestrial, there is another group of "stones that fall from the sky" that exists, the so-called tektites. For a long time, they were grouped in with meteorites. Today, however, it is proven that they consist of terrestrial matter. In contrast to meteorites, tektites do not have a crystalline or metallic structure but are recognized by the fact that they are made up of a mostly dark and glassy substance with a bubbly melted crust. Other than tektites, there are relatively few naturally occurring glassy rocks. First, obsidian needs to be mentioned here—a volcanic rock glass that forms through very fast cooling of a molten mass and whose sharp splinters were used in prehistoric times as knife blades or arrowpoints. Much-rarer finds are fulgurites, or fossilized lightning. They are formed when lightning strikes sand or sedimentary rocks and causes the grains along the impact zone to melt. This results in up to meter-long tubes of glass-like melted and caked sand. Goethe, fascinated by this rare phenomenon, owned—you guessed it—a fulgurite in his collection. Among the tektites, the subgroup of moldavites should be mentioned, whose beautiful, mostly bottle-green coloration catches the eye. Their origin and formation have been particularly well studied. Moldavites were formed exactly 14.3 million years ago, just at the time when the meteorite responsible for the formation of the Nördlinger Ries (diameter, 0.9 miles [1.5 km]) collided with the earth at a speed of about 60,000 mph (100,000 km/h). During huge impact events such as this, large amounts of matter are hurled into the atmosphere.

Figure 2.9 Allende meteorite. The round chondrules, yellow troilite, and irregularly formed inclusions (CAIs and amoeboid olivine aggregates) can be easily recognized. Planed cross section. Reflected light, image width 5.6 mm.

Figure 2.10 These "cosmic sand grains," chondrules from the Allende meteorite, are about 4.56 billion years old. Within, they are made up of crystals, mostly composed of olivine and pyroxene. Compiled by using thin sections. Crossed polarizers in transmitted light. Chondrule dimensions between 400 and 1,200 µm.

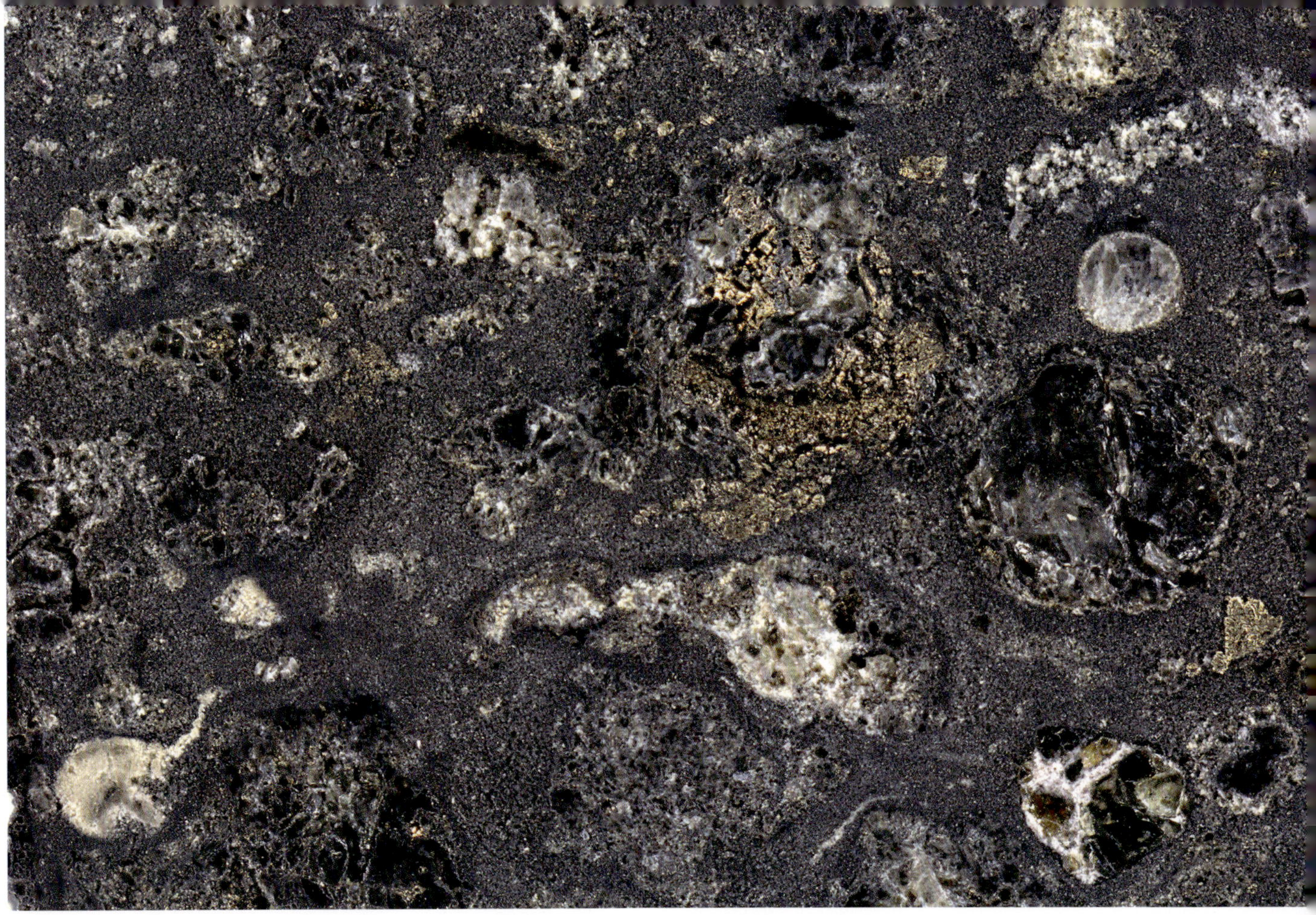

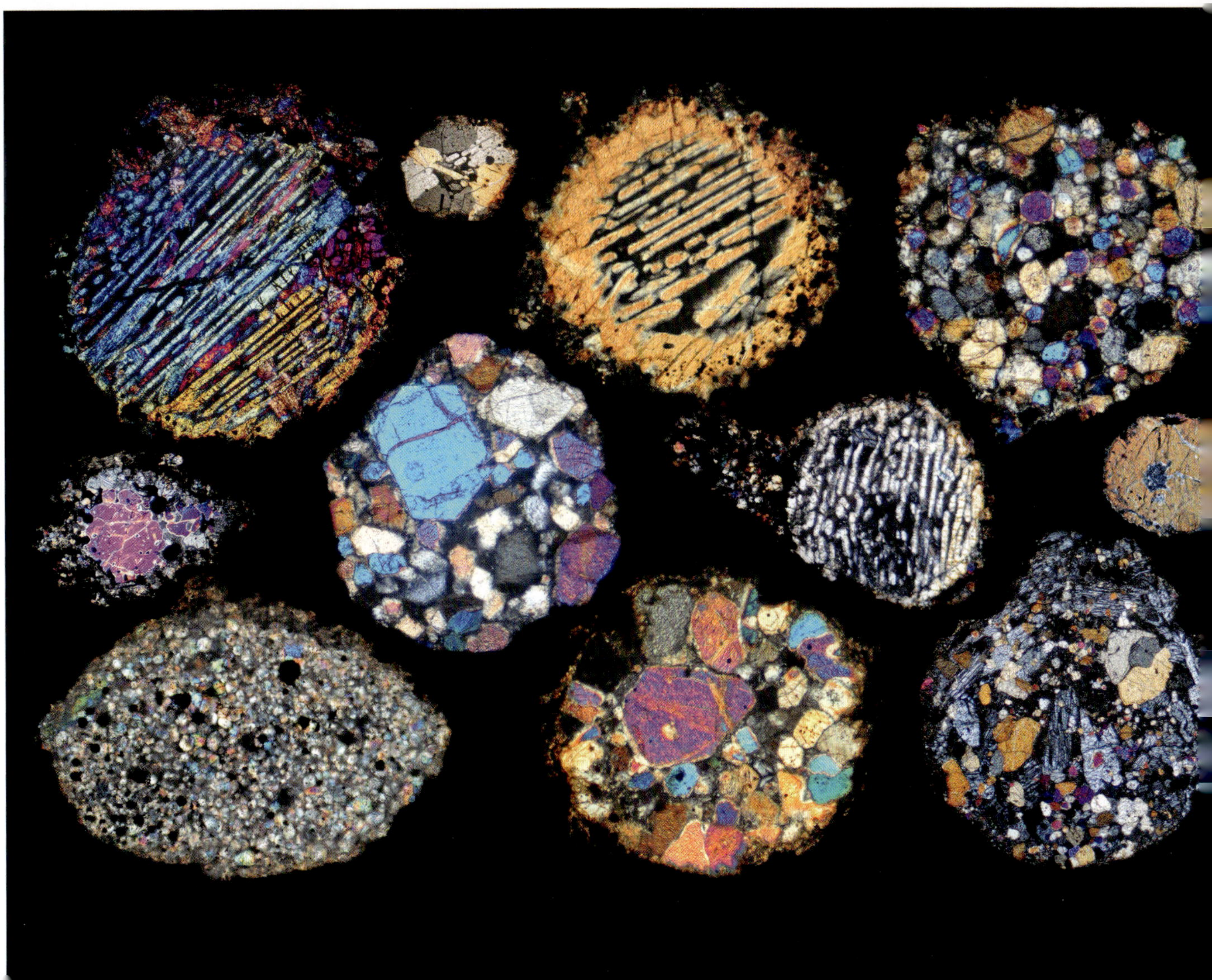

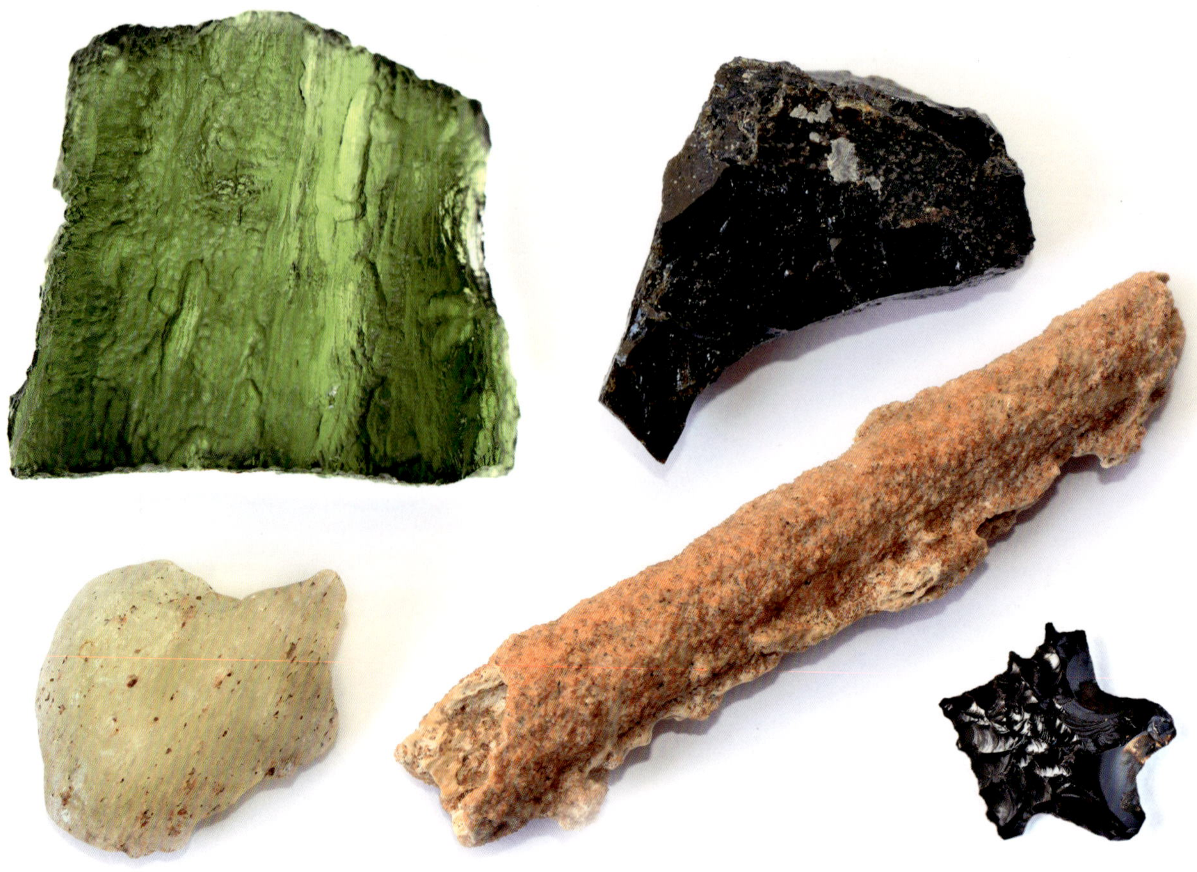

In the case of the Ries meteorite, the pressure and temperature were so high that a plasma beam several tens of thousands of degrees hot was generated from the sediments of the Upper Freshwater Molasse that covered the earth's surface at the time of impact.[20] And these sediments consisted predominately of sand. The sand plasma beam that shot up condensed under complex chemical processes in the atmosphere to glassy rock masses, the moldavites, which then scattered hundreds of miles farther east in the fields over Bohemia, similar to meteorites. Thus, we could call the moldavites "airborne beings" or "aeroliths": stones from the air.

Figure 2.11 Examples of igneous glass. Image width about 18 cm. *Top left*: Moldavite, Bohemia, Czech Republic. *Top right*: Volcanic glass (obsidian). *Bottom left*: Libyan desert glass, Sahara. *Bottom center*: Neolithic arrowhead of obsidian, South America. The bubbly melted texture and the typically bottle-green coloring can be well observed in moldavite.

Figure 2.12 Cosmic sand, composed of chondrules from the Saratov L4 meteorite, which fell over Russia in 1919. Each of these small grains is an unbelievable 4.565 billion years old. These unassuming grains show spectacular colors in polarized light, as you can see in fig. 2.10 as an example. Reflected light. Image width 4.3 mm.

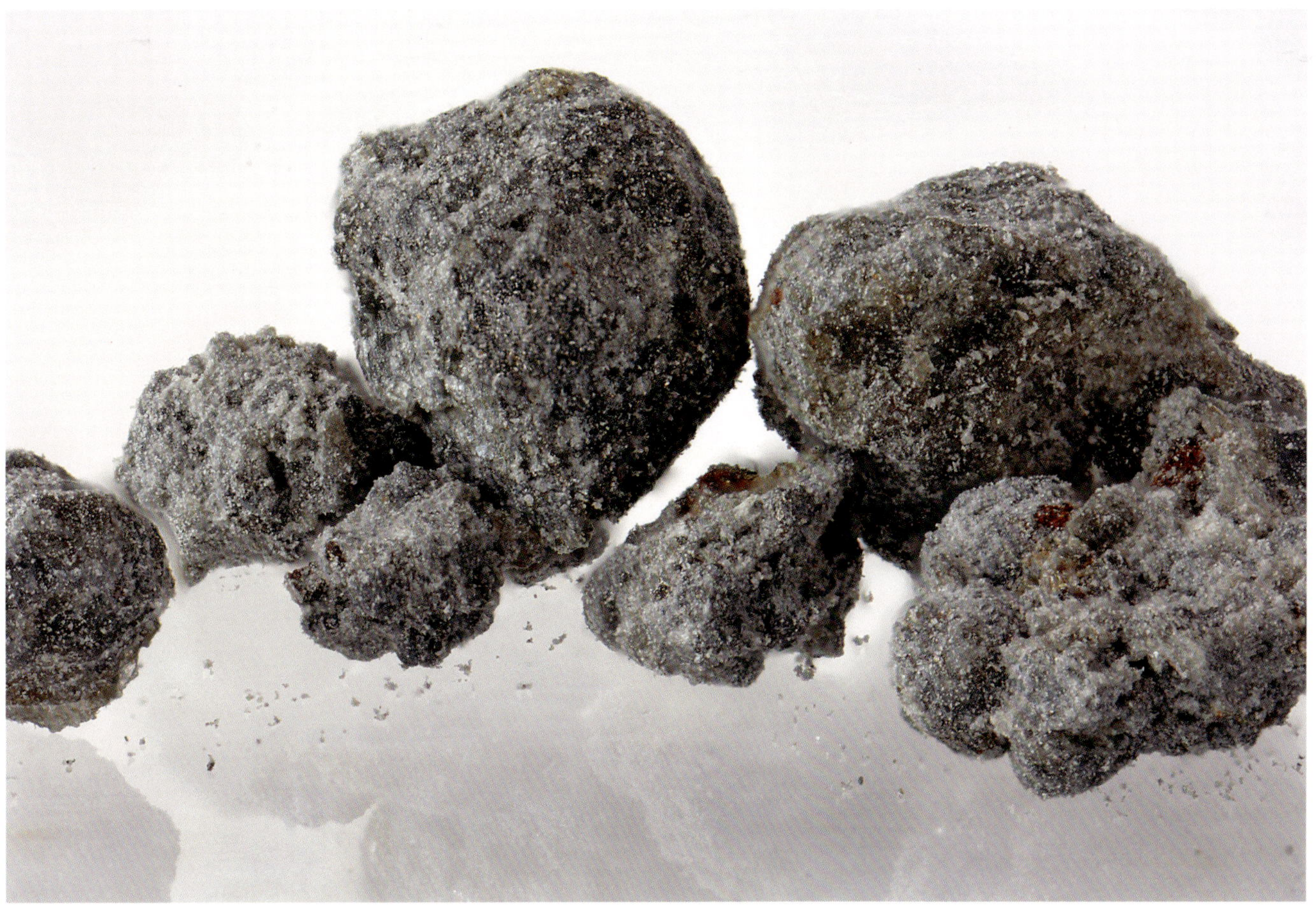

The Libyan desert glass (LDG) also appears to have come from a meteorite. Unlike the moldavites, the LDG has not been assigned to a crater that could be the source of the impact. And, as we have seen, a significant amount of the extraterrestrial matter that comes to Earth is the size of sand grains, which then mixes and connects to the sand of the earth. In 1970, Hoimar v. Ditfurth expressed the idea of a kind of cosmic material exchange.[21] Sand as a practically omnipresent natural substance can serve as a catalyst for thoughts that provide room for almost any discipline beyond pure natural science if one only looks carefully enough for suitable and fruitful connections.

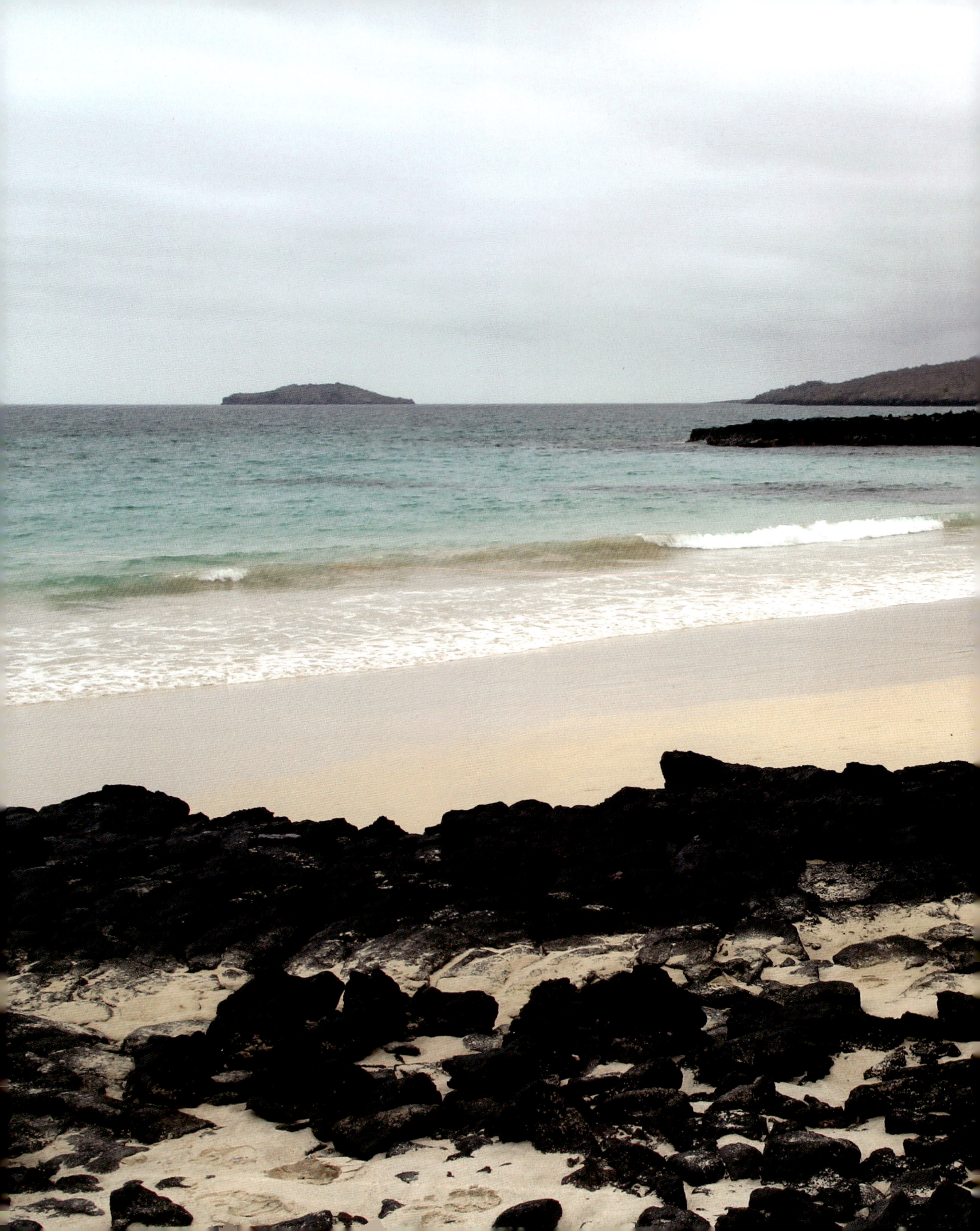

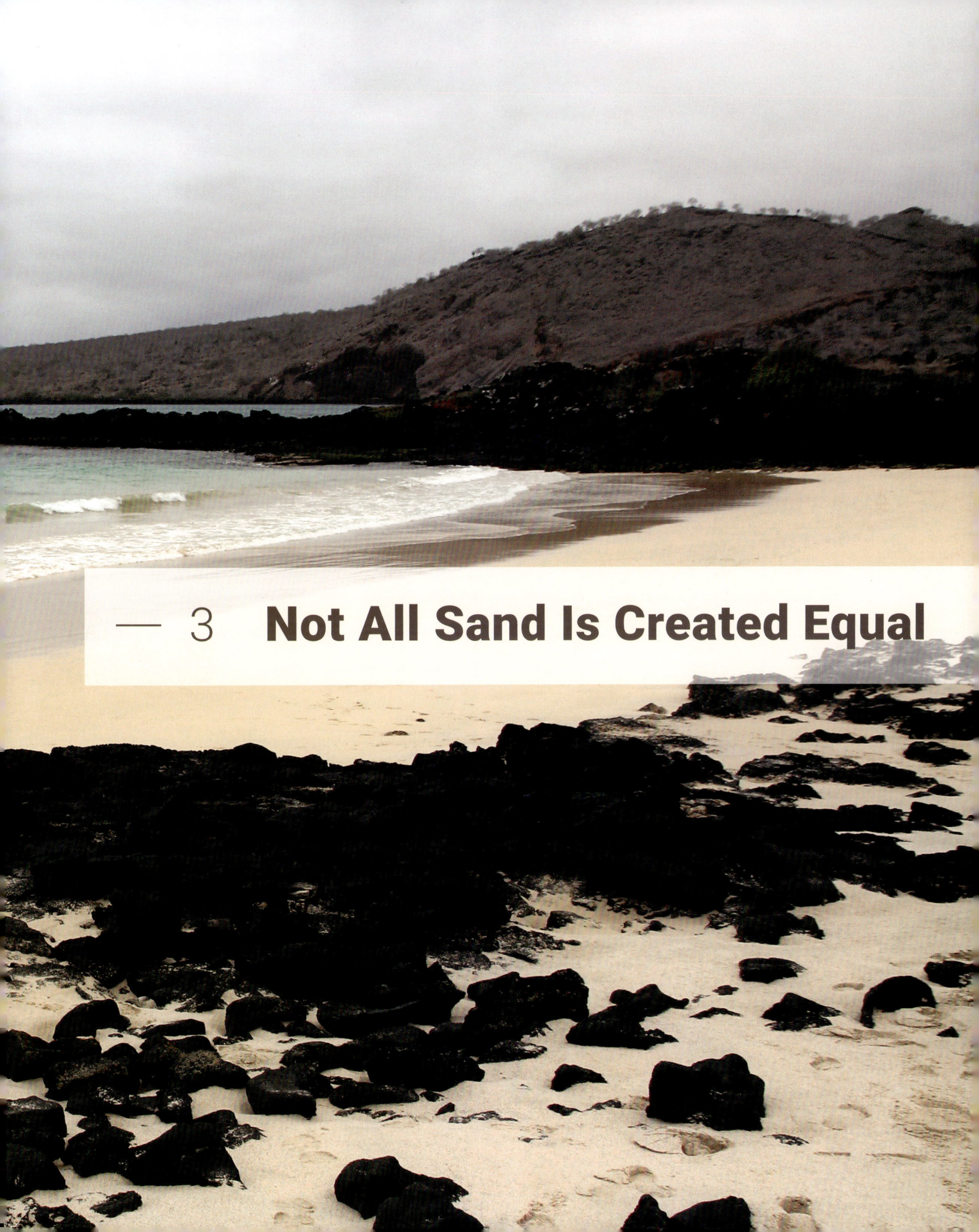

— 3 Not All Sand Is Created Equal

For in observation even the errors are useful, in that they call attention and give the sharp-eyed an opportunity to practice.

—*J. W. v. Goethe,* ***On Granite***

After taking a look at the extraterrestrial context, which we will briefly follow up on in this chapter, we will now deal directly with sand itself, its formation, and its components. Worlds hide behind the uniform beige color. We will become familiar with characteristics by which one can describe and distinguish sands / sand grains and try to understand which first conclusions can be drawn from the observation of sand samples. Along the way, we will learn amazing things about the study of crystalline sand grains, with the help of light.

"The main thing that seems not to be thought of when using analysis exclusively is that every analysis presupposes a synthesis. A heap of sand cannot be analyzed; but if it consists of different parts, such as sand and gold, then the washing is an analysis, where the light is washed away and the heavy is held back."[22]

The interesting part about this quote from Goethe is what is implied: sand is just sand. For distinction and analysis, a separate addition is needed; in this case, gold. Sand as an indistinguishable, uniform quantity. We will see that reality is quite different—not all sand is created equal. Each grain of sand has its own history. Over the course of time, external influences leave traces on the grain, which are erased and reshaped again and again. Grains of sand are individuals, and despite the immeasurably large number of them, there are probably no two that have lived through an identical history and are exactly alike in appearance. The uniqueness of each grain of sand is almost astounding. In contrast to living beings, whose appearance always remains similar, or at least develops in a certain, characteristic way from birth to death and accumulate different signs of age, matter particles, such as grains of sand, can change their appearance completely and abruptly as a result of external influences.

Previous double page:
Isla Floreana, Galápagos, Ecuador

In order to be able to differentiate and describe sand and individual grains of sand, we need to be able to assign them characteristics that are either measurable or observable for our purposes. This is not done here in the sense of a dichotomous key, as you would use for birds, plants, or insects. Flora and fauna are defined by clear and distinct characteristics, which are named and recorded according to strict taxonomic rules. Each new species that is discovered is given a two-part Latin name (for example, *Apis mellifera* for honeybee) and is deposited in a museum and preserved forever by means of a sample specimen, the so-called holotype, as a kind of primeval meter. By making comparisons to this holotype, scientists can determine at any time whether an animal belongs to an already-known species or if it possesses new characteristics that cannot be subsumed under the characteristics of the holotype. Such a sophisticated system does not exist for sand or rocks. This is because rock types can merge into each other, and minerals can sometimes form continuous series of mixtures that are not definable. A white stork and a chaffinch look the same in Germany and Italy, there are no connections between them, while granite and diorite are separated only by arbitrary definitions and can also very locally in appearance and composition.

Usually, there is nothing clear about the origin of sand and sand grains that can be inferred retrospectively on the basis of their characteristics. However, from a group of data and their respective coarseness, one can derive clues that allow interpretations about their formation, transport, and depositional conditions. Sand is a mirror of its environment and its history. In chapter 6, we will deal with this in detail.

By definition, sand is a collection of granulated minerals within a certain size class. A mineral is a chemically uniform substance of natural origin that appears in crystalline form, with just a few exceptions. For the most part, these mineral grains, at least in the case of continental sand, are composed of the decay products of compact rocks, which in turn are to be regarded as mineral aggregates. Rocks and the minerals they contain, along with biogenic materials such as shell remains of organisms or chemical deposits, form the basic material for sand. Over five thousand minerals can be found in the earth's crust and are principally available for use as sand components, but only a few of them take an exceptional role when it comes to descriptions of sand and rocks. In the simplest case, they can be brought into two groups: the group of the light, felsic or leucocratic minerals, and the group of the dark, mafic (artificial word from the initial syllables of magnesium and ferrum [iron]) or melanocratic minerals or rocks. The former includes quartz, feldspars, muscovite, and the zeolites, whereas the group of dark minerals is represented by olivine, pyroxene, amphibole, biotite, garnet, tourmaline, and magnetite.

A few words now on the origin of these minerals that are so important of the understanding of sand. Minerals have not existed since the beginning of the universe—they have formed during the course of history under quite-specific physical conditions. Only four elements existed immediately after the Big Bang: hydrogen being the most abundant at 90%, helium at 9%, and, finally, lithium and beryllium together forming approximately 1%. For the formation of twenty-four further elements, a fusion process was required, which takes place even today inside mass-rich stars, which, however, can develop only in a later phase of the universe. Only at temperatures between 10 million and 1 billion degrees Celsius can hydrogen protons overcome their mutual charge-induced repellence and fuse, emitting photons to form a helium nucleus, a new element. In this way, the sun converts about 560 million tons of hydrogen into helium and radiant energy every second. Despite this enormous turnover, the sun has used only about 20% of its hydrogen to date. Once all the hydrogen in a star has been converted into helium, such as will be the case for the sun in about seven billion years, new fusion processes begin, which depend very much on the total mass or size of the star. Because of its mass ratios, our sun will be able to form elements up to and including oxygen and carbon by fusion during its life. However, the mass of the sun is not sufficient for the fusion of elements with a higher proton number.

But how are other elements created with even more protons, such as gold or platinum, as well as elements that are commonly found in sand? As soon as a star with high mass (more than ten solar masses) is burned out, the outward radiation pressure that balances the inward gravitation force is removed, and gravity causes the massive star to collapse in less than a second without any significant counterpressure. The infernal explosion that subsequently occurs is called a supernova and is in fact crucial to the evolution of life in the universe. During a supernova explosion, extreme energy densities and temperatures of several billion degrees occur, conditions that existed probably only during the Big Bang. All further elements beyond the proton number of iron formed exclusively under these conditions, which then were scattered during the explosion. Our solar system consists of such scattered material from cosmic supernova events to a considerable degree, such as those of other stars, formerly far away, that exploded in ancient times. The solar primeval cloud of our solar system, mentioned in the previous chapter, was composed of 73% hydrogen and 25% helium. Only the remaining 2% of the heavier elements finally formed the basis for the solid components of the earth on which we live.

Figure 3.1 Sand samples from all over the world: *Top row (from left to right)*: El Condor, Argentina; Alice Springs, Australia; Galápagos, Ecuador; Vik, Iceland; Carlis Bay, England; Crete; Sossuvlei, Namibia; Lefkas, Greece. *Middle row*: Steinheim, Germany; La Gomera, Spain; Machalilla, Ecuador; Murzuk, Libya; Cotopaxi, Ecuador; Mauritius; Santa Cruz, Argentina; Gasenried, Switzerland. *Bottom row*: Torres del Paine, Chile; Lake Titicaca, Bolivia; Rustrel, France; Galápagos, Ecuador; *Île-des-Pins*, New Calcedonia; Green Sand Beach, Hawaii; Parga, Greece; Lomas de Arena, Bolivia.

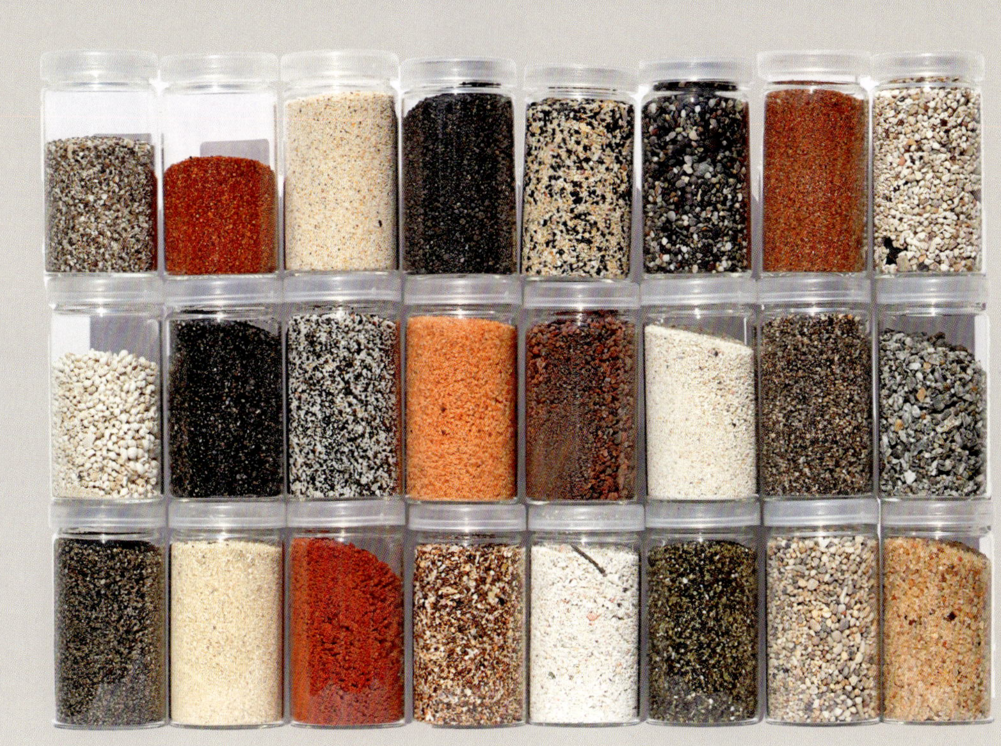

Without a doubt, every single atom in the gold ring on the reader's finger has its origin in a supernova explosion in distant and strange worlds—somewhere in the vastness of the universe that is inaccessible to our experience and our imagination. After these preliminary considerations about mineral formation, we return to our task to learn to distinguish different kinds of sand. In the section "Out with the Magnifying Glass," we limit ourselves to the properties of sand that can be determined with quite simple aids that enable first observations. The description of the inorganic components of sand then follows in the next section.

3.1 Out with the Magnifying Glass

Collecting sand, apart from the fact that it is simply fun, has a few special characteristics. Typical collectors usually plan on completing their collection. In some cases, this is quite conceivable, but never when it comes to sand. Even a single section of coastline can contain sands of very different types, depending on where and at what time the sample is taken, in spring or winter, upstream or downstream of the mouth of a body of water, and so on. No two samples will ever be identical; this is self-evident, but also fascinating. Another characteristic is that it is ridiculously easy to get sand samples.

Sand is everywhere, and apart from any sort of connected travel costs, it's free. Today, however, local regulations need to be considered. On some beaches, sand collection as well as any type of removal from the natural environment is prohibited. Excessive tourism and lack of responsibility on the part of the collectors necessitated this step—this is especially true in places with "beautiful" or unusual sand. At any rate, you should inform yourself sufficiently before taking samples (even in small quantities).

Let's pull out the magnifying glass and get started. Simple folding magnifiers with triple magnification are well suited. In addition, it is advisable to purchase a high-quality, multilens magnifying glass with 8–10× magnification, which is sufficient for most examinations and enables a color-fringe-free and leveled field of view. A little bit more money and we are rewarded with a much-better observation experience. If you can find a worn-out mechanical camera lens with a 50 mm focal length instead, you are also well equipped. It can be used as a magnifying glass; a good color correction is also to be expected with older models. Small excursion microscopes with magnifications up to 20×, which were popular in earlier times, also perform very well but deliver a mirror-inverted image that takes some getting used to and does not replace the magnifying glass. From my own experience, it is advisable, if possible, to label it on-site with a note or a sticker indicating where it was found, the time, and the collector. Otherwise, there is a risk of mixing up the samples after a short time, which would make them completely worthless. Small quantities, such as 2–3 tablespoons full, are usually sufficient and should be dried carefully to avoid mold growth. The use of simple newspaper folded into a small tub has worked well for this purpose. The moisture is usually absorbed quite quickly, and the sand can then be poured into a snap-top jar or something similar when dry and finally labeled. It is also advisable not to clean the sand samples but to leave them in their original condition. Once your friends and colleagues become aware of your collection, a whole series of vacation souvenirs from all over the world will accumulate over time. It is then advisable to record the data in a spreadsheet program. This makes it easy to sort through the collection by using defined groups of properties. Let us now go through the most-important properties of a sand sample, which we can observe and record quite easily. We will dwell on color a little longer at the end of the chapter, which is a very complex and interesting characteristic.

3.1.1 Grain Size

The size of the grain is the most important characteristic that a grain of sand can have when it comes to its behavior in the environment. As easy as it seems to measure the size of a grain of sand, it becomes increasingly difficult when dealing with a large amount of sand grains.

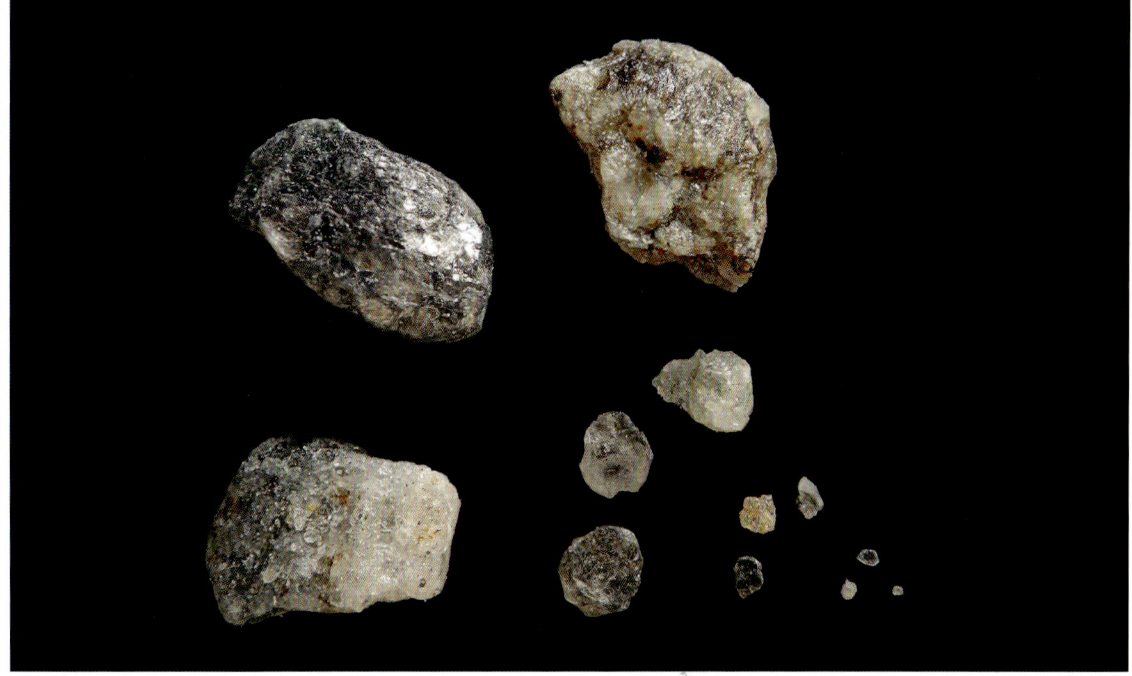

What do we mean when we refer to "grain size"? If each grain were perfectly round, you could simply use the diameter. But how do you define, for example, the size of a grain that is rod shaped? Here there are many different approaches. A value that represents the minimum mesh size of a sieve through which the grain of sand can just fall would be very practical. More than ten of these kinds of definitions are in use among sedimentologists.

When it comes to size classification, there is another difficulty that needs to be addressed: if we imagine enlarging the middle diameter of a pebble by 0.1 mm, it's plausible that its characteristics stay mostly the same. If, in contrast, we think about enlarging the diameter of a piece of desert sand by 0.1 mm as well, we will have doubled its size, through which we dramatically change its function in its environment. This is why, in the 1920s, geologist Chester Wentworth designed a nonlinear, geometric scale based on the ideas of Johan August Udden, which works in steps corresponding to doubling diameters and has proved so successful that it is still in use today and is certified according to DIN EN ISO 14688 norms. Here, sand is characterized into three categories: "fine" (0.063–0.2 mm), "medium" (0.2–0.63 mm), and "coarse" (0.63–2.00 mm). Technically speaking, common household sugar is therefore classified as "coarse sand," since the size classification does not consider the sample material. The vast majority of all sand grains in the world, by the way, belong to this size class, around 1 mm in diameter. It is interesting and astonishing that this finding can be attributed in origin to the underground conditions of solidification and crystallization of the molten rock that belongs to the granite group and produces a granularity of crystalline quartz, which in the end leads to the sand of seashores and deserts with which we are only too familiar.

Figure 3.2 Three sand grains in three different sizes, about 2.0 mm, 0.63 mm, and 0.063 mm, respectively (*from left to right*). These sizes refer to the distinctions between fine (0.063–0.2 mm), medium (0.2–0.63 mm), and coarse (0.63–2.00 mm) sand. Sample from Jungtal, Switzerland.

	Mud	Silt			Sand			Gravel		
		fine	medium	coarse	fine	medium	coarse	fine	medium	coarse
(mm)										
from		0.002	0.0063	0.02	0.063	0.2	0.63	2	6.3	20
to	0.002	0.0063	0.02	0.063	0.2	0.63	2	6.3	20	63
phi (circa)		8	7	5	3	1	0	-2	-4	-5

The typical size of quartz crystals in granite is approximately in this size class.

All common classifications refer to the (one-dimensional) diameter of the grains. If we look at the previous photograph of the different grain sizes (fig. 3.2), you get a prominent impression of the respective areas of the grains. A grain with a diameter of 2 mm has a diameter that is thirty times larger than a grain with a diameter of 0.063 mm and an area a thousand times larger. If we consider the volume or mass ratio, it's about 32,000. Since in many cases the physical behavior directly depends on mass as a parameter, so we must consider it in our calculation. The range between maximum and minimum grain sizes produces very different behaviors in their respective environments, such as in rivers, oceans, or dunes.

When it comes to scientific investigations and larger sample sizes, we don't work with the absolute diameter of a grain of sand, but with a reference and logarithmic quantity "phi," which has several advantages but is less intuitively accessible (phi = $-\log_2$ (d / 1 mm)). For size classes, sand ranges from phi = –1 (which means d = 2 mm) on the border to "fine gravel" to phi = +4 for very small grains on the border to silt (corresponding to d = 63 µm). For quantitative investigations of grain-size distribution in each sand sample, suitably graded sets of wired sieves can be used, which can be inserted into special vibrating machines. The sample is poured into the uppermost and coarsest sieve, which is then sorted into the various levels according to grain size during the shaking process (wet or dry).[23] In addition, methods are in use that measure the sedimentation velocity of the grains in liquids or, more recently, laser granulometry and computer-aided image-processing methods.[24]

In general, it can be said that the average particle size decreases with increasing distance between the supply area (i.e., the origin of the sand) and its deposition site. Smaller particles remain longer in the transporting medium during fluviatile (via rivers) and aeolian (via wind) transport.

Figure 3.3 Grain-size characterization of sediment according to Wentworth and DIN 14688. Diameters over 63 mm are classified as *rocks*.

The stronger the current, the larger the particles that can stay in it permanently (i.e., stay in suspension). Thus, a grain-size differentiation also takes place over the current distribution along the transportation routes. However, all attempts to create clear links back to where the sand originated have been unsuccessful. In this case, further parameters also must be taken into account.

3.1.2 Sorting

One speaks of well-sorted sand when all grains have a largely uniform diameter. This is rather rare. The degree of sorting is given on a mathematically defined scale and denotes the deviation from the mean grain diameter. Very good sorting carries the value 0.35, while very poor grading carries the value 2.0. The sand enthusiast and researcher Johan August Udden found a law called the "law of the chief ingredient" during his extensive studies on sand populations. It states that sand grains approximately equal to the mean diameter of the population occur most frequently.25 A number of conclusions can be drawn from the degree of sorting. Very rapid deposition usually leads to rather poor sorting, whereas long transport distances, violent water movements, and high wind speeds lead to better sorting. This is explained by the fact that the transport of sand in fluids such as water or air depends quite characteristically on its grain size. If, for example, we were to put a handful of sand with very different grain diameters into a very tall jar of honey, we would observe that over time the sand grains sort themselves strictly according to diameter. The longer we allow the sample to settle, the better the sorting will be in the direction of height, with larger and thus heaver grains in this example—unlike in nature—covering the longest distance by sinking the fastest. Of course, this applies only to the case where all grains have the same density (i.e., are monomineralic), which is the case for pure quartz sands, for example. Conversely, the density then becomes a sorting criterion, as can be seen in the example of heavy mineral soaps.

Furthermore, the starting material available in the supply area plays a major role. If the sand is a weathering product of an already well-presorted sediment, it has different starting conditions than if the sand grains originate directly from the erosion of a compact rock such as granite or basalt. Grains from weathered sandstone, for example, begin their existence in a state with higher degree of rounding and usually a better degree or sorting.

If you plot the diameters of the sand grains against their frequency or proportions in a diagram, important conclusions can be drawn about the depositional environment of the sediments.[26] You can also recognize if the sand sample has different populations of grain sizes, which can indicate changing sedimentation conditions.

However, none of the methods so far have produced clear results in the important distinction between river, beach, and dune sands. Beach sands tend to be very well sorted and have a higher proportion of smaller grain diameters. River sands are more poorly sorted and show a frequency distribution that is shifted toward larger grain diameters. Dune sands also tend to have larger diameters but are significantly finer grained than beach sands overall. Most the world's desert sands are in the "very fine" size class.

3.1.3 Density

The relative density of a substance is the mass of that substance compared to the mass of water of identical volume. The volume of 1 liter of water has a mass of 1 kilogram (and with it, a density of 1.0). The volume of 1 L (or 1,000 cm3) of quartz is 2.65 kg. This means that quartz has a density of 2.65. The density of minerals swings between 1 and 20, while the minerals that are predominately present in sand have a density of between 2 and 4. Minerals with a density of over 2.9 are characterized as heavy minerals. Along with the relative-density specification that is typical in mineralogy, there is also the physical density, measuring kilograms per cubic meter (kg/m3). It is easy to see that this unit is somewhat unwieldy for our purposes, which is why we use grams per cubic centimeter. .

3.1.4 Mohs Hardness Scale

Hardness refers to the resistance of a substance to mechanical intervention. The Moh's hardness scale, used here, has become generally accepted in the field of mineralogy because of its easy applicability for a quick and initial classification of the sample. The scale is divided into ten levels, which are constructed in such a way that each level is represented by a known and common mineral, and the mineral on the scale can scratch the mineral lower on the scale. Hardness 7 can thus score hardness 6. A fingernail scratches hardness 2; a copper coin, hardness 3; a pocketknife, hardness 5; and a better quality of pocketknife, *5½. However, it must be* considered that the levels chosen by Mohs have very different absolute hardness values (microhardness), so the scale is not linear.

The shape of sand grains, the grain morphology, is a very conspicuous feature that can be roughly estimated in the field. In addition to the grain shape, the degree of rounding and the sphericity of individual sand grains are considered. All three characteristics described below can be determined quite well, even quantitatively, with the aid of comparison tables.

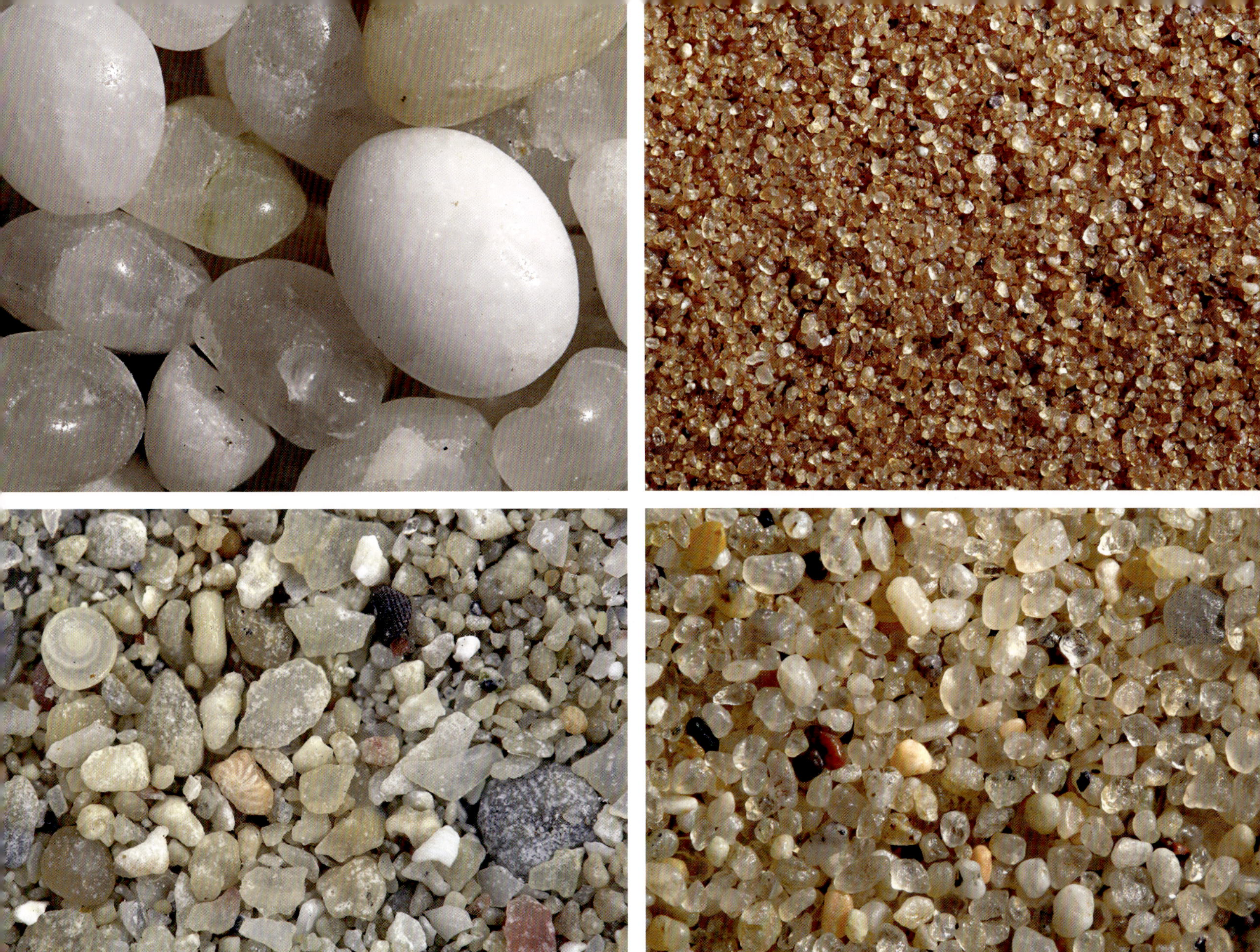

Figure 3.4 Coarse sand (bordering on fine gravel) that is well rounded and well sorted and has high sphericity. Italy, Sardinia, Sinis Peninsula, Is Arutas Beach. Image width about 8.5 mm.

Figure 3.5 At the same scale: fine sand with medium rounding, well sorted with medium sphericity, and with an average grain diameter of about 120 *μm*. The red coloring springs from iron oxide that has been deposited on the quartz surfaces. USA, Utah, Arches National Park, Pine Tree Arch. Arid inland desert. Image width about 7.7 mm.

Figure 3.6 Sand sample that is poorly sorted with different grain-size subpopulations. United Arab Emirates, city of Dubai. Image width about 10 mm.

Figure 3.7 Sand sample that is well sorted. The grains sizes are largely uniform. USA, California, Mojave National Preserve, Kelso Dunes. Image width about 9.6 mm.

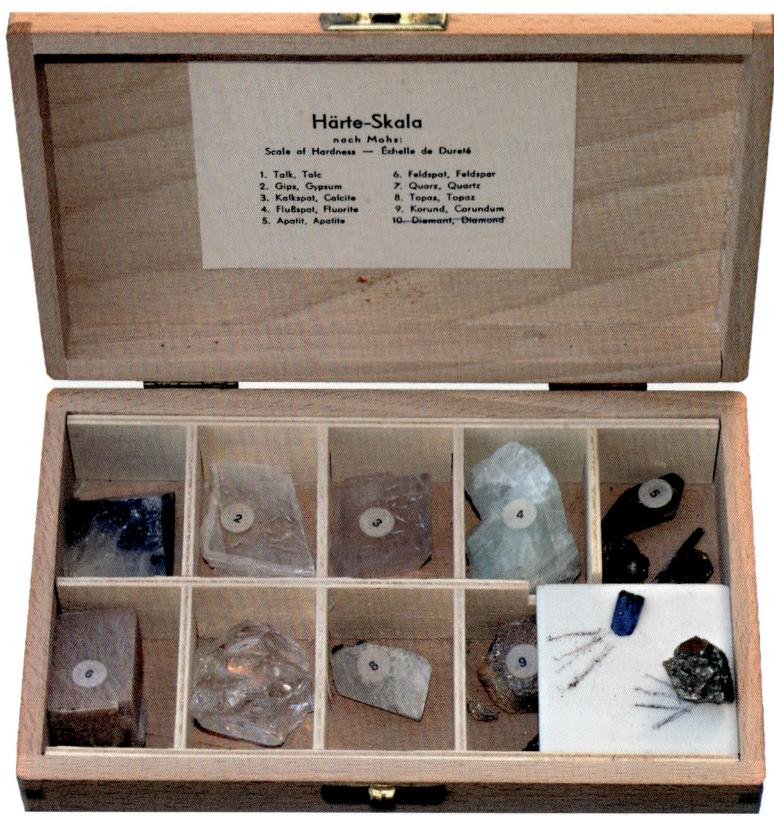

Mohs Hardness	Mineral	Vickers Hardness	
1	Talc	2.4	Can be scraped with a fingernail
2	Plaster	36	Can be scratched with a fingernail
3	Calcite	109	Can be scratched with a copper coin
4	Fluorite	189	Can be lightly scratched with a pocketknife
5	Apatite	536	Can be scratched with a pocketknife
6	Orthoclase	795	Can be scratched with a steel file
7	Quartz	1,120	Can scratch a windowpane
8	Topaz	1,427	
9	Corundum	2,060	
10	Diamond	10,060	Can be scratched only by itself

Figure 3.8 Mohs hardness scale with mineral samples and porcelain plate for establishing the line color. On the plate you can see black-blue hematite with a dark-red line color on the left and a gold-colored pyrite with a black line color on the right. Color and line color do not necessarily correlate.

Figure 3.9 Mohs hardness scale with the Vickers hardness test as a comparison. The Vickers test is a common technique in engineering, where a diamond pyramid is pressed into the material with a constant force and the depth of penetration is used to measure hardness.

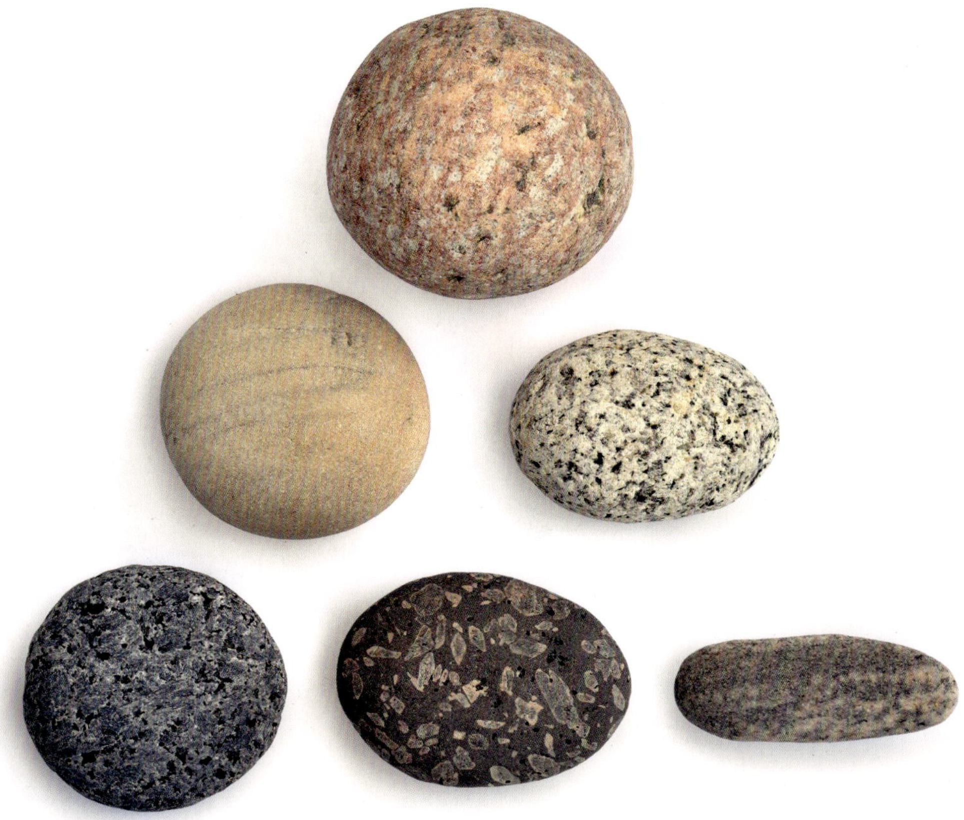

3.1.5 Grain Shape

In each grain of sand, one can imagine a rectangular coordinate system with three axes, simplified to length, width, and thickness. The grain shape is defined by the relation of these three axis lengths to each other. There are three basic shapes, which can be represented graphically as the corners of a triangle. One corner represents tabular shapes, disc shapes, and flat shapes; the second corner represents elongated, stemlike shapes; and the third corner represents the compact, cube, or sphere shape (fig. 3.10). Each grain of sand can be roughly classified into this scheme, although the transitions, of course, are fluid. This scheme is unsuitable for biogenic forms, with their special and more-complex structures and symmetries. The source rock and mineral structure have a lasting influence on the shape of the later sand grain. This dependence on the mineral structure is much less present in rocks, where factors such as texture, bonding material, schistosity, etc. dominate. The properties of the sand grains, on the other hand, usually directly follow those of the mineral from which the sand grain is made. For example, layered mica minerals such as biotite or muscovite form tabular grains as they wear along their cleavage planes. During wearing processes, geometrically similar grain shapes tend to be formed again and again.

Figure 3.10 Basic forms of grain morphology demonstrated with glacial debris from the west coast of Denmark. *Bottom left*: Flat form, two axes are the same, lower value for the upper axis. *Bottom right*: Elongated, angular shape; only one spatial axis pronounced. *Top of the pyramid*: All three spatial axes are of similar size (granite, diameter around 70 mm).

The behavior of the grains as they travel by water and wind is also influenced by the grain shape. Grain shape and transport therefore influence each other. For example, the transport of spherical grains tends to take place in a rolling manner, and that of tabular grains in a pushing manner. In contrast, materials such as quartz or garnet, which have the same properties in all directions (homogeneous and isotropic), can assume almost any shape; here, transport and depositional environment are more decisive for the respective resulting grain shape. If there are pronounced cleavage planes, as in the case of feldspars, angular grain forms with fractures are often found, which become more rounded through increasing erosion by wind or water.

3.1.6 Roundness and Sphericity

The degree of roundness and sphericity can be determined for individual grain sand and can be very informative. However, they can also be easily confused. Sphericity of a grain means how well it approximates the spherical shape (from sphere = ball). In most cases, a simple distinction is made between greater and lesser sphericity. Grain roundness, on the other hand, refers to the degree of roundness of all protrusions, recesses, edges, and structures existing on a grain, which is why an egg has a good degree of roundness but low sphericity. The scale according to Folk27 assigns the values 0–6 to grain roundness. It is very common and useful, which is why we reproduce it here in detail:

Well rounded (5–6):	All original structures of the grain of sand such as edges, corners, and protrusions have been destroyed though abrasion. The grain has soft curvatures without flat areas.
Rounded (4–5):	The original edges and corners become more uniform and are replaced by curves. Some flat zones or broken surfaces may be preserved.
Partially rounded (3–4):	Partially rounded. The original shape of the grain is still recognizable, although affected by erosion.
Subangular (2–3):	Angular, but free of sharp corners. Without gentle curves; the original shape and surfaces are partially preserved.
Angular (1–2):	Grain with sharp edges and corners. Erosion barely recognizable.
Very angular (0–1):	No erosion, freshly broken edges and corners

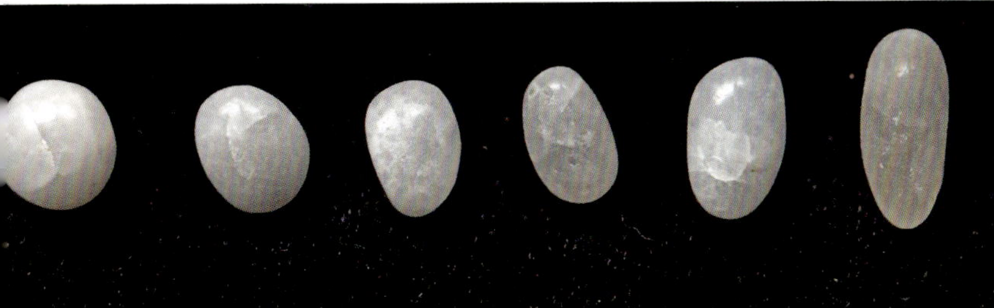

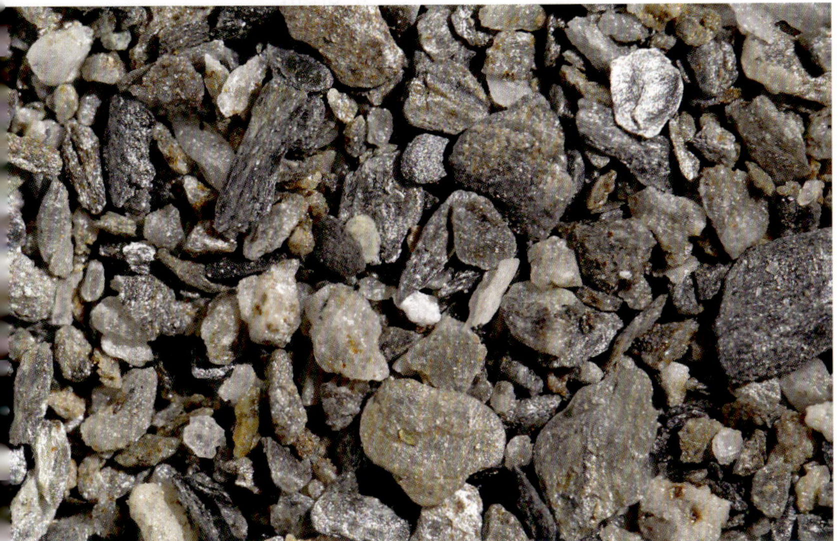

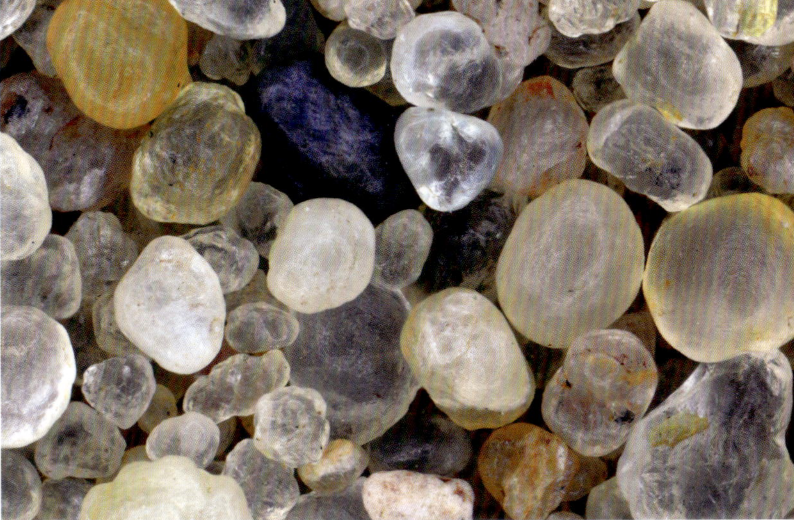

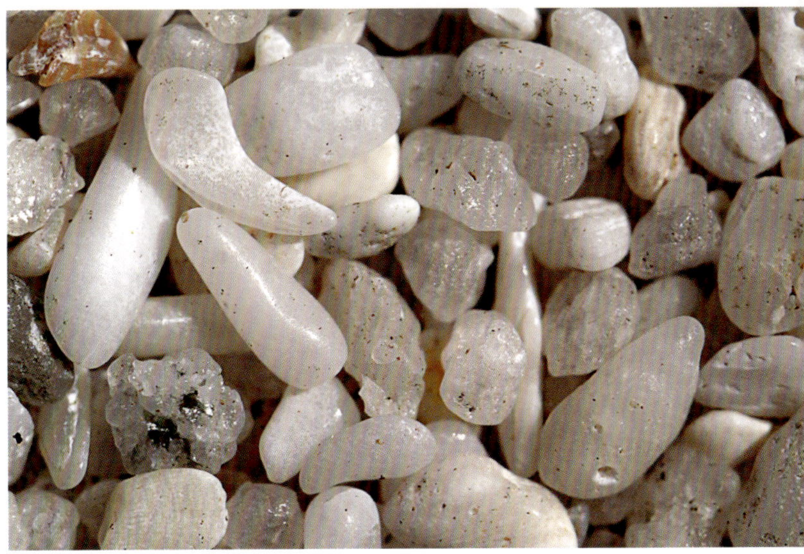

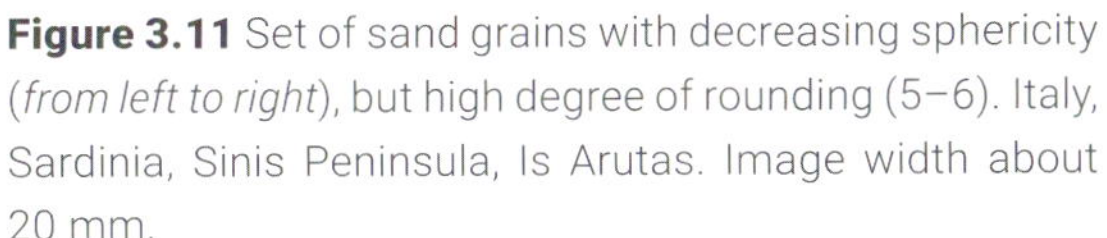

Figure 3.11 Set of sand grains with decreasing sphericity (*from left to right*), but high degree of rounding (5–6). Italy, Sardinia, Sinis Peninsula, Is Arutas. Image width about 20 mm.

Figure 3.12 Poorly sorted, immature river sand from Ubaye, St. Paul, French Alps. The grains are angular (sharp edged) to very angular and don't show any kind of sphericity. Image width about 8.5 mm.

Figure 3.13 Quartz sand, well rounded, and with differing, partially high sphericity and moderately sorted. Sand sample from the Bolivian inland desert of Lornas de Arena near Santa Cruz. Image width 7.0 mm.

Figure 3.14 Sand sample with very well-rounded grains, low sphericity, and moderate to low sorting. South Africa, Cape of Good Hope. Image width about 8.5 mm.

Sphericity and grain roundness are mostly independent from each other—you can easily imagine perfectly smooth grains that are long, as well as very circular forms that have coarse, angular structures. This can also be seen in the photo examples. The rounding process takes place very slowly, even slower than size reduction caused by erosion.[28] This means that larger grains of sand tend to be rounder than are smaller or miniature grains. Unfortunately, it is still impossible to directly infer the transport route from the degree of roundness. Nevertheless, grain roundness is a very important characteristic that can be used to interpret the origin and transport conditions of a sand sample. Direct comparisons should be made only between that samples that also have the same grain-size distribution and approximately the same mineral content.

3.1.7 Compositional Maturity

In contrast to the properties discussed so far, compositional maturity29 cannot be determined from individual grains; rather, it is the property of an entire sample. It denotes the proportion of quartz (and some other weathering-resistant minerals such as zircon and rutile) in the total sample. The higher the degree of maturity, the higher the proportion of quartz. For sands and sandstones, one observes a tendency for the degree of maturity to increase with the age of the sample. On the other hand, this process is time-controlled and is dependent on the supply of kinetic energy during the transport processes. The more mature a sediment is, the lower the proportion of clays, rock fragments, plant remains, and minerals with low mechanical and chemical resistance. Suspended particles are flushed out, chemically unstable components dissolve, and coarser components eventually break down into their constituent parts over time. Also, the degree of sorting increases with compositional maturity. Fresh sands in the process of formation, which have emerged on-site through weathering from rocks and have come together from different supply areas, are in a very immature state. The average content of erosion-prone feldspars in (secondarily formed) sandstones is only 10%–15%, whereas this proportion is about 75% in crystalline rocks. In contrast, coastal or desert sands that are well sorted by long aeolian or fluviatile transport routes and almost monomineralic when it comes to the quartz proportion represent the highest degree of maturity of a sediment. It is easy to observe that sands become lighter in color with increasing maturity. While immature and feldspathic sands usually appear beige or mottled, pure quartz sand appears white. In very simplified terms, it can be summarized that "the lighter and more quartz, the more mature and farther traveled."

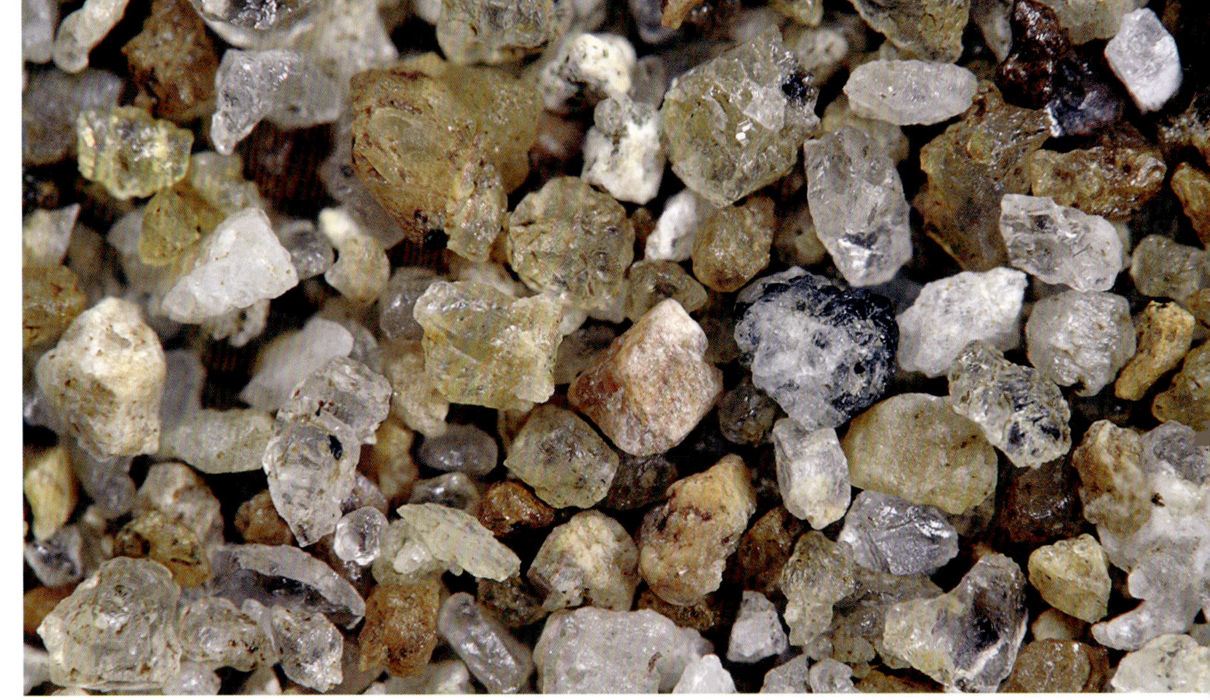

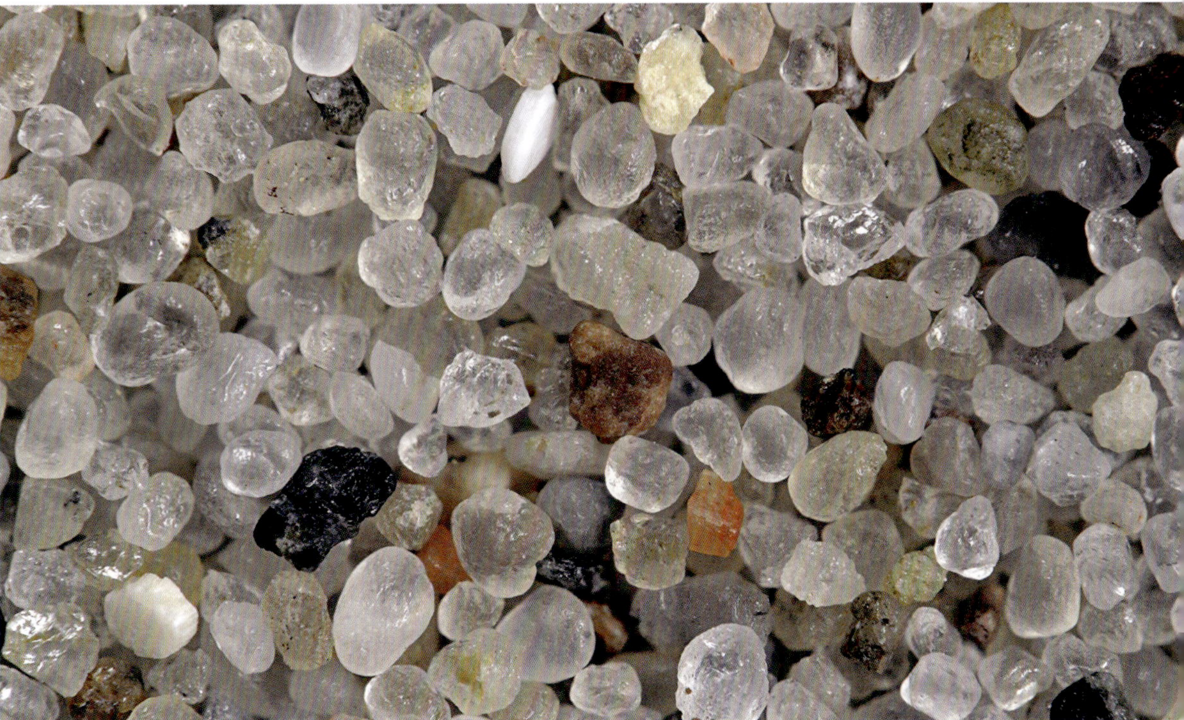

Figure 3.15 Sand sample from the highlands of central Norway (Valle) with low to moderate compositional maturity. The sand grains come from the edge of a boggy, originally glacial lake and show subangular to partially rounded structure with low sphericity. Main mineral composition: quartz, feldspars, mica. Image width about 8.3 mm.

Figure 3.16 Coastal sand from Fehmarn, Baltic Sea. The grains are mostly well rounded, partially with high sphericity. Quartz makes up the majority, so the sand has a high compositional maturity. Image width 7.8 mm.

3.1.8 Color

While the properties of "mixed-grain sand" presented so far were quite easy to describe, it is a bit more complicated when looking at the property of "color." We first want to clarify what is meant by "color" in general and what is meant by the term "color" in connection with minerals and, of course, sands. For this, we will go a little deeper into the matter, although even then we can only touch upon the very complex phenomenon and work with it in part. We preface with the following three statements about color, which are important and noteworthy:

1. Color does not exist without light (you can easily test this yourself).
2. Color is not a property that can be firmly assigned to a body.
3. Color is a subjective perception that takes place in the mind of the observer.

Goethe remarkably devoted a large part of his life to color and its effect on human beings, and ultimately also his most extensive work, the *Theory of Colours*. In its introduction we read the sentence that perhaps most beautifully summarizes the broad field of colors from his point of view:

> "The colors are acts of light; its active and passive modifications."

But what do these "active and passive modifications" look like? For this we turn toward physics. After all, physics isn't bad, but simply the description of nature and light, which seems so self-evident but is in fact a highly interesting and complex phenomenon. In the middle of the nineteenth century, the brilliant physicist Michael Faraday identified light as an electromagnetic radiation with wave nature. For the first time he described electricity and magnetism in a common concept. James Clerk Maxwell (1831–1879) finally summarized this concept in a set of equations. Maxwell's equations are still valid but are supplemented and extended by Einstein's theory of relativity and the so-called quantum electrodynamics.

As a result, we now know that light, amazingly, can exhibit the properties both of waves and those of a stream of particles (quanta). Waves (also water waves) generally possess a wavelength and an amplitude or wave height. Sunlight as white light can now be imagined as a bundle of very many wave trains with different wavelengths, which also oscillate with different amplitudes. However, the human eye is sensitive to only a limited range; namely, to the wavelengths between about 0.4 μm (blue) and 0.8 μm (red). The wavelength determines the color impression in the eye, and the amplitude determines the intensity. Wavelengths outside this range, such as radio waves or UV radiation, remain invisible to us.

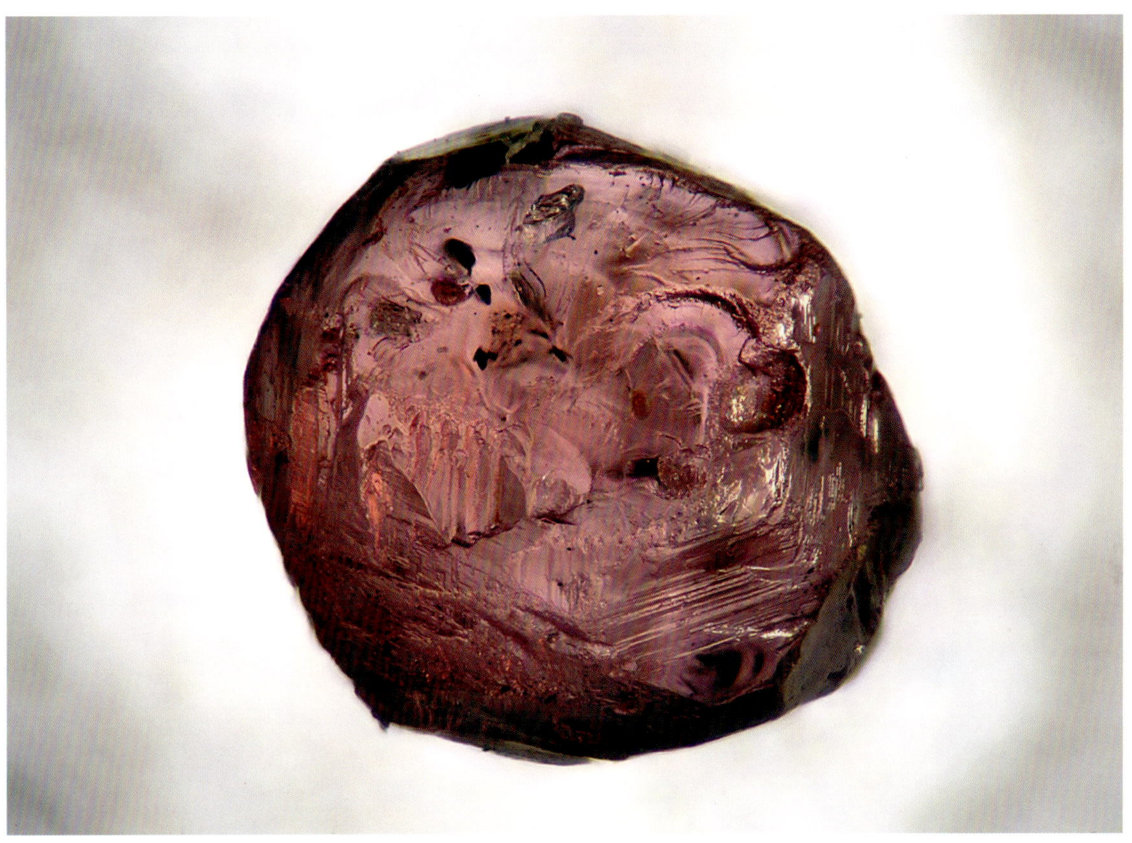

Let's illustrate this with an example. What exactly happens when a ray of light falls on a pinkish-red grain of sand made of garnet? Such grains are found in coastal sands. Let's travel along a light wave, from the place the light originates, the sun, to the sensory impression "pink" in our brain: the "white" sunlight surrounding us consists, as we described, of a mixture of all wavelengths or spectral colors. This light is produced by nuclear reactions in the sun and passes through the vacuum of space at the speed of light. The light takes only a few minutes to travel to Earth. When it hits the earth's atmosphere, it is scattered and finally falls on our grain of sand.

The crystal lattice of the grain of sand made of garnet absorbs certain green components of the incoming sunlight. These are swallowed up, as it were, and converted into heat. What remains is the counter color, in our case a pinkish-red hue, which results from the remaining rays of the originally white light.

Figure 3.17 Sand grain made of the garnet mineral almandine. Norway, sand beach at the Porsangerfjord. Grain diameter 630 *μm*. Reflected light with transmitted-light component.

At this point we have taken a big leap. "Is" the light red or does it only "seem" that way? After what we have mentioned previously, it's clear that the latter is the case. The grain of sand is not red, nor is the light itself. Without light, there is no absorption in the crystal, and without any absorption there is no change in the light spectrum. In the dark, all matter is colorless, as much as this seems to contradict our intuition. After all, to check you would just have to turn on the light!

Now we have followed the light from the sun to the grain of sand, but an essential building block is still missing in the phenomenon of "color": our perception. So how does a light ray with a certain wavelength become the sensory impression "red"? Why do I experience the color "red" at all? Part of the answer is already contained in the question: it is "only" a sensory impression, caused and generated in the human brain. The eye alone cannot see! It is part of the perception apparatus in the brain. Nobody knows how a fellow human being experiences the color red.

The world around us is projected through the lens of the eye to the retina in the back of the eye, like onto a sensitive screen. There are two different types of light-sensitive cells (technically speaking: sensors), called rods and cones. The approximately 125 million rods are extremely sensitive to light, but they distinguish only light and dark, not colors. The wedge-shaped cones, of which there are about seven million, are responsible for this. They are less sensitive and occur in variants that respond only to the wavelengths for red, green, and blue. This is the key to color perception. In the case of the garnet grain, more cones of the red-sensitive type are activated in the retina, and their impulses are transmitted to the neocortex, the multisensory part of the cerebral cortex, where a series of downstream signal-processing steps, a kind of image processing, takes place, which, however, remains hidden from our consciousness. The brain performs extraordinary feats in this process. The actual perception of color therefore takes place in the brain and not in the eye. It is the same with all other stimuli. So much about the phenomenon of color.

Sand never appears in pure (spectral) colors—it always appears to us as a mixture of the colors of all sand grains reflecting light, as we have seen in the example of the garnet grain. How does the mixture of colors work now? Here, we can differentiate between additive and subtractive mixing. In additive color mixing, lights of different colors are projected on top of each other. This can be easily seen by using flashlights and color filters. The more lights or lamps that are added, the brighter the projection becomes as a result. If lights with the colors red, green, and blue are placed on top of each other, the result is white light as a mixed color. This is easy to understand from the explanations given above, since each of these colors addresses one type of cone, with their sum causing a visual perception of white. If one mixes only two lights, you get magenta with blue and red, cyan with blue and green, and yellow with green and red.

It's astonishing that we perceive the color yellow, for example, even though there physically is no light frequency that directly corresponds to the color yellow.

Subtractive color mixing is more relevant for the color perception of mixed sand grains. This can be illustrated by holding different-colored foils against a bright window: in our case, yellow-, magenta-, and cyan-colored filters are held one behind the other, resulting, of course, in a darker light beam passing through. Each filter absorbs its respective complementary color. Thus, the yellow filter absorbs the color blue, the magenta filter absorbs the color green, and the cyan filter absorbs its complementary color red. If blue, green, and red are removed from the light by absorption, we perceive black as a mixed color because no cone on the retina can then be stimulated.

We find these conditions with color powders used as pigment mixtures, like sands, which consist of different-colored components. However, it is not possible to get black powder by mixing three color powders: the closest you get is something like gray. This is mainly due to a lack of an unreachable spectral purity of the individual components, as well as refraction phenomena. Black sand always consists of black sand grains and cannot be obtained by mixing colored sands.

Throughout his life, Goethe never got used to the idea that white light "consists" of all colors, which can be "divided" through a prism, and that conversely, the combination of spectral colors "results" in white light. His approach to perception, description, and explanation of phenomena, including color phenomena, was completely different. He considered the given phenomena in the following way. He understood it as a whole and that it was simply presented in this way to the senses. For him, the act of perceiving color stood alone: the pure reception via the senses was to be separated from the assignment of a subsequent theoretical meaning.

If you continue to follow this thought, both perception and meaning are linked in the customary process of empirical activity. For the observer it usually remains hidden that every act of perception already contains a kind of directionality and thus a mental assignment of meaning. If I catch sight of a tree, I assign the meaning "tree" to the sensory stimulus pattern almost simultaneously, unreflectively, and almost inseparably. This can be understood very well by casually looking at the picture shown below (fig. 3.19) with black and white areas and at first recognizing nothing but these very areas. Suddenly the pattern of a giraffe might become recognizable.[30] However, the purely sensory activity on the retina has not changed at all during the switch from indifferent pattern to the actual giraffe.

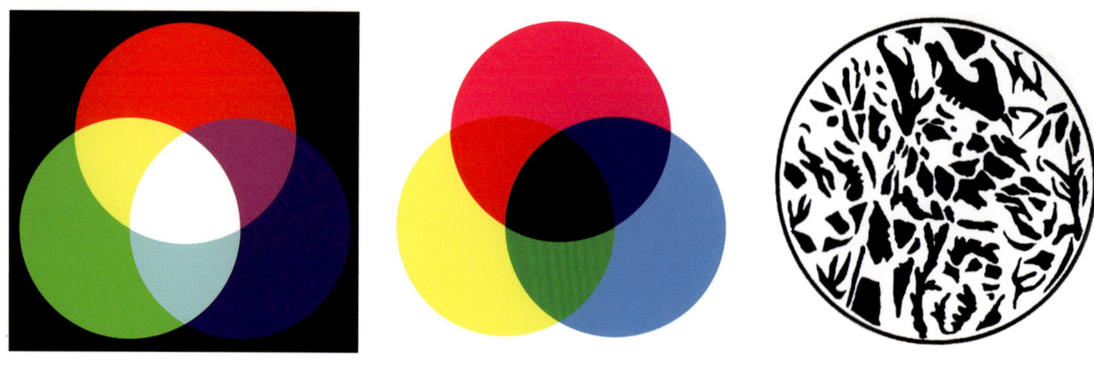

The brain assigned the pattern from one moment to the next the meaning "giraffe." However, we still do not see a giraffe, but a pattern with which we couple the stylization of a giraffe in the extended act of perception. In the case of the example shown, another strange effect occurs. Once the meaning "giraffe" is assigned, it is very difficult to observe the pattern neutrally—the giraffe continuously appears as a linked perception.

Painters, photographers, and artists generally know that things under different lighting situations can evoke quite different color impressions regardless of their actual color. A forest edge in the light of the setting sun appears in different colors than it does in the bright light of midday. The human brain can separate a large part of these external influences from the perception of the object. An object remains identical to us, even if it takes on different colors under changing light situations, which are quite verifiably and measurably different. A red tomato in the living room under artificial light remains red for us even if we bring it into the bright sunlight, without the color change being immediately noticeable or even causing the impression of a change in the object itself. The brain possesses other constancy phenomena other that the one described, which, for example, can also affect the size of an object when viewed from different perspectives.

What has been said applies both to surface colors of solid objects as well as to colors of transparent objects such as crystals or liquids. The completely different color perception of a turbid medium (e.g., a few drops of milk in a glass of water) with changing illumination in front of changing backgrounds is very revealing, caused by effects of frequency-dependent scattering. Goethe studied this extensively and derived from turbid mediums a kind of color scale that resembles prismatic colors and is directly observable in atmospheric phenomena.

Figure 3.18 *Left*: Additive color mixing done by projecting lights of different colors on top of each other. *Right*: Subtractive color mixing through laying different color filters on top of each other.

Figure 3.19 Pattern, see text.

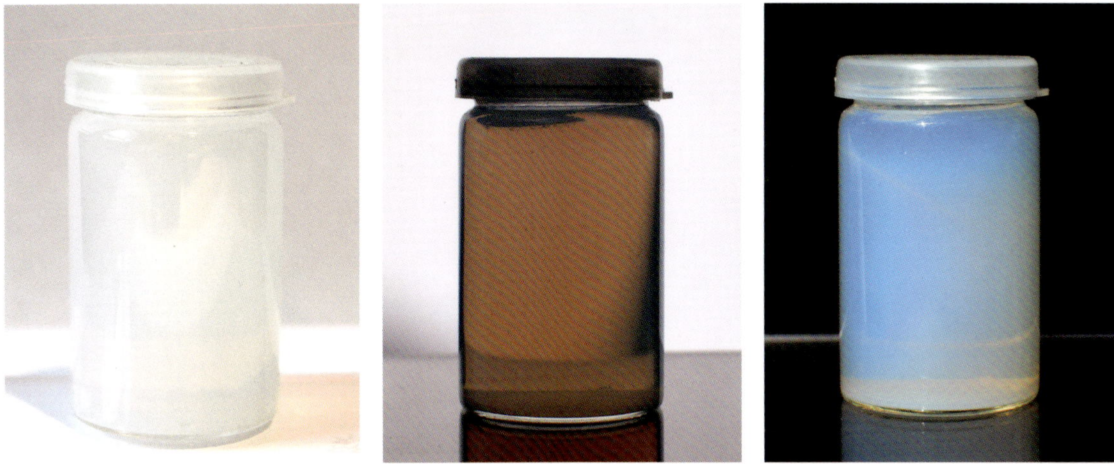

The turbid medium of the atmosphere changes the light-yellow color of the sun into orange and red colors over longer distances (sunrise and sunset). When you're looking toward the zenith in the high mountains, the sky appears deep blue to dark blue, almost an extenuation of the black space behind the atmospheric layer. Longer distances over air layers, on the other hand, result in sky blue, and finally, when you're looking toward the horizon, a whitish, cloudy pale blue or light gray.

While the brain does most color adjusting automatically and in a way that is unnoticeable to us, when it comes to photographic images, care must be taken to ensure that the colors are displayed correctly, meaning in a way that corresponds to the way in which they are perceived. For digital cameras, this is usually done through a function called "automatic white balance," which is based on the image analysis. The problem that is solved here is that a white sheet of paper appears white to us as an observer, at twilight, in the shade, or in the sun, but not at all to the camera sensor.

There is another aspect that is unique to photographing sand: the question of liquid exposure. Mineral mixtures and stones change their color impression significantly as soon as they are wet, causing the reflection behavior of the surfaces to change as a result. In most cases, the colors become clearer and stronger underwater (fig. 3.22). This has been experienced by anyone who has picked up a pretty stone along the shore, only to be disappointed once it dried and the pretty colors became hardly recognizable.

The majority of all naturally occurring sands can be classified within a very subdued color range. Apart from monomineralic sands, there is always a mixture of minerals and thus also a mixture of color shades. Earth tones, beige tones, and all kinds of gray shades predominate. The proverbial white-sand beach is only really white in the rarest of cases. Pure white can generally be achieved only with pure quartz sand that is free of impurities but always tends toward light gray.

Figure 3.20. Color perception is highly dependent on the lighting. If a turbid, colorless liquid (*far left*) is lit in front of a dark background, it appears to be blue (*far right*). The same liquid, when placed in the shade in front of a light background, appears orange colored (middle). All three pictures were taken of the same sample glass.

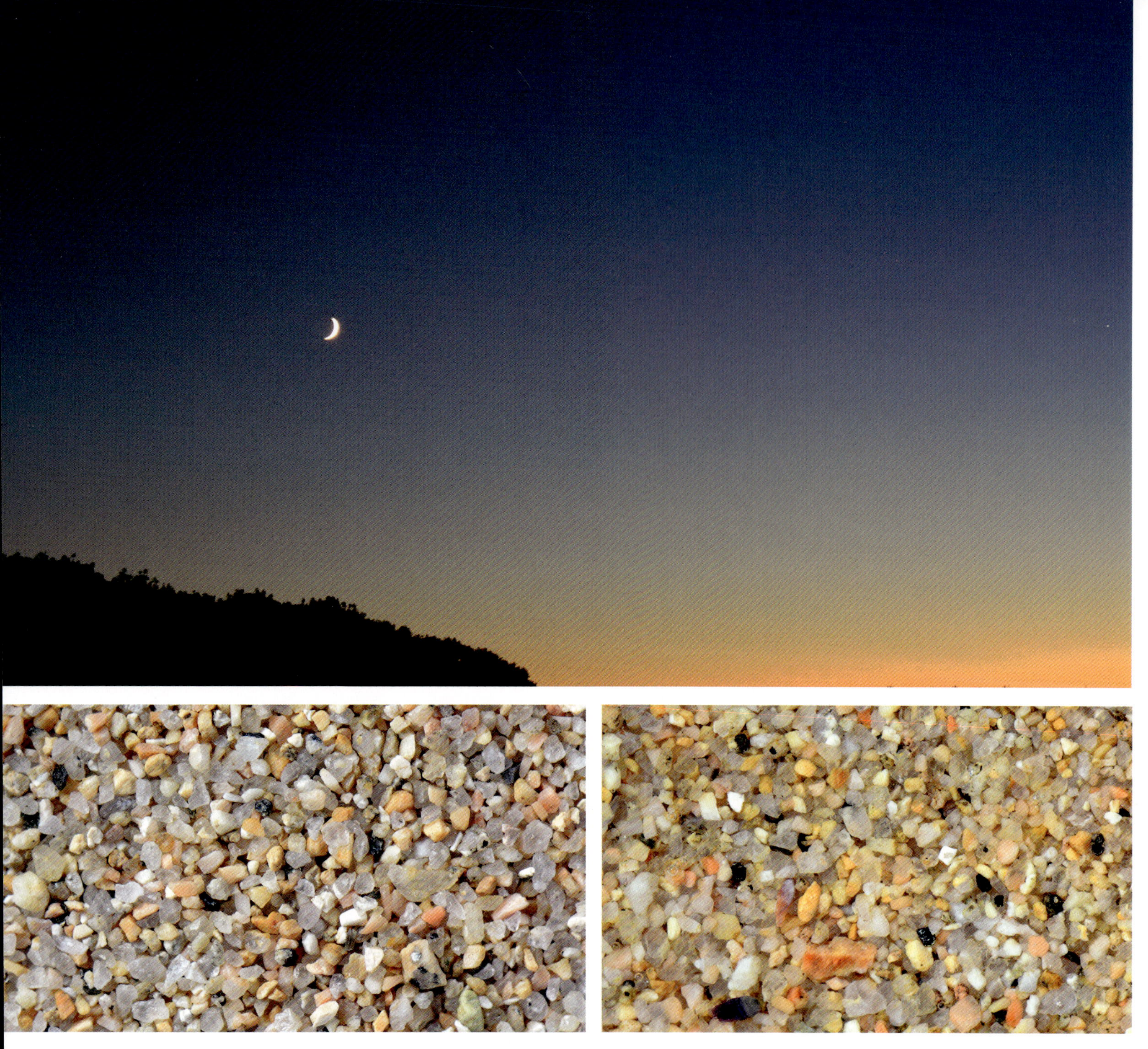

Figure 3.21 Spectrum of colorful impressions from orange to dark blue in the evening sky, influenced by the atmosphere between the light-giving sun and the darkness of space

Figure 3.22 Identical sand sample (origin: Norway), dry on the left and moistened on the right. The camera settings and white balance are identical. The more intense, yellowish coloration of the wet sediment is clearly visible.

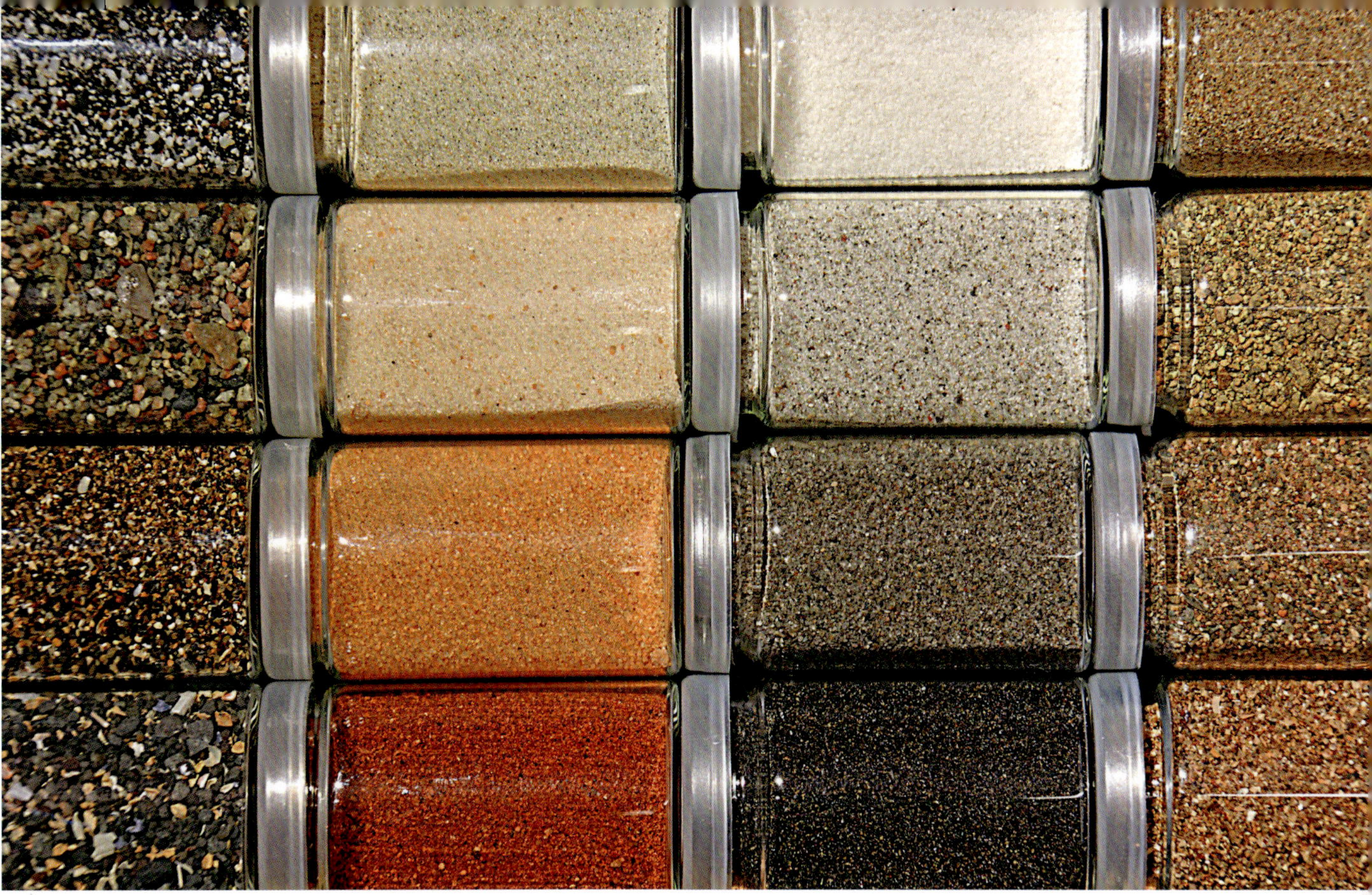

In comparison, biogenic sands made of white coral appear even whiter, also without significant additives. The beige coloration of most continental beach sands is caused by feldspars or tinted silicates, while in biogenic sediments, it is more likely to be caused by shell remnants. The more immature the sand, the more beige it appears. Rarer, greenish colorations derive from mostly volcanically formed minerals such as olivine or pyroxenes. Otherwise, volcanic sands are matte black to glossy black. Dull coloration predominates when there is a microbubbly grain texture, while glossy black is achieved by dense, closed, and polished grain surfaces. Metallic black coloration indicates the presence of magnetite or ilmenite, which are opaque in transmitted light, meaning they do not let light through and can also easily be extracted from the sand with a magnet. The latter, together with red garnets, are the main ingredients of many placers.

When sands of different origins come together, whether from geological causes or local sediment behavior, such as coral fragments, shell remnants, and lava sand, they can cause the sand to look sprinkled with contrasting coloring ("salt and pepper"), where the individual components remain noticeable and do not become lost in a mixed color such as beige or gray.

Figure 3.23 Sequences of typical color shades found in sand. The places of origin are listed from top to bottom. *Left*: Contrasting mineral mixtures including mixtures of shell fragments. Machalilla, Ecuador; Kirkenes Norway; Jeju, Korea; Snaefjellsjökull, Iceland. *Left center*: Sequence from beige to red through increasing proportion of hematite or additives of flesh-colored minerals such as feldspar. Sands such as this are often found inland, such as in inland deserts. Goa, Caravela Beach, India; Murzuk 4, Libya; Murzuk 2, Libya; Alice Springs, Australia. *Right center*: Sequence of gray tones as they are typically found on continental beaches. Alamogordo White Sands, Mexico; Sehlendorf Beach, Germany; Toronto's Lake Ontario, Canada; W-Limbe, Cameroon. *Right*: Earth-colored sands: Coloring with very different origins. Iron oxide to flesh-colored feldspar and portions of dark minerals and organic fragments can lead to this type of coloring, but they do not appear very often. Colakli Beach, Turkey; Landmannalaugar, Iceland; Gobi Desert, Mongolia; Darwin, Australia.

Quite often, one finds gradations of red, which, as already mentioned, is explained by thin hematite layers that have settled around the sand grains. In rare cases, garnet additives are responsible for the red coloration.

In addition, mineralogists describe what is called the "streak color" of minerals. In this process, the specimen is rubbed over a plate of rough, unglazed porcelain. The abrasion, the "streak," has the complete characteristic coloration of the mineral powder and is diagnostically relevant. Surface color and streak color do not have to match. For example, black hematite has a blood-red streak color while the golden-yellow pyrite has a black streak color.

We will now include a picture gallery with colored sand grains and some special-colored sands. Actually, there are as many colored sand grains as there are colored minerals, so we will give just a few examples.

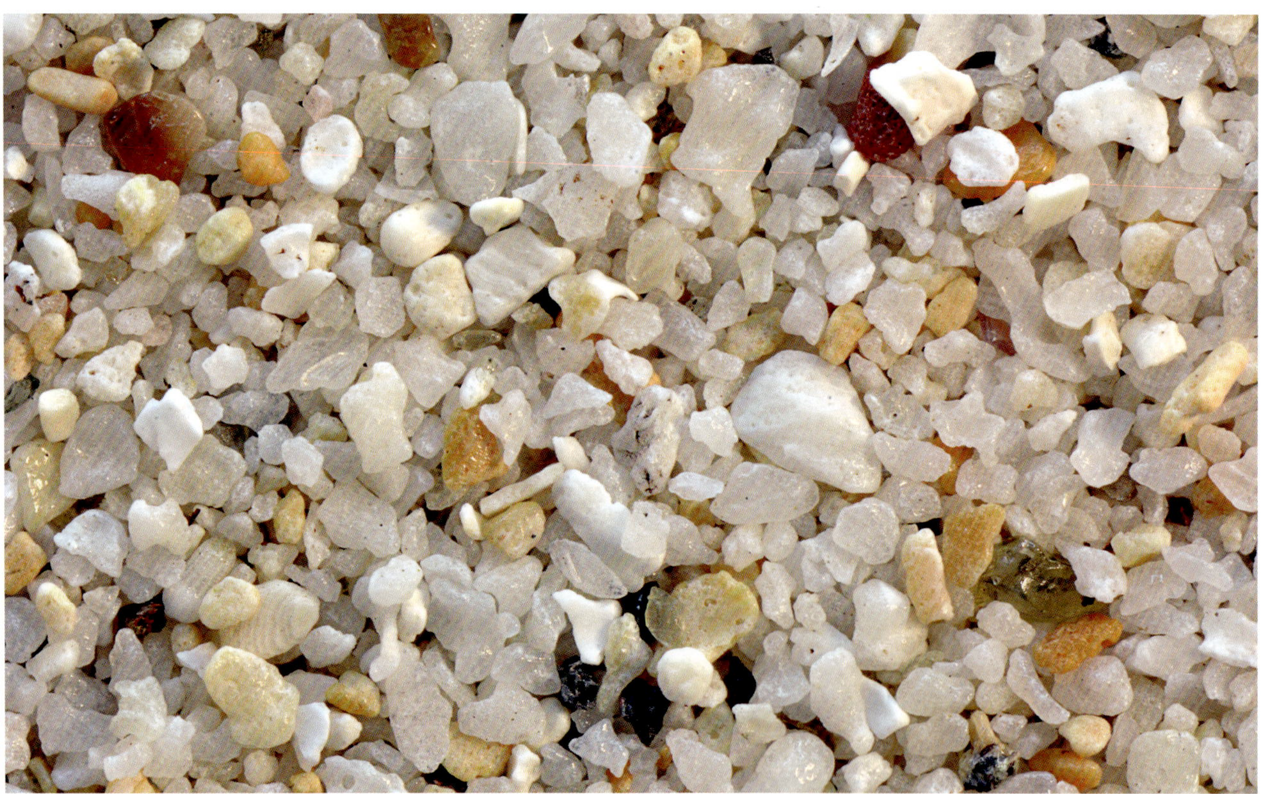

Figure 3.24 Sand that appears to be cream colored from the Galápagos Islands, Isla Floreana, Punta Cormoran, Ecuador. The beige color appearance comes from the mix of white and honey-yellow-colored shell fragments. Image width 5.5 mm.

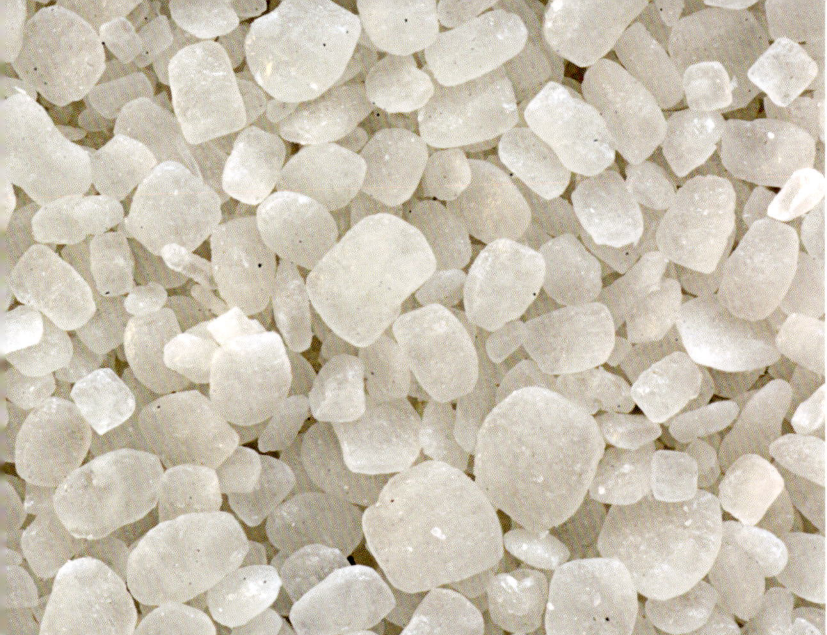

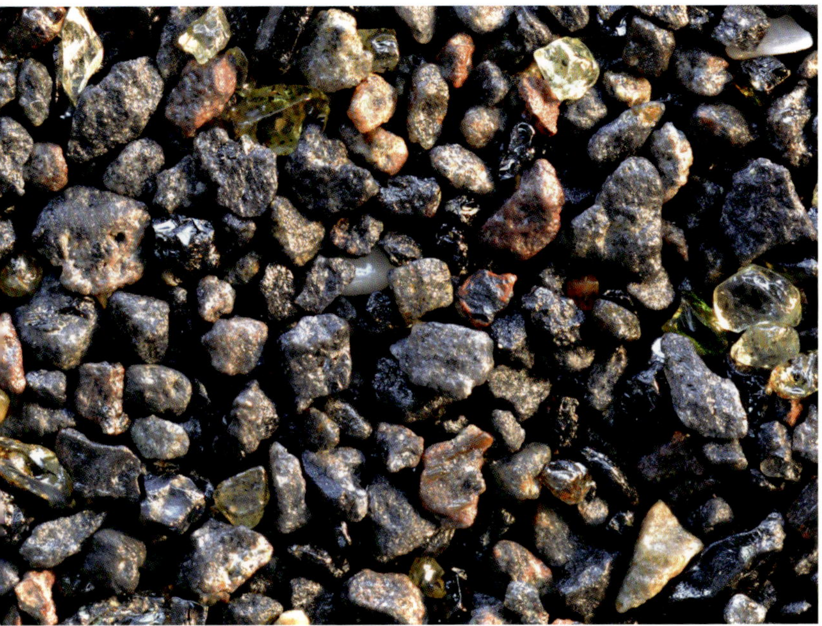

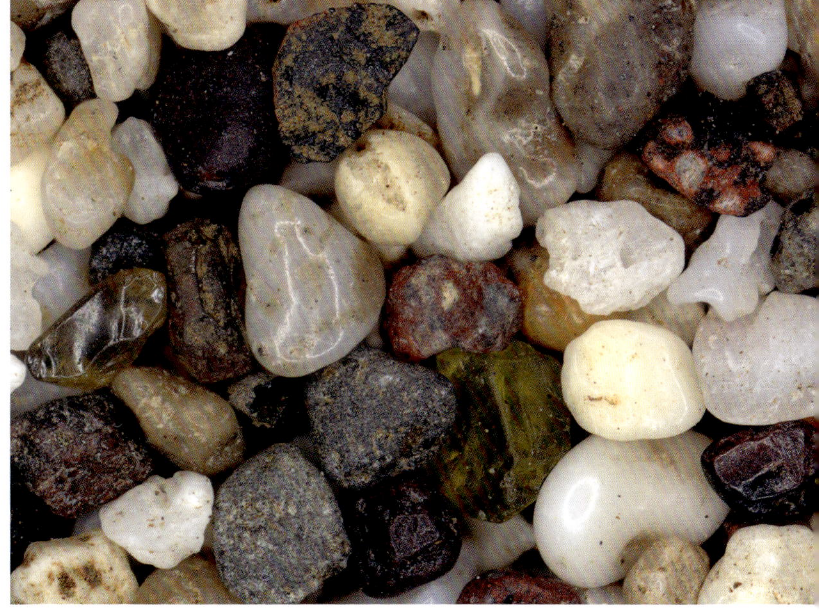

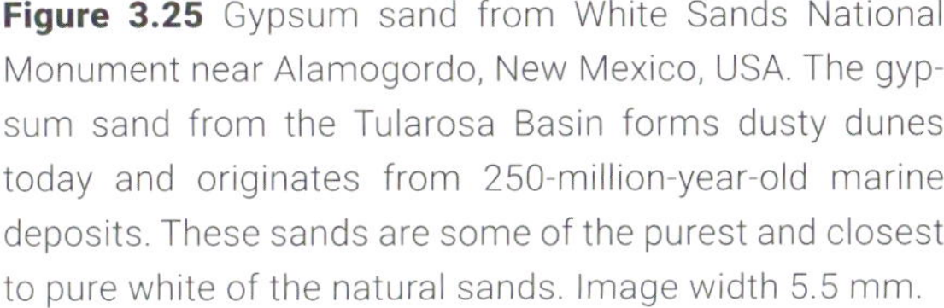

Figure 3.25 Gypsum sand from White Sands National Monument near Alamogordo, New Mexico, USA. The gypsum sand from the Tularosa Basin forms dusty dunes today and originates from 250-million-year-old marine deposits. These sands are some of the purest and closest to pure white of the natural sands. Image width 5.5 mm.

Figure 3.26 Puerto de la Cruz, Tenerife, Spain. Very black-looking volcanic sand that gains structure and tonal range only after enlargement and illumination. Among the natural sands, you can rarely find sand that is a deeper black—the only brightness you see comes from light reflections. Image width 5.5 mm.

Figure 3.27 Sand from Plage Grand Ansa, La Reunion, Seychelles. A typical sprinkled "salt and pepper" sand. The effect comes from the combination of dark volcanic minerals and white biogenic fragments. The grains are well rounded and polished. Image width 5.7 mm.

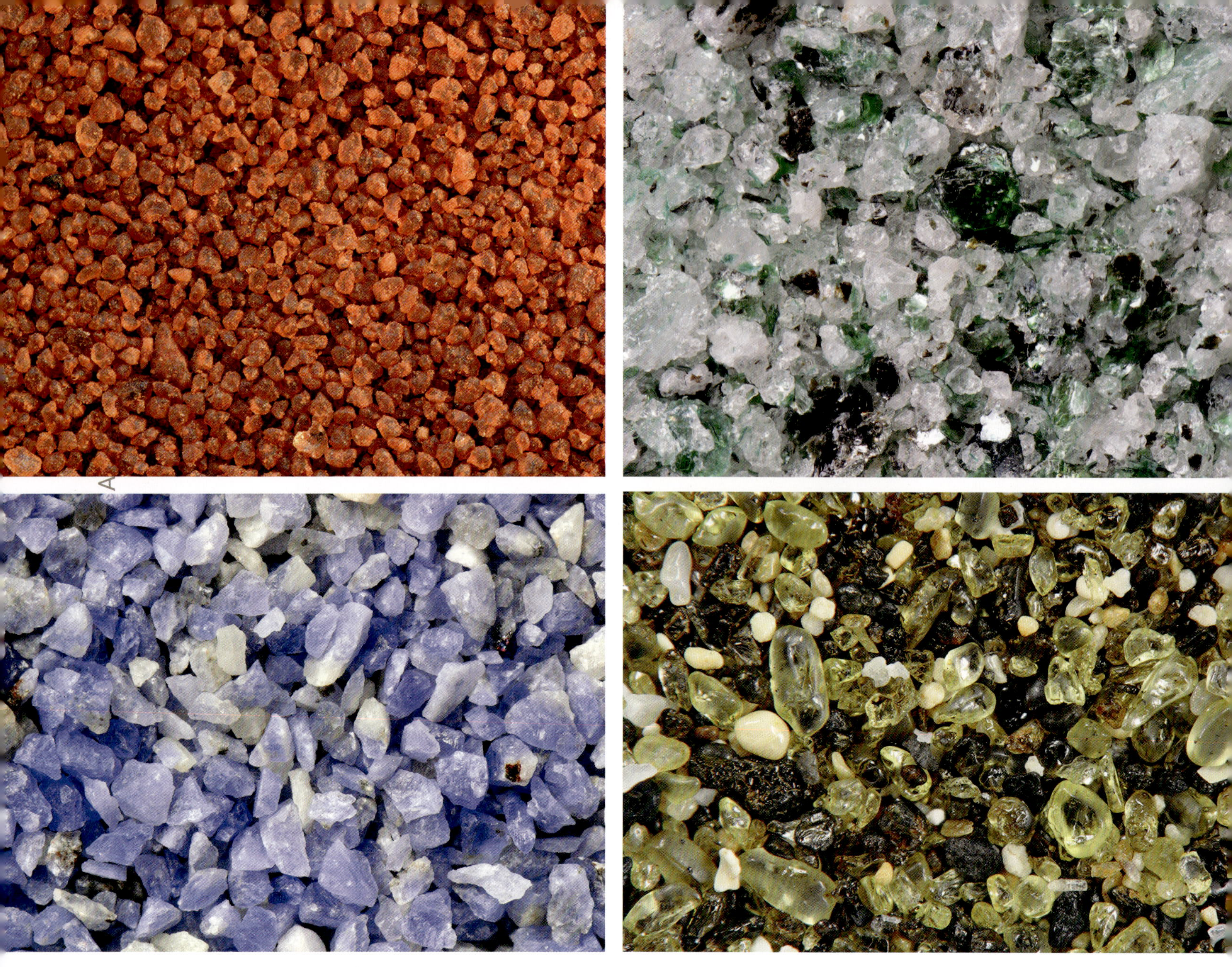

Figure 3.28 Sand from the ocher quarry in Rustrel, France. The colorful, very fine ocher-colored sands from Rustrel are well known and were coveted by artists for mixing paints. Image width 4.9 mm.

Figure 3.29 Blue sodalite sand from mine tailings in Kunene, Namibia. Sodalite, such as hauyne and nepheline, belongs to the feldspathoids or feldspars. Image width 4.9 mm.

Figure 3.30. Masi quartzite is the raw material of this cold-green sand. The green grains consist of the chromium-containing mica mineral fuchsite. Northern Norway. Image width 17.5 mm.

Figure 3.31 The Green Sand Beach, Hawaii, USA, is one of only four known beaches worldwide with green sand. The heavy mineral mixture of volcanic origin is composed of olivine, hornblende, clinozoisite, aegirine-augite, and other materials. Image width 8 mm.

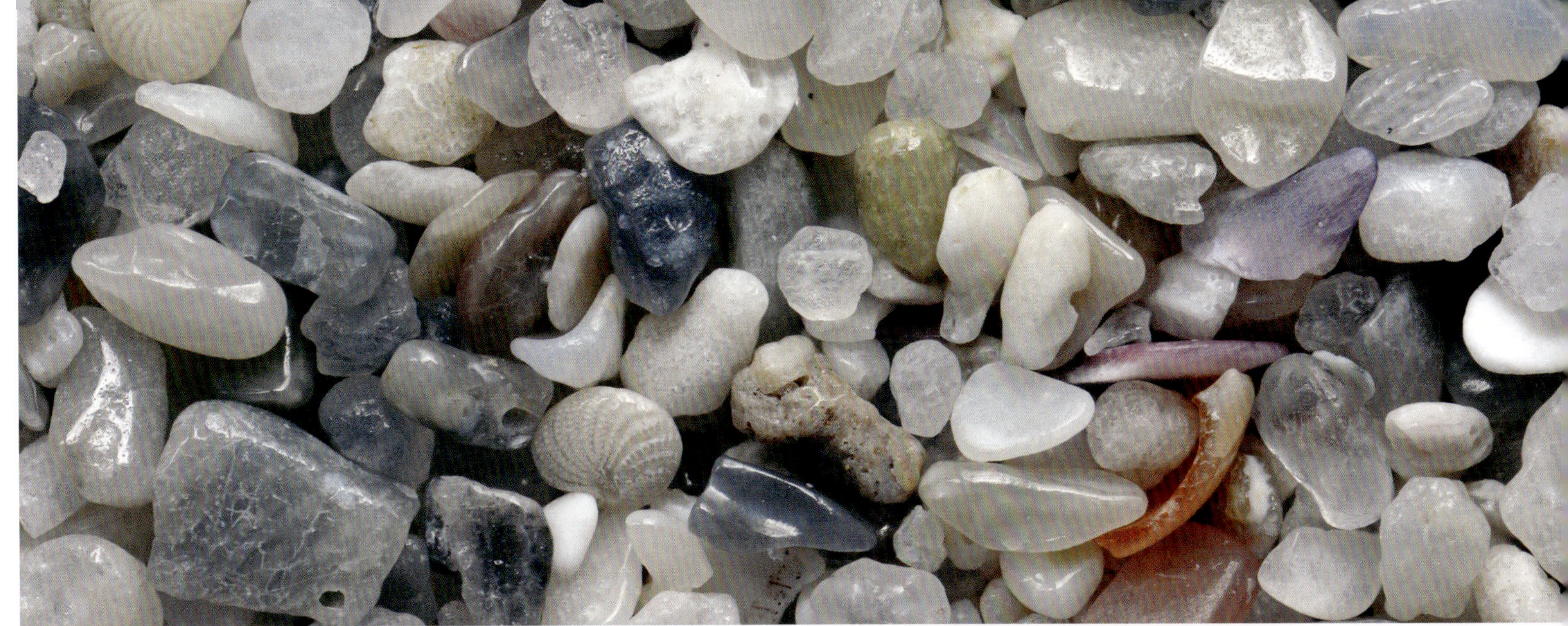

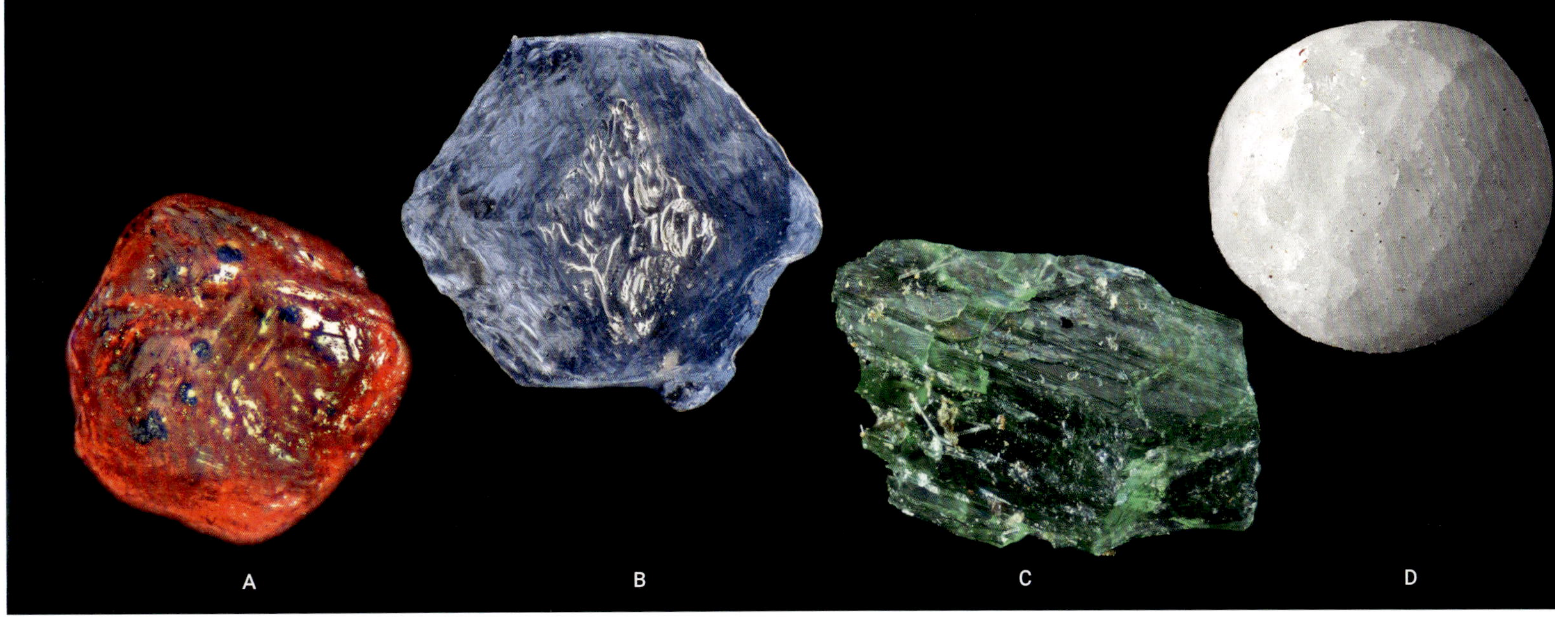

Figure 3.32 Sand with a very rare light-gray coloration. The overall gray impression increases with increasing distance: the colored grains fade into the background.

Figure 3.33

A Red grain of sand made of granite. Diameter 850 *μm*. Porsangerfjord, Norway.

B Blue idiomorphic grain of sand made of hauyne. Diameter 500 *μm*. Southern Tenerife, Spain.

C Green grain of sand made of Ca-clinoamphibiol (actinolite). Width about 1,000 *μm*. Finero, Italy.

D White polycrystalline grain of sand made of quartz. Diameter 2,600 *μm*. Is Arutas, Sardinia, Italy.

3.2 Components of Inorganic Sand

Until now, we have dealt with sand and its characteristic as a mixture of grains, and now we want to take a closer look with a few specific examples. Just as a cook must know his ingredients, we should learn to understand the main ingredients of average sands. We will look at individual sand grains and understand them as mineral particles that can be examined and identified through suitable methods. We will not go into the corresponding physical principles, the research methodology, or the necessary equipment technology. Chapter 7 is dedicated to biogenic sand grains (fragments of mussels, snails, etc.).

Of the more than five thousand minerals, we are particularly interested in the most-important rock-forming minerals, which are also the main components in the sands of the world. Among them, the silicates take a prominent position. The two most common elements in the earth's crust, silicon and oxygen, combine here on an atomic level to form a tetrahedron and the basic building block of all silicate minerals on Earth.[31]

— *Framework silicates*: The building blocks are strongly connected, like a scaffold. Quartz is the purest form of the framework silicates, which also include the feldspars. The latter form spatial arrangements between SiO_4 and AlO_4 tetrahedra. Framework silicates belong to the most-common silicates—they are structurally very hard.

— *Layered silicates*: The building blocks are connected in layers. The layers among each other are only loosely bound. Water is always incorporated. Examples: mica, chlorite, clay minerals.

— *Ribbon silicates*: The building blocks are connected as chains or ribbons. Rodlike to fibrous minerals consisting of tetrahedral double chains. Amphibole family with hornblende as the main representative. Medium hard.

Since silicates play such a decisive role in the composition of sand and account for over 90% of the volume of the earth's crust, we will now introduce their main representatives in more detail: quartz, feldspar, and mica minerals.

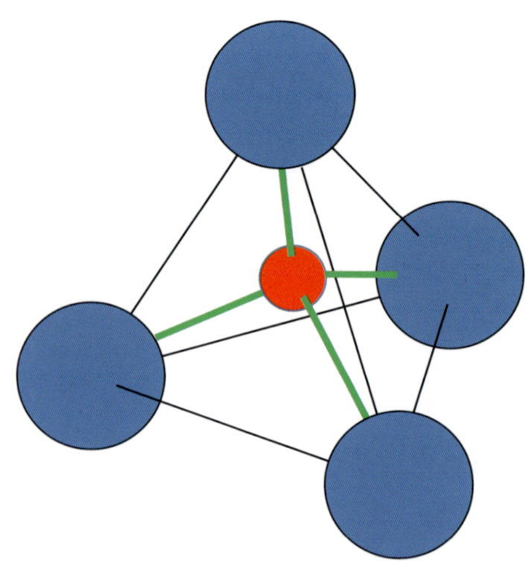

Figure 3.34 Basic building block of silicate minerals: an atomic $[SiO_4]^{4-}$ tetrahedron consisting of four oxygen atoms (blue) and a centrally arranged silicon atom (red) in the tetrahedral gap. The oxygen atoms are with a ratio of 1.27 to 0.34 angstroms, much larger than the silicon atoms, and would, if represented dimensionally, be closely adjacent to each other.

3.2.1 Quartz

Formula:	SiO_2
Mohs hardness:	7
Density:	2,65
Crystal system:	trigonal

Quartz (SiO2), as the most important mineral of the earth's sands, consists, as described, of the two most abundant elements in the earth's crust by mass percentages, silicon and oxygen. Although it's chemically not quite correct, geologists speak of silicic acid. Consequently, rocks consisting predominantly of quartz are called "acidic," and, in the absence of quartz with all conceivable transitions, "basic." Rocks rich in silicon melt at temperatures around 750°*C* –800°C, whereas rocks low in silicon do so only at much-higher temperatures, around 1,100°C.

Quartz in its pure, crystalline form is completely colorless and transparent, is relatively hard, and fractures without any preferred direction in a shell-like form. The known quartz varieties such as amethyst, rose quartz, citrine, etc. get their coloration from added or incorporated metal ions, or, in the case of smoky quartz, from crystal defects caused by radioactive cosmic radiation. Quartz is extraordinarily stable against chemical and mechanical influences.

The chemical and physical stability of quartz is the reason why many sands consist almost exclusively of quartz grains, and quartz also occurs as the most common mineral in sandstone, with an average of 65%.[32] For all its chemical resistance, quartz is soluble in water, albeit to a very small extent, and conversely can reattach to sand grains as a silicate layer if appropriate conditions exist. Amorphous silicates precipitated from such solutions are known as opal, chalcedony, agate, or flint, depending on the formation conditions.

Of geological importance is a small series of organism groups that build their skeletons from opal, a kind of silica gel in which water is additionally incorporated. In addition to diatoms, which inhabit seas, rivers, and lakes in impressive diversity, radiolarians and siliceous sponges should also be mentioned, the latter of which we will discuss in chapter 7, "Living Sand," in more detail.

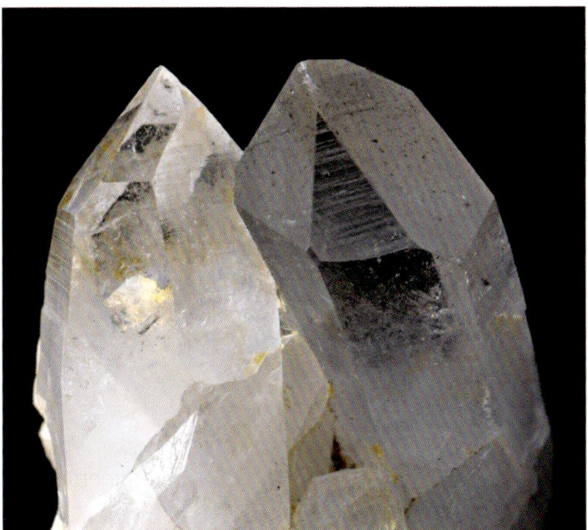

Figure 3.35 Quartz stage, low quartz. Image width 33 mm.

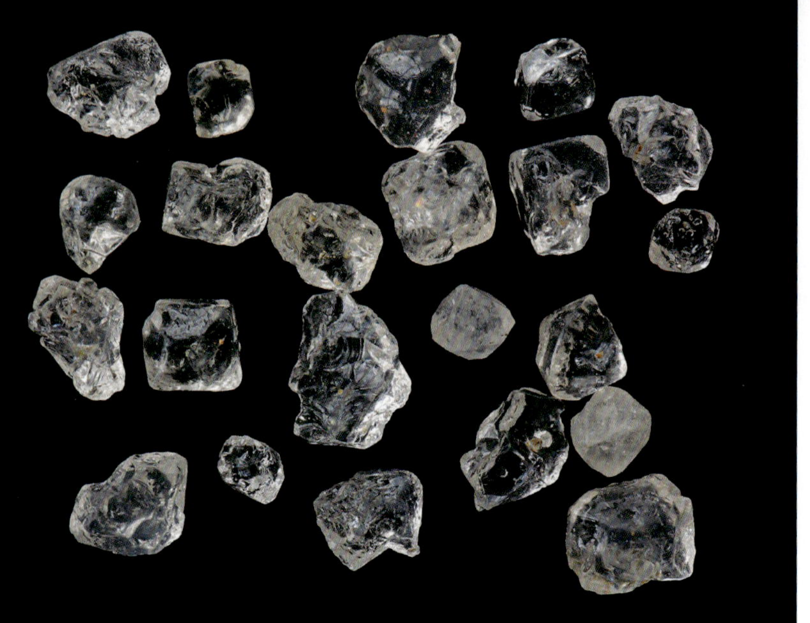

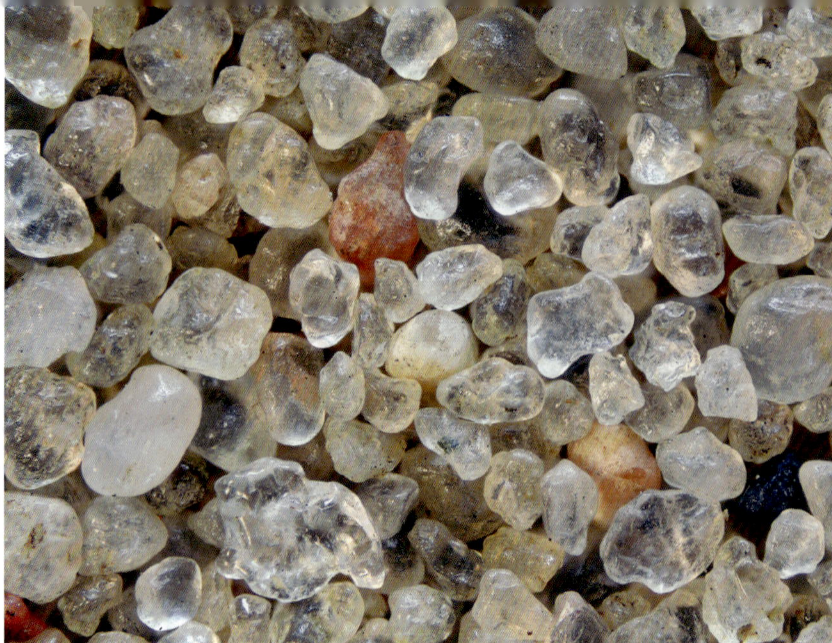

All three groups of organisms are of considerable importance for the formation of mostly marine sediments and are also found in marine sands. The annual production of siliceous skeletons is 70 billion tons according to studies by Calvert (1974). However, only about 1% of this is preserved in the ocean—the rest is dissolved again.[33]

Diatoms (also called phytoplankton), in particular, are a fascinating group of species that perform photosynthesis and can be found most about anywhere there is light and water. Diatoms have dimensions ranging from a few thousandths of a millimeter to a millimeter and are divided into two groups according to their symmetry: centric (circular, marine) and pennate (more boat shaped, and common in fresh water). Their diversity is impressively high, with an estimated 15,000 species. Many diatom species are interesting border crossers: during the day, they float in or on ocean water, only to sink in the evening and be washed up on the beach by the waves. There, they adhere to the grains of sand with sticky mucilage, from which they detach in the morning to be carried away again by the waves into the open sea. Diatoms can even be found in sea ice: they protect themselves with a mucilaginous substance that acts as an antifreezing agent down to 14°F (–10°C).

Chalcedony is crypto- to microcrystalline quartz with very fine fibers: chalcedony is therefore not its own mineral. Chalcedony forms dense masses, which are mostly colorless to cloudy translucent but sometimes appear colored due to chemical additions, and then these bear special names: red colorings are carnelian; green colorings are chrysoprase. If layers deposited on top of each other are rhythmically and differently colored, they are called agate. The latter is very popular because of its decorative appearance. Unfortunately, you find artificially colored specimens for sale quite often that are not always easy to identify but are often conspicuous by their much-too-bright colors. Jasper is a very intensely colored, completely opaque type of chalcedony.

Figure 3.36 Crystalline quartz sand grains polished by the waves of the surf at Playa Santa Clara, Panama. Reflected light, grain diameter around 1 mm, image width around 3.6 mm.

Figure 3.37 Sand from an inland dune at the Besenhorster Sandberge near Geesthacht, Germany. The sand is almost exclusively composed of quartz grains that were deposited during the last ice age in the glacial valley of the Elbe. Image width 3.8 mm.

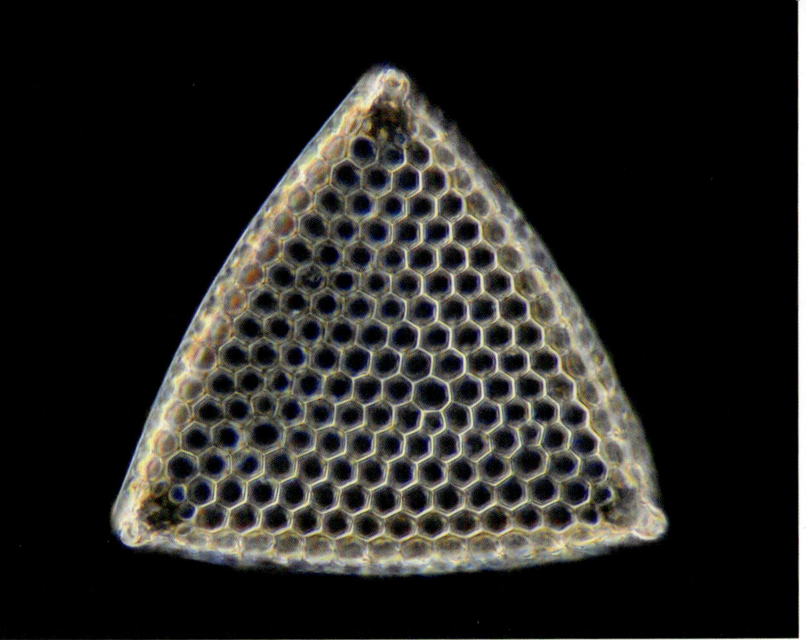

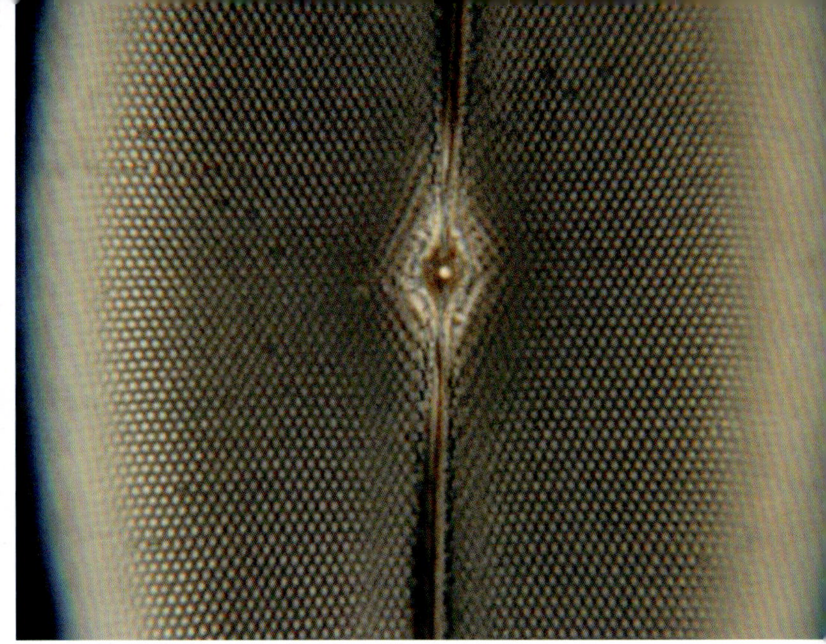

Here, too, several additional denominations appear when available for purchase, but microcrystalline quartz is always their basic material. Chalcedony is often formed by the crystallization of silicate aqueous solutions at not very high temperatures. This forms cavity fillings in volcanic rocks, or general fissure fillings in any kind of rock. The quartz concretions in sedimentary rocks are particularly striking and conspicuous, which are due to siliceous skeletal parts of the previously mentioned diatoms and other microorganisms. These skeletal components, consisting of a water-containing and microcrystalline quartz form (skeletal opal), dissolve preferably at slightly elevated temperatures and pressures and enrich the pore water in sediments with silica. The quartz-containing solutions circulate in the pore network, crystallize amorphously, and finally form lenticular or nodular concretions.

Quartz appears in two different crystal formations: low quartz, which crystallizes trigonally up to a temperature of 573°C, and the so-called high quartz, which crystallizes hexagonally at temperatures between 573*°C and* 870°C. At even higher temperatures, tridymite is formed, and at temperatures over 1,470°C, cristobalite is formed. In addition, there are the two high-pressure modifications coesite and stishovite, which are formed at pressures above around 35,000 bar (coesite) and 100,000 bar (stishovite). If you know the crystal formation, you can make inferences about the origin conditions. Pressure at those kinds of levels occurs almost exclusively during impact events of cosmic matter, so the presence of a quartz high-pressure modification indicates meteorite impact. This is a very important indicator when talking about these types of events that occurred long ago.

Both coesite and stishovite have been found in the meteorite craters in Arizona and Nördlinger Ries. Coesite crystals have also been discovered in metamorphic rock, which indicates that the accompanying parent rock sank to depths of over 62 miles (100 km) before they rose back to the earth's surface through mountain-forming processes.[35]

Figure 3.38 *Left*: Diatom shell from *Triceracium favus* (North Sea, Germany). The hexagonal shell elements have the highest structural integrity. Transmitted-light dark-field illumination. Image width 340 *µm*.
Right: Diatom shell from *Pleurosigma angulatum* (La Rochelle, France), whose regular shell structures (grating constant about 0.5 *µm*) are close to the optical resolution limit of microscopes. *P. angulatum* was therefore often used as a test specimen for checking the quality of objectives. Transmitted-light dark-field illumination. Image width 48 *µm*, total magnification around 1,600×.

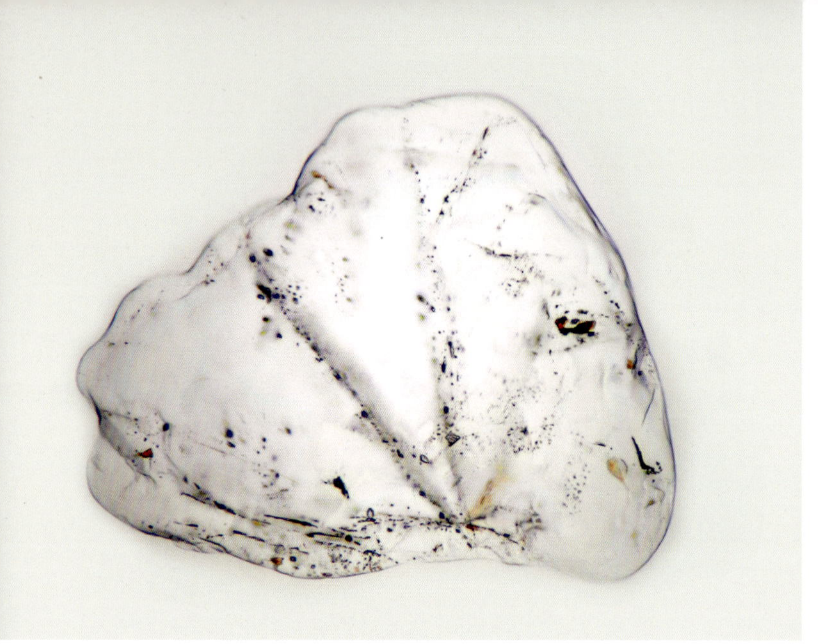

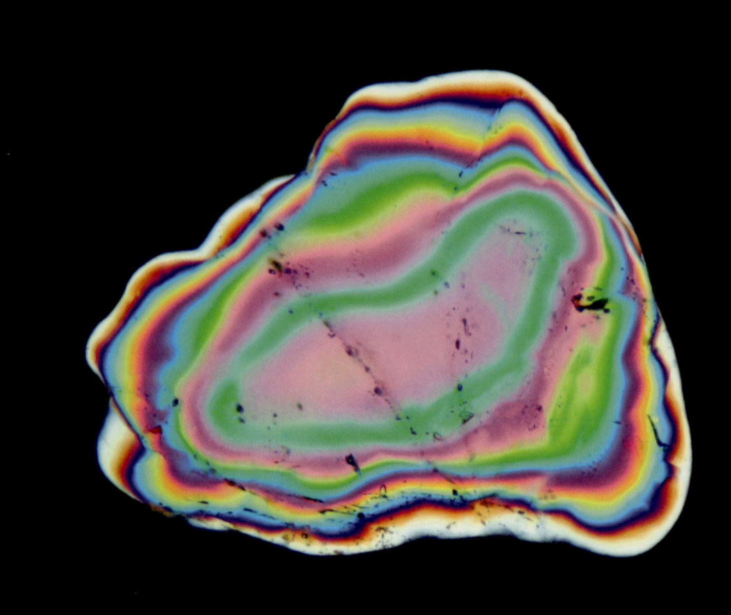

Especially in volcanic areas, small bipyramidal, crystal-clear grains of high quartz can be found in sand, which, together with possibly present inclusions, can give clues about the area it originated and formation conditions (see fig. 6.24). Quartz occurs only extremely rarely (in eucrites) in meteorites themselves.

As mentioned, the colors of quartz minerals can vary, which makes the identification of sand grains using only a magnifying glass difficult and can lead to confusion with other minerals. Often, for example, in the case of desert sands, this is the reddish hematite. Depending on the thickness of the layers, the grains appear anywhere between light and dark red. The following list for the coloration of the quartz grains does not refer to these layers but to substances that color the crystal body itself:[36]

Figure 3.39 Typical grain of sand made of quartz in transmitted-light bright field. The grain comes from Copacabana, Rio de Janeiro, Brazil. The edges are well rounded with low sphericity; many small inclusions are grouped in rows. Grain 560 × 465 *μm*.

Figure 3.40. The same grain, with the picture taken between crossed polarizers. From the outside in, the sequence of interference colors up to the fourth order of red can be seen. The color series becomes paler as the grain thickness increases inward. From the inference phenomena, clues to the mineral present can be deduced.

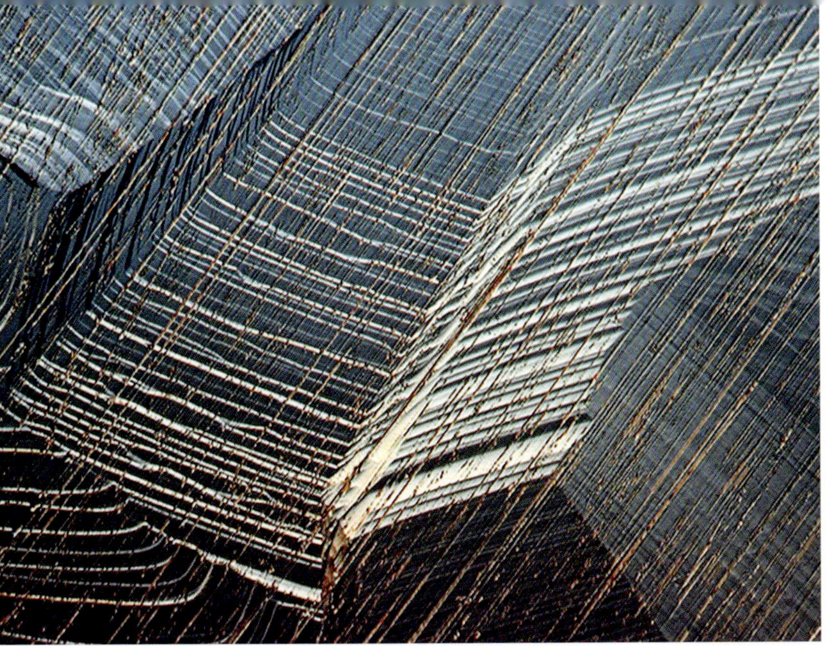

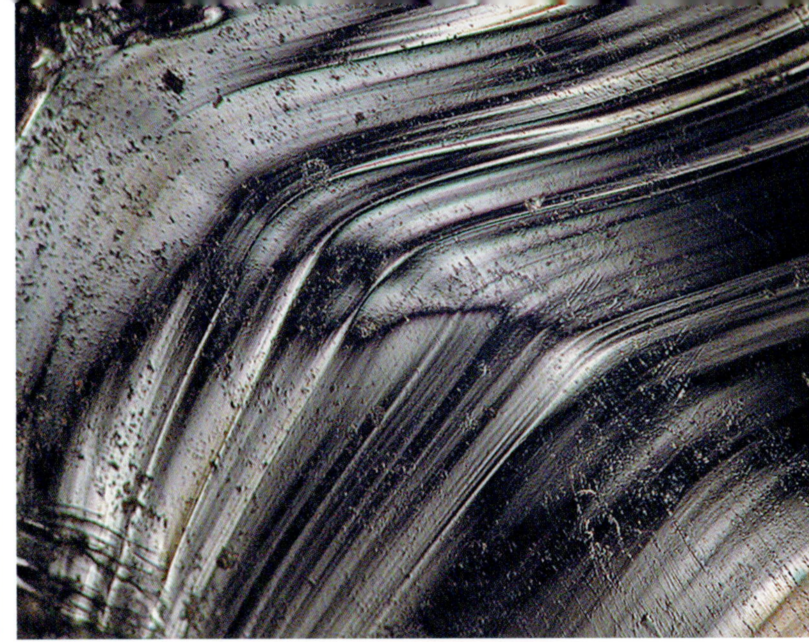

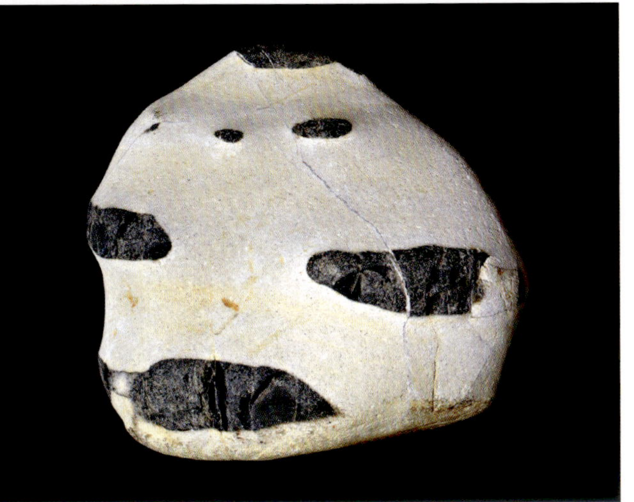

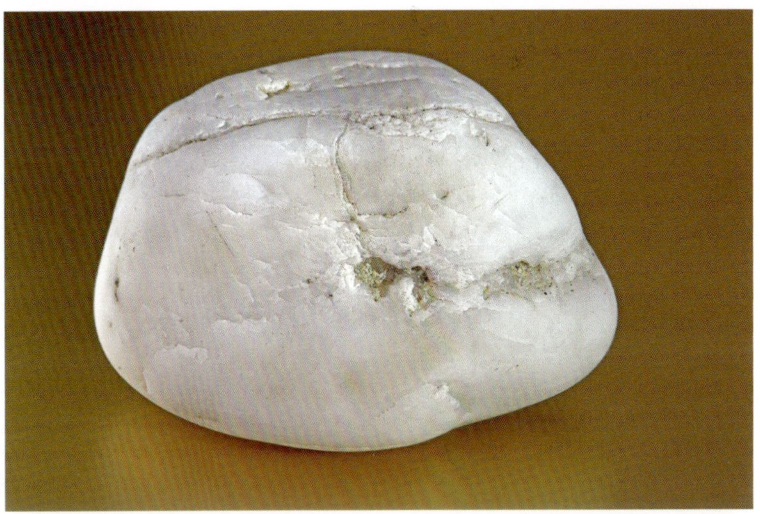

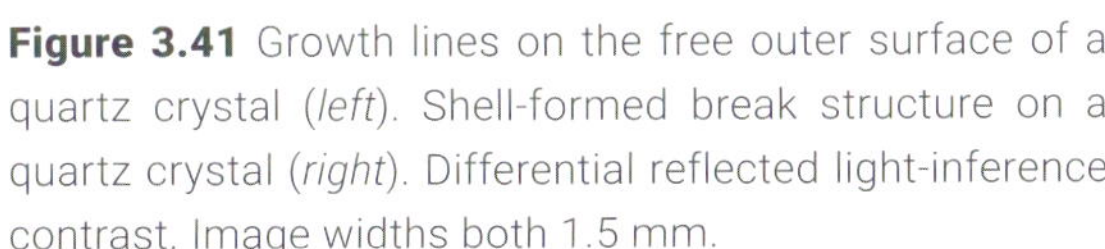

Figure 3.41 Growth lines on the free outer surface of a quartz crystal (*left*). Shell-formed break structure on a quartz crystal (*right*). Differential reflected light-inference contrast. Image widths both 1.5 mm.

Figure 3.42 *Left center*: Black chert concretions of quartz also in fine crystalline sediment. The lenticular chert concretions are harder than the surrounding material, causing them to protrude. The name results from the hornlike dense appearance of the nodules. Rhine sediments from a gravel pit near Kehl, Germany. Rock size 60 × 60 mm.

Figure 3.43 *Right center*: Milk quartz, which gets its white coloring from microscopically small gas or liquid inclusions. Quartz pebbles from a gravel pit near Kehl, Rheinebene, Germany. Rock size 45 × 35 mm.

Figure 3.44 Smoky quartz from a pegmatite vein in Vang Granite Quarry, Bornholm, Denmark. The middle granite pegmatite is composed mostly of coarse crystalline, pale-colored potassium feldspar, and the glassy dark smoky quartz crystals. Image width around 150 mm.

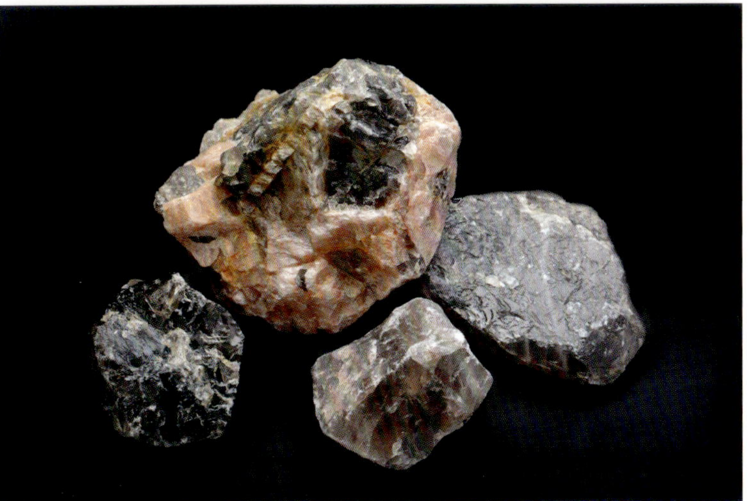

Red:	Inclusions of biotite or hematite, incorporation of metal ions ("rose quartz")
Brown:	Goethite inclusions
Yellow:	Iron hydroxide inclusions or metal ions
White:	"Milk quartz," liquid inclusions or gas inclusions
Gray:	"Smoky quartz," radioactive exposure (also man-made), mica or graphite inclusions
Green:	Chlorite, biotite, or hornblende inclusions
Blue:	Rutile inclusions
Pink:	Manganese compounds, inclusions of rutile or hematite

The very low but measurable solubility of quartz in water has already been pointed out. During his years in Frankfurt, Goethe was engaged in, among other things, alchemical experiments, in which the easily accessible quartz played a role. He was particularly interested in the potential transition of the mineral to the vegetal or organic. These areas, also in the tradition of the alchemists, were thought to be separate in principle. Goethe's early experiments in his specially assembled laboratory attest to his urge at the time for practical knowledge rather than the application of strictly scientific methods.

> "But what busied me most, for a long time, was the so-called *Liquor Silicum* (flint juice), which is made by melting down pure quartz-flint with a proper proportion of alkali, whence results a transparent glass, which melts away on exposure to the air and exhibits a beautiful clear fluidity. . . . I had acquired a peculiar dexterity in preparing this *Liquor Silicum*; the fine white flints which are found in the Maine furnished a perfect material for it."

In fact, Goethe apparently annealed silicone dioxide with sodium oxide and obtained a chemical substance called water glass, from which a gelatinous mass of silicic acid can be separated. This substance, slimy in consistency, reminded him of the life-giving "primeval slime" that had to be found.

Incidentally, the origin of the word "quartz," which has been attested to since the fourteenth century, has not been definitively clarified. It is probably a Middle High German familiar form of "querch," dwarf. In early times, miners attributed the occurrence of worthless, hard rocks between the ore veins to evil mountain spirits or dwarves. Thus, the name of the element cobalt also comes from the goblin, the mountain dwarf. Water-clear quartz crystals, on the other hand, have always impressed people.

In ancient legends, it was said that nymphs and fairies live under the mountains in palaces made of sparkling crystals. In times far before the alchemists, the Greeks already named the crystal-clear quartz crystals *krystallos*—the ice. And Plinius reported that the crystal is found only where "winter snow brings the greatest cold." It was actually believed that quartz crystals consisted of fossilized ice, which could not be melted even by heat. This view was widespread for a long time. Even in 1689, the chemist Johannes Kunkel still reported in his *Vollkommene Glasmacher-Kunst* ("Perfect art of glassmaking") that the stone was nothing but "ice coagulated together." At that time, however, Robert Boye had already proved in his *Essay about the Origine & Virtues of Gems* that quartz is much heavier than ice and therefore cannot be identical to it. From about the end of the seventeenth century, the term "crystal" was finally used independently of quartz crystals and was generalized to its current meaning.[39]

Even today, quartz has considerable technical significance. Quartz owes this to a physical phenomenon known was the "piezoelectric effect." This means that when a directional mechanical stressor is applied, for example by applying pressure to the crystal, it generates an electrical voltage. This also works the other way around: if you apply an electrical voltage to a suitable crystal, it changes its length. The effect was discovered with tourmaline crystals in 1880, but it took until the twentieth century for a useful mathematical-physical model to be developed for it. The first applications consisted of the construction of so-called oscillating crystals, which stabilize an electrical frequency through resonance, which was used, for example, to manufacture precise clocks: quartz clocks. Today, the corresponding quartz elements are produced artificially or replaced by other substances and combinations of substances. Quartz, and thus quartz sand, has become an indispensable raw material in many areas of technical and industrial applications.

3.2.2 Feldspar Group

Mineral:	Orthoclase (example)
Formula:	$K[AlSi_3O_8]$
Mohs hardness:	6
Density:	2.53–2.56
Crystal system:	Monoclinic

How does salt end up in the sea? We can answer this frequently asked question by looking at feldspars. While quartz is a single mineral, the term "feldspar" encompasses an entire mineral family that is also very complex in structure. In addition to silicon and oxygen, feldspars contain aluminum and another metal (potassium, sodium, or calcium) so that a distinction is made between potassium feldspars, sodium feldspars, and calcium feldspars, depending on the presence of the respective additional metal.

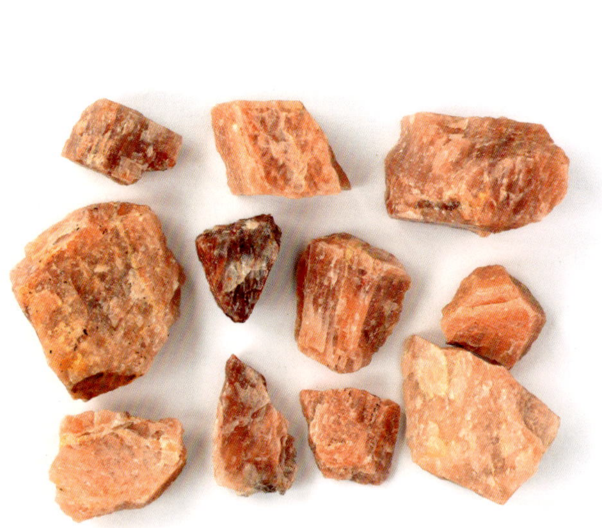

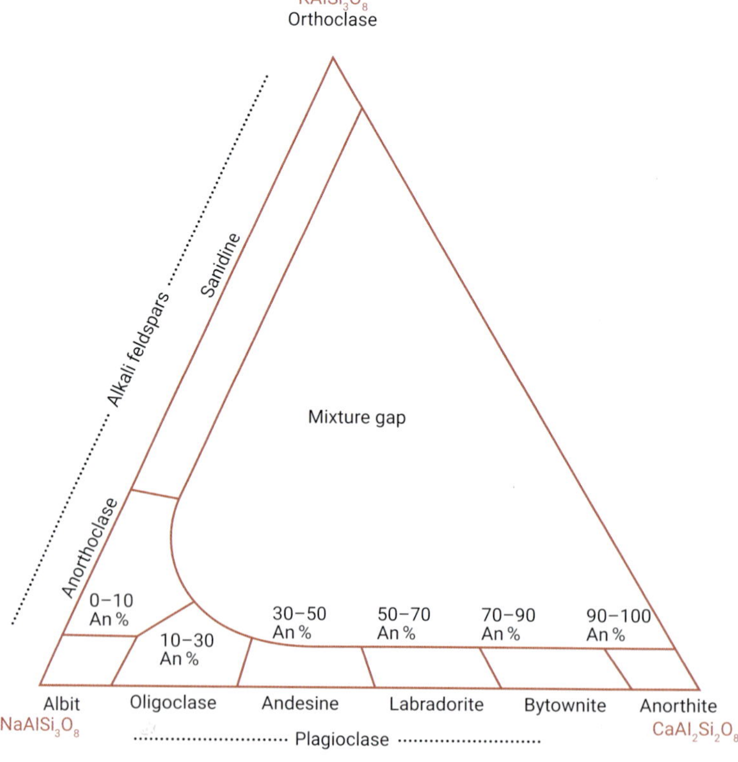

Figure 3.45 *Left*: Potassium feldspars (orthoclase) from the tailings pile of the Vang Granite Quarry, Bornholm, Denmark. Image width 12 cm. *Right*: Simplified ternary depiction of the trimaterial system of feldspars.

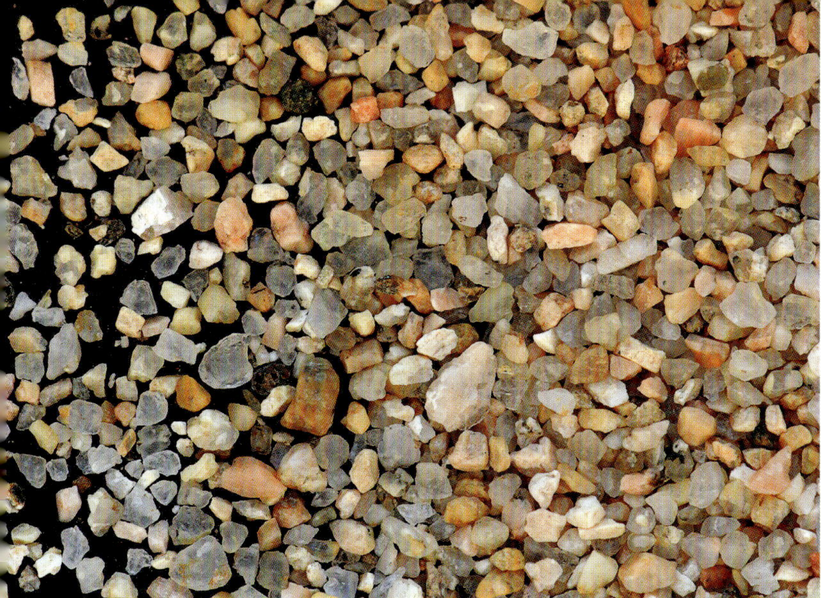

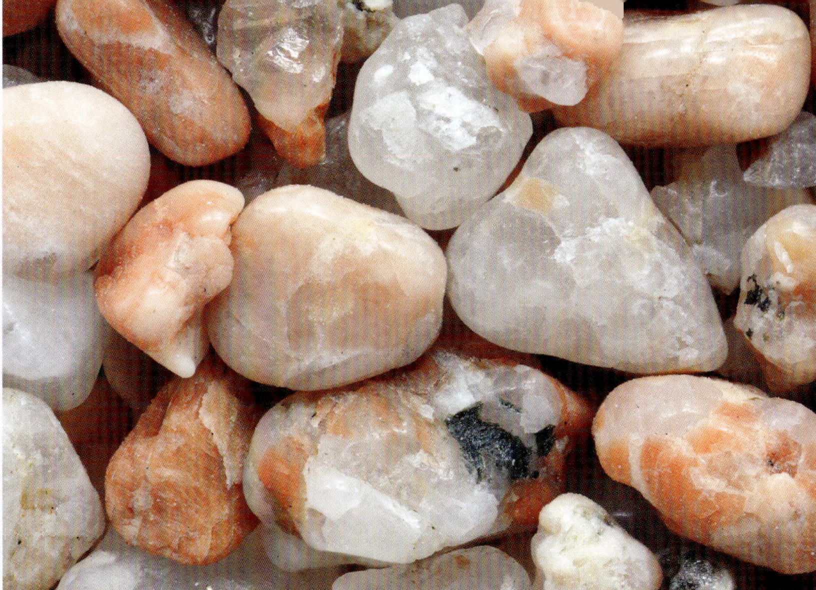

Calcium feldspars are usually referred to by the mineralogical name anorthite, sodium feldspars by albite. Between these two end members any ratio of sodium to calcium is possible—this binary mixed crystal series is called plagioclase. Ternary feldspars contain sodium as well as calcium and potassium. As mixed crystals they stand between albite, anorthite, and orthoclase. They are formed in magmatic rocks at high temperatures, with subsequent slow cooling. Ternary feldspars of certain compositions are also called anorthoclase. All these varieties can be found in sand.

Potassium feldspars occur very frequently in granitic rocks and are, due to their abundance and pervasiveness, important components of sands and soils. The erosion stability is higher for potassium feldspar than for plagioclase. With about 60%–65% (volume percent), feldspars are by far the most abundant mineral in the earth's crust. Soils rich in feldspar are very fertile, in contrast to sandy, quartz-containing soils, because the weathering of feldspars produces kaolin, and the released potassium is then absorbed by plants, which have a high demand for this mineral. Sodium, on the other hand, is needed by plants to a much-lesser extent, so it remains in the soil and is eventually washed out by rainwater and transported toward the sea. This explains the high salt content of the world's oceans, in which predominantly sodium ions, and not the potassium ions taken up by plants and extracted from the erosion products of the rocks, accumulate over millions of years.[40]

Feldspars are easily eroded and very easily split, which is indicated by the suffix "spar." Their coloration is characteristic, although not unique. Potassium feldspars show a spectrum of red, pink, brown, and violet, often also with flesh-colored tints with rather larger crystal dimensions.

Figure 3.46 Immature and rich in feldspars, only lightly rounded and relatively fresh. Fietjetjörn, Valle, Norway.

Figure 3.47 Immature, highly to moderately rounded sand rich in feldspars from Cala di Principe, Sardinia, Italy. The direct origin from the nearby granitic stone can be easily seen. The grain below the middle is a miniature stone fragment out of potassium feldspar (orange), quartz (white), and biotite (black). Image width 8.5 mm.

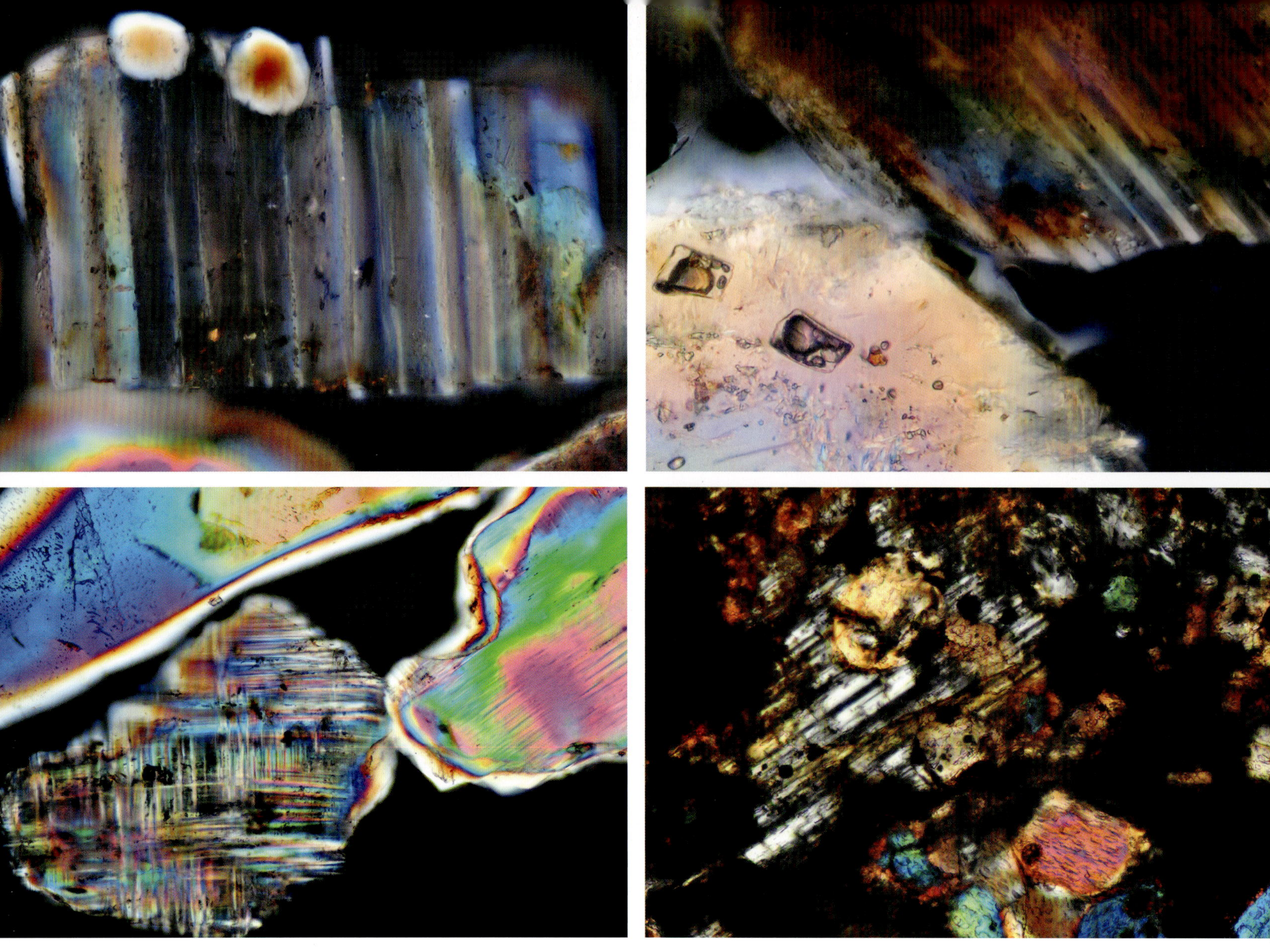

Figure 3.48 Sand grain of feldspar. The crystalline twinning is clearly seen under crossed polarizers. Fietjetjörn, Valle, Norway.

Figure 3.49 *Bottom left*: A grain of sand made of microcline with typical clathrate twinning (450 × 250 μm). *Right*: A feldspar with lamellar twining. *Top*: A grain of sand made of quartz. Fietjetjörn, Valle, Norway. Transmitted light, crossed polarizers.

Figure 3.50. *Top right*: A grain of sand made of plagioclase, recognizable by sharply defined black and white twin lamellae. Two diatoms adhere to the lower sand grain. Crossed polarizers. Fietjetjörn, Valle, Norway. Image width 270 μm.

Figure 3.51 Extraterrestrial plagioclase from the Allende meteorite. It is fascinating to see that the extraterrestrial material is composed of the same basic components as our earthly materials. Many of the earthly minerals can also be found in space, such as the plagioclase crystal in the figure shown with typical black and white twin lamellae. The colorful grains are composed of olivine. Crystal size around 200 × 150 μm. Thin section, crossed polarizers.

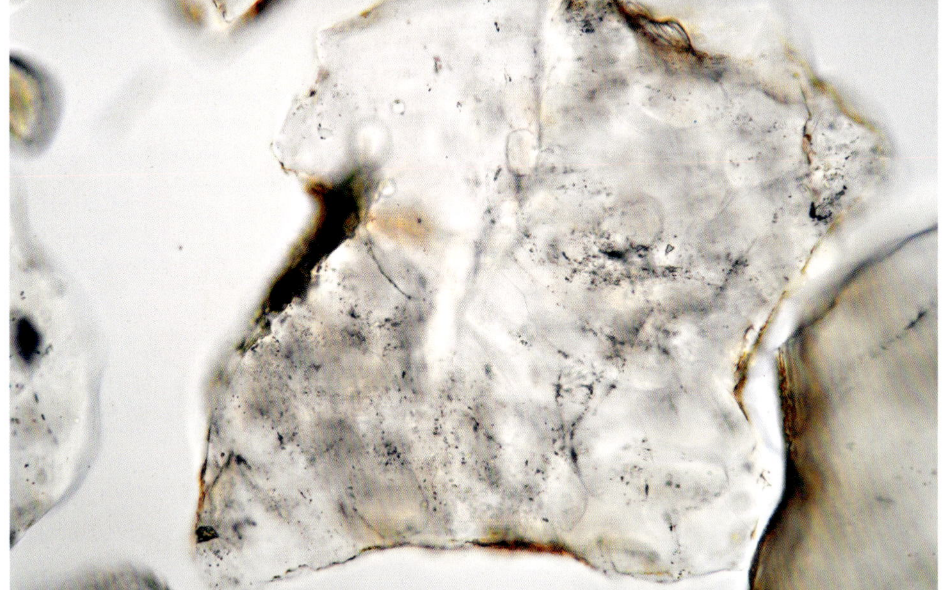

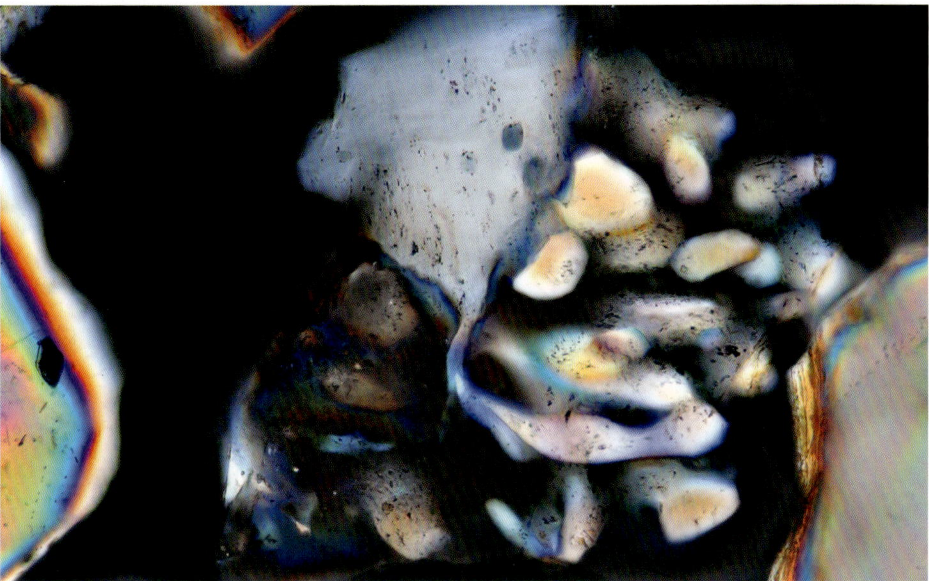

Plagioclase tends to be made up predominantly of white, yellow, and green colors with—compared to potassium feldspars—smaller crystal dimensions in the rock formation. The characteristic tendency of the feldspars to form crystal twins can be seen in the microscopic image, sometimes even with the naked eye when looking at freshly broken rock.

Alkali feldspars are miscible without limitation at high temperatures, as described. By increasing and slow cooling of magma, under 600°C, exsolution takes place in the solid state. In the process, exsolution structures are formed, which are called perthites. Perthites can be recognized by exsolution spindles and striations, which are formed from albite and are located in the alkali feldspar as host crystal. Quartz can form intergrowths with feldspars, which produce a wormlike or hieroglyphic appearance.

Figure 3.52 A so-called myrmekite is when quartz appears as a wormy intergrowth in plagioclase. In the bright field (*above*), there is hardly a recognizable difference between the quartz and feldspar (plagioclase) components. Only under crossed polarizers (*below*) is growth structure of the sand grain revealed. Valle, Norway. Transmitted light. Grain dimensions 380 × 430 *µm*.

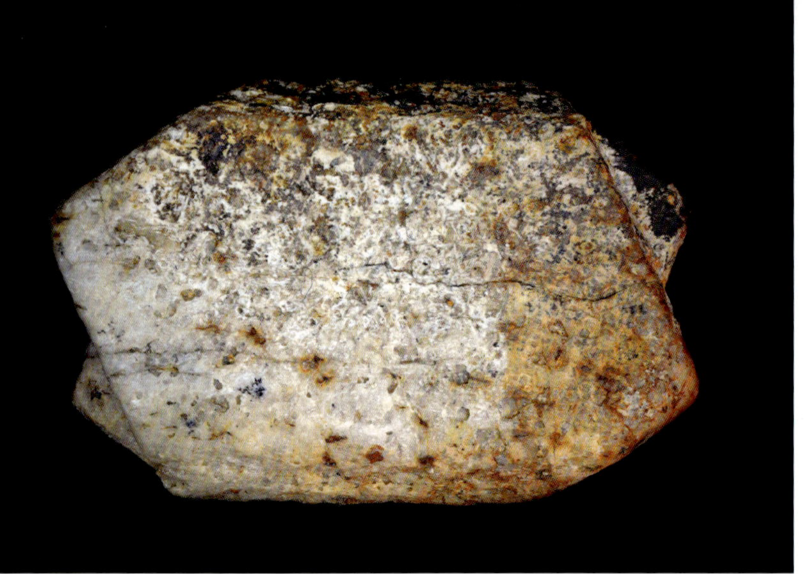

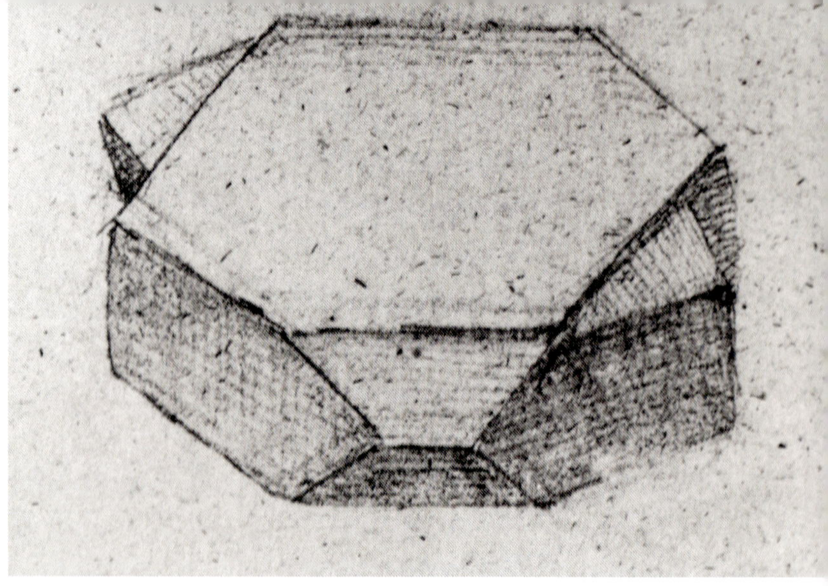

Occasionally, in sand grains, such a structure can be detected only under crossed polarizers. Intergrowths of this type are called symplectites. If quartz is intergrown with plagioclase, it is a myrmekite.[42]

Feldspar crystals form, as described, twins relatively often, by which they can be easily recognized and determined under the microscope. The twinned lamellae are particularly obvious in transmitted light with crossed polarizers, since the adjacent lamellae are not simultaneously in solution, thus showing a light-dark contrast. The twin formations according to the Carlsbad law, called Carlsbad twins, are particularly striking with potassium feldspars, whereby only two individual pieces are grown together as twins, while for plagioclases, many lamellae appear alternating.

Goethe, who liked to stay in the fashionable Bohemian spa town of Carlsbad for relaxation and social occasions as well as to collect minerals and rocks, described this special feldspar twinning as early as June 1807 in the following way:

> "There are large masses of Carlsbad granite in which one finds perfect crystals of very complicated formation. They are double crystals, which seem to consist of two interlocking crystals, but one cannot think of one without the other. Their form is not easy to describe to the imagination, but one can imagine them as two rhombic plates, one inside the other."[43]

Figure 3.53. Potassium feldspar twin crystal according to the Carlsbad law, a so-called Carlsbad twin. Crystal from the Enfola Peninsula, Elba, Italy. Length 38 mm. *Right*: Goethe's sketch of a "left twin" from July 20, 1807, during his description of Joseph Müller's mineral collection. (taken from J. W. v. Goethe, *Writings on General Natural History, Geology, and Minerology*. Frankfurt Edition [FA] 1. Department, Complete Works Book 25)

To the essay "Zur Kenntnis der Böhmischen Gebirge," from which this quotation is taken, Goethe added a precise pencil sketch of the crystal and continued as follows:

> "But we would not have a clear idea of their peculiar formation if the granite that contains them did not sometimes weather in such a way that the surroundings crumbled to sand and crumbs, but the crystals themselves came to freedom firm and unchanged; whereby, however, it must be observed that they must soon be picked up, because they also crumble or at least become brittle through time and weathering."

Orthoclase, otherwise known as potassium feldspar, where this twin form appears, is, as Goethe accurately observed, actually somewhat more erosion resistant than the other feldspar varieties. Seven years later, Goethe received a Carlsbad twin from Heinrich von Treba, the mining officer from Freiburg, which Treba described as a "witch crystal" in the accompanying letter (fig. 3.53). At that time, the classification of the rocks was still unclear and heavily discussed in the context of the Neptunist controversy.

3.2.3 Mica Group

Mineral:	Muskovite
Formula:	K $Al_2[(OH, F)_2 AlSi_3O_{10}]$
Mohs hardness:	2–3
Density:	2.78–2.88
Crystal system:	monocline

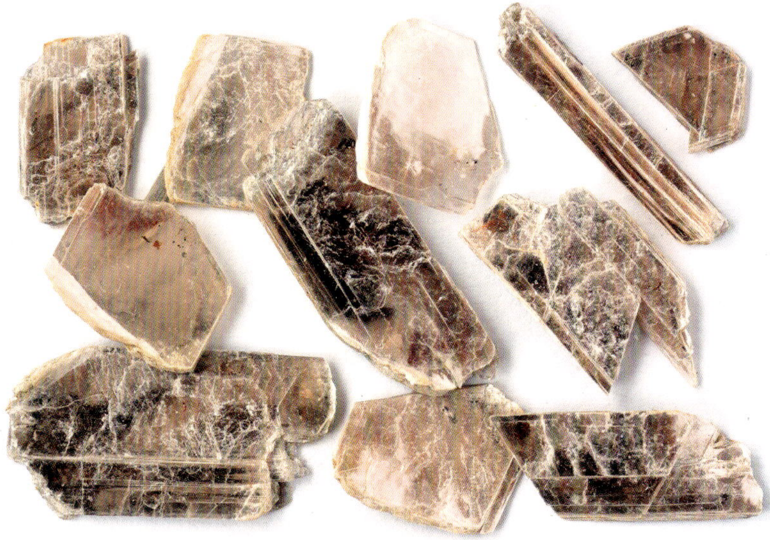

Figure 3.54 Muscovite crystals. Image width about 90 mm.

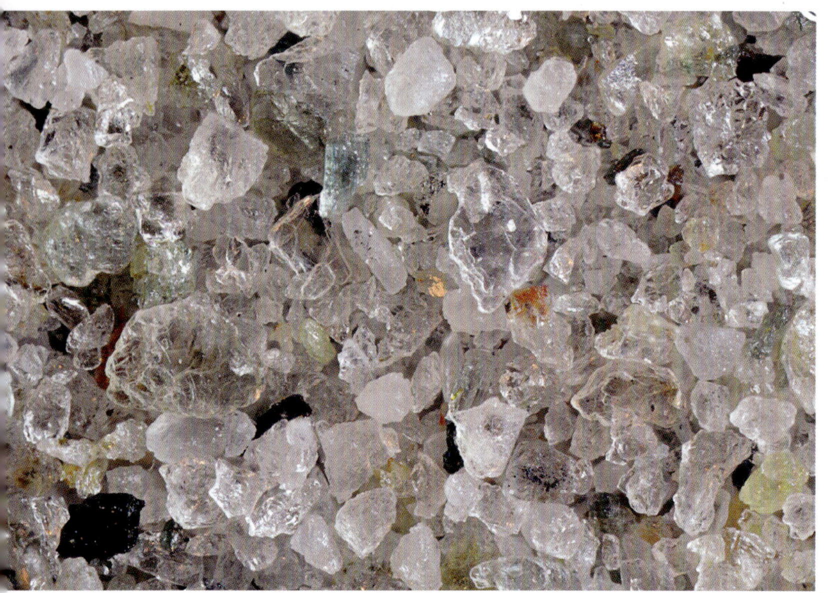

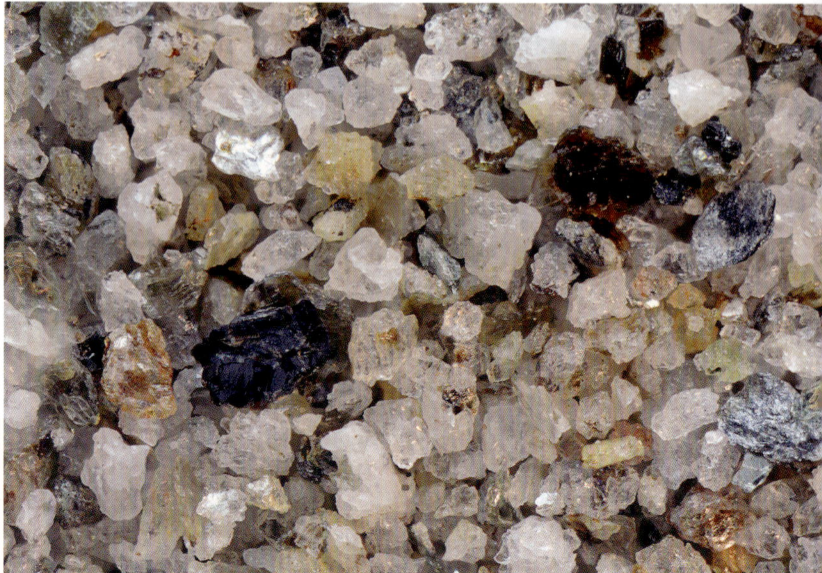

Quartz and feldspar belong to the silicate group. Mica is the third important rock-forming mineral group that is counted among the silicates; more precisely, the phyllosilicates. Muscovite and biotite are the most-frequent representatives of this group. Chlorites, also phyllosilicates, resemble mica with respect to preferential cleavage but are not as elastically flexible and are always dark green in color.

The mineral muscovite, otherwise known as mica, exhibits almost perfect cleavage. Individual flakes can be easily detached with a knife and are transparent and pliable. The formerly used name "muscovite glass" comes from the fact that in the Moscow area, larger mica sheets were used as a glazing. However, the flakes are very soft, with a Mohs value of 2, and are susceptible to scratching. Muscovite occurs in numerous acidic magmas, is found in granites of various types, and is the main component of mica schist, which gets its good cleavability form the layered orientation of the muscovite platelets. If a granite contains light mica (i.e., muscovite), it can be concluded for geochemical reasons that the corresponding melt must have solidified at a depth of at least 7.5 miles (12 km).

Figure 3.57 Grain of sand of muscovite. Hinterrhein in Nufenen, Switzerland. Transmitted light, bright field in linearly polarized light with slight inference colors, caused by the separation of thin split particles from each other. These are not comparable to the inference colors that occur in polarized light but correspond to Newtonian rings as seen on oil puddles. Dimensions 510 × 950 *μm*.

Figure 3.58 Grain of sand of biotite. Hinterrhein near Nufenen, Switzerland. Transmitted light. Dimensions about 400 × 800 *μm*.

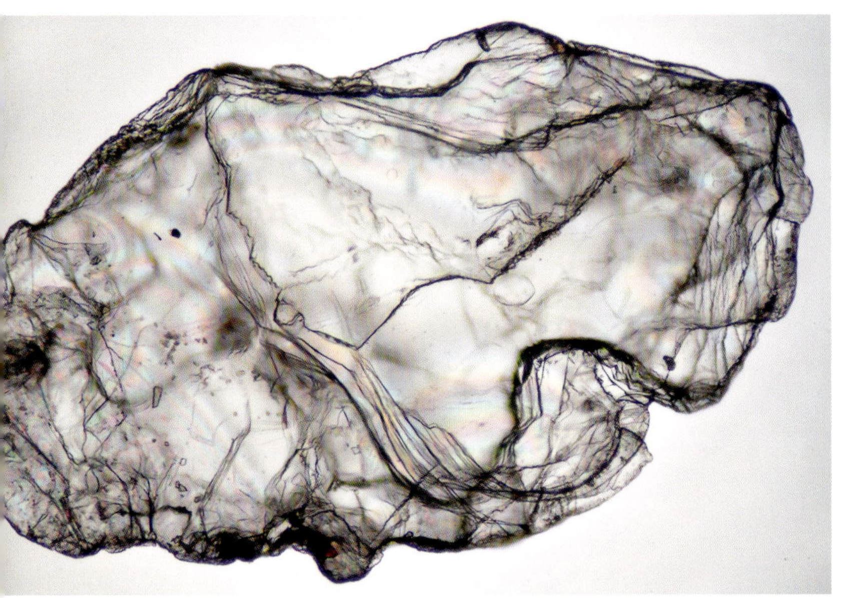

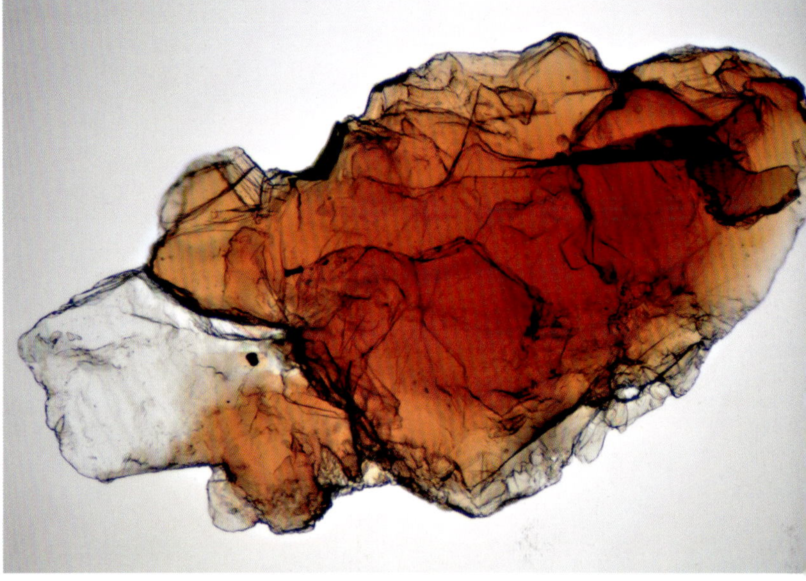

Muscovite is also an important component of metamorphic rock (see next chapter). From the presence or absence of mica, inferences can be drawn about the degree of metamorphism and the composition of the origin rocks. Its shimmering, partly pearly luster is caused by light reflections on the thin, foliated layers. Muscovite is extremely erosion resistant, whereas biotite or dark mica is less resistant and increasingly fades with erosion and can take on a gold-like, iridescent hue.[44] The latter, unlike muscovite, is also stable at higher temperatures and therefore is also found in volcanic rocks.[45] Overall, biotite is the most common mica. In sediments, however, muscovite occurs about four times as frequently as biotite because of its higher durability. The dark, mostly gold-colored glittering crystals of biotite are particularly easily recognizable in sands, while the partly delicate muscovite flakes are often overlooked and catch the eye only because of light reflections. Mica components are easily washed out of the sediment by water and transported far as a suspended load by floating. Along with clay minerals, they are responsible for the milky appearance of glacial waters. The boundary to the chemically and structurally related but much-smaller clay minerals is set at a size of 20 *μm*. Clays themselves have by definition grain sizes below 2 *μm*, consisting of minerals such as illite, smectite, kaolinite, etc., which are indistinguishable even under the light microscope.

Abb. 3.57 Sandkorn aus Muskovit. Hinterrhein bei Nufenen, Schweiz. Durchlicht Hellfeld im linear polarisierten Licht mit leichten Interferenzfarben, hervorgerufen durch Ablösung der dünnen Spaltplättchen untereinander. Diese sind nicht vergleichbar mit den Interferenzfarben, die im polarisierten Licht entstehen, sondern entsprechen den Newtonschen Ringen, wie sie auf Ölpfützen zu sehen sind. Abmessungen 510 × 950 μm.

Abb. 3.58 Sandkorn aus Biotit. Hinterrhein bei Nufenen, Schweiz. Durchlicht Hellfeld. Abmessung ca. 400 × 800 μm.

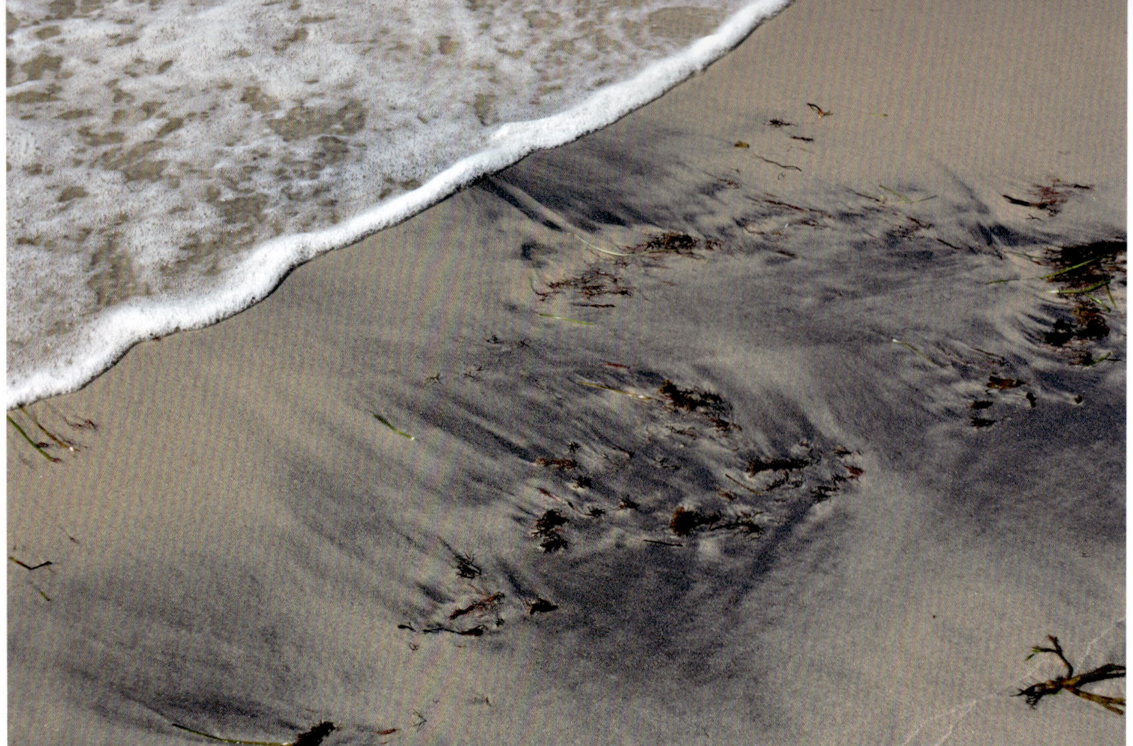

3.2.4 Heavy Minerals

Through the influence of wind and water, sand is not only transported but also sorted. Above all in water, the periodic movement of waves leads to a sorting of the materials that have different densities.

Depending on the source rock, many sands contain a number (0.01%–1% in the case of sandstones) of accessory (i.e., additional mixture) components, which are specifically heavier than quartz and can withstand weathering similarly well. These heavy minerals then accumulate in initially small quantities at the wash margin of larger rivers, in recesses, and at the waterline of coasts. The conspicuous dark band, which can be observed some days while walking along the beach, consists of such heavy minerals, which on closer inspection allow a glimpse into the world of very attractive, sometimes brightly colored miniature gemstones. Tourmaline, zircon, rutile, garnet, topaz, spinel, and many others can be found, as well as the opaque (nontransparent) metallic materials hematite, ilmenite, and magnetite. The latter are strongly magnetic and therefore somewhat easier to identify.

If, in general, there is a local accumulation of specifically heavier or mechanically or chemically more-resistant minerals, they are referred to as heavy-mineral sand deposits. Depending on the type of enrichment process, different types of heavy-mineral sand deposits can be distinguished.

Figure 3.59 Heavy-mineral sand deposits (placers) on the beach of Hasle, Bornholm, Denmark

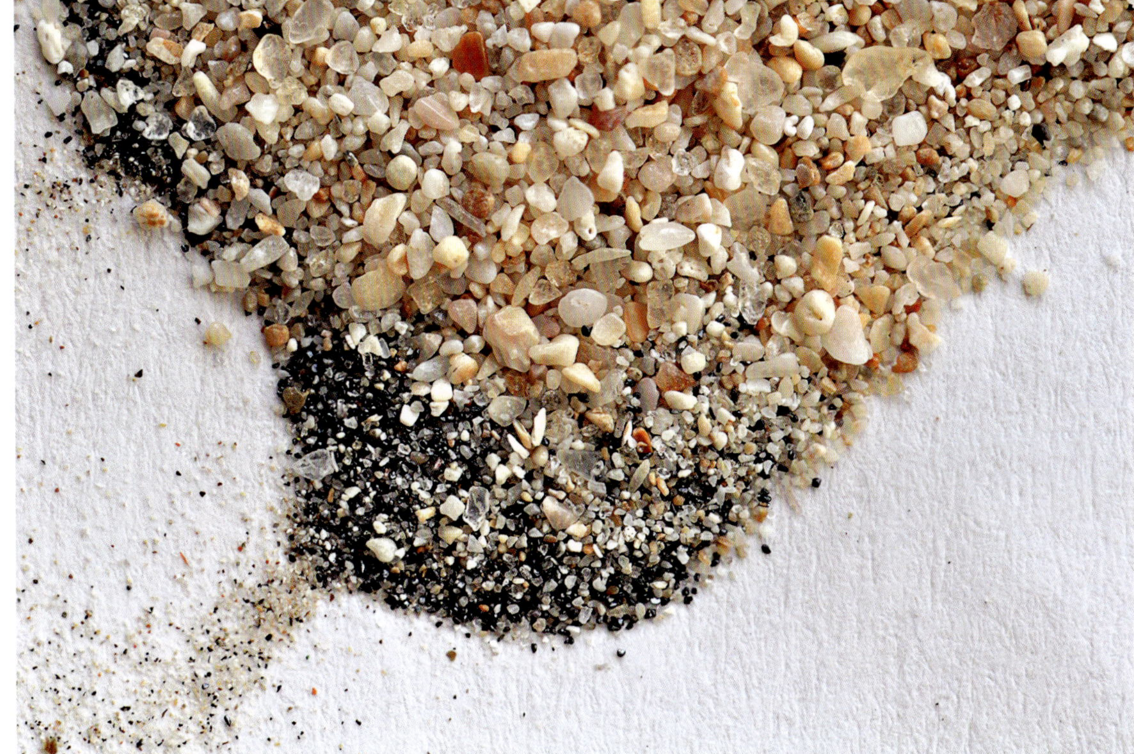

— Residual placers, which occur through erosion on-site
— Fluviatile or alluvial placers, which are formed by enrichment in rivers or come about by rain floods
— Marine placers, which occur through enrichment from surf zones
— Aeolic placers, formed by blowing wind and separation by air currents

The American gold rush was based on rich gold discoveries in fluviatile placers. At that time, pans had to be laboriously washed out to separate the lighter minerals from the heavier ones through gravity and current. Heavy mineral deposits reach mineable thickness in some places. Marine placers in particular can extend for many miles along coasts and in some cases reach enormous sizes. On the eastern Australian coast, beach placers with enrichments of zircon, ilmenite, and rutile extend over 560 miles (900 km). The onshore wind, which usually prevails on coasts, also leads to the inland transport of the minerals and to enrichments many miles from the coast. For gold, tin, tantalum, and titanium, the share of production from placers worldwide is up to 70%. The diamond soaps in the Namib Desert are of particular importance, as are the tin soaps in Southeast Asia.

By analyzing the content and components of heavy minerals, important conclusions can be drawn about the formation conditions and the origin of sands and sedimentary deposits (also see chapter 6). For example, the successive erosion of the French Central Massive over long geological periods could be reconstructed by analyzing the heavy-mineral sequences in the surrounding sedimentary basins (Rhone Basin, Paris Basin).[47]

Figure 3.60. Sand from Unawatuma, Sri Lanka. The concentration of dark, metallic heavy minerals can be seen on the lower edge. The light sand is made, in contrast, mostly of biogenic material with few rock fragments. Scattered on the paper with a light shaking. Image width about 50 mm.

For scientific research, the heavy-mineral content is removed from a sediment sample by separating processes, and the type and quantity are determined. In addition to magnetic separation methods, either high-density liquids are used, and the lighter minerals are separated by floating to the top, or centrifuges are used, which cause separation through mass inertia. The highly toxic liquid bromoform has been used in the past for separation, with its density of 2.85 g/cm^3. Everything that sank in the separating vessel was a heavy mineral. Over one hundred heavy minerals have been recorded, but in total they account for only about 0.1%–0.5% of the content of sediments.[48]

Due to their higher specific weight, heavy minerals are transported in their respective environment with quartz grains, whose diameter, however, is on average significantly larger (in phi units: 0.5–1.0). As a result, the average diameter of the heavy minerals is smaller than the average diameter of the sediment in which they are found. This difference is known as the *hydraulic ratio*.[49] Furthermore, the size distributions of heavy minerals also differ among themselves. Heavy minerals are mechanically and chemically stressed during their transport; after their deposition and the beginning of diagenesis, the mechanical stress ceases, whereas chemical influences continue. If a further sedimentation cycle follows, its influences override the first cycle and so on.

During the diagenesis phase, minerals can reach complete dissolution—their substances passing into the sediment pore water.50 This process, called int*rastratal solution*, has the effect that minerals that are still present in upper layers become absent with increasing depth. A gradual depletion of older formations of heavy minerals is observed. For example, investigations in the Paleozoic era of the North Sea showed that amphibole disappeared at a depth of 1,700 feet (600 m), epidote at a depth of 3,600 feet (1,100 m), disthene at 5,900 feet (1,800 m), and taurolite at 7,900 feet (2,400 m), while garnet, rutile, tourmaline, and zircon could still be found at depths over 9,200 feet (2,800 m). Garnet develops pronounced and increasing etch figures.[51] It is known that garnet's chemical stability in eroding waters with small pH values around 3.6 (acidic range, comparable to apple juice) is very low and increases with rising pH values. At pH values from 8 to 10, comparable to household soap, garnet belongs to the chemically most-stable heavy minerals of all.

Figure 3.61 Heavy mineral deposits on the beach of Hasle, Bornholm, Denmark (see fig. 3.59). Along with the light quartz grains, opaque, metallic, shiny minerals such as ilmenite and magnetite can be seen as well as pink- and orange-colored, round granite crystals. Image width about 7.5 mm.

Figure 3.62 Sandwich Harbor, Namibia. Beach placer, consisting of granite (orange color: grossular, red and pink almandine/pyrope) and dark ore minerals (magnetite). Image width 5.5 mm..

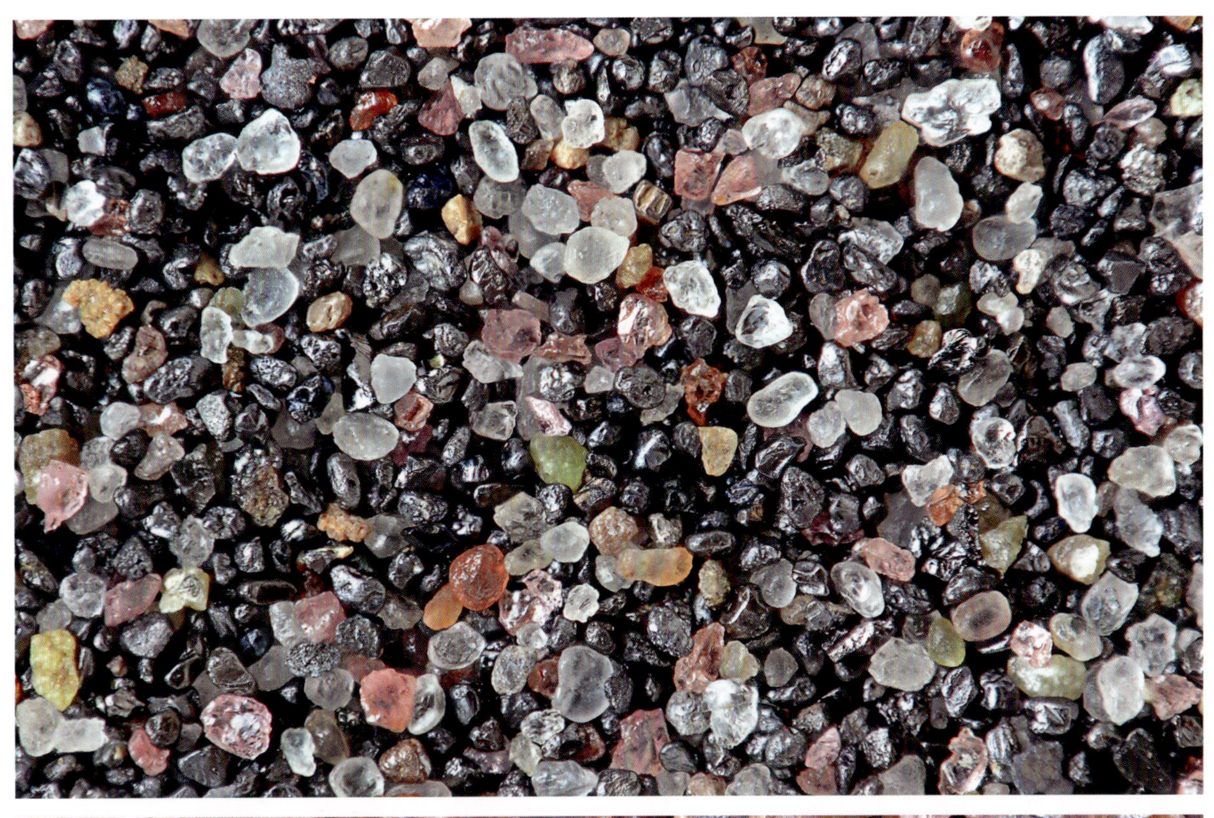

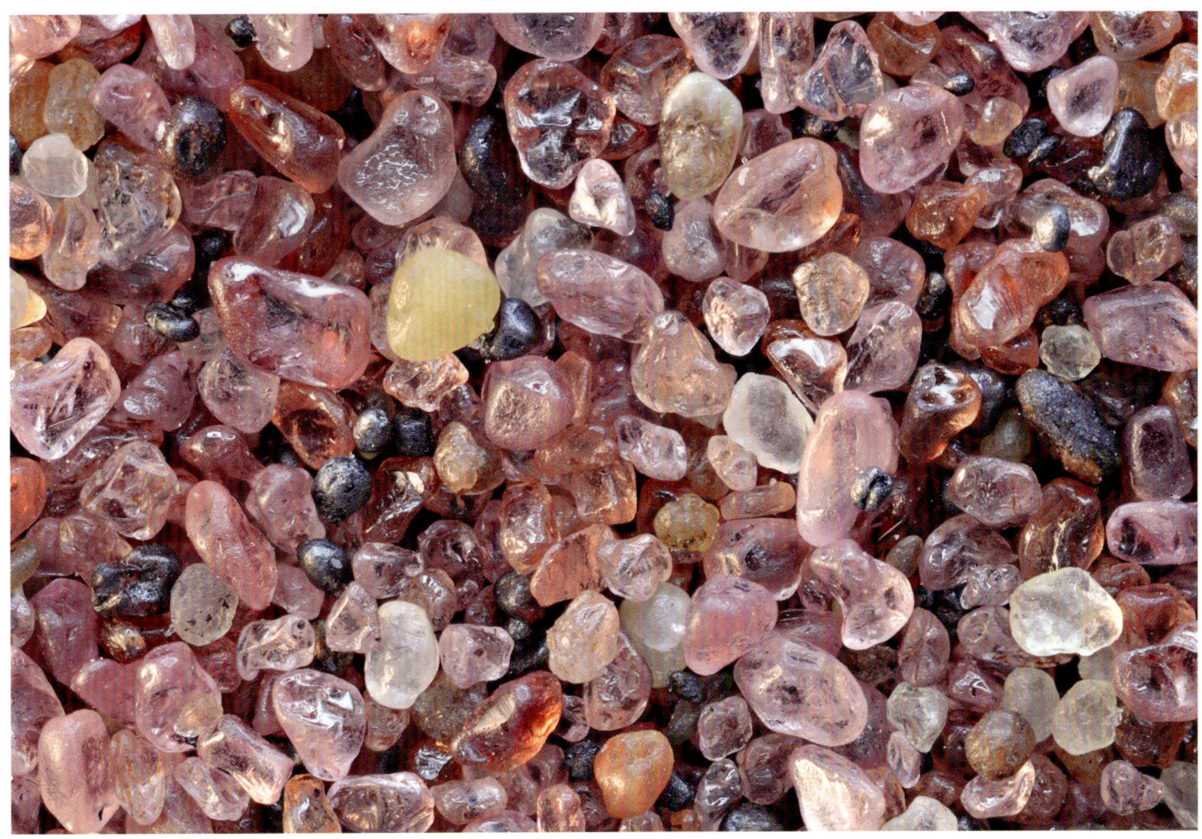

Granite's stability is exceeded by tourmaline, rutile, and, above all, zircon. In contrast, olivine is one of the very unstable heavy minerals, which is why it is only rarely found in placers. Due to its special meaning and because of its comparably very high proportion of sand content worldwide, we will take zircon as an example for the diversity of heavy minerals and explore it in more detail.

Zircon

One of the especially interesting minerals that tends to be highly present in placers is zircon (Persian: zar = gold and gum = color). When it comes to erosion, it is very durable, almost indestructible. In experiments in which natural transport processes were simulated, no signification rounding of the grains could be detected even after more than 2,800 miles (4,500 km) of simulated transport distance.[52] Zircon can also survive high-pressure environments over the course of a rock metamorphosis, or even the melting of the host rock, largely unchanged. Therefore, zircons from former sedimentary rocks can be found in metamorphic rocks as well as in igneous rocks, which are rounded due to their previous life, whereby grain rounding can also take place during metamorphism and therefore does not provide a clear characteristic of erosion processes.

Abrasively rounded grains can even regrow metamorphically into complete crystals, which then show a shell-like zoning.[53] To extract zircons from granite rocks for analytical purposes, the rock lumps are crushed and ground into a fine powder. This powder is then treated with highly aggressive hydrofluoric acid, which dissolves out any minerals until only the zircons remain. The tiny zircon crystals survive this procedure so undamaged that they can be used for subsequent analysis.[54] Even in very mature sediments, which consist of over 99% quartz, zircons can be found that have survived a very eventful transport history. Conversely, geologists can infer from this the degree of erosion of sands and sandstones. They summarize the three most stable minerals—zircon, tourmaline, and rutile, which have similar properties in this respect—into a "ZTR index," for this purpose.

Zircons occur predominantly in magmatites as well as in acidic volcanic rocks and form an important basis for geological age determination. The crystals usually contain traces of radioactive uranium isotopes, which decay into lead in very determinable but long periods of time: lead itself is not deposited. The 238U and 235U uranium isotopes suitable for this purpose decay into the lead isotopes 206Pb and 20Pb after half-lives of 4.468 and 0.704 billion years, respectively. By measuring the ratio of remaining uranium to newly formed lead, the decay state of the uranium and thus the age of the sample can be calculated. For the core areas of zonally built-up zircons from the Jack Hills in Western Australia, an age of 4.404 billion years could be determined through this method, meaning that these zircon crystals

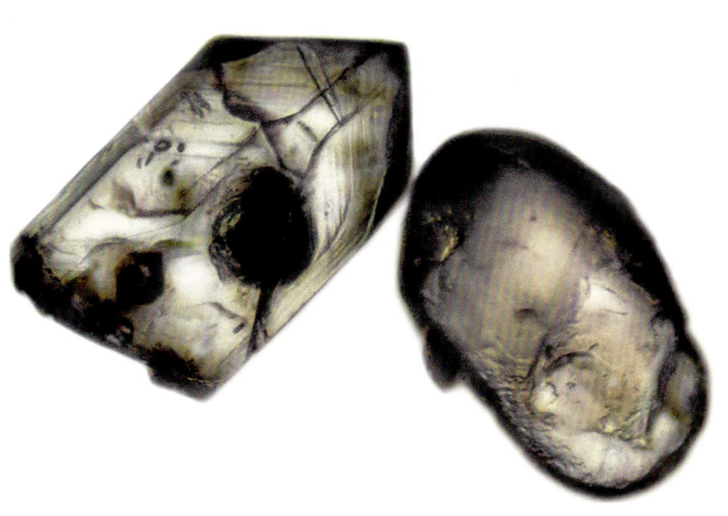

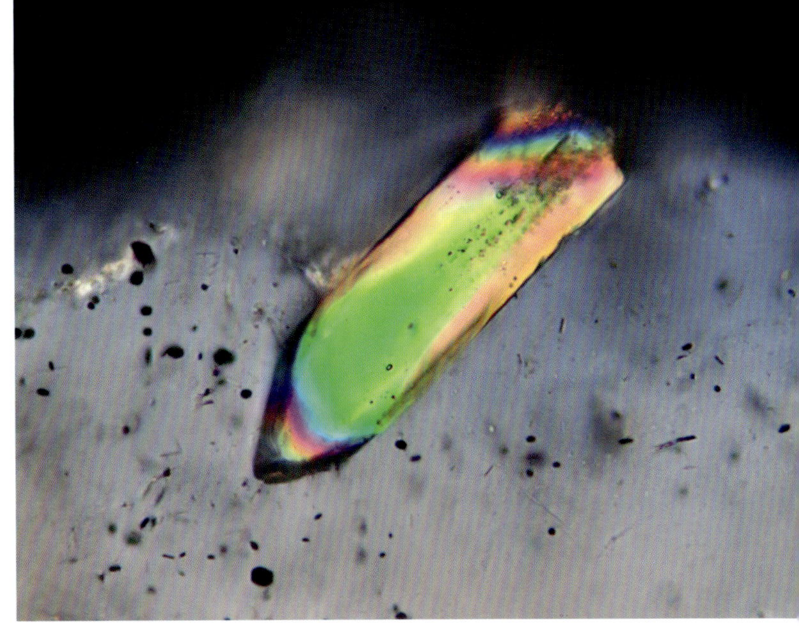

probably belong to the few known witnesses of the primary continental crust of the earth (also see chapter 6) that survived the Hadean eon.

Their zonal structure, similar to tree rings, provides the chronicle of the entire continent. The age of each individual accretion layer can be determined with high precision. Additionally, zircon sand grains often contain inclusions of various kinds, including glassy remnants of the melt from which they were once formed. The study of these inclusions is a science in its own right and provides insights into, among other things, the formation conditions at the time of the solidification of the surrounding rock.[56]

While sand grains from zircon crystals can be easily recognized by their shape and size with a little practice, identifying other minerals in the sand is very difficult without aids. Also, the attempt to assign sand grains to certain minerals on the basis of their color does not lead to a clear result. For example, a beautiful greenish grain could be actinolite, hornblende, epidote, olivine, augite, aegirinaugite, or aegirine, to name only the most-important ones. Similarly, not every crystal-clear grain is quartz. For precise identification, polarizing microscopes or even more-elaborate methods are needed. For the sand lover, such information is interesting but not essential. For the scientist, however, they are indispensable. Here, we show a selection of ten monomineralic sand grains, which are often found in the placer deposits of beaches, in order to give at least an impression of the variety and beauty of these placers.

Figure 3.63 Sand grains composed of zircon crystals, from a placer on the beach of Hohwacht, Baltic Sea, Germany. The left grain is a zonally built, idiomorphic crystal without visible signs of abrasions, while with the grain on the right, the crystal forms have been mostly rounded. With the left grain, there are multiple visible inclusions of different kinds (melt inclusions large and round, angular mineral crystal inclusion). Bright field image, width of the left crystal crosswise to the axis 100 *µm*.

Figure 3.64 Zircon crystal as a small inclusion within a grain of sand. Dimensions of the zircon 24 × 95 *µm*. Tigaki, Kos Island, Greece. Transmitted light, crossed polarizers.

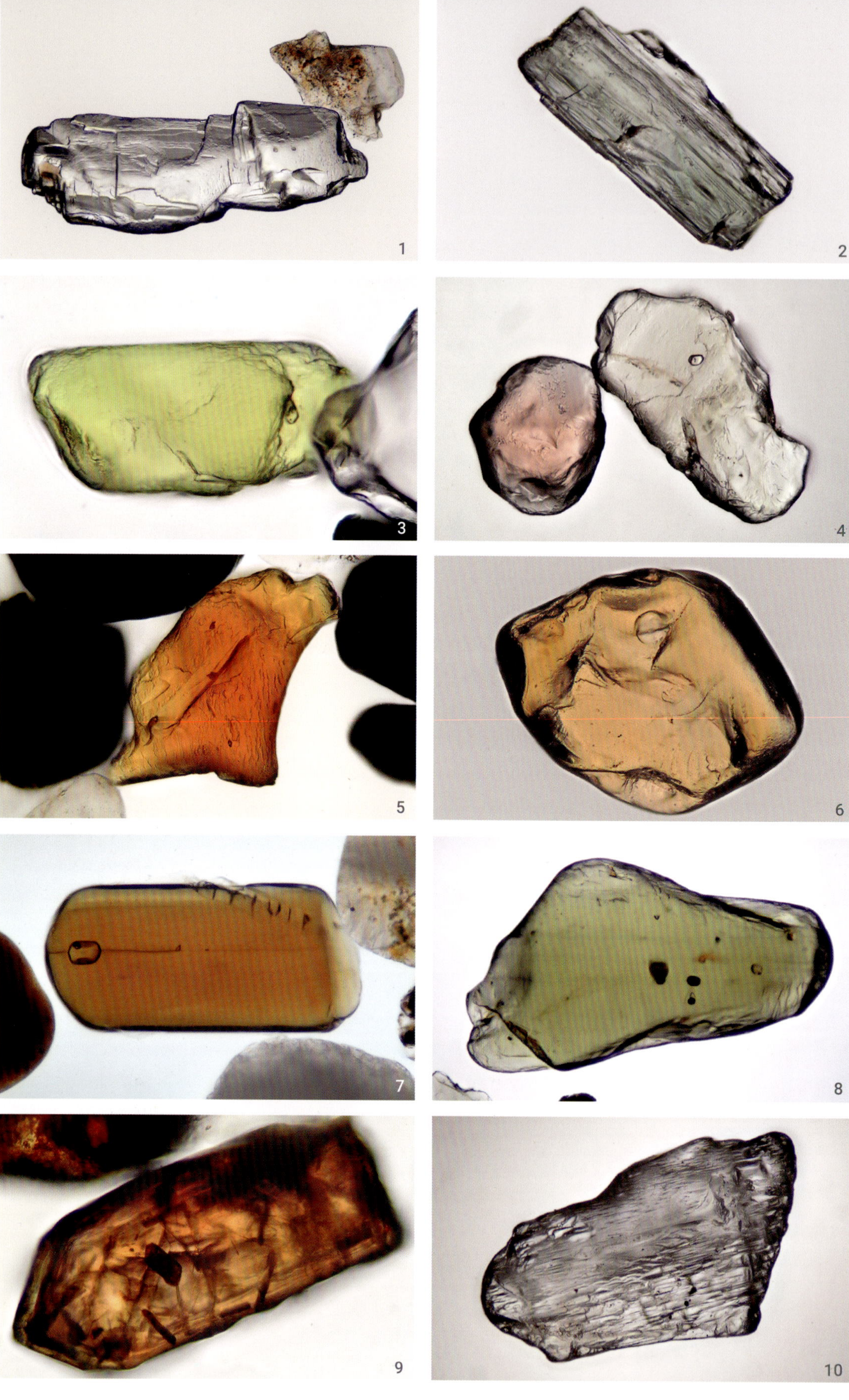
1
2
3
4
5
6
7
8
9
10

Ore Minerals

All the previously introduced heavy minerals are transparent and therefore can be observed and researched with transmitted light. That is not possible with ore minerals: ore minerals are opaque when the size of sand grains, meaning they are nontransparent. This is why reflected-light methods are used to analyze them, for which specially equipped ore microscopes are used. Here, we are interested only in the two most important ore minerals in sand, magnetite and hematite, which are good to be able to differentiate for the pure sand lover.

Magnetite or magnetic iron (Fe_3O_4), together with ilmenite, forms the main mass of all the oxidic ore and is the most important iron ore mineral. It is very widespread and can be found as an additive part in many rocks. Magnetite is even embedded in the body of bacteria of the genus *Magnetospirillium* in the form of fifteen to twenty tiny crystals, which enables them to orientate themselves on the basis of the earth's magnetic field lines and to move accordingly. Magnetite has a high melting point and is therefore deposited very early from rock melts in small, complete crystals, which we find later as opaque, black-glossy, and erosion-stable grains in the placers of fluviatile or marine sands. Above a material-dependent temperature, the Curie temperature, the magnetism disappears, and it returns only after cooling.

There is a reference to a shepherd named Magnes as early as in the writings of Pliny the Elder, whose hobnails are said to have stuck onto a stone on Mount Ida. The stone then received the name of the shepherd. Less mythical interpretations of the name are more probable, but it is necessary to realize the peculiarity of an object that, without being touched, exerts a force on another object. This was an altogether unique and mysterious phenomenon to our ancestors. According to Lucretius (ca. 99–55 BCE), stones with magnetic properties were found "within the borders of the people of Magnesia"[57] and named accordingly. The ingenious and clear-sighted Roman thinker, who showed through logically sound thought experiments that matter cannot be indivisible, and deduced from this the existence of atoms, attempted an interpretation of magnetic action in his work *On the Nature of Things*. He assumed that all matter is "porous" and interacts with other bodies via a permanent exchange of particles.[58] This is quite an amazing insight. If one substitutes "atom" for "primeval element" in his writing and understands the pores as the actual "empty" space between all atoms of matter and thinks of the exchange not through air but through the modern concept of the field, we are not far away from today's state of knowledge. An incredibly perceptive observation more than two thousand years old, purely won by the power of thinking. Even today, the permanent magnetism of solid bodies is still astounding.

Figure 3.65 Further examples of grains made of heavy minerals: (1) Kyanite, Rio Amarco, Brazil; (2) Actinolite, Nufenen, Switzerland; (3) Epidote, Hohwacht, Germany; (4) Olivine (re) and garnet, Hochwacht, Germany; (5) Staurolite, Hasle, Denmark; (6) Tourmaline, Kerala, India; (7) Brown hornblende, Lovina, Bali; (8) Augite, Lovinia, Bali; (9) Rutil, Valle, Norway; (10) Sillimanite, Kerala, India. All photographs were taken in a bright field with transmitted light.

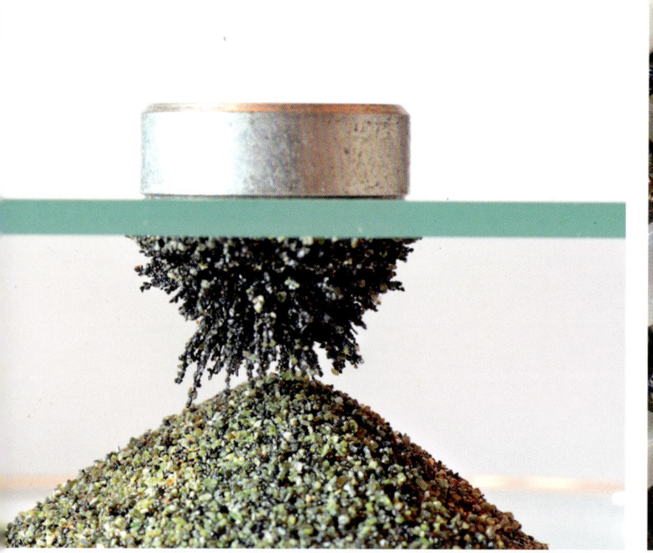

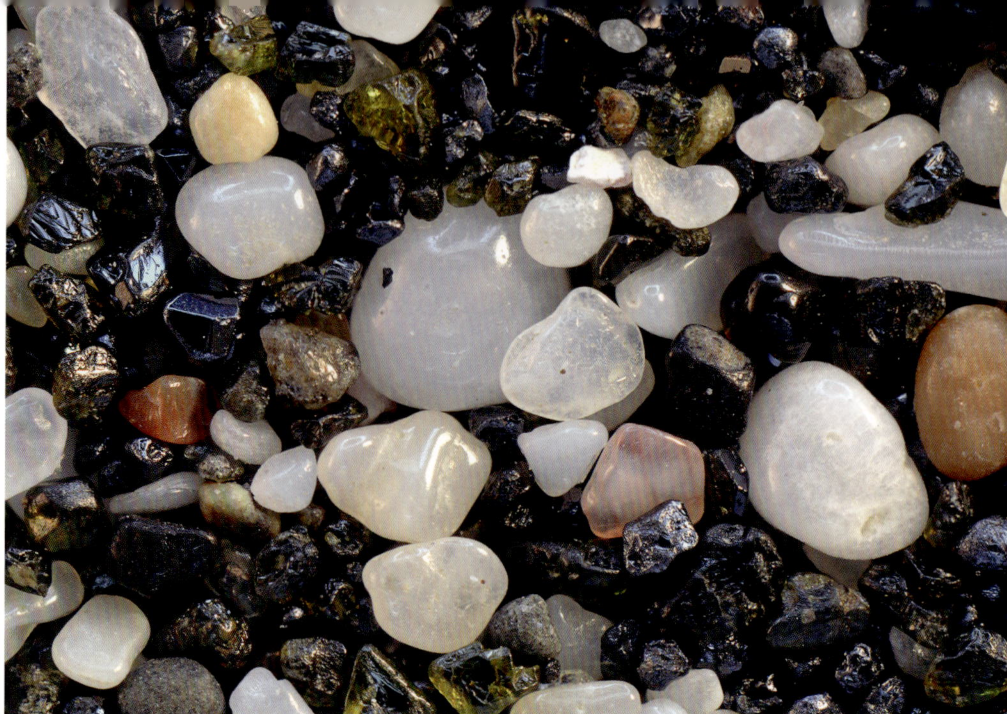

To date, it has not been possible to detect magnetic single poles. No matter how often one divides a magnetic substance, a positive and an opposite negative pole always exist on the same body, down to the elementary-particle level.

Now we are going to investigate why many sands, especially in desert areas, have red to orange-red coloration. It is due to a mineral that does not occur in sands as larger crystals, but only as a pigment layer consisting of the iron oxide hematite.

Hematite, or red iron ore (Fe_2O_3), is an important ore that is found as a byproduct in metamorphic rocks, less frequently in igneous, and otherwise in hydrothermal veins as a volcanic-exhalation product.[59] Hematite is grayish black to reddish brown and only moderately magnetic. It occurs in many different shapes alongside thin films in coarse, spherical masses ("glass head"), in metallic shining crystals, and in fine, translucent flakes ("iron mica"). As a detrital mineral, hematite is found, as already mentioned, only rarely in placers and is at most locally accumulated. It is quite stable against erosion but eventually transforms into limonite, goethite, or siderite. As the most common color pigment in sediments such as sand, sandstone, and claystone, hematite is encountered practically at every turn. The typical reddish to deep-red sands of the Sahara, Namib, or even the great Australian deserts consist of quartz sands around whose grains a fine film of hematite has been deposited from solution.

Figure 3.66 Sand sample from Limassol, Cyprus. A glass plate has been secured over the sample, which lies on a strong magnet. The sand grains experiencing the pull are made up mostly of magnetite and arrange themselves along the magnetic-field lines. The green color of the sand comes from an uncommon clustering of amphiboles and pyroxene.

Figure 3.67 Sand from Machalilla National Park, Los Frailles Beach, Ecuador. Alongside the white shell fragments of clams and snails as well as green volcanic olivine, the black, shiny, metallic sand grains made of magnetite stand out. To the left of the middle, an idiomorphically developed crystal of magnetite can be easily seen. Image width 5.5 mm.

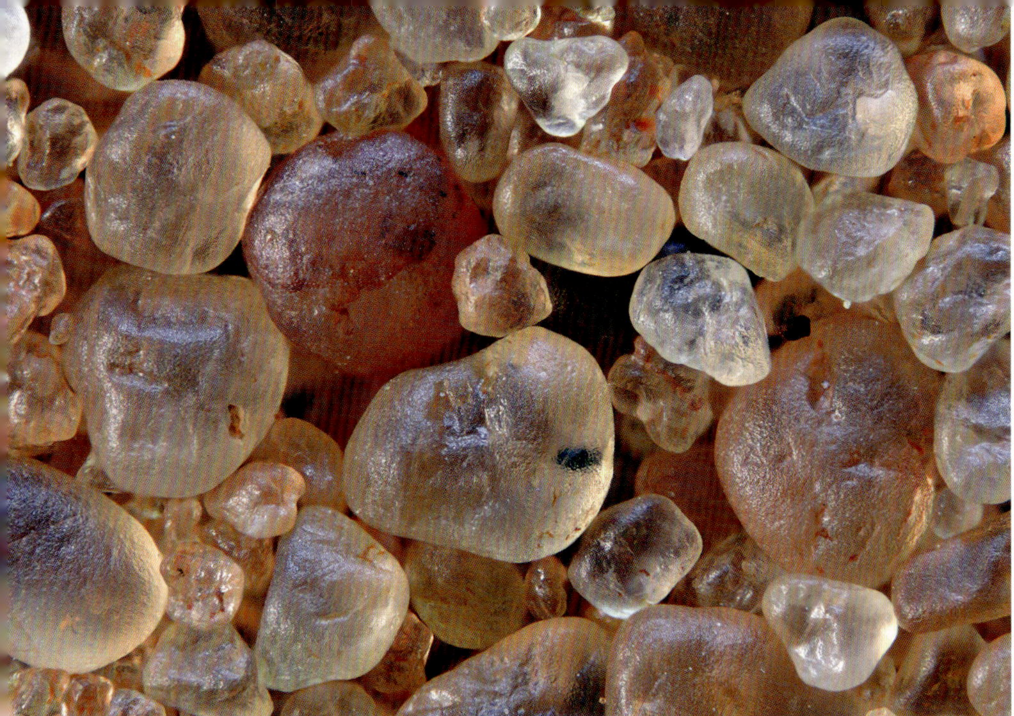

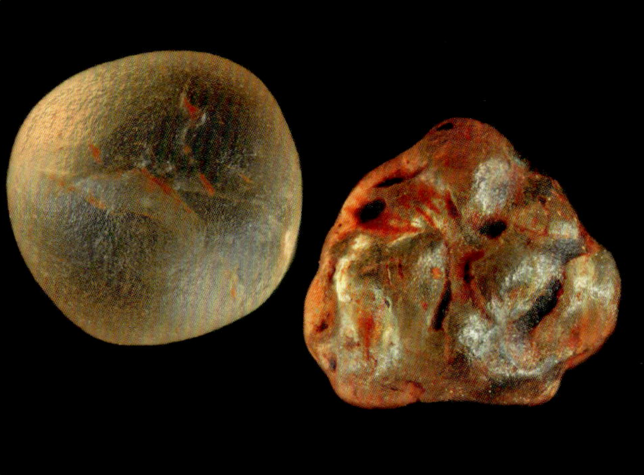

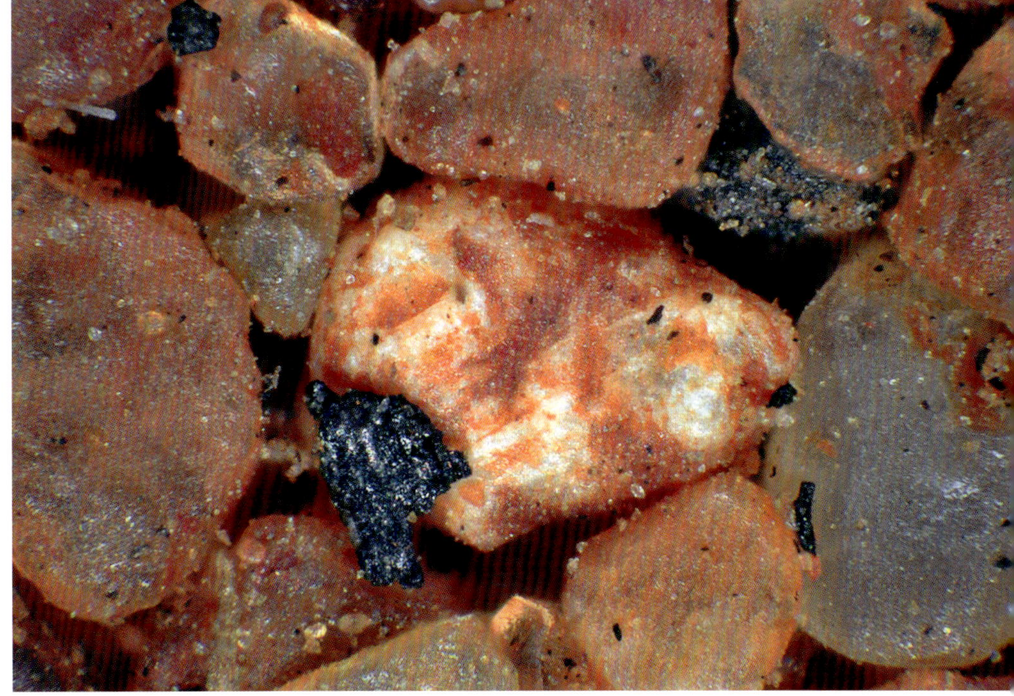

Figure 3.68 Desert sand from the Sahara near Murzuk, Libya. The red color of the quartz grains comes from the accumulation of ferric oxide. The tarnished surfaces of the sand grains are typical of deserts. Reflected light. Image width 3.8 mm.

Figure 3.69 Quartz grains from desert sands: *to the left*, the Sahara near Murzuk, Libya; *to the right*, Alice Springs, Australia. There is a more intense coloring that has remained in the furrows and depressions of the grains. Abrasive processes from wind transportation work against the accumulation of ferric oxide layers. Reflected light. Grain diameters each about 730 *µm*.

Figure 3.70. Sample of fine desert sand from Ayers Rock, Australia. Many grains from desert sands have a ferric oxide casing, which, for the middle grain, is partly broken through. Black magnetite grains probably contributed to the iron in the sediment. The sand has many adherent microparticles. Reflected light; the image width is only 720 *µm*.

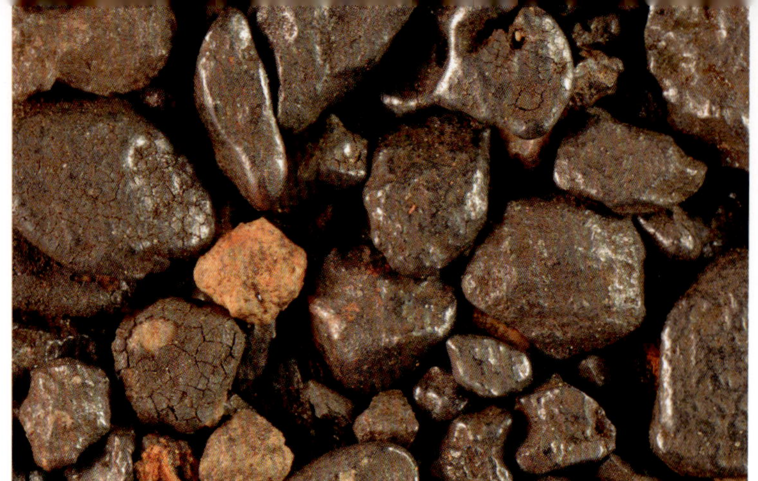

The red sandstone from the Triassic period also got its color this way. However, the process of sediment reddening has not yet been completely explained and is still controversially discussed. It has been discovered that the reddening of sands is limited if the sediment is transported by wind at the same time. In any case, a semiarid to humid climate is conducive to the reddening process.[60] The iron oxide originates from the erosion of iron-bearing minerals in otherwise pure quartz sands of the arid dunes or is added by adjacent rocks via wind transport. According to studies by Pye,[61] the red pigment of the sand grain coatings of coastal dunes consists mainly of a mixture of weakly crystallized hematite and goethite, whereas well-crystallized hematite is rare. Certain climatic conditions certainly favor the process of reddening in sediments. It always requires the addition of a ferrous substance, which can occur by external input or by the chemical transformation of exiting minerals. Tröger (1969) found that in sandstones older than the Middle Tertiary period, more than 80% of all opaque heavy-mineral grains consist of hematite, while in younger sediments, as we have already seen, magnetite and ilmenite clearly predominate.[62] In this context, it should be mentioned that colorful, layered reddening can also occur as a result of subsequent processes, such as the seepage of fluids containing iron oxide into sandstone layers. These do not always have to be oriented parallel to the layers of the original sedimentation. Effects of this kind are known as Chiasma sandstones or Liesegang rings.[63]

Hematite occurrences date back to the earth's early history. In the time span from 2.5 to 1.8 billion years ago, the oxygen newly formed by cyanobacteria in the atmosphere combined with the iron content of the world's oceans to form oxides that sank to the bottom. Powerful formations of so-called banded iron ores of hematite, magnetite, and silicates were formed. Today, these ores form the basis of the worldwide steel industry. They are also witnesses to an early dramatic change in the surface chemistry of our planet in the early Proterozoic period. Unlike other rocks, these ores cannot be newly formed—they are unique.[64]

Figure 3.71 Sand from Triebtal near Jocketa, Saxony, Germany. The sand grains are made of limonite or show a partly crumbled limonite crust (see small grain at the bottom left). Near Jocketa, iron ore mining was carried out on the Eisenberg in earlier times. Image width 7.6 mm.

Figure 3.72 Grain of sand with cracked, eroding crust of limonite. Bergsee, Valle, Norway. The images of the surface topography are made with increasing magnification in the reflected-light dark-field method. Image width: top: 2,550 *µm*, middle: 880 *µm*, bottom: 450 *µm*.

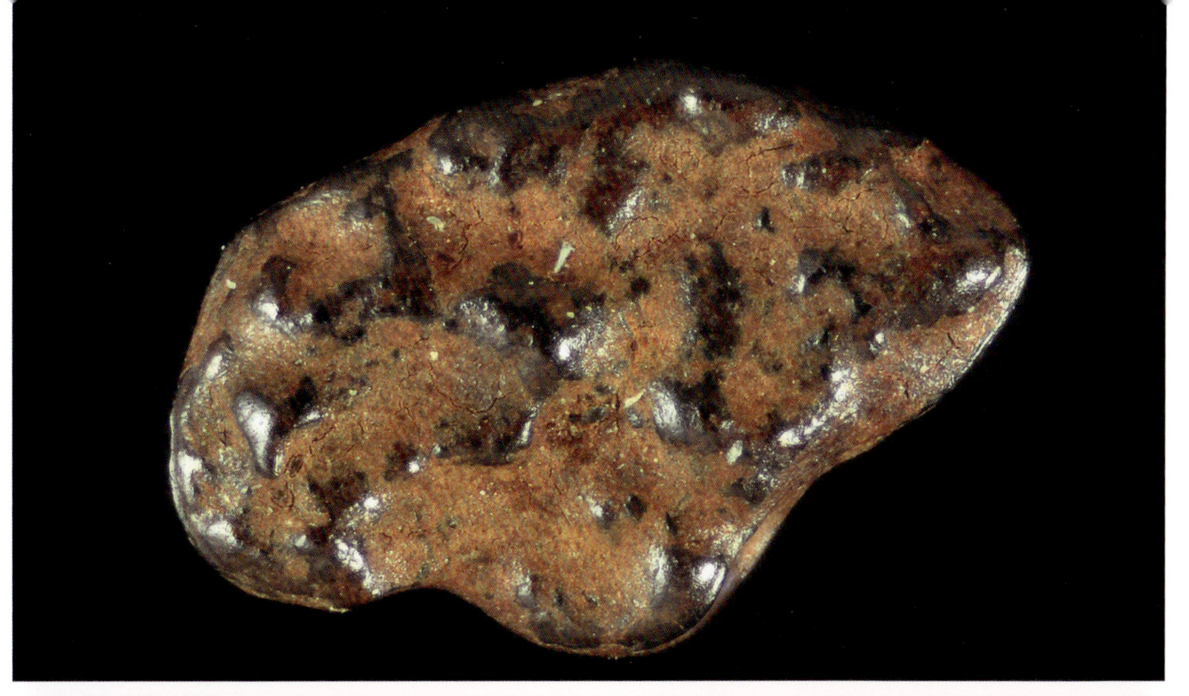

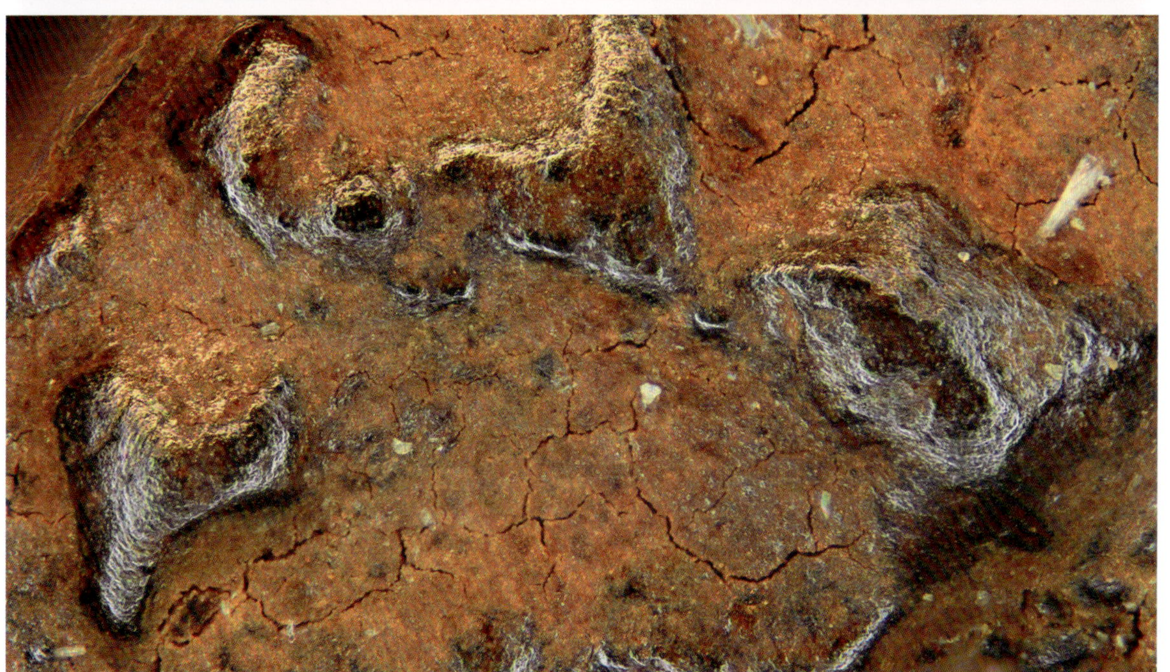

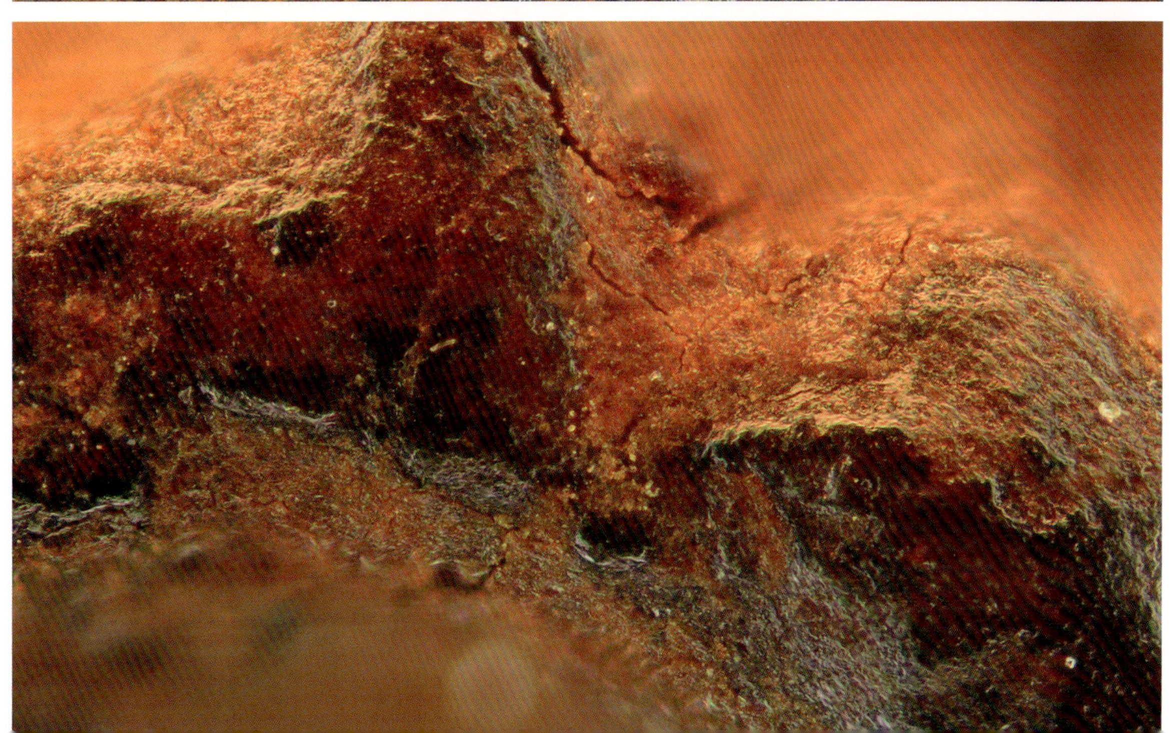

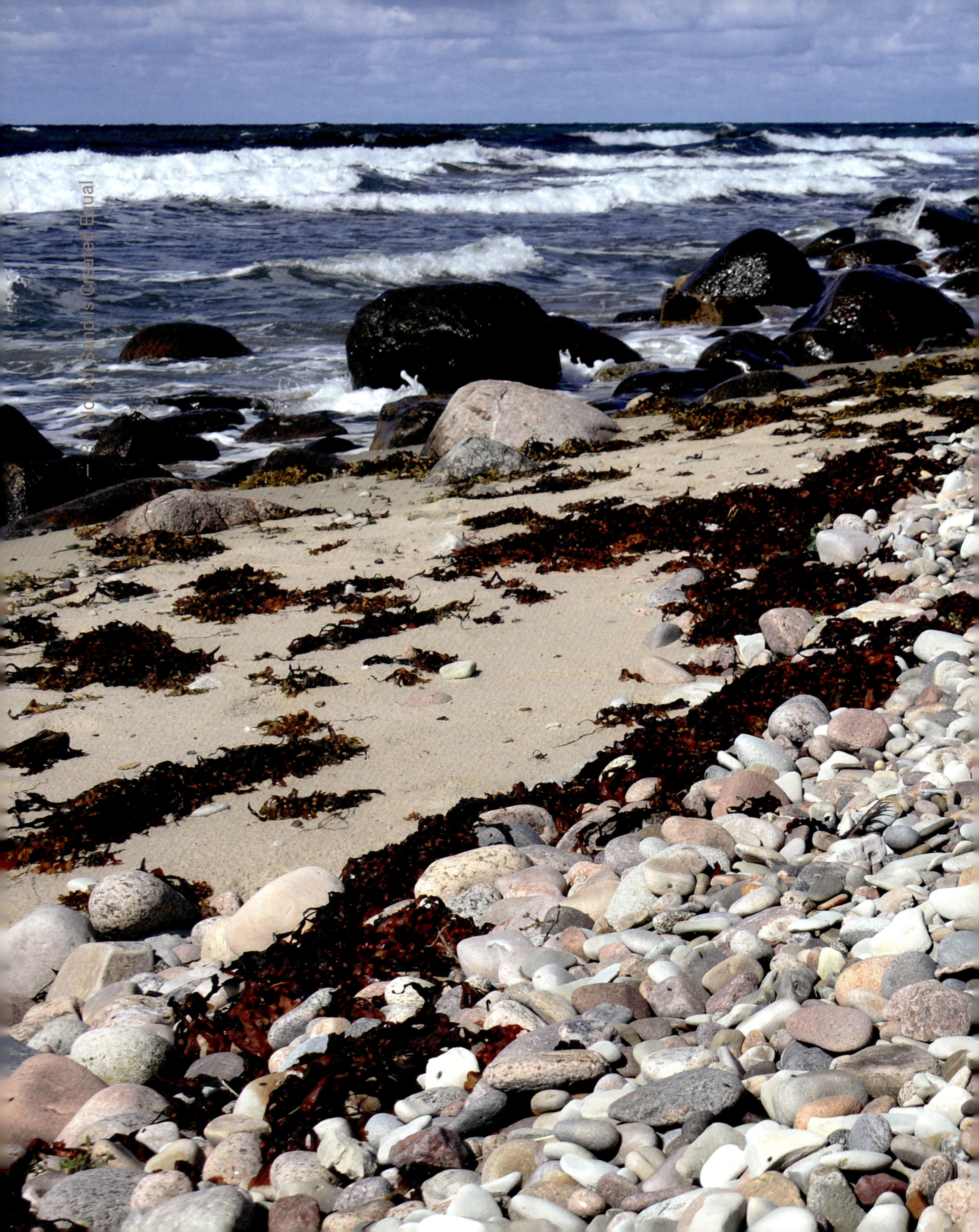

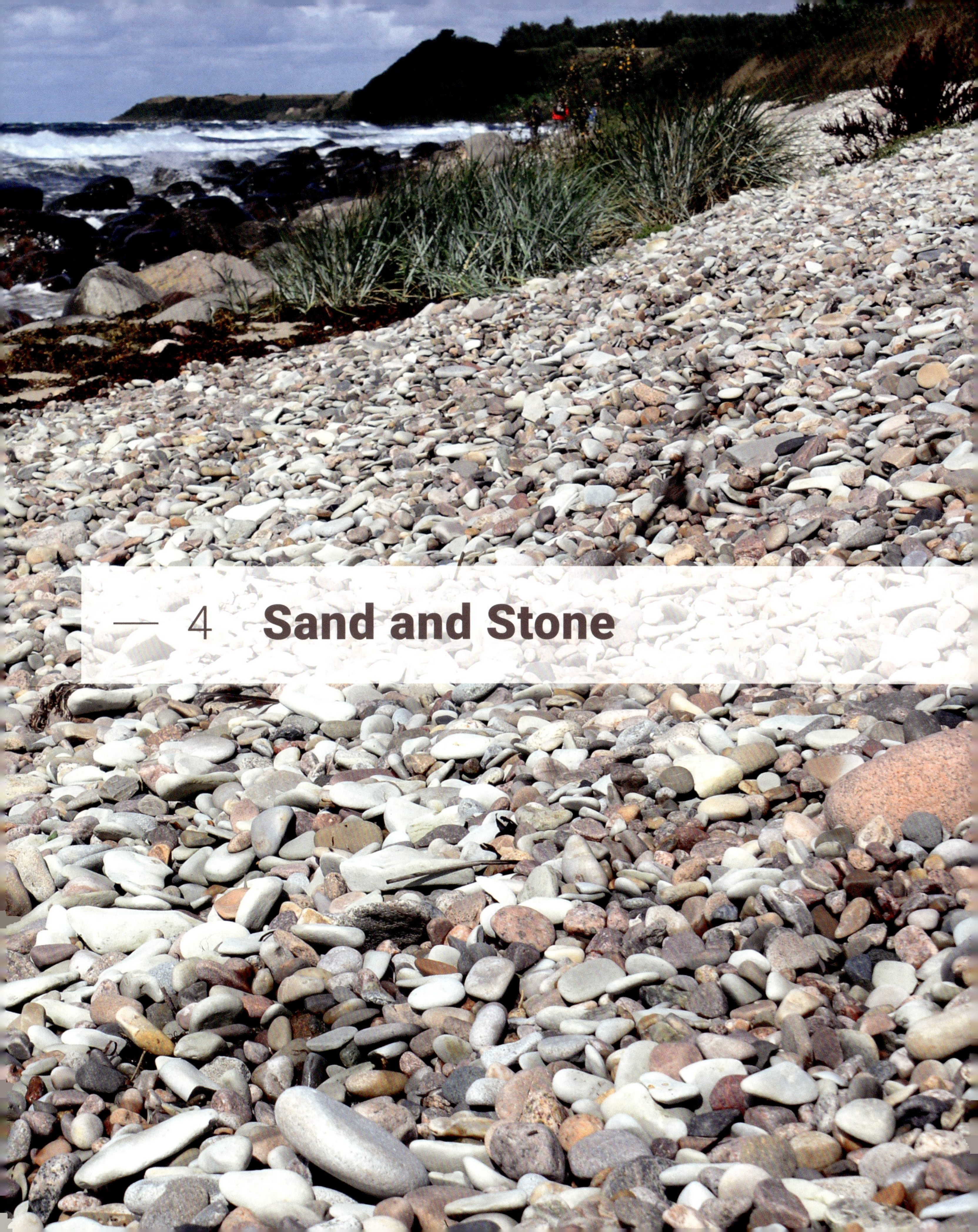

4 Sand and Stone

Stones are mute teachers; they silence the observer, and the most valuable lesson we can learn from them we cannot communicate.

—*J. W. v. Goethe*, ***Maxims and Reflections***

As we have seen, sands can differ from each other significantly. Now we will pursue the question of where sand comes from and how it is formed. We will get to know sand as a mixture of the remains of organisms and the broken-down product of the rocks of the earth's crust. But where do the rocks themselves come from, how are they formed, and how can they be classified and labeled? To clarify, sands and rocks are closely connected. Later we will deal with the organic remains in detail.

"Palermo, Wednesday, April 4, 1787
Still more singular did our guide deem me, when at all the shallow places, of which a great many are left dry by the stream, I searched for pebbles, and carried off with me specimens of each sort. I again found it difficult to make him understand that there was no readier way of forming an idea of a mountainous district like that before us, than by examining the nature of the stones which are washed down by the streams; and that in so doing, the purpose was to acquire a right notion of those eternally classic heights of the ancient world."

—*J. W. v. Goethe*, ***Letters from Italy***,
Weimar Edition [WA] I, 31

In the previous chapter, we dealt in detail with the properties of sand and with its composition as a heterogenous, granular mixture of crystalline matter. As it turned out, a given sand sample can be described by mineral composition, grain size, sphericity, color, degree of rounding, etc. But where does the sand come from, how was it formed, how was it subsequently transported, and what will become of it? To already give it away: It is very rare to be able to discover the origins of a given, unknown sand sample by using simple means. Much less is it possible to reconstruct the place of origin of individual grains. In some places, mineral composition, sorting, and grain size can already change within an area of 330 feet (100 m).

Previous double page:
Bavnodde, Bornholm, Denmark

Exceptions to this would be, for example, very locally specific mineral compositions or the locally restricted occurrence of certain organisms whose shell remains are found in the sample.

In the forensic geosciences, such knowledge is used and elaborate statistical methods are employed alongside equally elaborate physical/chemical investigations. In the case of the Italian prime minister Aldo Moro, who was assassinated in 1978, sand residue was found in his clothing, and after a geological analysis, a 93-mile-long (150 km) coastal strip was identified as a possible area of origin.[65] From this area, ninety-two comparative samples of sand were taken and analyzed in detail for inorganic and organic indicators.

As a result, the potential area of origin could be narrowed down further to about 7 miles (11 km). Unfortunately, however, the subsequent search of this stretch of coastline failed to yield any clues to the perpetrators. After their subsequent arrest, they claimed to have intentionally placed the sand samples in the vehicle and on the victim's clothing as a diversionary tactic. However, it is believed that this claim is false.

Group 1	Inorganic sands that are formed mainly through the erosion of crystalline rocks and therefore have a continental character
Group 2	Inorganic sands that are predominantly composed of the products of limestone erosion (notwithstanding that limestones themselves are often organic in origin)
Group 3	Inorganic sands consisting of grains formed by chemical or physical processes. They usually have only local and limited occurrence. Examples: ooids, gypsum sands, sands formed pyroclastically (volcanically).
Group 4	Sands generally of biogenic origin. The sand grains consist of shells, casings, or fragments of living organisms such as shells, snails, foraminifera, corals, sea urchins, and sponges. The sands have a marine oceanic or, more rarely, fluviatile character.
Group 5	Anthropogenically produced sands or sand fractions. Mostly, these are industrial wastes such as slags from blast furnaces, coal, plastic residues and granulates, glass fragments, and, above all, microplastics.

As has already been shown, each type of sand differs from other types through a variety of characteristics. But we still want to bring some sort of order to this variety and therefore split sand into five groups, which, of course, naturally flow into each other but are useful for a first look:

In this chapter, we will first focus on group 1, which represents the largest proportion of the world's sands when looking at volume. This is sand that has directly or indirectly resulted from the erosion of continental, crystalline rock masses. In order to gain a deeper understanding of this sand group, it is necessary to take a look at the different rock types and their formation history.

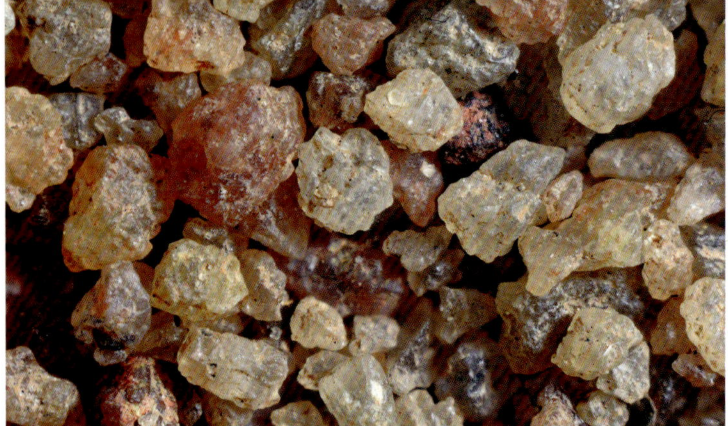
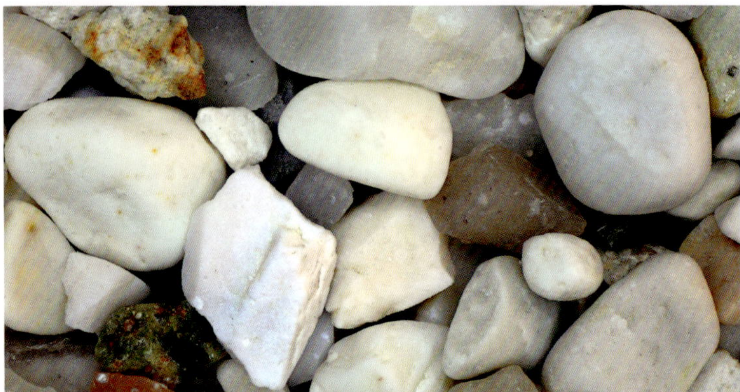
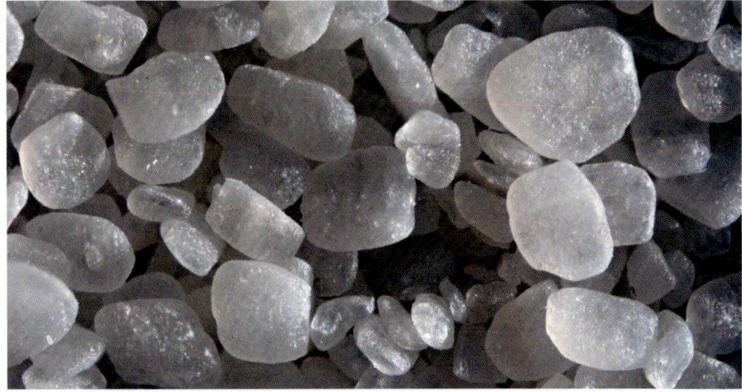
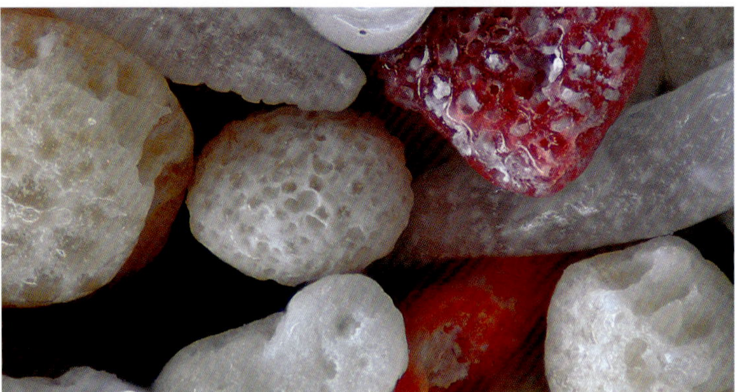
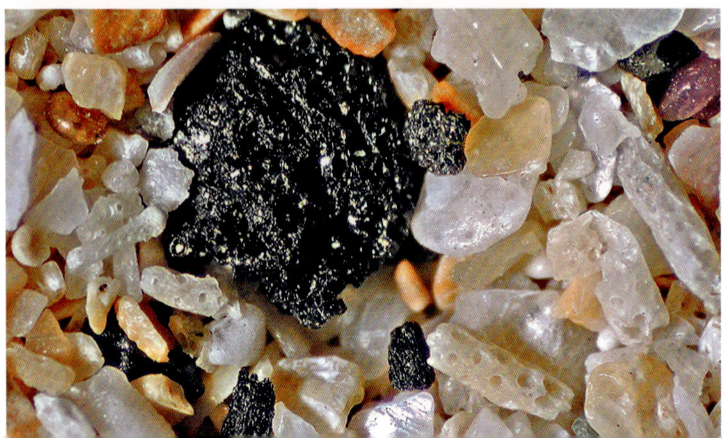

Figure 4.1 Sand examples from the five different groups:

Group 1: Inorganic sand, chapter 4. Binao-Bousooue, Ivory Coast. Quartz-dominated land sediment. Edges only rounded.

Group 2: Inorganic sand made of limestone, chapter 7. Lefkas, Greece. Mostly well-rounded limestone grains as well as a few colorful minerals along with traces of marine life.

Group 3: Sand formed by chemical or physical processes, chapter 4. Alamogordo, White Sands National Monument, New Mexico, USA. Monomineralic, fine-grained gypsum sand. Image width 4 mm.

Group 4: Sand of biogenic origin, chapter 7. Michaelmas Cay, East Australia. All the sand grains are made of the shell remnants of marine organisms.

Group 5: Sand with anthropogenic traces, chapter 9. Carbis Bay, southern England. Marine sand with slag residue and coal fragments, apparently freighter cargo that has gone overboard.

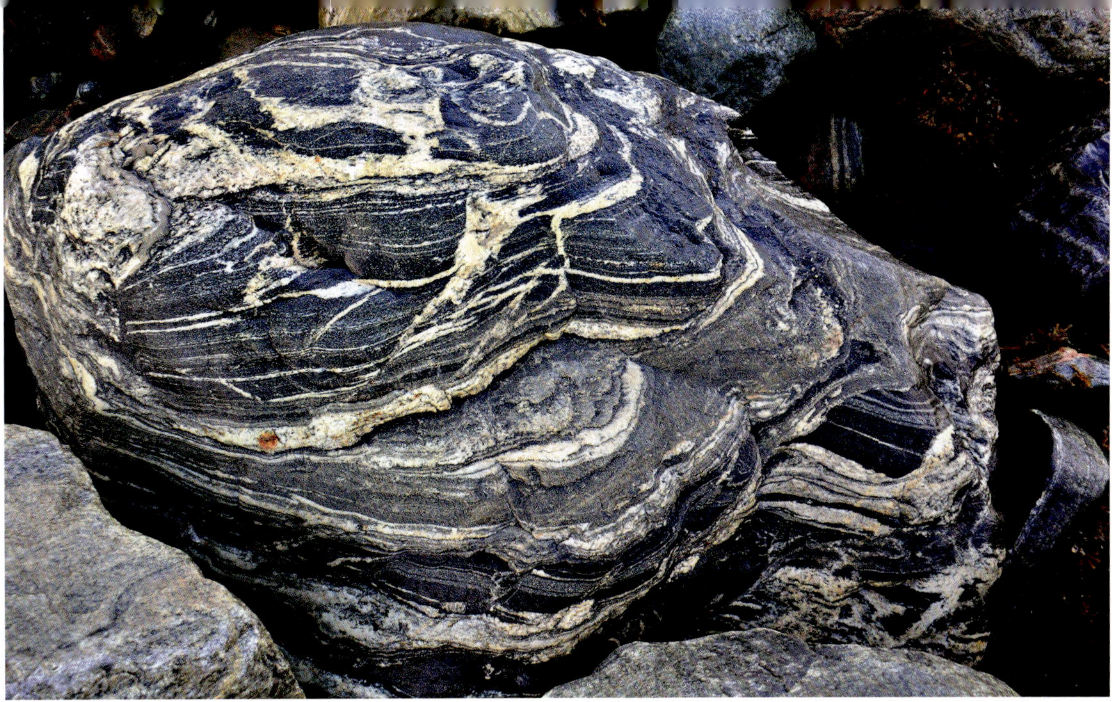

How was the earth formed, and how were all the different types of rock created that form the basis of the sand we see today? Where did matter come from? For that, we must go back to the distant history of our universe.

More than 8.5 billion years following the Big Bang, a cloud of matter formed within our galaxy, the Milky Way, that consisted of 98% gas and only 2% heavy elements. From this rotating gas cloud, as we saw in chapter 2, our sun was formed. The four inner, Earth-like planets Mercury, Venus, Earth, and Mars condensed from the heavier elements. They initially formed as rather loose associations of dust, chondrules, and CAIs that gradually solidified through partial melting. The decay of radioactive elements later heated the early earth to such an extent that its surface was characterized by vast magma oceans. These oceans of molten rock were more than 620 miles (1,000 km) deep.[66] About 4.5 billion years ago, an event took place that had a crucial impact on the fate of the earth and became known as the giant impact theory.[67] At this time, a gigantic body, which was about the size of the planet Mars, struck the early earth. It was given the name "Theia." This was a grazing impact, therefore at a very acute angle, whereby the angular momentum of the earth increased and the earth rotation accelerated. In contrast, a frontal impact would have completely destroyed the earth, and we would not exist. The impact shortened the day/night rhythm of the earth to levels that were better suited for the development of higher life. The impact energy of the protoplanet Theia was so high that it evaporated itself to a large extent and melted a large part of the rock masses of the earth again. The ejected matter, mainly lighter crustal material, formed the moon. This is why the moon is lighter than the earth. The impact also tilted the earth's axis to the value of about 23.5 degrees relative to the earth's orbit, which still exists today. Without this tilt, there would be no summer/winter cycle and probably no higher lifeforms. Apparently, we have Theia to thank for a lot—sand included.

Figure 4.2 Archaic, metamorphic crust rock about 2.8 billion years old. Kirkenes, Norway. Image width about 1 m.

As the earth's crust melted, heavier elements, mostly iron but also other metals as well as probably the remaining metal core of Theia, sank toward the earth's crust. There, a liquid iron core gradually formed, which still exists today and is responsible for the vital dynamo effect that produces the earth's magnetic field and protects us all from deadly solar wind and cosmic radiation.

During the following years, the earth's core, mantle, and crust developed through complex processes. Depending on the respective composition, heavy basic magmas sank, lighter silicate magmas rose, convection cells formed, and plate tectonics were initiated. Chemical processes led to the separation of the combined crystal water, and the oceans were formed. Volcanic activity brought gases and water vapor to the surface, a first thin atmosphere covered the earth as it further cooled, and the first major continental masses formed about three billion years ago.

Core parts of this archaic crust can still be found today on almost all continents. The oldest stone that is accessible here on the earth's surface is approximately 4.03 billion years old. It is a so-called metatonalite, a gneiss stone, which can be found in the Canadian Western Territories (Acasta gneiss). Even this stone includes even-older components, such as zircon crystals, which were dated back 4.2 billion years ago, making them tiny witnesses to the earth's crust that has long since disappeared. In northern Scandinavia, on the western side of the Kola Peninsula in Kirkenes, there are more very old rocks from the ancient earth's crust, which were created 2.8–2.9 billion years ago. About 2.5 billion years ago, free oxygen entered the earth's atmosphere, and geochemical cycles and weathering processes began. A long period on the way to the sand as we know it today.

Before we take a closer look at these important components of the earth's crust, rocks, it is important to clarify what exactly is meant by rocks. We will refer to Rothe:[68]

> "Rocks are monomineralic or polymineralic aggregates of minerals, which represent independent, self-contained areas of the earth's crust."

One must add that an igneous rock is formed by definition at the time of solidification—time starts here. If we hold a piece of rock that is four hundred million years old in our hand, this means accordingly that the melt from which the rock originated solidified this period of time long ago. This does not take into any consideration that the molten mass itself consisted of a molten rock from an older time. For sedimentary rocks, the time point of sedimentation is used as the age indicator.

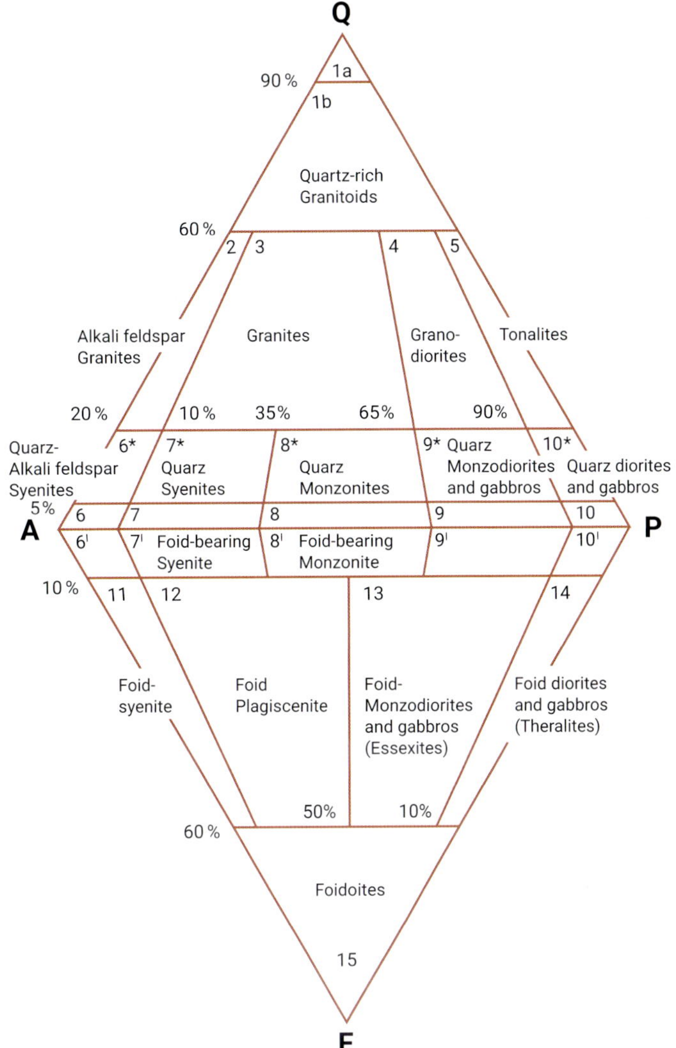

How can the rocks, or "mountain species" as they were called in Goethe's time, be distinguished and, above all, named? Unlike, for example, the classification of plants or bird species, a naming system for rocks can never be complete or exact. Each formation of rock bodies is in principle unique and dependent on very many factors. Transitions are fluid and difficult to discern for the samples at hand without precise analysis or contextual knowledge. Many of the names in use as well as colloquial rock names are derived from place-names, so-called type localities. These are concise but usually have narrowly limited occurrences such as kinzigite, gabbro, syenite, or bauxite, which are named after cities.

Figure 4.3 QAPF diagram for plutonic rocks. M = 90%–100%. Q: Quartz, A: Alkali feldspars, P: plagioclase feldspars, F: feldspathoids (foids), M: mafic minerals. Data in percent by volume. The corresponding corners identify 100% of the individual components. Source: Rothe 2010.

Rocks and minerals were also named for famous researchers or the initial descriptors (forsterite, goethite, etc.). None of this is systematic, and some of it is even wrong. According to new findings, syenite, which is present near and gets its name from Aswan (Egypt), is actually red granite. If one looks at commercial names, it gets even more confusing. In commercial settings, the systems that have been introduced have moved far from the scientific systems or never used them in the first place.

A systematic classification system has been developed by the International Union of Geosciences (IGUS), which we briefly present here. The system works with the relative mineral content of the rocks and is based on the idea of Albert Ludwig Streckeisen, a Swiss petrograph. Using a so-called QAPF diagram (alternate name: Streckeisen diagram), both plutonic rocks and volcanic rocks can be represented by measuring the volume content of the characteristic materials of alkali feldspar (A), which includes potassium feldspar and plagioclase albite; plagioclase (P), the mixed crystals of the plagioclase series without albite; quartz (Q); and mafic minerals (M) under the microscope on thin rock sections (see below). Mafic or dark minerals contain a lot of *ma*gnesium and iron, *f*errum; hence the word. The common base of the double triangle is the feldspar line A–P, which reflects the ratio of alkali feldspar (A) to plagioclase (P) in the rock. In the upper triangle, rocks are represented with relevant proportions of quartz (Q), the percentage of which is plotted above the feldspar line in each case. The fourth, lower apex of the double triangle considers the so-called foid (F) or feldspar representatives. They include minerals such as leucite, nepheline, and sodalite. Foid (contraction of the longer term feldspathoid) is formed when there is too little silica in magma to form feldspars. For this reason, quartz and foid never occur in the same rock. The aspect gets put to good use in the classification system.

For classification according to Streckeisen, only these four groups (quartz, alkali feldspar, plagioclase, foids) are used and summarized under the term "felsic" or light minerals. "Mafic" or dark minerals are not considered as long as their volume proportion does not exceed 90%. Among them are mica, epidote, magnetite, olivine, orthopyroxene, clinopyroxene, and others. However, if their volume fraction exceeds 90%, the rocks are counted among the ultramafic igneous rocks and treated separately in their own classification.

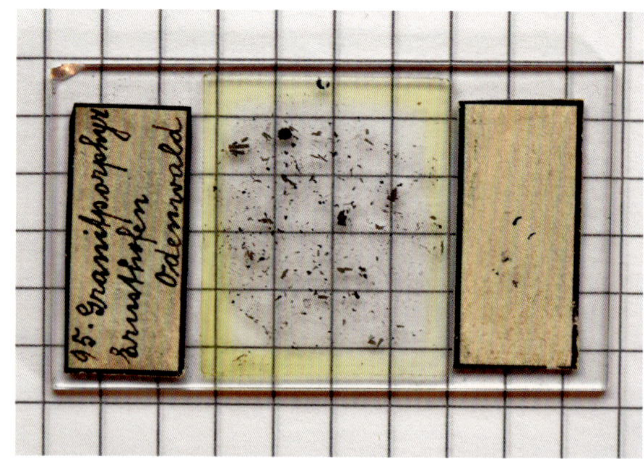

Figure 4.4 Historical thin rock sections of granite from the Odenwald (around 1900). A piece of granite (*center*) is stuck to a piece of glass and ground so thinly (0.03 mm) that you can see through it, so it can be looked at closely with a microscope. In the picture, you can easily see the checked grid through the glass plate with the rock. The glue substance has since yellowed. The dark spots on the glass plate are brown-black biotite grains in the granite.

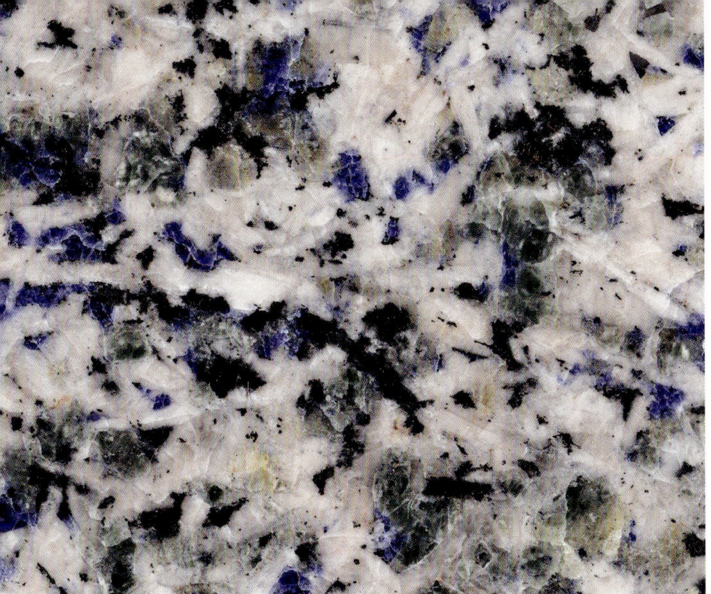

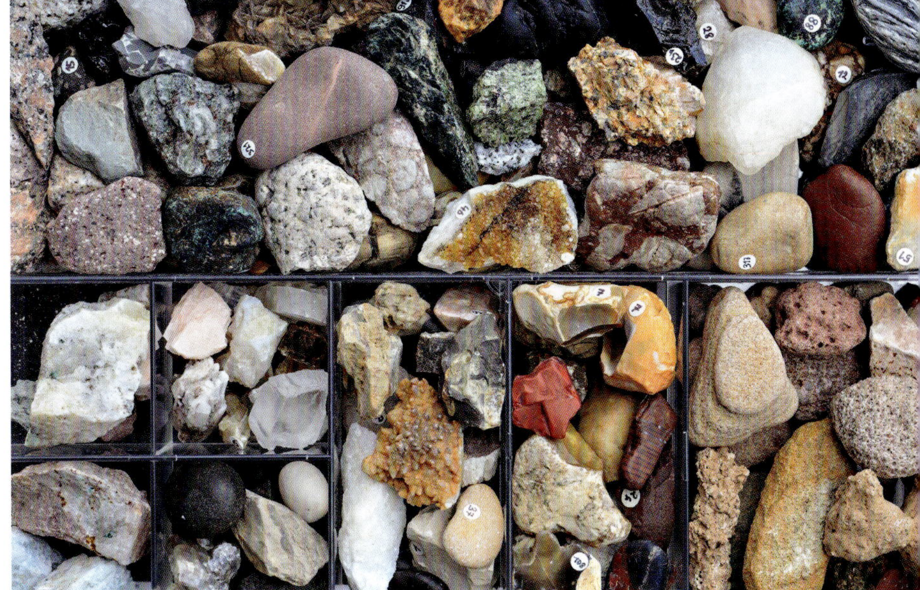

When looking at rocks containing quartz, the content Q + A + P is set to 100% and plotted in the upper triangle, and alternatively, in the case of rocks containing foid, F + A + P = 100% and plotted in the lower triangle. The relative ratio A:P is plotted on the baseline, and in the case of pure "A" on the left point, above which the quartz proportion is then plotted upward or the foid proportion downward. The point resulting from the measurements lies in a field of the distance triangle, to which a rock name is assigned according to its definition. If, for example, A and P are present in equal amounts in a sample and the quartz content is less than 20%, then it is a quartz monzonite (see graphic); if the plagioclase content is zero with the same quartz content, then it is a quartz alkali feldspar syenite. The QAPF diagram, however, says nothing about rock formation—rocks are classified, but arbitrarily.

Generally, rocks are split into three different types:

- igneous rocks or magmatites (plutonic and volcanic rock)
- sedimentary rock or sedimentites
- metamorphic rock or metamorphites

Figure 4.5 Foid-containing igneous rock, foid-syenite (field 11 in the QAPF diagram), here: sodalite-containing nepheline-syenite. Green-gray nepheline, blue sodalite, white alkali feldspar, and black biotite. Neither quartz nor plagioclase is included. Polished plate, image width about 70 mm.

Figure 4.6 In this small collection, there are many important types of rock that can be found on Earth today, including igneous, sedimentary, and metamorphic rock. There are many different minerals represented in the same stone, here in wallstone (German: *Wallsteine*), coming from the ocean's wash margin: peridotite from the earth's interior, desert gypsum rose, high metamorphites, obsidian (from which our ancestors made knives), rock salt, and a loess doll and lava from prehistoric times whose pores are filled with crystallized minerals. Dimensions: 20 × 30 cm.

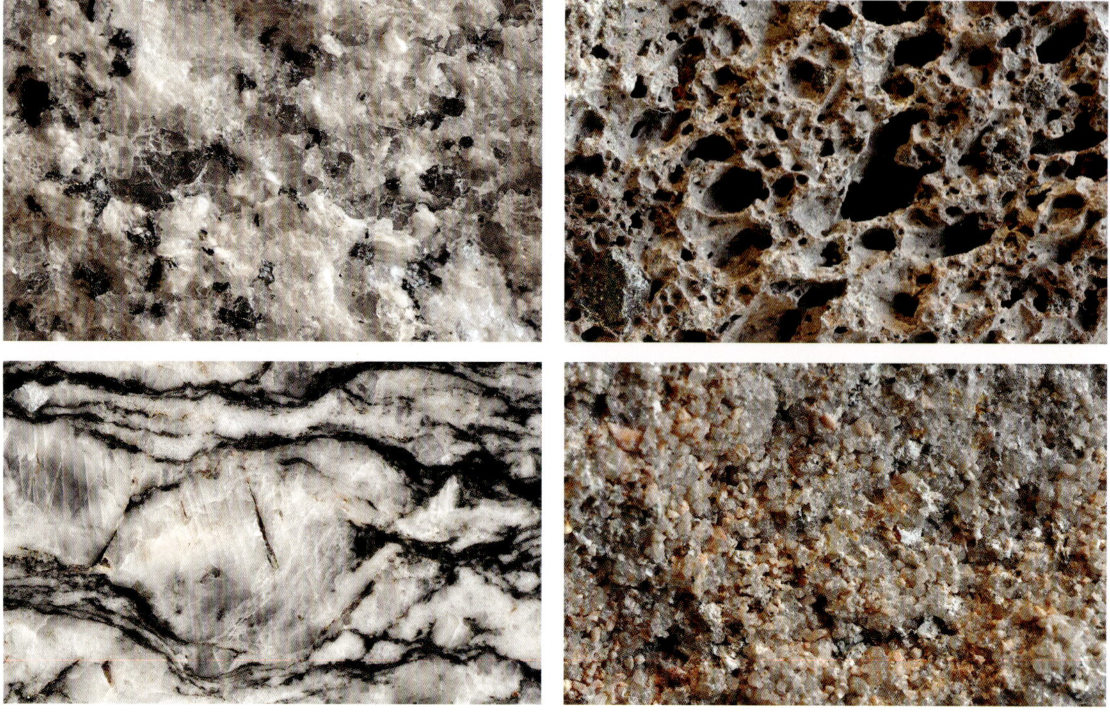

While igneous rock (65% of the earth's crust) is made up of solidified melted rock that has cooled down in the earth's crust (plutonic rock) or aboveground (volcanic rock), metamorphic rocks (27% share) are formed by the transformation of any rocks under pressure up to more than 10,000 bar and temperatures up to more than 800°C within the earth's crust in depths up to 22 miles (35 km). Metamorphic rocks are therefore mostly layered or schistose in structure and very compact. Due to the unimaginably high pressures and temperatures, new very characteristic minerals form, and the rock is transformed into a dough-like, plastic state, from which it can then be folded and even compressed like paper 180 degrees through mountain-building processes. Metamorphic rocks, however, must not be confused with sedimentary rocks, which are also often layered. Sedimentary rocks or sedimentites form without pressure on the surface of the earth or on the seafloor, where products from mineral erosion, after being deposited in layers, are formed into new rocks through physical or chemical processes (diagenesis).

Figure 4.7 An overview of the most-important families of rocks. The image width is 30 mm, respectively. *Top left*: Igneous/plutonic rock: granite with white feldspar, black biotite-mica, and glassy-gray quartz; *top right*: igneous/volcanic: lava rock with a significantly bubblier texture; *bottom left*: metamorphic: augen gneiss from Switzerland with layered texture and augen feldspar, polished sample surface; *bottom right*: sedimentary: fluvial sandstone (from the Triassic period) from Hohenhaslach, southern Germany.

4.1 Igneous Rock

4.1.1 Plutonic Rock

Plutonic rock forms the largest portion of the earth's land mass in terms of quantity. These rocks were named after Pluto, the god of the underworld. With a share of almost 90%, granite rock is its most prominent and common representative. Today, there is no doubt that granites are formed from originally liquid magmatic melts of very different compositions, which crystallize relatively slowly in the deep interior of the earth in so-called plutons (so-called I-type granites, from "igneous").

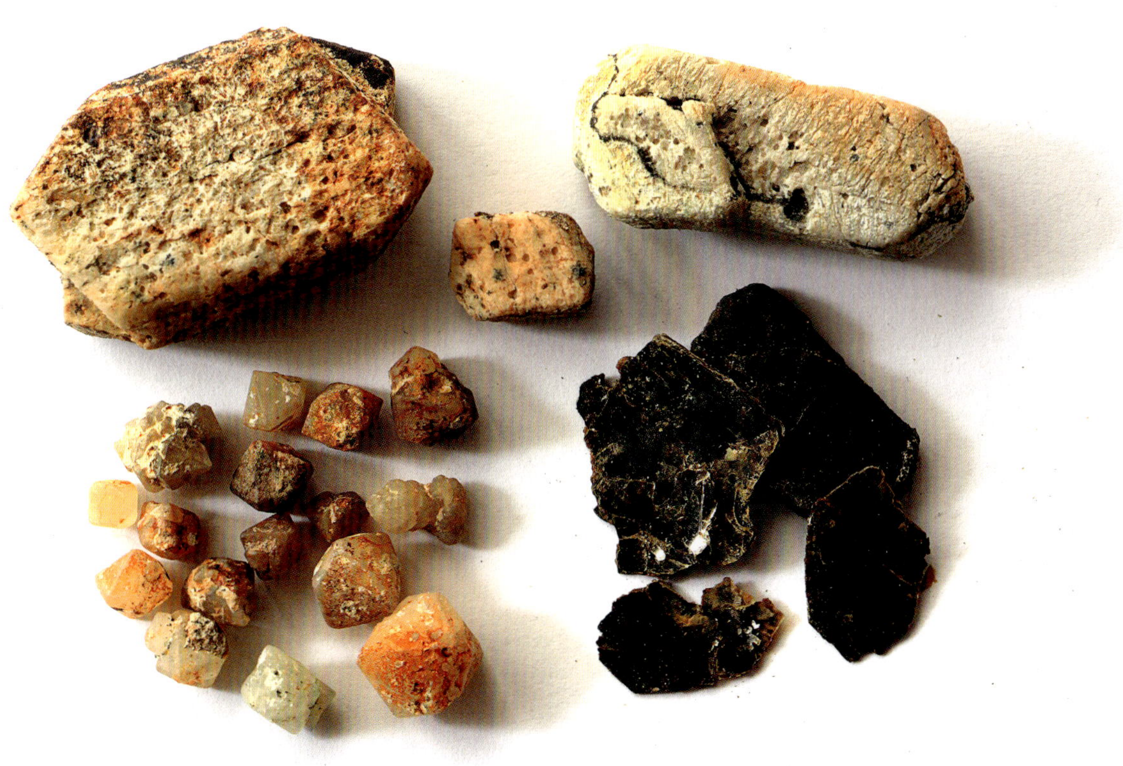

Figure 4.8 Some of the most common rock-forming minerals, here the main species involved in the formation of granite: feldspar (±65%), quartz (±27%), and mica (±5%). *Above*: Feldspar crystal from Elba, Italy; *far left*: Carlsbad twin, like Goethe described, which has been weathered out of the rock formation. *Bottom left*: Quartz crystals eroded out of porphyritic granite, Enfola Peninsula, Elba, Italy; *bottom right*: Biotite from Norway.

Plutons typically have diameters between 6 and 62 miles (10–100 km), while batholiths can reach sizes of more than 100,000 km^2; for example, in South America or along the North American western coast. Furthermore, granites can be formed by the melting (anatexis) of sedimentary source rocks (so-called S-type granites, from "sedimentary"). Melting processes of this kind take place during the collisions of whole continents at great depth. The melting of aluminum-rich sediments thereby leads to the formation of muscovite-containing granites. Muscovite can therefore, as already described, be regarded as a kind of depth indicator, since it remains stable during melting only in depths greater than 7.5 miles (12 km). If we hold a piece of granite with muscovite in our hand, we have to bear in mind that it was once formed at great depths before it reached the earth's surface, where it will most likely eventually break down into sand.

Today, we know that granite is by no means the oldest rock. The oldest rocks, which are still accessible today, emerged from even-older sediments, which in turn represent decomposed rocks of even-older origin. Probably the "original rock" was a formation from basaltic lava masses. European granites, like the Scandinavian ones, can be anywhere between 1.4 and 1.8 million years old. But there are also granite formations that are much younger, as found, for example, on Sardinia, Elba, or also in the Alps. At about thirty million years old, they are relatively young. Granite can be formed in principle at any time: a general age does not exist.

No one has yet been able to observe the process of granite formation directly or even indirectly.[69] We derive our knowledge from the plutons exposed through erosion, which are visible on the earth's surface and make the rocks available for analysis. In the 1950s, experimental petrology succeeded in transforming graywacke and clay shales into granite-like melts during experiments under high pressure with the addition of water and drew important conclusions from this.[70] The chemistry of magmatic melts proved to be quite complicated; this became apparent when investigating the spatial and temporal sequence in which minerals crystallize out of a melt during slow cooling and settle or float. It becomes even more difficult when it must be taken into account that the magma is locally altered in its composition through the melting of adjacent rocks. This gives an idea of the great variety of plutonic rock types.

In addition to granitic rocks, there is the important family of gabbroic rocks. Gabbro is a depth equivalent of basalt and builds up the oceanic portion of the lithosphere, the outer layer of the earth's body. Gabbro consists of basic plagioclase and half of mafic minerals, mostly of the mineral augite. Other plutonic rock families are diorites and tonalites, which are conspicuous by their black-and-white coloration (like pepper and salt). Syenites and monzonites look very similar to granite at first glance but have much lower quartz content (less than 5%). It stands to reason that the different rocks then produce differently composed sands during erosion.

The meanwhile largely secured knowledge about the formation of igneous rock and, above all, granites was very controversial for a long time. In Goethe's time, the scientists were divided two ways: the Neptunists and the Plutonists. While the Plutonists already assumed that granites, or igneous rocks in general, originated from molten masses from the earth's interior, the Neptunists decisively rejected this view. They saw purely sedimentary processes at work. Granite—according to their view—was the product of a very slow depositional process from a primordial ocean spanning the entire earth, with increasing compaction from the overlying material. Goethe came very close to this interpretation, since his deeply antiaggressive attitude toward the things of the world and toward nature could not accept seeing the volcanic activity of the earth's interior, which was perceived as aggressive, as responsible for the emergence of the granite rock, which he valued so much as the primordial rock of the world. In a conversation with Eckermann on April 17, 1825, he remarked,

> "You know how happy I am about every improvement that the future holds in store for us. But, as I said, every violent, erratic thing is repugnant to me in my soul, because it is not natural."[71]

This attitude, which was also a result of the traumatic impression the French Revolution, he pursued consistently throughout his life. It is easy to forget that it was he, after all, in his function as minister of state, who largely disarmed the Duchy of Weimar militarily.

Goethe was scientifically close to the then leading geologist Abraham Gottlob Werner, one of the spokesmen of the Neptunists. Werner explained the visible volcanic activity, the escape of magma, and also the formation of basalt with a smoldering fire of coal beds hidden underground, a hypothesis that Goethe then tried to support with his own observations. In the autumn of 1789, he presented a small paper titled "Vergleichs Vorschläge die Vulkanier und Neptunier *über* die Entstehung des Basalts zu vereinigen" ("Comparison proposals to unite the Plutonists and Neptunists on the formation of basalt"),[72] which admittedly never found circulation or recognition. In contrast, his essay "On Granite," in which he falsely presented granite as the oldest rock on Earth, gained some attention. In *Faust*, he had Mephisto present the views of the Plutonists. In later years, when the evidence against the Neptunists became overwhelming, he relativized his views and even acquiesced when he received an essay in 1823 on the effects of volcanoes by Alexander von Humboldt, whom he admired:

"The most diligent study of these few pages, according to the literal and the sense, shall help me to solve an important task, shall encourage me when I try to think like such a man, which, however, is only possible when his counterpart becomes the counterpart for me, towards which I must work with all my strength. If this succeeds, then it will not be a shame, but rather an honor for me to place my rejection of the old and my acceptance of the new doctrine in the hands of such an excellent man and tested friend."[73]

What is remarkable is the willingness shown to concede and join the factual arguments of the scientific opponent. In his other great struggle, the theory of colors, Goethe unfortunately never entered this phase and held on to his theses until the end. Incidentally, the term "granite" can be traced back to the Latin *granum*, which means nothing other than grain; we are not far removed from the grain of sand here, since typical quartz sand grains correspond approximately to the size of quartz crystals in the grain structure of granite.

Pegmatites have a special structural feature that we will return to in another context. Like their fine-grained counterpart, aplite, pegmatites occur predominately as vein-shaped rock bodies. Characteristic is their extremely coarse-grained structure with pronounced crystal formation. The size of feldspar crystals here can range from a few centimeters to several meters. Potassium feldspar is usually represented by microcline, which often shows shapes indicating segregation. If rapid and simultaneous cooling of quartz and microcline occurs, angular interpenetration and intergrowth of the crystals may occur at fairly regular intervals, resulting in an appearance of cuneiform characters or runes, depending on from what direction they are observed. This so-called graphic granite is very decorative and popular with collectors. Pegmatite veins often carry rare accompanying minerals such as beryl, spodumene, and lepidolite, which formed from the rare elements enriched in the residual melt.

Gabbro and diorite are sometimes difficult to distinguish since the distinction is made by the content of anorthite (above 50%: gabbro; below 50%: diorite). However, this can be determined only with a microscope. On the glass plate, it is easier to distinguish the darker parts of the mixture. Other than the obligatory pyroxene mineral augite, gabbro possibly contains olivine. Diorite, on the other hand, contains mainly amphibole (hornblende) and possibly biotite. After simply determining the presence of biotite or olivine, a classification can be made. In addition, the diorites possess only about 30% of mafic-mixture parts and are thus usually clearly brighter than gabbros. In the Streckeisen diagram, both types of rocks occupy field 10.

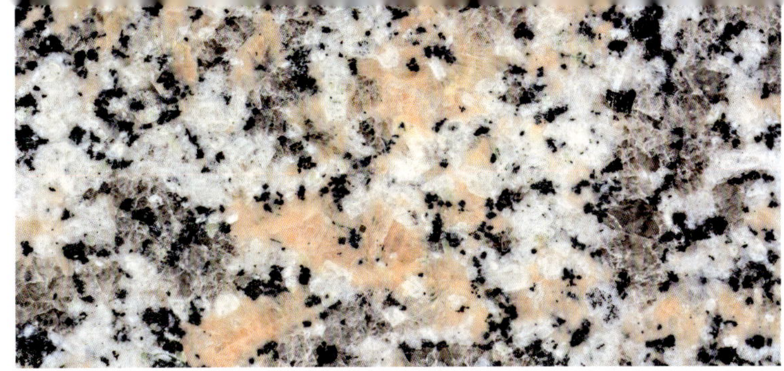

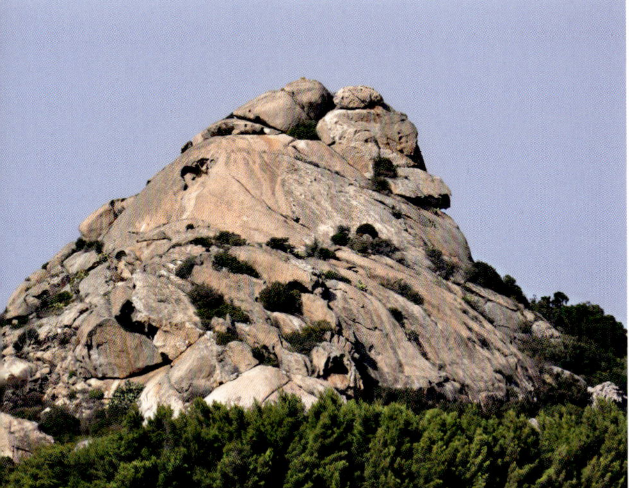

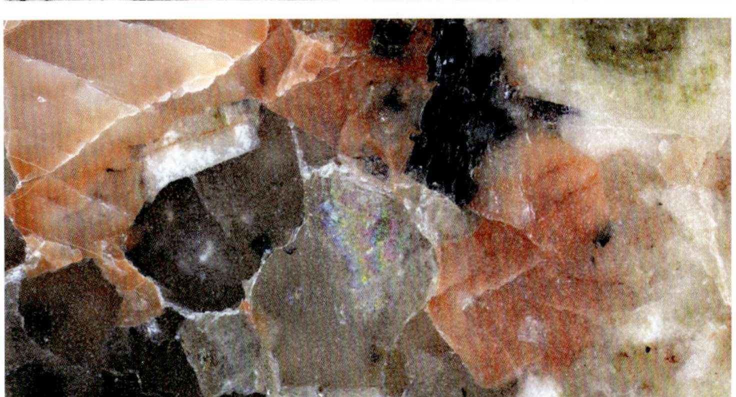

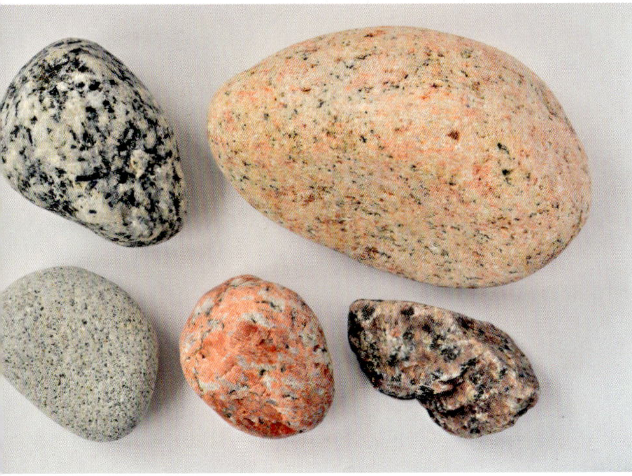

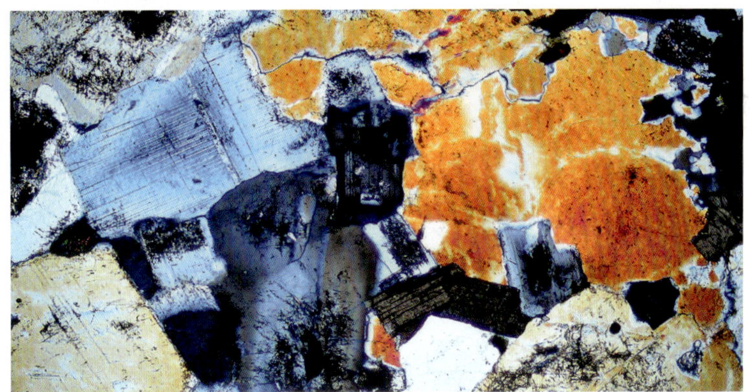

Figure 4.9 Weathered-out granite pluton with easily recognizable spheroidal weathering. Northern Sardinia, Italy.

Figure 4.10 Beach debris from Bornholm: granite of different types and origin. *Starting from the left and moving clockwise*: Sala granite from Uppland, Sweden, 1.9 billion years old; Almindingen granite, Bornholm, Denmark, 1.4 billion years old; Hammer granite, Bornholm, Denmark, 1.4 billion years old; Red Gravenfors granite, *Östergötland*, Sweden, 1.7 billion years old; Stockholm granite, Uppland, Sweden, 1.8 billion years old. Image width 200 mm.

Figure 4.11 So-called Baveno granite from the Permian period, Italy. The sample clearly shows gray quartz, pink-colored potassium feldspar, porcelain-white plagioclase, and black biotite. Polished granite surface in reflected light. Image width 90 mm.

Figure 4.12 Polished granite surface (detailed) in reflected light. The gray quartz, red/pink-colored potassium feldspar, green and porcelain-white plagioclase, and black biotite can be easily recognized. The cloudlike structure in certain areas is because of unfinished polishing. Beach debris (slice) from Bjerregard, Denmark, cut and polished sample. Image width 7 mm.

Figure 4.13 Thin slide of granite. The minerals can be classified on the basis of their interference colors. Quartz grains are orange, at top left is a feldspar crystal in light blue with twin lamellae (plagioclase), and just below right center is a biotite that is brown black in color. Reflected light, crossed polarizers. Image width 3 mm.

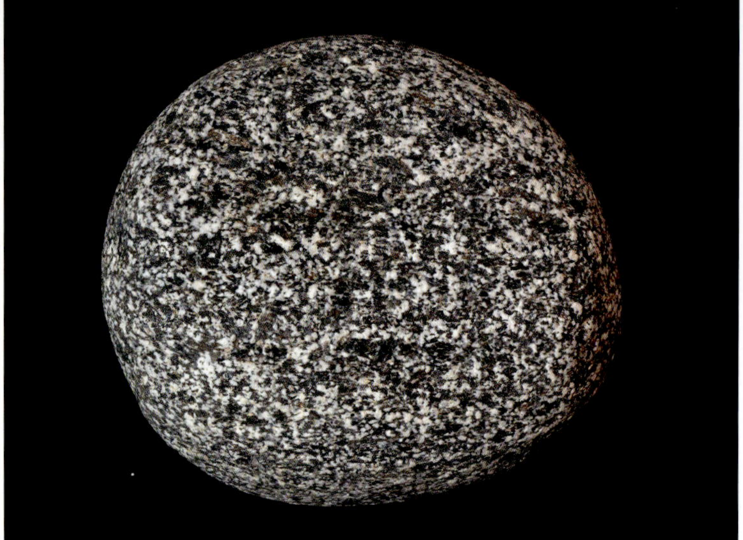

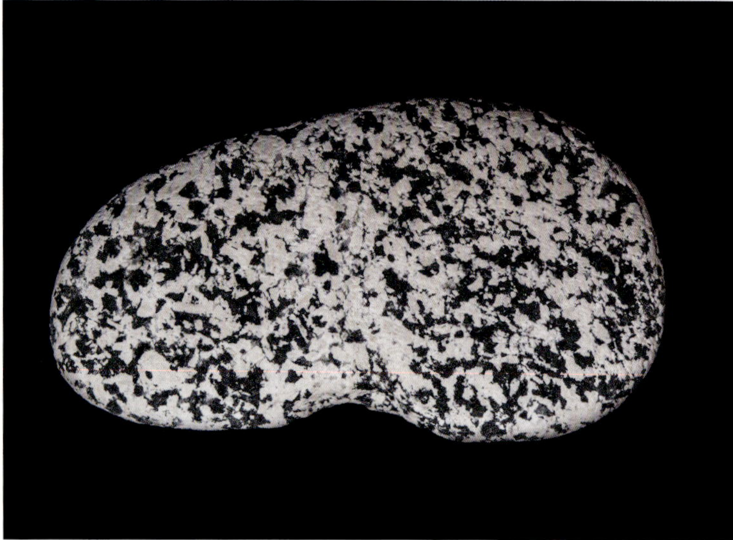

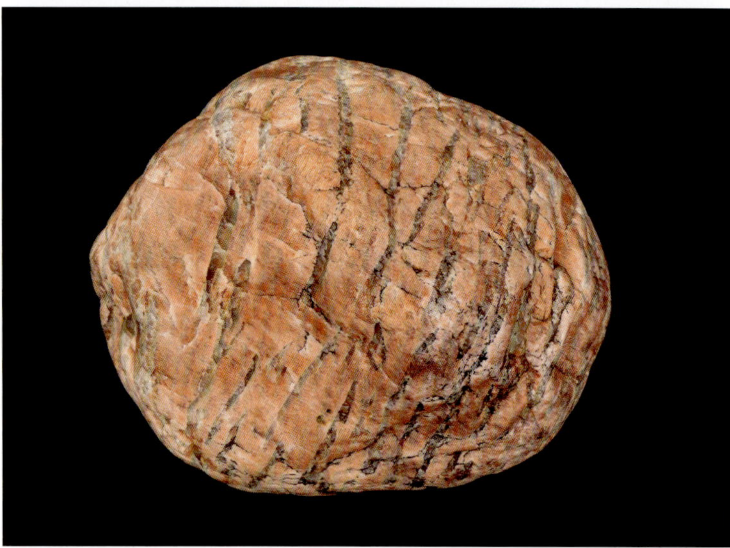

Figure 4.14 Gabbro, made of light plagioclase as well as hornblende and augite (dark), beach debris, Baltic Sea. Rock size 43 × 49 mm.

Figure 4.15 Diorite, made of white plagioclase and dark hornblende, Rhine gravel near Kork, Germany. Rock size 26 × 52 mm.

Figure 4.16 Pegmatite veins in the example of a graphic granite from Norway. This stone has been clearly rounded through glacial transport. The narrow strip-shaped quartz crystals are clearly visible—the base consists of feldspar (microcline). Sandvatnet, Valle, Norway. Rock size 90 × 110 mm.

4.1.2 Volcanic Rock

We already distinguished earlier between light, felsic minerals and dark, mafic minerals. From this rather broad division into two groups, there are further important properties to be derived, each of which is specific to each group. Rocks with predominantly light mineral compositions lead to acidic, silicate-rich magmas when melted, and those with predominantly dark minerals lead to basic, silicate-poor magmas. The most-important dark minerals can in turn be divided into the two main groups, pyroxenes and amphiboles, with augite being the characteristic representative of pyroxenes and hornblende being the main representative of the amphiboles. Hornblende contains hydroxyl groups in its crystal structure and therefore, in contrast to augite, does not tolerate particularly high temperatures. The lighter minerals tend to have a lower melting point than the darker ones.

Granite belongs to the lighter rocks, from which we can deduce from previous remarks that they solidify at comparatively low temperatures (650°*C*–800°C) and have an acidic chemical composition. Hornblende can often be found as an accompanying mineral, but augite hardly ever. Conversely, rock melts that solidify at high temperatures around 1,000°C tend to contain dark minerals such as augite and have a basic chemical composition. Basalt is the main representative here. What gabbro is to plutonic rock is what basalt is to volcanic rock. Together with andesite, basalt is the most important casting stone on Earth. We can consider basalt to simply be a mineral association of pyroxene and plagioclase, often with olivine (fig. 4.17). These minerals are then also consequently found in the respective local sands.

Plutonic rock solidifies in the earth's interior completely thermally isolated and under pressure from the surrounding and loading rocks. This results in a very slow cooling rate, which gives the minerals time to crystallize at intervals according to melting temperature to form well-defined crystals. In contrast, volcanic rock solidifies very rapidly on the earth's surface. The rapid cooling at the surface does not allow time for distinct crystal formations, so volcanic rock is always very fine grained. For many volcanic rocks, therefore, the classification in a Streckeisen diagram can be very difficult in practice when using conventional methods, or sometimes not possible at all. Classification is then done through a chemical analysis and then with the help of a special diagram in which the SiO_2 content is plotted against the sum of the alkali oxides (TAS or total alkali silicia diagram, fig. 4.18).

When magma emerges from the earth, it is called lava. Magma with an acidic chemical structure has a much-higher viscosity than the thin-bodied basic magma. Dark, basic melts can therefore cover very large areas in a short time after they emerge, whereas lighter melts form only short flows. Experts can already tell from the slope of the volcanoes whether the lava is acidic (large slope) or basic (shallow slope).

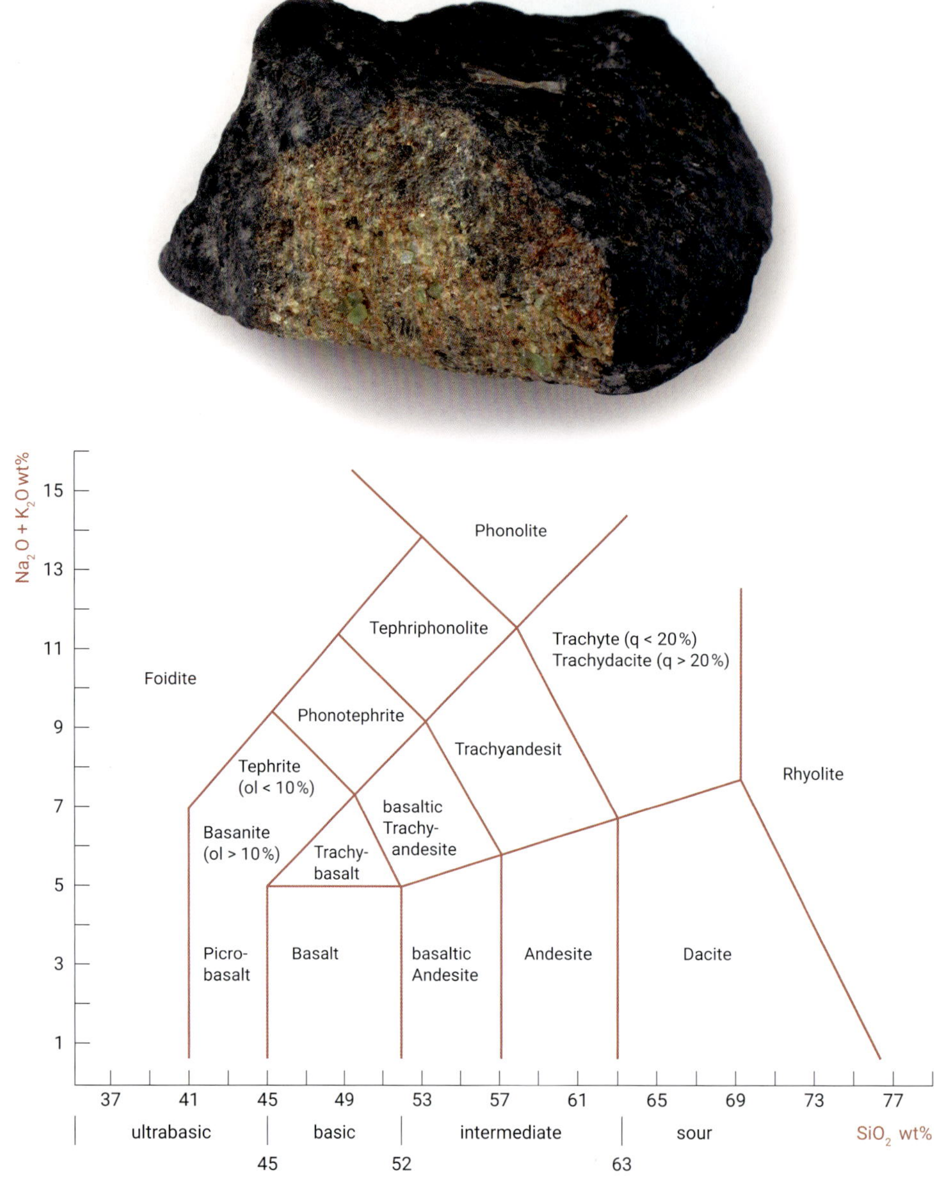

Figure 4.17 Piece of basalt with large foreign inclusion made mostly of olivine. Here, we are looking at a peridotite nodule. The upper mantle is composed of peridotite. Fragments of this rock can be brought to the surface when magma rises. Peridotites are also formed, for example, in intruding objects from gabbroid melts through the sinking of the heavier olivine crystals, which belong to the early crystals. Banks of the Rhein near Düsseldorf, Germany. Rock width 60 mm.

Figure 4.18 TAS diagram (total alkali silica) according to IGUS2004. Classification of volcanic rocks based on chemical analysis. The SiO_2 content is recorded against the total amount of alkali oxide. According to Maresch 2014.

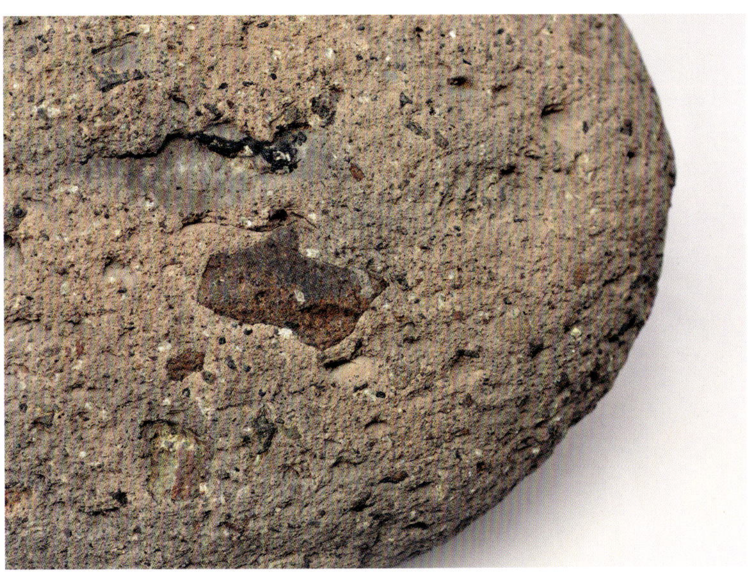

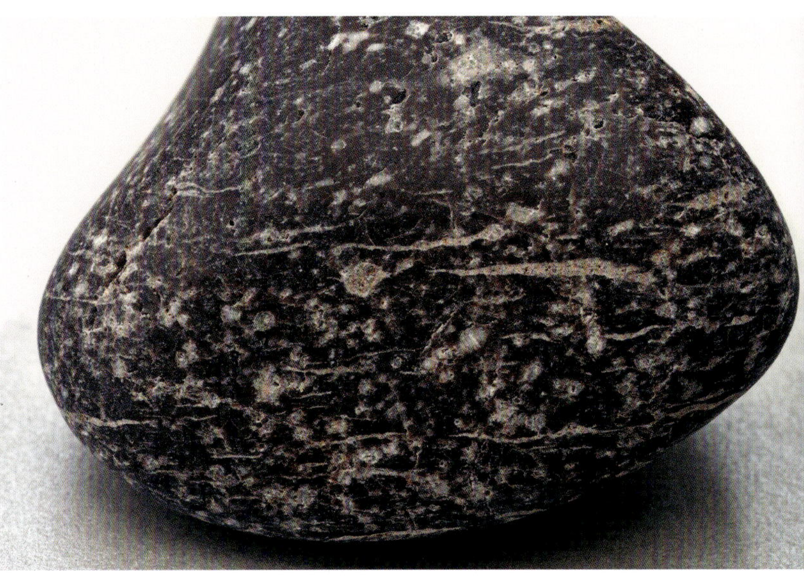

If the magma cools down even faster, rock glasses such as obsidian, from which our ancestors knew how to make razor-sharp blades, are formed. In addition to ordinary magma eruptions, there are also volcanic activities that can cause catastrophic effects in the near and distant environment through explosive eruptions. In the rising underground column of magma, volcanic gases are carried upward and partially dissolved in magma. When their dissolving capacity is exhausted, large gas bubbles form, which can lead to an explosive release of lava melt particles and gas. In this process, at the mouth of the vent of the volcano, conveying velocities are reached that in some cases exceed the speed of sound and cause eruption columns that can rise up to several miles. Depending on the supply from the vent, such eruption columns collapse again and then pour into the surrounding area as pyroclastic flows (from the ancient Greek *pyrien* = fire). Such pyroclastic flows can reach peak velocities of up to 435 mph (700 km/h) and temperatures of up to 700°C. All life along the way is abruptly destroyed. Finally, deposits are formed that are called "ignimbrite" (from Latin *ignis* = fire and *nimbus* = cloud). In general, deposits from volcanic eruptions are called "pyroclastics" and are differentiated according to grain size and mineral content. Sand can also have pyroclastic origin and composition.

Circulating mineral solutions can subsequently crystallize in the bubby cavities of volcanic rocks and fill them completely (fig. 4.22). Furthermore, a number of ore minerals and minerals from the zeolite group are often found in the vicinity of volcanic activity.

Figure 4.19 Ignimbrite rock from Sardinia, Italy. Beach debris. The large foreign inclusion can be easily recognized (width 35 mm) in the center of the picture. This is a piece of rock that was carried up with the magma out of the vent. Close to the top, you can see a long, black filament, a so-called fiamme ("fiamma"). Fiamme are melted rock glasses that are partially layered in the ignimbrite.

Figure 4.20. *Älvdalen* ignimbrite. Beach debris from Bornholm, Denmark; origin Dalarna, Sweden, transported by a glacier during an ice age. In contrast to the comparably young ignimbrite from fig. 4.19, this one has been compressed to a compact, tough rock. Fiamme are easily recognizable and layered. 1.6 billion years old. Image width 70 mm.

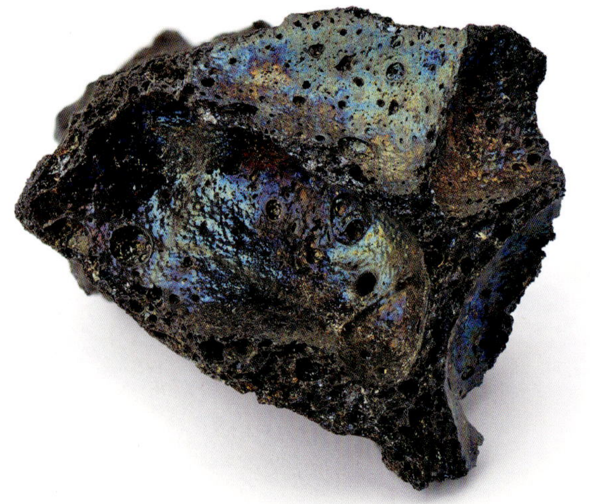

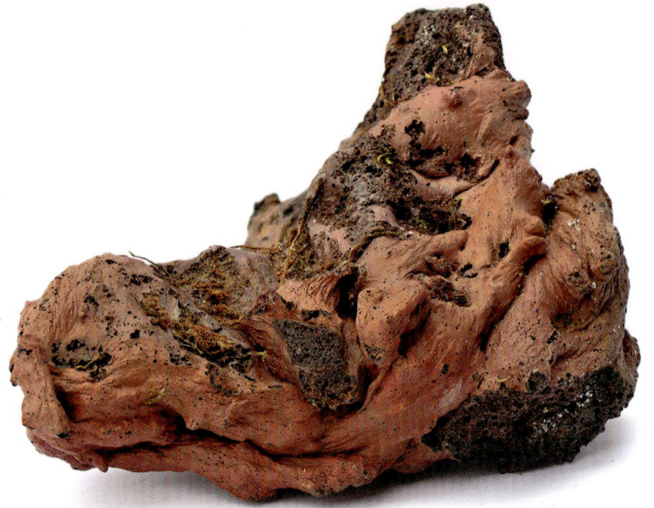

Mostly, we can see that the coastal regions of volcanic islands have black sand. Volcanic sands are consequently rich in mafic minerals such as olivine, amphibole, pyroxene, and dark magnetic ore minerals. However, many of the minerals in the volcanic coastal areas, which rarely occur elsewhere, are not particularly chemically stable and are susceptible to erosive processes, such as olivine. Their occurrence indicates rather young sedimentation or freshly formed sand.

Another group of volcanic rocks are the porphyries. The term "porphyry," however, can be found and used in a few different contexts. When referring to a porphyritic structure, one understands the coexistence of two grain-size classes. In a very fine-grained mass, which is not separable with the naked eye or even a magnifying glass, a few larger crystals float, mostly quartz, plagioclase, and potassium feldspar. The porphyritic debris of the North Sea and Baltic Sea coasts is composed mostly of rhyolites or trachytes, but because of its porphyritic structure, it's called porphyry but with the area of origin (e.g., Bredvad, *Älvdalen*, and Dala porphyry). In addition, these porphyries, such as many of the ones that come from the Swedish Dala area, could have been formed by pyroclastic events, which is why they can also be classified as ignimbrites.

Porphyries can be traced back to SiO_2 magmas and have either a direct volcanic origin or are made from magma that got stuck just below the surface and cooled there. The latter is called granite porphyry. In both cases, the cooling occurred much faster than that of plutonic

Figure 4.21 *Left*: Solidified, bubbly lava from an eruption of the Krafla volcano in 1995, Iceland. Iridescent oxide layers are visible. Rock width 45 mm. *Right*: Solidified lava from the last eruption of the Krafla volcano, Island, 2017. A clear flow texture is recognizable. Rock width 80 mm.

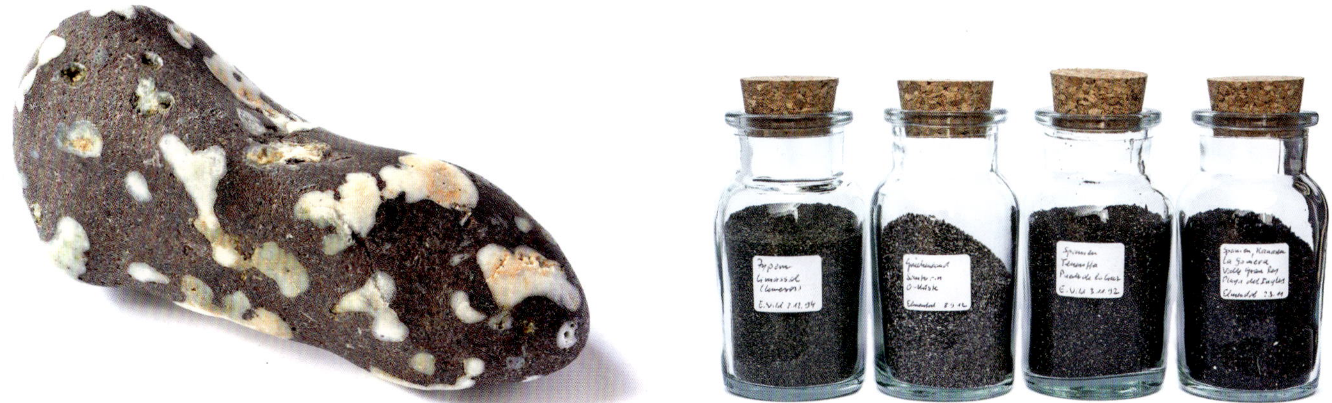

rock, which leads to the fine-grained groundmass. The large inclusions previously crystallized deeper in the earth's interior at a slower cooling rate and subsequently ascended with the magma. Porphyritic structure thus results from a change in cooling rates. Most of the red color comes from finely distributed hematite, which was formed in the early cooling phase by circulating and oxidizing water.[74]

Porphyritic debris has combinations of characteristics that make it easily sorted according to its place of origin and is therefore suitable to be used as a guide.

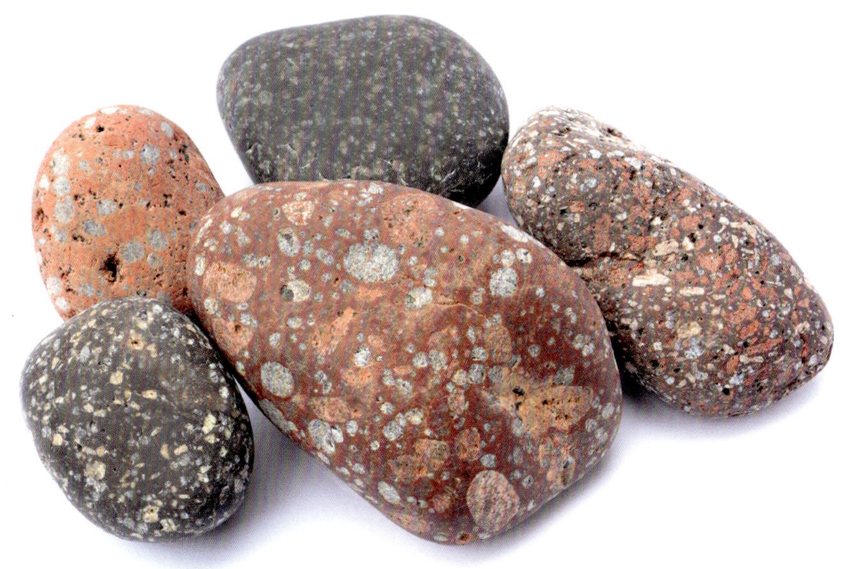

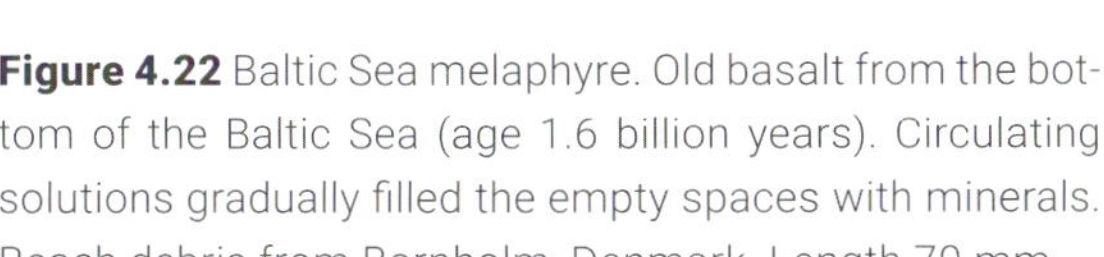

Figure 4.22 Baltic Sea melaphyre. Old basalt from the bottom of the Baltic Sea (age 1.6 billion years). Circulating solutions gradually filled the empty spaces with minerals. Beach debris from Bornholm, Denmark. Length 70 mm.

Figure 4.23 Different dark coastal sands from the volcanic islands of Cyprus, Santorini, Tenerife, and La Gomera (*from left to right*).

Figure 4.24 Porphyritic debris from Roenne, Bornholm, Denmark. In the foreground, an Aland quartz porphyry that's about 1.6 billion years old. Rock width 55 mm. The finely grained groundmass can be easily seen, where you can also notice the round, gray quartz and the pink-colored potassium feldspar crystals. *In the front left*, there is a porphyry with a darker groundmass but a comparable structure.

4.2 Sedimentary Rock or Sedimentite

In contrast to the igneous rocks discussed so far, sedimentary rocks have a secondary formation history. Sediments consist of decomposed, crushed, or disintegrated rocks. If sediments solidify due to pressure, mineral solutions, crystallization, or other processes, sedimentary rocks are formed. The process of assembling individual sediment particles into a rock is called diagenesis. In general, sediments can be divided into clastic (and pyroclastic), chemical, and biogenic sediments, which is a similar classification to what we already discussed for sands. Here, however, we want to deal only with the clastic sediments in more detail, which are sediments consisting of crushed parts.

Source rocks for the formation of sediments can be all kinds of rocks, including sedimentary rocks themselves. Apart from local residual formations, sediments and their individual particles each have an individual transport path behind them. We distinguish them between the formation of the sediments, its transport conditions (the transport regime), and the depositional or sedimentary milieu. The latter refers to the place where the sediment is ultimately no longer transported. Continental sedimentation environments can be classified and named as follows, everywhere typical sands can be found.[75]

Alluvial environment:	Debris pockets on the edge of mountains or mountain hollows
Aeolian environment:	Deserts and desertlike broad river valleys in arid zones
Limnic environment:	Deposits in marine areas and small-scale coastal deposits
Glacial environment:	Glacial deposits, moraine debris, meltwater deposits
Littoral environment:	Sea coasts (marine environments) or large river deltas
Bathyal environment:	Sedimentary deposits on the continental shelf and deposits of the suspension currents of the continental slope
Pelagic environment:	Mud, fine sand, and silt in deep-sea areas

Sandstones are formed from sand through the process of diagenesis. A distinction can be made among grain, matrix, and cement. "Matrix" refers to the finer-grained material that is located between the actual sand grains and was deposited simultaneously with them. The binder comprises matrix and cement, the latter of which may be siliceous or carbonate in nature and consists of chemical precipitates in the grain or matrix spatial gaps. Depending on the composition of these components, silica sandstone is formed from pure sand grains and a siliceous binder with a slightly pronounced matrix, or calcium silicate sandstone, in which the cement is carbonate in nature.

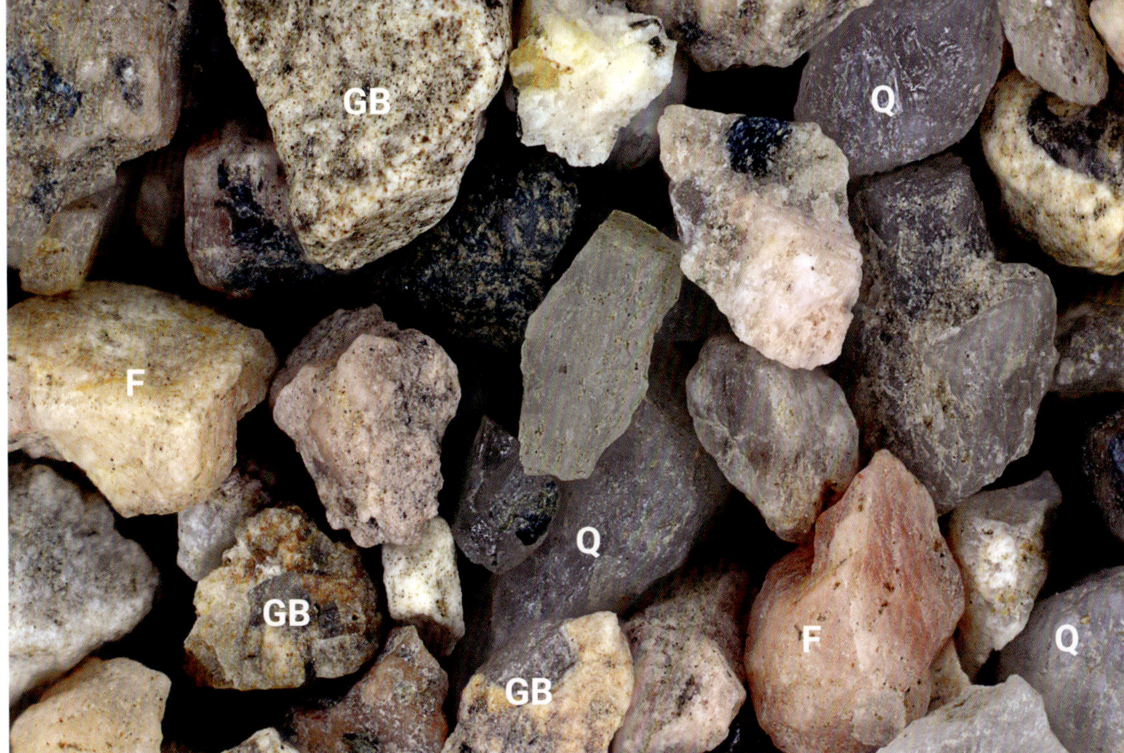

The old terms "arkose" and "graywacke" have been used as field names for the first classification. The term "arkose" is used to describe sandstones that, in addition to quartz grains, contain at least 25% feldspar. "Graywackes" are sandstones whose pelitic (clay) matrix combines sand grains and rock fragments as well as clay components. This combination is clearly different from pure sandstones, which is why graywackes are also called "dirty" sandstones.

The classification of sandstones is not uniformly defined. However, the relative proportions (volume percentages) among quartz, feldspar, rock fragments or rock clasts, and matrix are always used. According to Füchtbauer (1988), the following classification applies::

1. Quartz sandstone	> 90 % quartz
2. Sandstone (without additional descriptor)	50–90 % quartz
3. Feldspar sandstone	> 50 % feldspars
4. Rock fragment sandstone (litharenite)	> 50 % rock clasts
5. Feldspar-rich sandstone (arkose)	20–50 % feldspars
6. Sandstone with many rock fragments (litharenite)	25–50 % rock clasts
7. Feldspar-bearing sandstone	10–25 % feldspars
8. Sandstone with rock fragments	10–25 % rock clasts
9. Weakly feldspar-bearing sandstone	< 10 % feldspars
10. Sandstone with few rock fragments	< 10 % rock clasts

Figure 4.25 Sand from the Black Forest from Lake Schluchsee, Germany. Fresh, unsorted sediment lacking compositional maturity and without rounded edges. The different crystalline components quartz (Q), feldspar (F), and rock fragments (GB) are easily visible, which are labeled here as an example. Image width 7.5 mm.

If a sediment consists of components that have diameters exceeding 2 mm, angular fragments form breccias and round components form conglomerates. Breccias usually contain very different-sized fragments, because the components have had only small transport distances; for example, through landslides. In the case of crushed breccias, the source rock has sometimes even been crushed and compacted on-site. If breccias or conglomerates consist of only one type of rock, they are called "monomict"; otherwise they are called "polymict." Unlike breccias, the components of conglomerates have been subjected to a longer transport route. Many conglomerates are of fluviatile origin and were deposited in the no-flow zone of rivers, where the transport energy decreases (river gravels). However, very good rounding usually indicates a marine origin from surf debris. Examination of the distribution and nature of the components can provide information about the former delivery area. This is of particular interest if the conglomerate is the last witness to an eroded mountain range. Pelitic sediments with grain sizes under 0.063 mm lead to loess or siltstones, and with even-smaller diameters to mudstones or clay shales in the consolidated state.

Diagenetically consolidated formations can be divided into two groups. In one group, consolidation is caused by cement formation or chemical changes as described, whereas in the other group, consolidation is caused by compaction. In this process, the grains, mostly quartz, literally weld together. The thicker the overlying sediment layers become over time, the greater the pressure on the individual grain contact points. The pressure, together with micromovements, leads to dissolution (pressure dissolution) and welding of the contact points. Compaction structures are formed in the grain bond as the load increases, usually before cementation, in which the remaining gaps are filled.

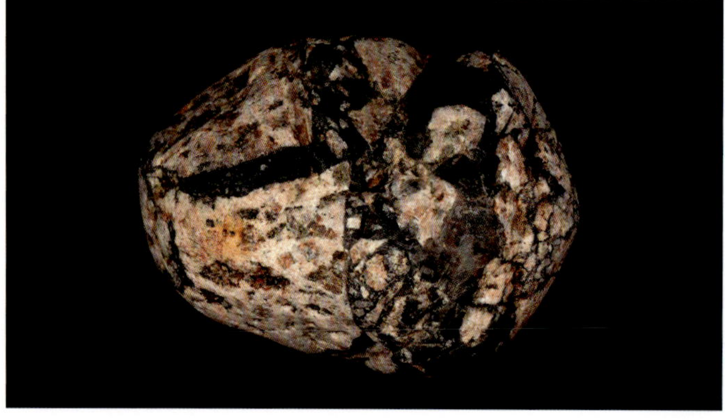

Figure 4.26 Breccia. The angular fragments of granite and the black, siliceous binder are clearly visible. The fragments do not show any rounding, which indicates a very short or even absent transport path.

Figure 4.27 *Right side, top*: Alluvial debris pits in a cirque—a depression excavated by glaciers. The sliding sediments that collect at the bottom of the cirque are clearly visible. To the right of center, we see a stream that apparently drains the highlands and carries sand and other sediments into the sea. Near Hammerfest, northern Norway.

Figure 4.28 *Right side, bottom*: Sedimentary layers that were placed vertically during the formation of the Caledonian mountains between the Ordovician and Silurian periods

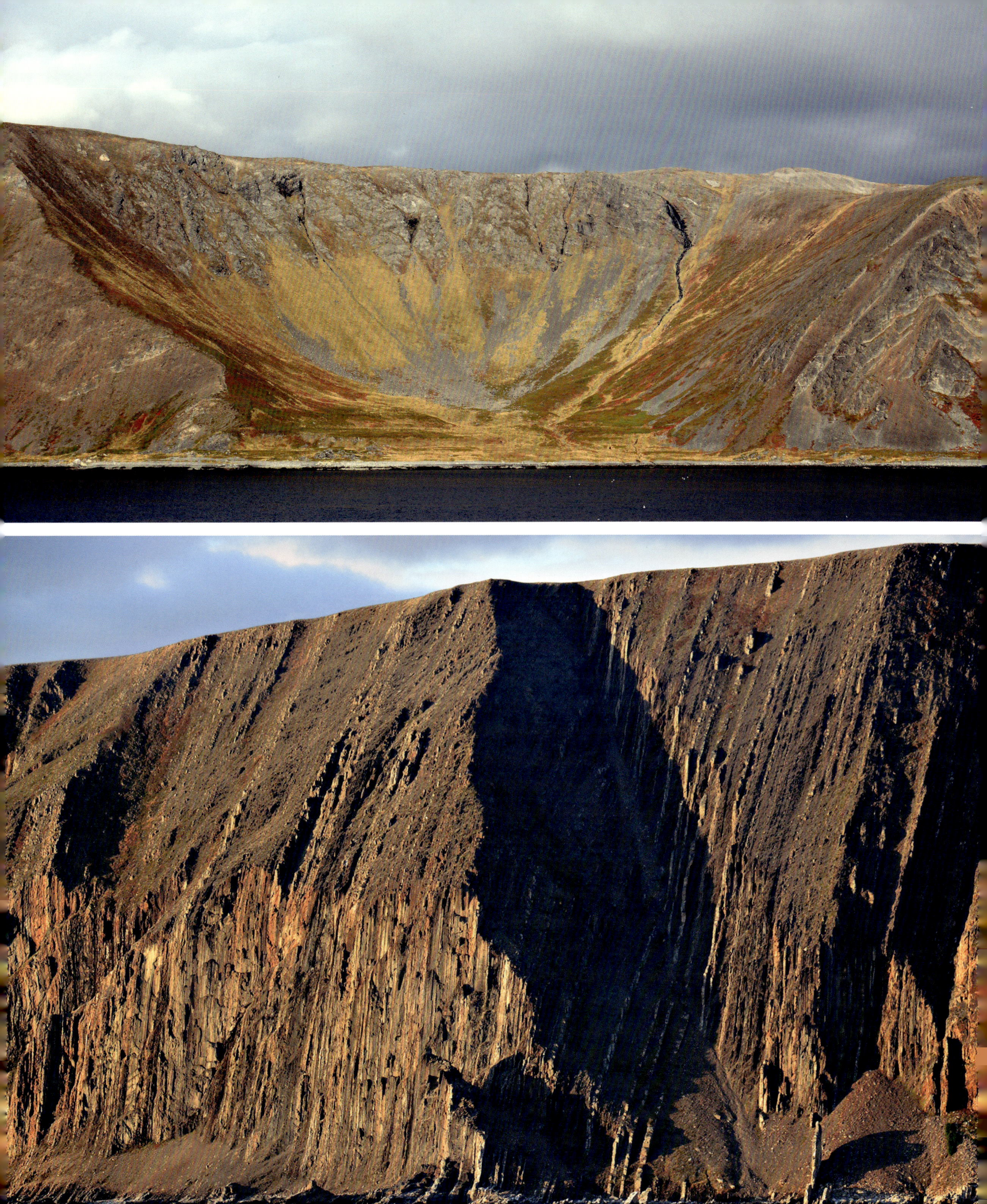

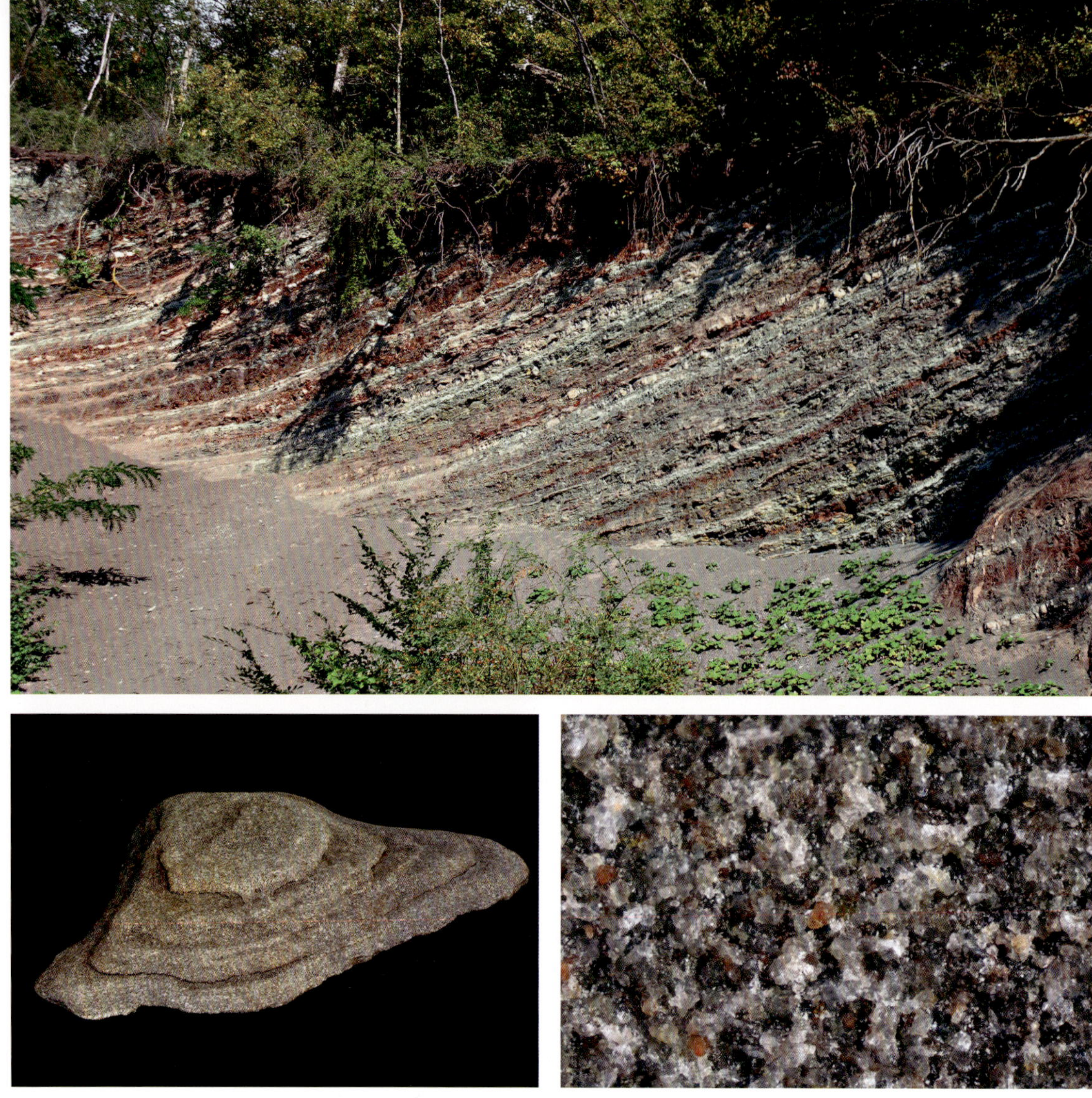

Figure 4.29 Layered texture in the sediments of the Upper Keuper, which was deposited about 210 million years ago (Bunte Mergel). In this period, dinosaurs lived in middle Europe, whose bones and traces can be found in Triassic deposits. The Keuper belongs to the Triassic geological formation and consists of differently colored clay sediments with sand beds in between. The incline of the stratigraphic sequence should be noted. At the foot of the outcrop, the formation of a small-scale sandy bed can be observed. New sediment is formed from old sedimentary rock.

Figure 4.30. Heavily weathered, slaty sandstone from the outcrop of Vardo, Varangerfjord, northern Norway. The Varanger Peninsula is composed of protozoic sediments, which are sediments that date back to the Protozoan period and are some of the oldest that can be found in Europe. Length 110 mm.

Figure 4.31 Detailed view of the Vardo sandstone from fig. 4.30. Grains are rounded, with low matrix proportion. Medium sorting. Reflected light. Image width 4.2 mm.

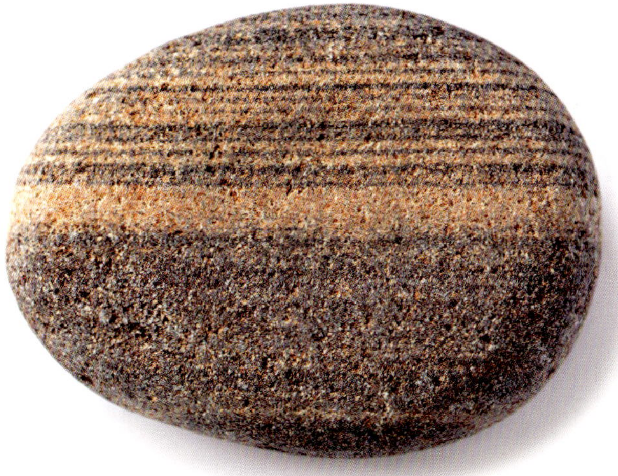

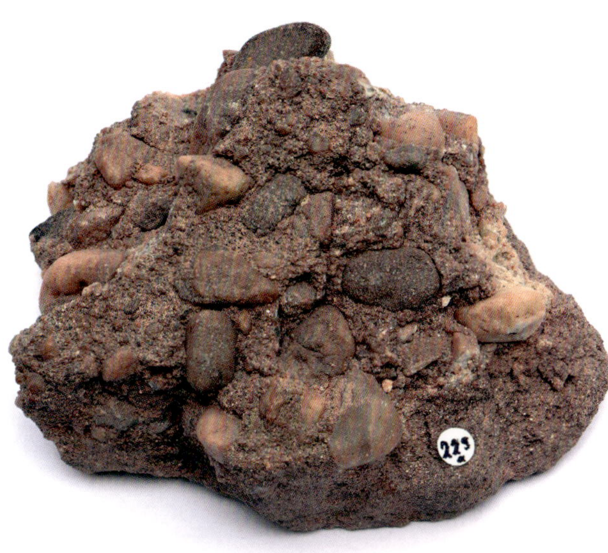

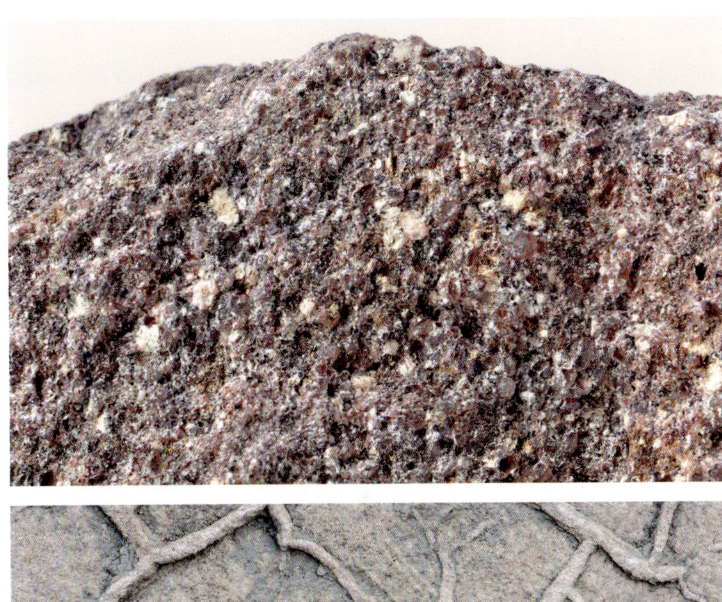

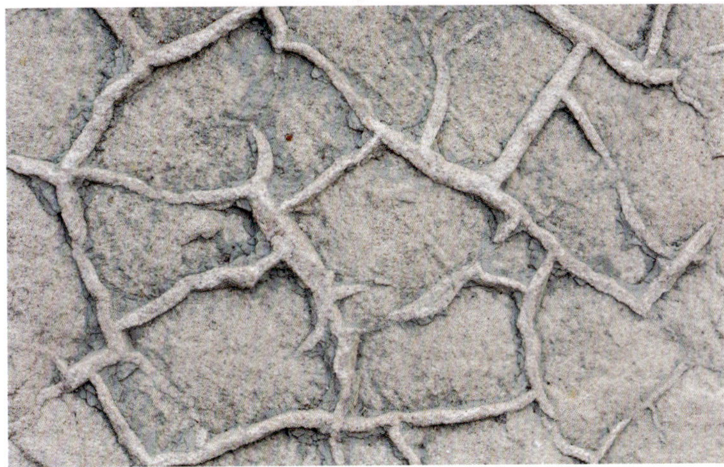

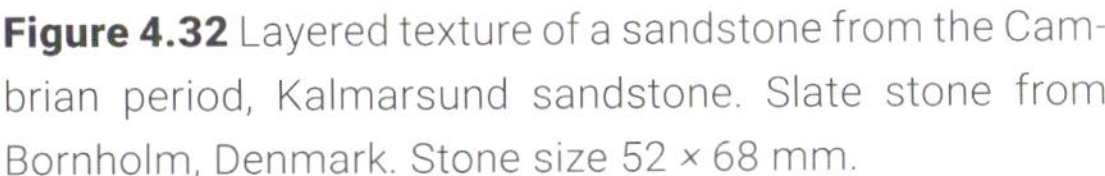

Figure 4.32 Layered texture of a sandstone from the Cambrian period, Kalmarsund sandstone. Slate stone from Bornholm, Denmark. Stone size 52 × 68 mm.

Figure 4.33 Colorful sandstone from the Black Forest, Germany. Conglomerate with a sandy matrix. The red color comes from iron hydroxides. The roundness points to a long transport route. Stone size 75 × 90 mm.

Figure 4.34 Nexø sandstone from South Bornholm, Denmark. The detailed picture shows quartz and feldspar with sharp-edged grains. It is an arkose from a fluviatile environment. The Nexø Sandstone was deposited during the Lower Cambrian period directly above the old bedrock base.

Figure 4.35 Dry cracks on a sandstone from the Keuper (Stubensandstein). Drying mud puddles ruptured and were subsequently covered by sand sediments. The sand penetrated the cracks and preserved them. The green marl remnants, so to speak, testify to a fossilized puddle from over two hundred million years ago. Hohenhaslach, Germany. Image width 11 cm.

Many sedimentary rocks are fossil bearing because, unlike igneous rocks, they form under so-called normal conditions. This means that the continental sediments are at least deposited at an ambient temperature and pressure, meaning that living organisms can exist in the depositional environment. Their traces eventually fossilize through the diagenetic process.

Due to their formation history, sedimentary rocks have a more or less pronounced layer texture. This can be seen even in a hand-sized sample but becomes even more clearly apparent in terrain outcrops.

When it comes to quantity, marine sediments exceed all other deposits and consequently form the group of sedimentary rocks with the largest volume. The continental sediments, including sand, are transported through gravity by rivers toward the sea. Coarse, clastic sediments such as gravel and sand remain near the coast or around the continental shelves. With increasing distance from the mainland, the grain size of the sediments decreases from sand to silt to clay. Clays, which remain suspended in the water for a long time due to their very small grain size, are widely transported and represent the main mass of marine sediments, with a proportion of 85%. Interesting special cases are so-called dropstones. These are stones that were transported across the sea on icebergs until the ice melted. This means these glacial erratic sediments end up in quite unexpected places, but in the best cases they can still be classified to their place of origin.

In the deep sea, there is no longer any significant proportion of sand. However, a rhythmic buildup in graded stratification can be observed there. Coarse grains are overlain by increasingly finer grains usually ending with clay, which in turn are followed by coarse material and start a new rhythm. This phenomenon is caused by sudden landslides of sometimes gigantic dimensions. From the steep continental slope, turbulent, high-density sediment suspension currents, called "turbidity currents," slide toward the deep sea. In the resulting rocks, the turbidites, very different patterns can be seen, which are characteristic and bear names such as "Bouma sequence," "Stow sequence," or "Sparks sequence." Experts can measure and identify the respective sequences on the rock. Investigations of turbidite sequences are particularly important in the context of oil exploration. An increased interest in the exploration of sands and sediments in general after the Second World War is also due to the global search for fossil oil deposits.

By far, the largest part of the seafloor consists of biogenic muds (i.e., the mass accumulation of shells and shell fragments of marine organisms). These can be distinguished according to the material from which their skeletal or shell parts are constructed.

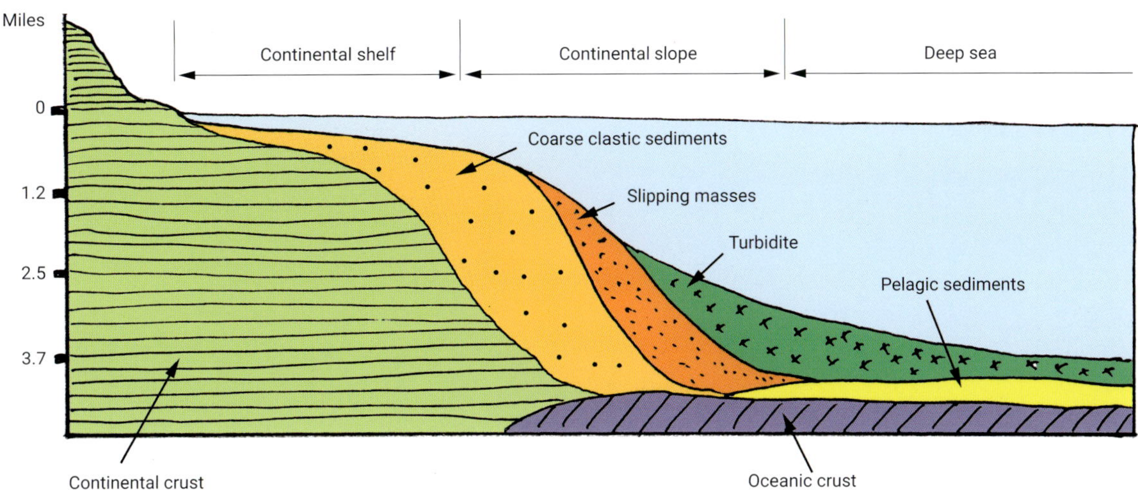

Essentially, these are either calcareous shells (i.e., shells that have calcium carbonate as their main component) or siliceous skeletons, which consist of silicon dioxide (silicic acid). Both substances have different solubilities in seawater. While calcareous-organism fragments have a relatively high solubility, that of silicious fragments is much lower. However, the solubility of lime increases—quite counterintuitively—as the water temperature drops.

Other than sponge needles of siliceous sponges and diatoms, radiolarians in particular are organisms that have a siliceous skeleton and occur in rock-forming abundance. Radiolarians have maximum diameters of 0.5 mm and live mainly near the surface of the water. During the sinking of shells and skeletons, silicic acid especially accumulates in deep-sea sediments, where the calcium carbonate almost completely dissolves in the cold deep-sea water layers. Thus, there is a sea depth below which the (pelagic) sediments remain calcium free. This limit, called CCD (calcite compensation depth), is about 15,700 ft. (4,800 m) but has fluctuated over geological time. If the CCD is known, sediment researchers can estimate the respective sea depth belonging to a stratification on the basis of the calcite content.

Large parts of the deep sea are covered with radiolarian mud, which, due to increasing loading and diagenesis, solidifies into a rock called radiolarite, which can take on a reddish, green, or black color, depending on the additives.

Figure 4.36 Regions of marine sediment deposits. In the continental crust, there are mostly crystalline rocks, and the oceanic crust is made up mostly of basalt and gabbros. Coarse clastic sediments include sands, clays, and carbonates, while the 4-degree inclined continental slopes are dominated by graded stratified deposits of turbidites (Rothe 2010).

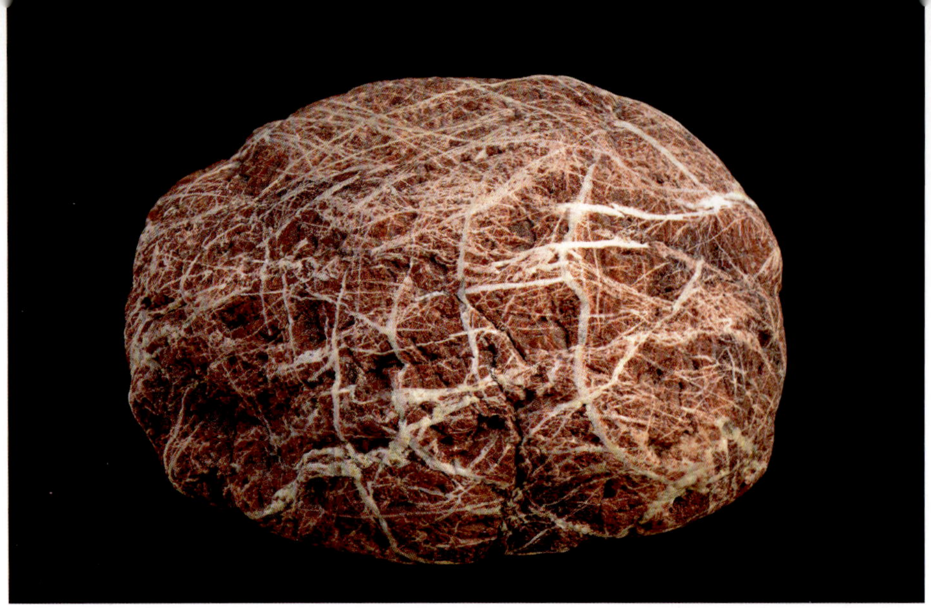

The accumulation rate is about 1 mm per 1,000 years. From the Alps, red Jurassic radiolarites are known, which can be easily identified among the gravels of the Rhine by their beautiful coloration and alteration. Radiolarite is hard and brittle and fractures under aggressive compressive stress as part of tectonic processes during mountain building. Penetrating and crystallizing mineral solutions cement the fractures together and form the characteristic white veins that are often found (so-called "tortured" rock). Our ancestors made very effective and useful tools from hard, sharp-edged radiolarite pieces.

A very well-known rock that is also made of amorphous silica (chalcedony) is flint. Flint is usually very homogeneous, tough, and hard; breaks conchoidally; and is best suited for the production of very sharp blades, arrowheads, and knives if the appropriate skills are mastered. By the Stone Age, there was a lively trade in this raw material, which was so important for the hunters of that time. The name "flint" indicates that flints in the "shotguns" of earlier times provided the ignition spark when striking pyrite. Flintstones are secondary formations of biogenic opal, which can be found in bulbous form mostly in deposits from the Cretaceous period in large quantities. There they originate from dissolved silicic acid of various marine organisms such as coccoliths or sponge needles. Circulation mineral solutions are separated again after being transported onto suitable rocks again and form the occasionally bizarre forms that are gladly collected by beach hikers.

Still to be distinguished from the pebbly nodules of flint are the no-less-bizarrely-shaped concretions of loess deposits. Like the former, these are usually up to palm sized and often very sculpturally formed, which has earned them the name "loess dolls." Loess is the most important fossil dust sediment and contains mainly quartz with a limestone content of maximally 30%. The grain size is in the range of coarse silt, mostly between 0.02 and 0.05 mm.

Figure 4.37 Radiolarite rock, colored red due to iron additives. After fragmentation, cemented with quartz through mountain-building processes. Rhine gravel near Kehl, Germany. Rock size 60 × 74 mm.

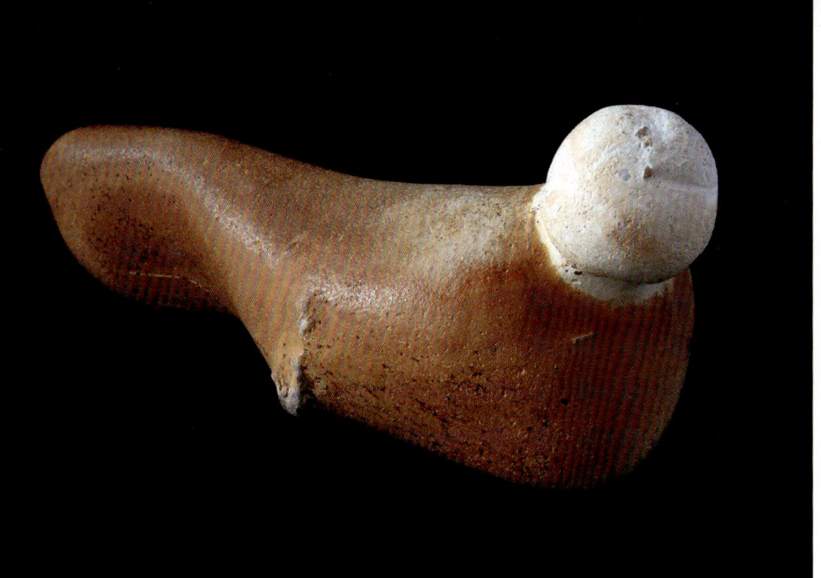

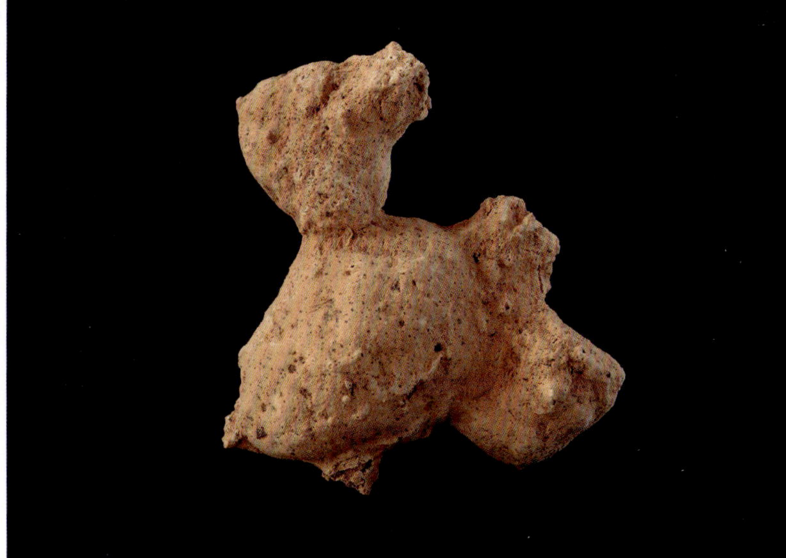

The largest layers of loess came to be during the Pleistocene period through wind blowing and deposition in the lees of large river valleys and mountains, yielding very fertile soils. Circulating calcareous solutions in the sediments accumulated in layers on the root remains or other foreign matter.

Before we look at metamorphic rocks, we should deal with a very remarkable group of sand sediments, the ooids, which belong to sand group 3 according to our classification. "Ooids" are roundish or ovoid grains with dimensions about the size of a grain of sand, consisting of a core and shells of limestone (calcite or aragonite) grown concentrically around it. This shell-like structure leads with each additional shell to an increasing rounding of the original core, which may consist of a quartz grain or other particles. Ooid sands preferentially form by rolling on the ocean floor, whereby certain saturation conditions of lime and salt in the water must be maintained. The shallower the ocean depth and the greater the water movement, the thicker the shells become with permanent movement. Measurements of ooids from Cat Cay in Florida demonstrated an age of 225 years on the outer 10% of the hull layer, whereas the inner layers were much older at about 2,530 years. In the Persian Gulf, it can be proved that ooids apparently remain at rest for years before they are shifted once again into areas of more strongly moving water, where the continued growth of the outer hull is favored.[76] One can imagine the growth of ooid shells similarly to the growth of a rolling snowball, to which shell after shell is added as it moves. Other than these rather technical aspects, ooid sands are particularly striking and their uniformity is strangely beautiful. They are found not only in marine environments, but also in some inland lakes (lacustrine deposits). Famous are the locations in the Bahamas, Florida, the Gulf of Suez, and the Persian Gulf. "Oolilte" refers to sedimentary rocks composed of ooid sands that are bonded.

Figure 4.38 *Left*: Concretion of circulating silicic acid, flint. Hirtshals, Denmark. The different colors come from differing exposures to erosion. Length 70 mm. *Right*: Concretion of circulating calcareous solution, "loess doll." Mettertal near Sachsenheim, Germany. Width 70 mm.

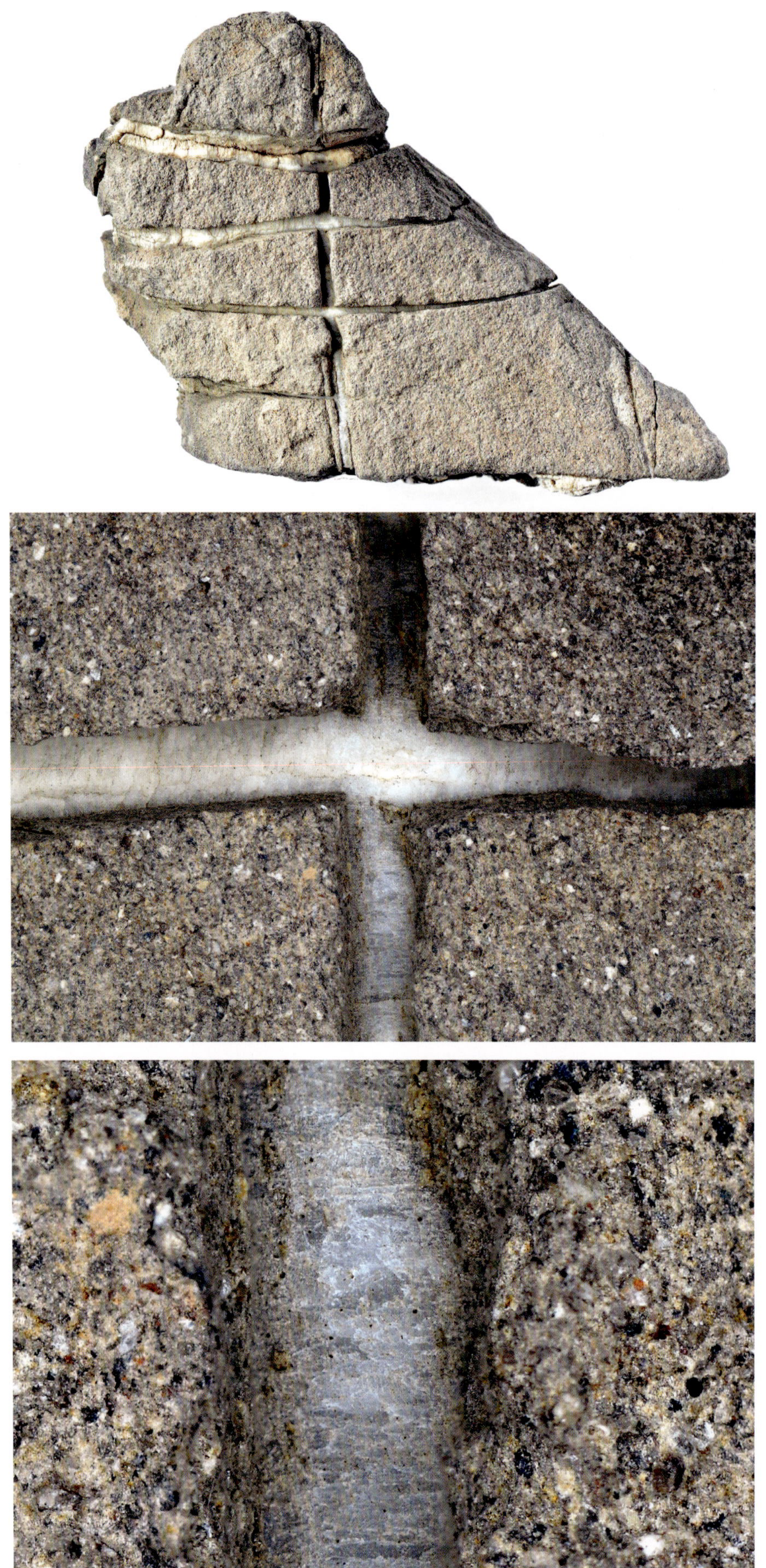

Figure 4.39 Sandstone from the Flysch zone of the Walmendiger Horn in Kleinwalsertal, Austria. The step-by-step zooming in shows the partly eroded, light gap filling, which in this case is composed of crystallized calcium carbonate. The sandstone itself is composed of barely rounded grains that are poorly sorted with nonhomogeneous composition (immature sand as its basis). Stone size 97 × 140 mm.

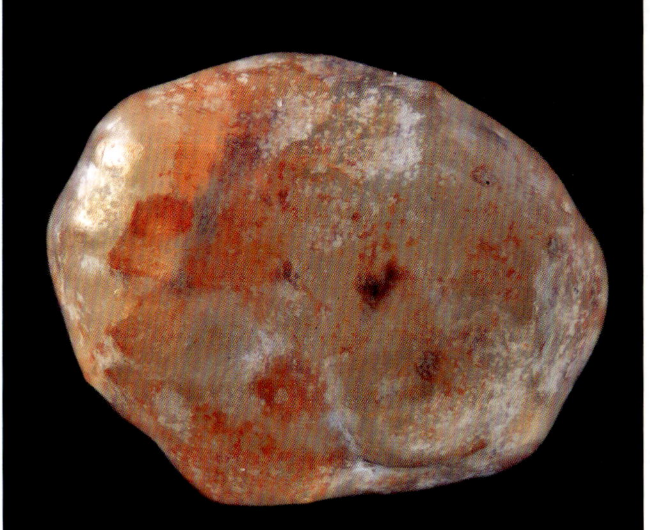

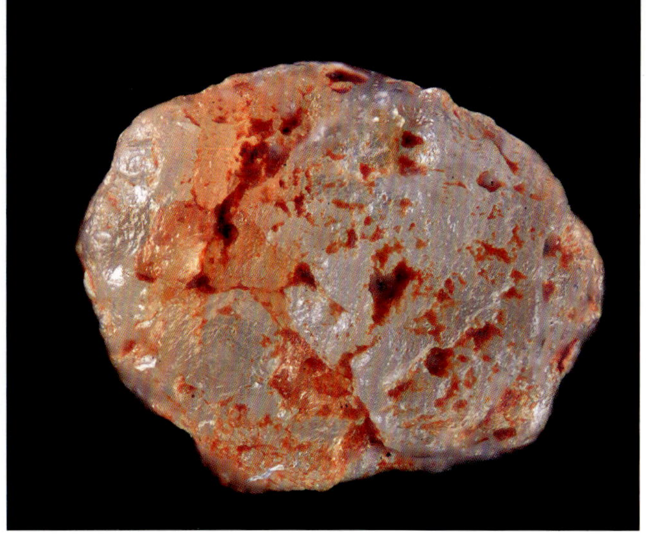

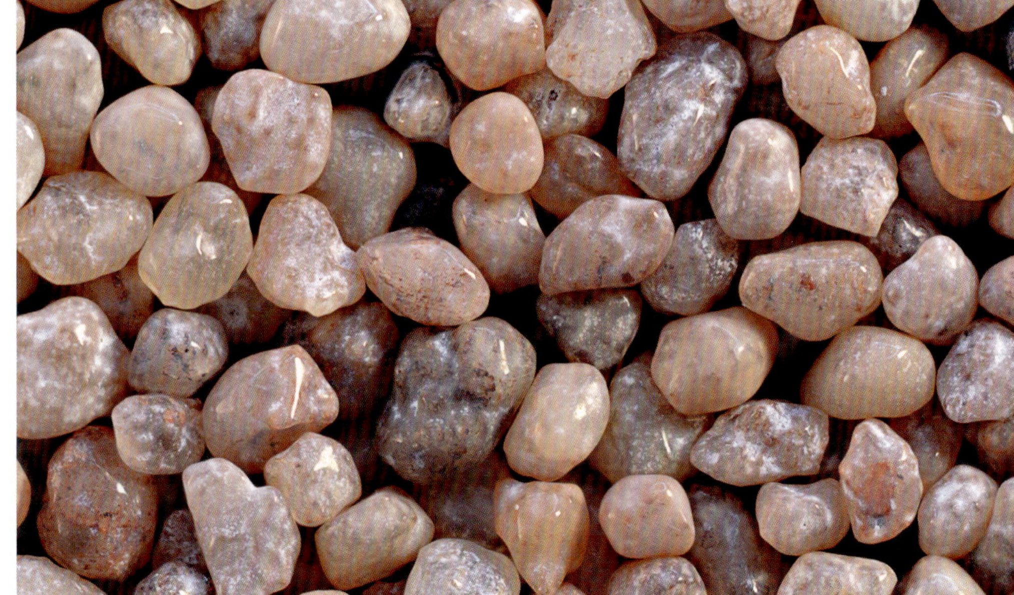

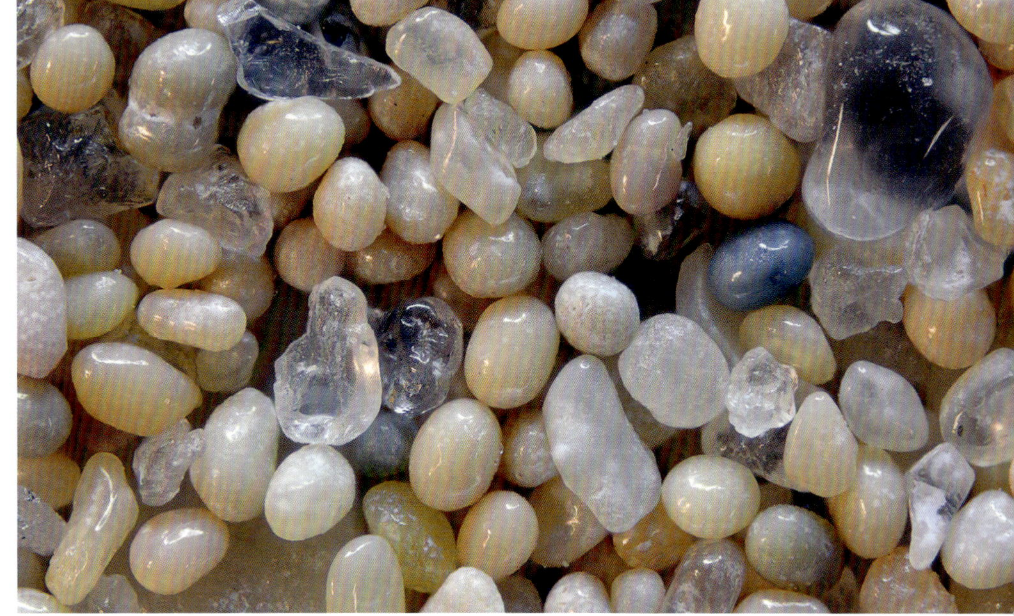

Figure 4.40. Ooid sand grain from Lake Tanganika near Kigoma, Tanzania. Both pictures depict the same grain of sand. In the left picture, the grain can be seen in its original condition, while in the right picture, the grain has been treated with watered-down hydrochloric acid. The hydrochloric acid has dissolved the outer calcareous shell of the ooid with violent bubbling. What remained was the quartz grain with red iron oxide deposits in the depressions. Obviously, the grain was subjected to a weathering process on land before it reached Lake Tanganyika, where it received its coating of lime. In the left picture, the red iron oxide areas are still visible under the transparent lime layer. The rounding of the grain edges is clearly visible. Reflected light. Grain size 935 × 1,120 *µm*.

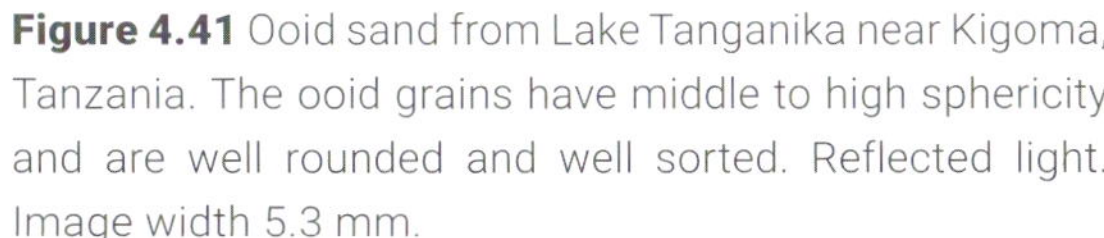

Figure 4.41 Ooid sand from Lake Tanganika near Kigoma, Tanzania. The ooid grains have middle to high sphericity and are well rounded and well sorted. Reflected light. Image width 5.3 mm.

Figure 4.42 Marine ooid sand from the Golf of Suez near El Ein Bay, Egypt. The grains are mostly perfectly rounded and have medium to high sphericity. Reflected light. Image width 3.6 mm.

4.3 Metamorphic Rock or Metamorphite

Last, we will now consider the family of metamorphites or metamorphic rocks. The term "metamorphosis" comes from the Greek and means "change, transformation" and is found in many branches of science: in geology, zoology, and botany. The metamorphosis concept pertaining to rock science follows the classical definition according to Winkler (1967). It reads as follows:

> "Metamorphism is the mineralogical alteration of rocks while retaining their solid state as a result of physical and chemical conditions that have prevailed outside the realm of weathering and diagenesis in the depths of the earth and that are different from those conditions in which the rocks were formed."

All stones that we have talked about so far can go through the process of metamorphosis and be transformed into other types of rock. For this type of transformation, high pressures and temperatures are required that lead to the original rock being recrystallized in a solid state without being melted. There are different types of rock metamorphosis, which can make things complicated. For example, depending on the formation conditions, basalt can produce a wide variety of metamorphic end products, such as greenschist, amphibolite, or eclogite, to name just a few. During the transformation, conditions are sometimes so extreme that rock layers that are miles thick are folded like paper during mountain formation processes.

The two main types should be discussed. During regional metamorphism, we mainly deal with mountain-building processes. For example, about fifty-three million years ago, the African plate began to move directly northward, triggering complex shifting and folding that led to the formation of the Alpine arc, with its typical metamorphic rocks. The beginning of the mountain building was well before that and continues to this day. Each year, the African plate moves 2 inches (5 cm) north and pushes the Alps up about 5 mm. Similar conditions can also be found in the Caucasus and Himalayan Mountains.

If liquid lava penetrates rock formations and cools within the earth's crust (magmatic intrusion, pluton formation), changes occur in the surrounding rock due to pressure and temperature. Such contact metamorphism is localized and can vary from a few inches to a few miles. Usually, low pressures but very high temperatures are effective. For example, limestone is transformed to marble and sandstone to quartzite. "Skarns" (calc-silicate rocks) are what rocks are called that form in the contact area of quartz and carbonate-containing source rocks, mostly sedimentary rocks, through contact metamorphic overpressure by intruding magma and additional mass transfer with chemically reactive, hot solutions (metasomatosis).

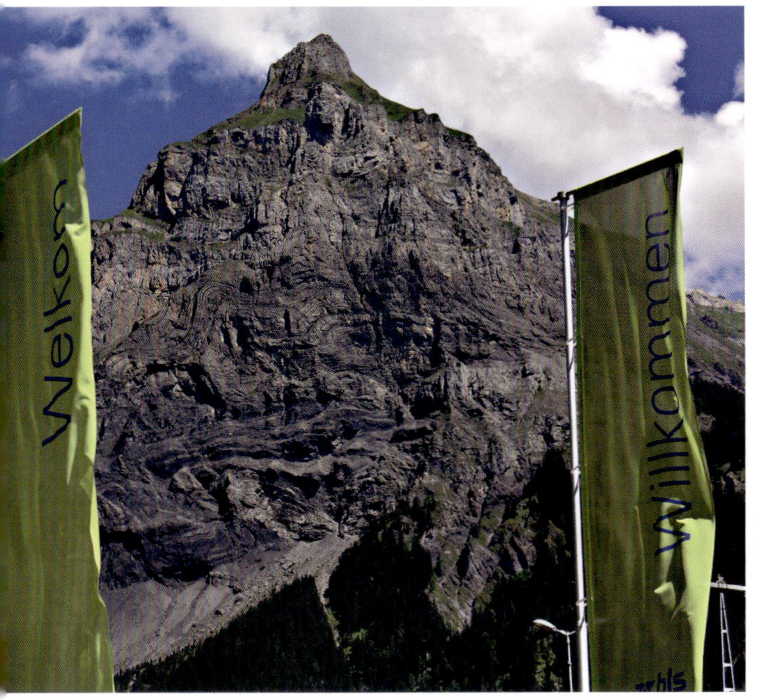

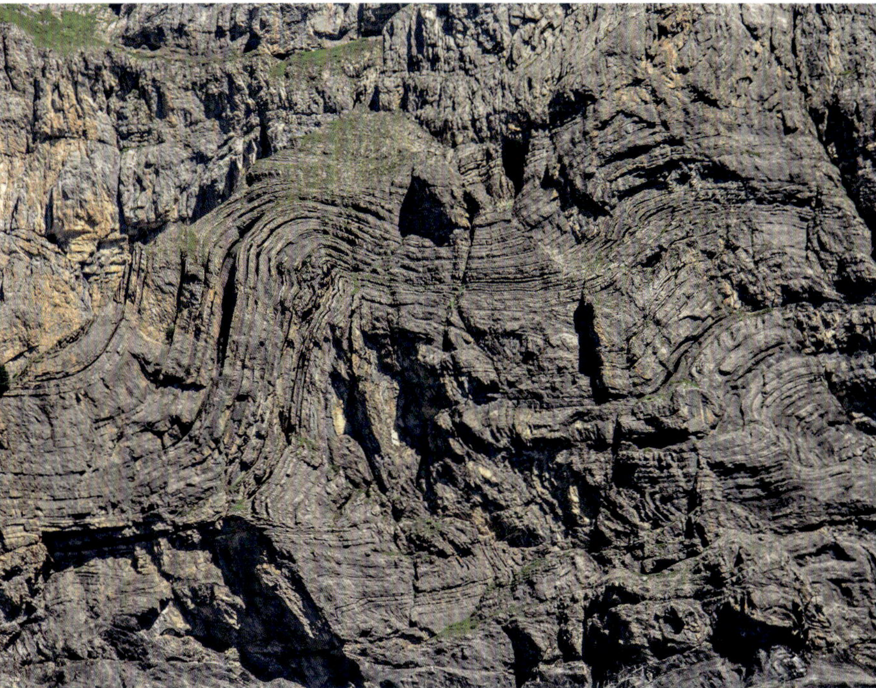

This results in a variety of interesting new mineral formations, which are formed in the contact area in a ring around the intrusion source and then are also found in the products of erosion: the sand. Depending on the composition of the melt product and the overlaying rock, quantities of rare minerals and ores worth mining can be produced (for example, Rio Marina on the island of Elba).

At plate edges or subduction zones, plates push on top of each other, with one of the plates, usually the heavier, oceanic one, being subducted (i.e., pushed under the other, lighter, usually continental plate). This causes considerable pressure and shear stresses and a rise in temperature. It has been demonstrated that during the subduction of the oceanic plates, parts of them are scraped off, so that accretionary wedges of oceanic sediments form at the edge of the continental plates in the area of subduction (tectonic erosion).[77]

Metamorphic facies are fields with defined pressure/temperature ratios, such as eclogite facies located at temperatures around 600°C and pressures around 20,000 bar. Series of metamorphic facies and their sequence provide important clues to the respective geotectonic position of the rocks.

Figure 4.43 *Left side*: Kandersteg, Switzerland. Folded layers of a mountain massif, and, *to the right*, an enlargement of a section. Mesozoic sedimentary rocks of the Helvetic period.

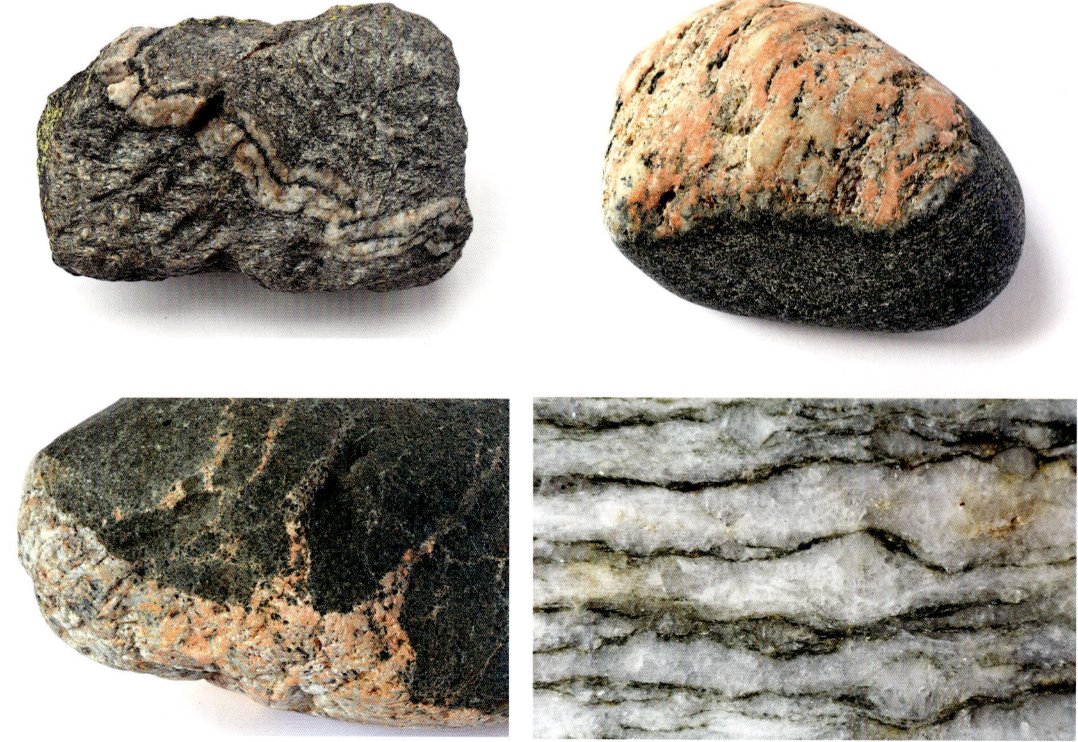

During metamorphism, certain minerals emerge depending on the chemical composition of the rocks brought into contact. Typical minerals of metamorphic rocks include sillimanite, disthene, garnet, andalusite, staurolite, and epidote. We have already encountered some of these in the group of heavy minerals in sand placers. From the occurrence of these minerals in the sand, conclusions can be drawn about formation conditions in the rock, as well as about the occurrence of the corresponding rocks in the delivery area of sand.

If very high temperatures are present during metamorphism, the rocks involved can partially melt under certain conditions. The resulting new rocks are called migmatites. The melting process does not go so far as to completely melt the rock and recrystallize, which would be counted as a magmatite, but accounts for partial melts and their crystal. Characteristic is the structure of intertwined bands and streaks of lighter (leucosome) and darker (melansome) parts. There does not have to be a stringent parallel course of the structures; mostly folds and undulations of different wavelengths are developed, which result in a ductile deformation of the rock under pressure from all sides.

Figure 4.44 *Top left*: Metamorphic rock from the Swiss Alps, Jungtal, Wallis. Rock dimensions 7.5 × 11 cm. The layered texture and the mica-rich layers are clearly visible. In the middle, a folded quartz vein. This is a muscovite-rich paragneiss—the original stone was a sedimentary rock. *Right*: Beach debris from Bavnodde, Bornholm, Denmark. The rock shows the transition between granite (*on top*) and a mica-rich greenschist. The intrusion of the granite melt left behind an easily recognizable contact zone to the old slate rock, which is of metamorphic nature. Rock size 53 × 73 mm.

Figure 4.45 *Bottom left*: Contact zone between granite magma (light) and a gabbro. One can easily see the penetration of the formerly thin melt into the dark rock during the erosion. A contact zone is not clearly pronounced in this case. Beach debris, Vang, Bornholm, Denmark. Size of the cutout is 20 cm. *Bottom right*: Gneiss, Gasenried, Wallis, Switzerland. Typical rolled-out, layered structure with quartz and feldspar crystals; layers in between of chlorite-containing, green mica layers. Ground hand piece, image width 12 mm.

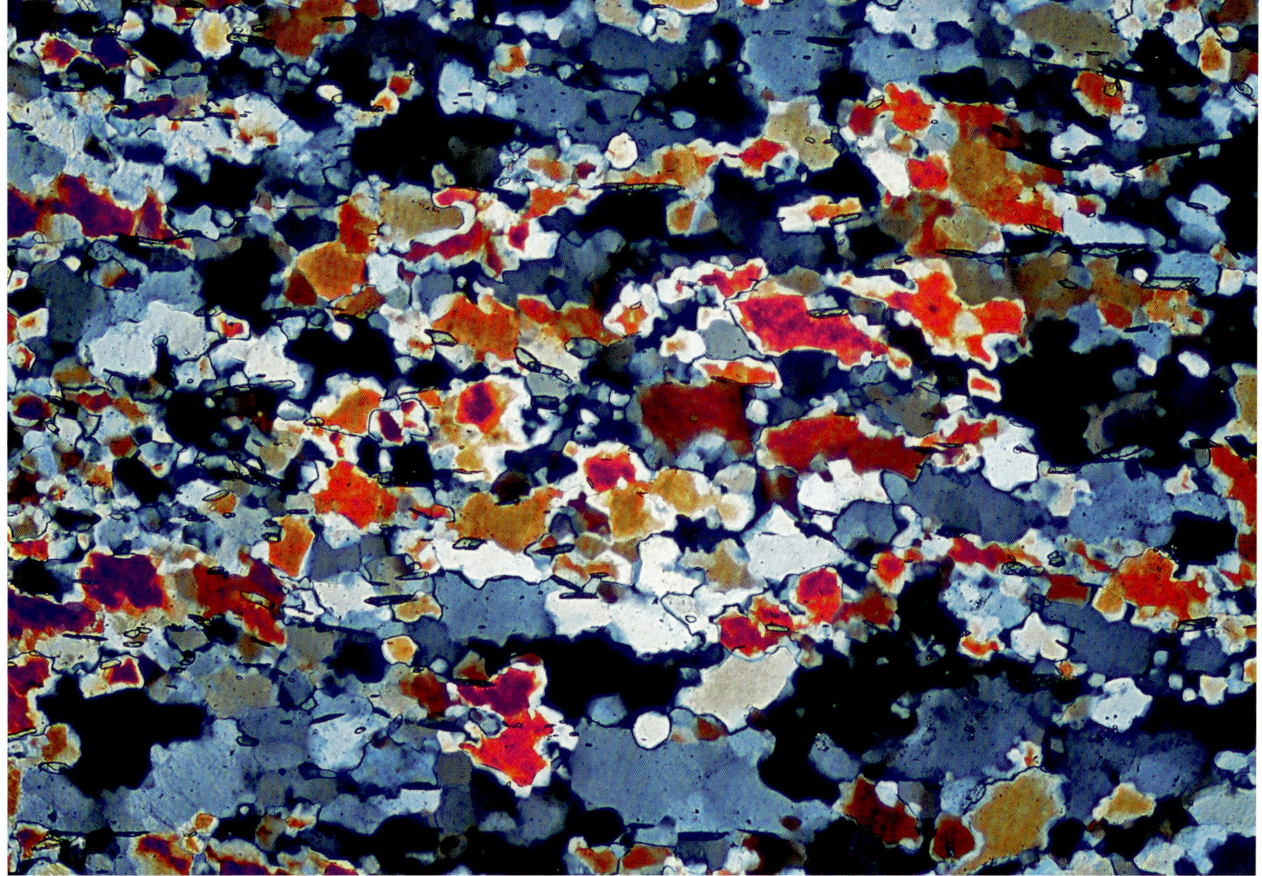

In contrast, in gneisses, the foliation planes between lighter and darker layers are always parallel. In migmatite, whose structure can be distinguished from gneisses only in somewhat larger dimensional ranges, the lighter and thus more-acidic areas melt first and form a granitic composition after cooling. The darker areas consist of biotite, hornblende, and rarely garnet and are not melted. Portions of the unaltered source rock are usually present (see fig. 4.2).

Migmatite can be found (often, but only) in areas with old, protozoic and eroded bedrock areas, where they were formed in the base regions of mountain building under high-grade metamorphism. Today, they can be seen as glacial debris from the area of the Baltic Shield in the Baltic Sea region and are witness to their remarkable past.

Figure 4.46 Quartzite is created from the metamorphism of pure quartz sandstone. The picture shows a thin section of such a piece of quartzite with a significant layered structure and interlocking quartz grains. The original, probably much more strongly rounded quartz grains experienced a partial recrystallization during the metamorphism. During weathering, the grain structure, which is clearly visible in the picture, dissolves along the grain boundaries and forms individual sand grains. Reflected light. Sample between crossed polarizers. Gasenried, Switzerland. Image width 3 mm.

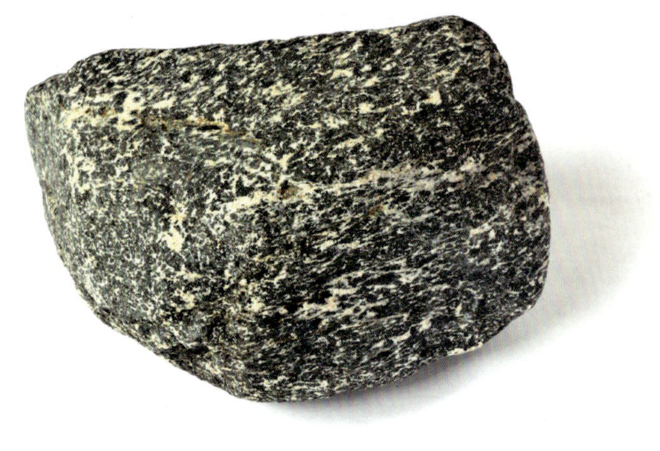

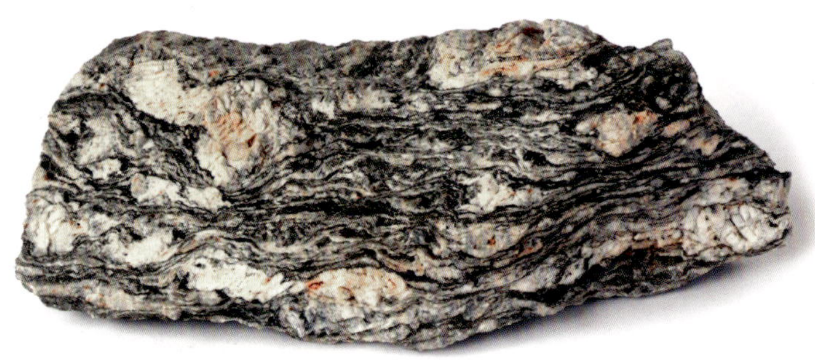

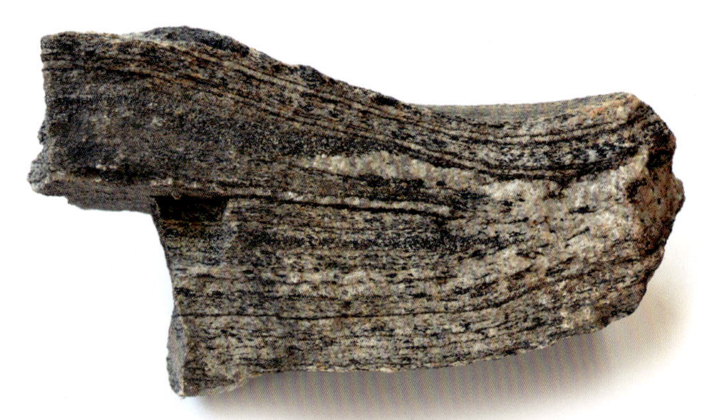

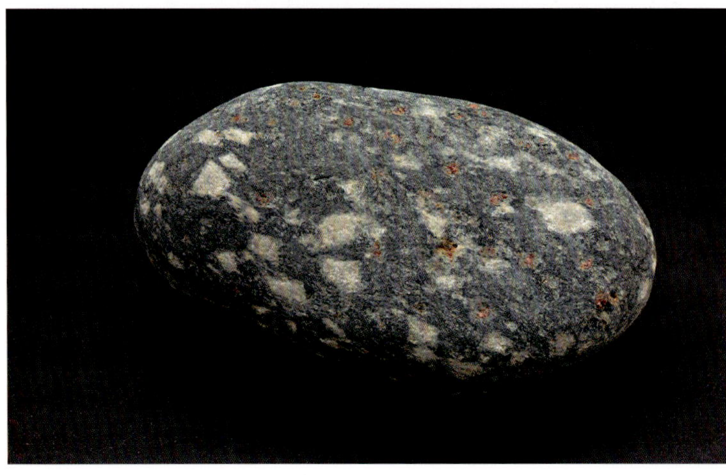

Figure 4.47 Amphibolite, Jungtal, Switzerland. The amphibolite is made of amphibole and plagioclase as well as some admixtures and is an important representative of the metamorphic facies of the same name. Rock size 85 × 120 mm.

Figure 4.48 Augen gneiss, Gasenried, Switzerland. The light potassium feldspar crystals ("Augen") were created during the metamorphism. The crystal on the top left has also rotated during the adjustment of the layers through a shearing motion. Length 70 mm.

Figure 4.49 A 180-degree folded stone made of banded paragneiss. Dark mafite and biotite-rich areas alternate with light, mafite-poor areas. Highland west of Rysttad, Valle, Norway. Dimensions 260 × 180 mm.

Figure 4.50. Eclogite. Debris from the Rhone near Salgesch, Switzerland. Eclogite is metamorphic rock from the eclogite facies of the same name. Eclogites are very hard and tough rocks that are very dense and are typically composed of pyroxene, omphacite, and granite. Admixtures often include quartz, disthene, glaucophane, zoisite, muscovite, and rutile.[72] In order to create eclogite, pressures over 12,000 bar are needed. Changing basalt rock from the ocean crust into eclogite requires a tectonic immersion depth of at least 22–25 miles (35–40 km). This means, on the other hand, that tectonic movements in the border areas from the earth's crust and the earth's mantle must have taken place. The sample shown here most likely comes from the Zermatt-Saas zone, which is the largest ophiolite zone of the Alps (more about this term can be found in chapter 7). The sample shows white zoisite pseudomorphs after the prograde mineral lawsonite (according to Meyer 2019). In addition, the red granite with traces of erosion can be easily seen. Rock size 86 × 52 mm.

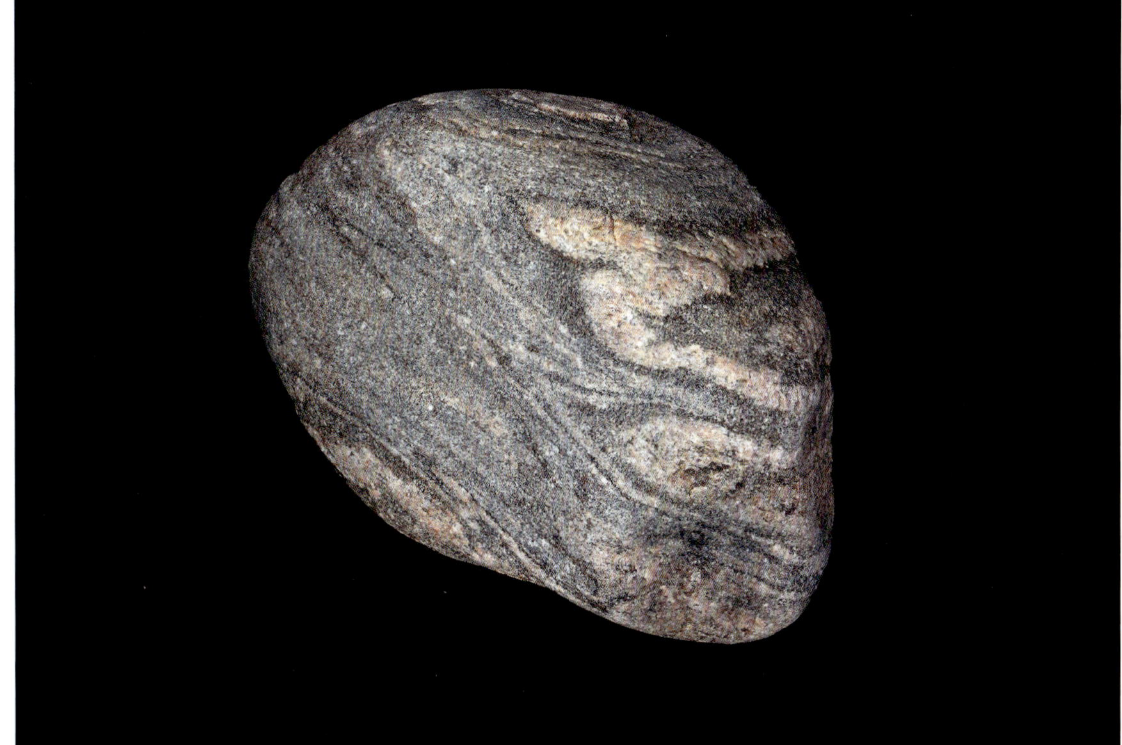

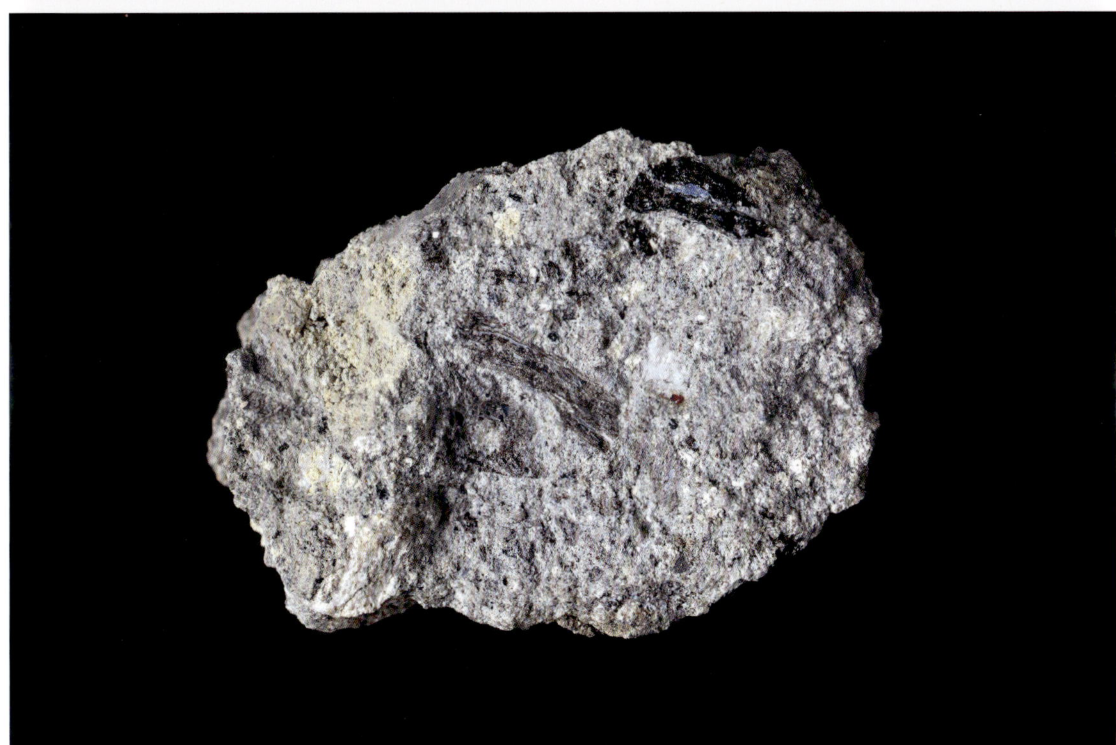

Figure 4.51 Migmatite. Stone fragment from Vang, Bornholm, Denmark. Rock size 130 × 165 mm.

Figure 4.52 Suevite or an impact breccia, N*ördlinger* Ries, stone quarry near Aumühle, Germany. In the middle of the picture, there is a piece of glass with a bubbly structure. *To the top right*, there is a dark piece of glass that looks similar to obsidian and is completely melted through and smashed. Sample size 37 × 45 mm.

A special and very extreme form of metamorphism is displayed in shock-wave or impact metamorphism, which can occur only in connection with a meteorite impact. The maximum speed that a meteorite can have upon hitting the earth lies between 43,500 and 124,000 mph (70,000 and 200,000 km/h), depending on if the meteorite is moving in the same direction as the earth or in the opposite direction. Depending on the mass of the meteorite, pressures can occur in the center of the collision up to 1,000,000 bar at temperatures over 10,000°C. Upon impact, the kinetic energy of the meteorite is transferred to the surrounding rock, which either melts, undergoes metamorphosis, or is crushed. The crater from the Nördlinger Ries has a diameter of 15.5 miles (25 km) and a depth of 2,000 feet (600 m), making it the largest accessible crater in the world. As described in chapter 2, a 0.6-mile-long (1 km) stone struck here 14.6 million years ago and completely evaporated. Drillings have shown that the crater is filled with a special impact rock, suevite, with a layer thickness of up to 1,300 feet (400 m). This suevite consists of molten rock in the form of glassy flakes, in addition to fragmented rock debris. Furthermore, high-pressure modifications of minerals were found, which could have been formed only under the aforementioned conditions. In the case of the Ries crater, this can be taken as evidence for the impact theory and is accordingly also considered as evidence for meteorite impacts where suevite also occurs on Earth. It was previously believed that the Ries crater had volcanic origins, but this theory is considered disproved today.

There is another meteorite crater in Europe that is very accessible but much lesser known. It is the Gardnos crater in Hallingdal near Nesbyen in Norway. It was not until the 1990s that evidence of a meteorite impact was found. In contrast to the quite recent giant impact, the impact event at Gardnos crater lies very far back, which explains the strong overprinting of the crater structure, which therefore remained undiscovered for a long time. The crater has a diameter of 3.1 miles (5 km) and features a small central mountain. The impact event is dated to the late Precambrian period and took plate about 500–600 million years ago, depending on sources. At this time, a rock body of about 820–984 feet (250–300 m) in diameter impacted the old bedrock of granitic gneisses, quartzites, and amphibolite present at the site. As a result of the impact, two typical impact metamorphic rock types were formed. In the central area of the crater, suevites are found, as they are also known from the Nördlinger Ries. Ring shaped around the suevite zone lies a belt of Gardnos breccia, named after the site. This is a lithic impact breccia that, in contrast to the suevite, has no molten rock components. In contrast, the Gardnos breccia consists of clastic mountain base fragments composed of gneisses and quartzites. Many of these fragments show significant strain deformation. They are embedded in a black matrix of pulverized rock, which exhibits significantly lower rigidity.

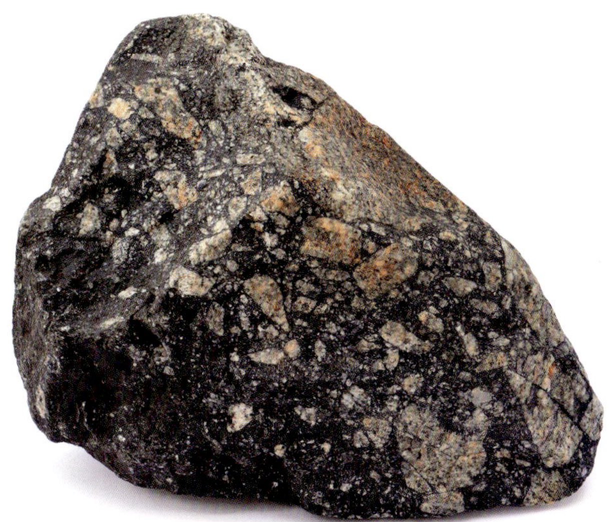

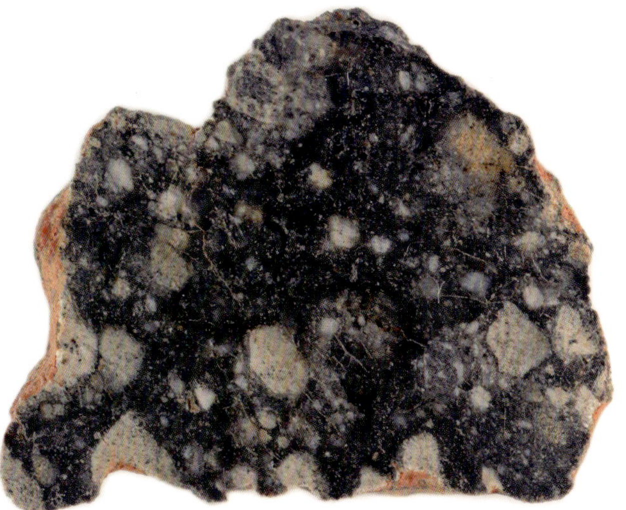

Lithic impact breccia isn't found only on Earth. If one imagines that the bombardment with meteorites takes place over long periods of time and is not cushioned by an atmosphere, it quickly becomes plausible that on a celestial body an ever-thicker layer of debris material accumulates and gradually condenses to a breccia. Exactly these conditions we find on our Earth's moon, a circumstance that was confirmed by the geological investigations of the Apollo missions. The moon is covered with rock strata, the so-called regoliths or regolith breccias, next to the flood basalt areas, the mares. The debris is composed of basaltic material originating from the mare basins. The latter formed on the moon about 3.5 billion years ago through enormous meteorite impacts, which then likewise led to enormous lava eruptions. If you would like to be a little more comfortable and forgo the trip to the moon, you can examine these rocks in peace in the form of moon meteorites, which unfortunately are found very rarely as earth science treasures on Earth (fig. 4.54).

Figure 4.53 Impact breccia from Gardnos near Nesbyen, Norway. The black matrix, made out of the rock pulverized on impact, is visible, which was later solidified, as well as the light broken pieces of the original rock. Rock size 110 × 135 mm.

Figure 4.54 Regolith breccia from the moon. The different fragments of rock debris are clearly recognizable, as is the finely grained black basic mass. The white clasts are composed of anorthite, a plagioclase. The red outer layer is an earthly erosion crust. Polished piece of the moon meteorite NWA 11273, Algeria. Dimensions about 17 × 15 mm.

4.4 The Rock Cycle

As we have now seen, no stone remains on the other: the three big rock families are related, merge into each other, and transform. Let us skip the orders of magnitude and look at how this appears on a large scale. In a later chapter, this examination will be able to tell us where all the sand masses remain, which are created daily through erosion and other processes all around the world.

The lithosphere of the earth's crust consists of the lighter continental crust, which is formed mainly by granites, and the heavier oceanic crust of basalt and gabbroic rocks. The lithosphere floats on the asthenosphere below it, which is indented where the continents are due to their weight to a depth of about 62 miles (100 km). However, the earth's crust is by no means homogeneous and closed but fragmented very early into a puzzle-like structure of individual plates, the lithospheric plates. These play a decisive role in the formation of rocks, since they are permanently in motion due to the thermal processes in the earth's interior and can move relatively freely on the viscous malleable asthenosphere.

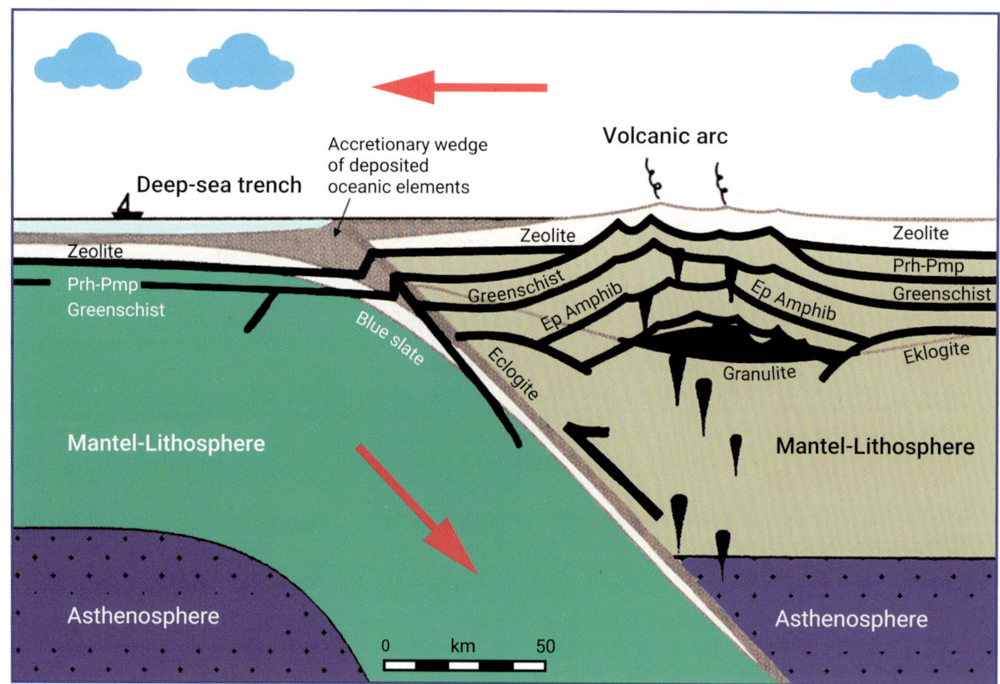

Figure 4.55 Profile of converging plates. The oceanic lithosphere (*green*) is subducted under the continental plate (*brown*). The blue asthenosphere forms about the 1,000°C boundary. Named are the facies series from the zeolite to the eclogite facies. The black drops symbolize ascending magma intrusions, formed as a result of heating and pressure rises. The wedge of scraped oceanic sediments is easily visible (according to Okrusch 2009, adapted).

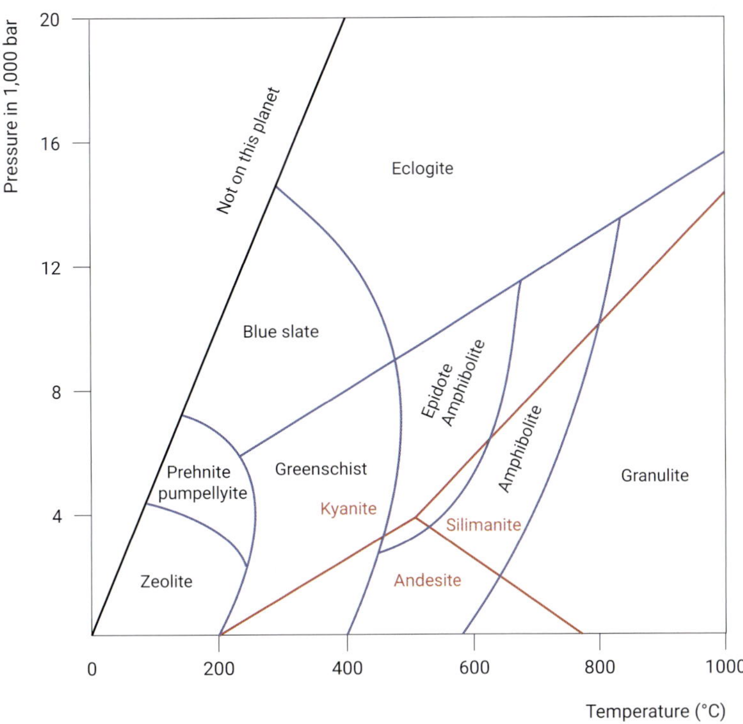

It's a huge station for shifting—driven by convections in the earth's interior. The shifting of the whole plates and thus whole continents (continental drift) started about 3.8 billion years ago. The all-decisive geological processes for the formation of continental rocks take place at the edges of these plates.

When two lithospheric plates move toward each other, the heavier plate is pushed under the lighter one (subduction) and finally melts in the asthenosphere. A formidable process: whole continents are largely melted in the earth's interior. One speaks very vividly of a destructive plate boundary; the Andes are mentioned as an example. Subductions are typically accompanied by speeds of up to 3.5 inches (9 cm) / year, which in connection with friction leads to mutual tensions of the plates, which at some point can jerkily loosen and cause severe earthquakes. Besides volcanic activities, which are caused by the melting of the enormous rock masses, a metamorphosis of the involved rocks basically takes place at colliding plate boundaries. When plates collide, the subduction process can also come to a standstill. Then, the oceanic part sinks into the asthenosphere after rupture, and both continental plates drift on each other, under thickening and folding. The Alps and Himalayas were formed in this way.

Figure 4.56 Pressure and temperature diagram of the metamorphic mineral facies. The borders are not exact but overlap. In black are the typical rocks, and in red is the border between the indicative and typical minerals of metamorphic rock. According to Okrusch 2009.

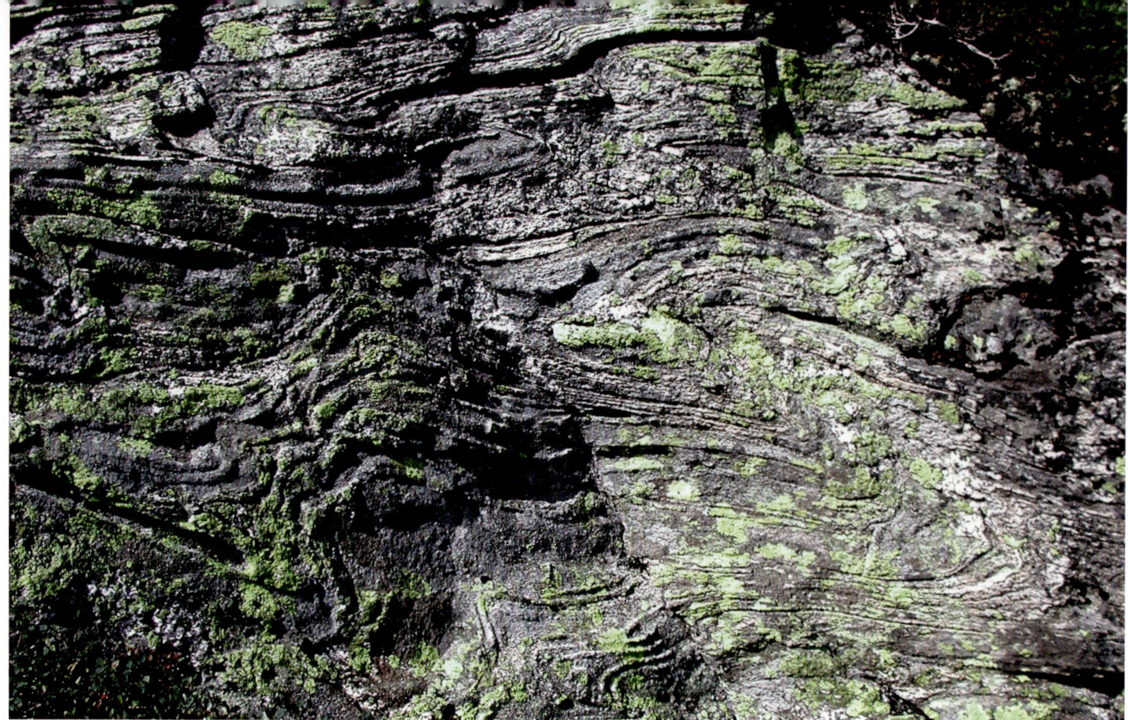

If two plates drift away from each other, which, for example, is the case at the midoceanic ridge, we speak of constructive or diverging plate boundaries. The space freed by the drift is filled with rising material from the mantle, and the (oceanic) crust is successively re-formed at this point.

Over the course of the earth's history, there have been a large number of mountain-building processes (orogenies), most of which are still visible today as a direct result of plate tectonics. The continuing movement of the continents for over three billion years has led to a complete repositioning of the continents relative to the earth's poles, apart from the building up of mountains. Alfred Wegener discovered this around 1915 and made the argument plausible by looking at the almost perfect fit of the African and South American coastlines. Much later, his theory was proven through surveying the rock, looking at fossil remains, and other evidence.

Today, the former position of rock masses can be determined quite precisely by using an interesting technique. It makes use of the fact that magnetic particles (e.g., the mineral magnetite, which is also common in sand) are frozen in almost all rocks. Such particles have the magnetic orientation of their place of origin at that time, which is stored like on a hard drive. This information can be read out and related to the current orientation of the earth's magnetic field. So, for example, if a researcher measures the orientation of a miniature magnet relative to today's North Pole and its angle of incidence on a piece of sandstone from the Paleozoic period, he can determine exactly where on Earth the piece of sandstone was when it solidified, given the known age of the sample. This is the principle of geomagnetic research or paleomagnetism.

Figure 4.57 Rock folding to the time of the Caledonian orogeny, in the course of which the primeval continents of Laurentia and Baltica met about 450 million years ago. The age of the stone is about 1.5–1.8 million years. Highland west of Rysttad, Valle, Norway. Image width about 2 m.

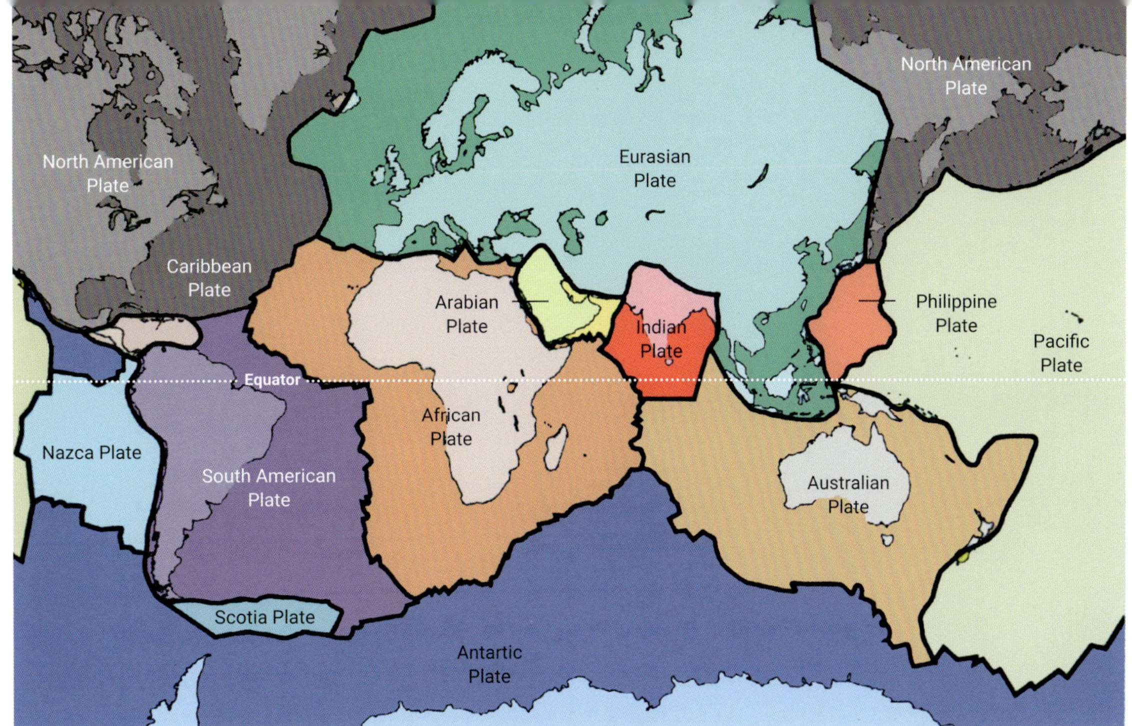

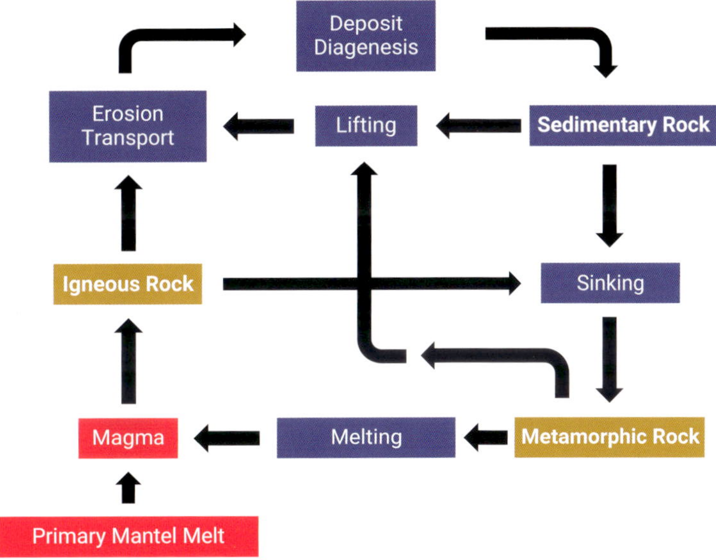

Figure 4.58 Current split of the lithosphere plates, according to N. Welsch et al, *Erde und Leben* (Berlin 2017)

Figure 4.59 Rock cycle, simplified. The three main rock families of metamorphic, sedimentary, and igneous rock find themselves undergoing permanent change over geological time periods. Lifting, sinking, erosion, transport, and melting cause the rock to transform into each of the families, depending on the dominant process. We start with the primary melting of the mantle and the resulting basalt and igneous rock, which then, depending on the erosion, transport, and diageneses, compacts into sedimentary rock. If the rocks are then subjected to high temperatures and pressures through tectonic processes, metamorphic rocks are formed, which are then once again weathered down according to tectonic-lifting processes.

After the formation of the first solid continents and with the beginning of plate movements, a process of transformation of existing rocks into other rock types also began. Water played an essential role in this process. Water affects the cycle of rocks in many ways. On the earth's surface, water is the driving force behind erosion and the removal of erosion particles. In the earth's interior, it is dissolved in the melts or bound to minerals. Water lowers the melting point and viscosity of magma, which can contain up to 10% water.

Let's summarize the rock cycle: The weathering and erosion of any rock forms sediments, which after compacting and squeezing water out of the pore spaces become sedimentary rocks and are eventually transported via plate tectonics to subduction or collision zones. Pressure and high temperatures at the contact zones eventually turn the sedimentary rocks into metamorphic rocks. The rocks formed in this process can be uplifted and eroded, then eroded and sedimented again over the force of the rock formation processes during plate collisions. Some of the rocks are further subducted (i.e., pushed toward the asthenosphere into the earth's interior and melted there). Molten rock, liquid magma, eventually rises again at plate edges toward the earth's surface. Slow solidification of the magma while still within the earth's crust produces plutons and plutonic rock, which will also one day fall to erosion once the plutons are exposed. If the magma reaches the earth's surface through volcanic activity, igneous rocks are formed. Also, these fall to erosion and become sediments and sedimentary rocks. The cycle of weathering, transport, deposition, compaction, subsidence, deformation, and uplift begins again, a mechanism that has been going on for more than three billion years up to today.

Figure 4.60. A segment in the rock cycle. Sand (*top and left*) as an erosion product becomes compacted into a sandstone (*middle*) through diagenesis. The individual grains of sand are still recognizable in the layers, and there is still a relatively large pore volume. Likewise, mostly layered stratifications are present, corresponding to the sediment-depositional conditions. If the sandstone is now subjected to metamorphic conditions (i.e., high pressure and high temperatures), a rock called quartzite is formed (*below and right*), which consists mainly of quartz and no longer has any significant pore volume. Mostly in the quartzite, no more layered stratifications are pronounced. All detailed photos in the same scale, with image width of 6 mm. The quartz sand comes from N-Fehmarn, Germany; the Lower Cambrian Nex*ö* sandstone from Bornholm, Denmark, contains along with quartz many feldspars too, thus originating from an immature sand. The quartzite from Bjerregard, Denmark, is free of secondary aggregate parts and consists entirely of metamorphically connected quartz grains. The latter has an age of about one billion years and probably originates from Norway. Image width of main picture is 17 cm.

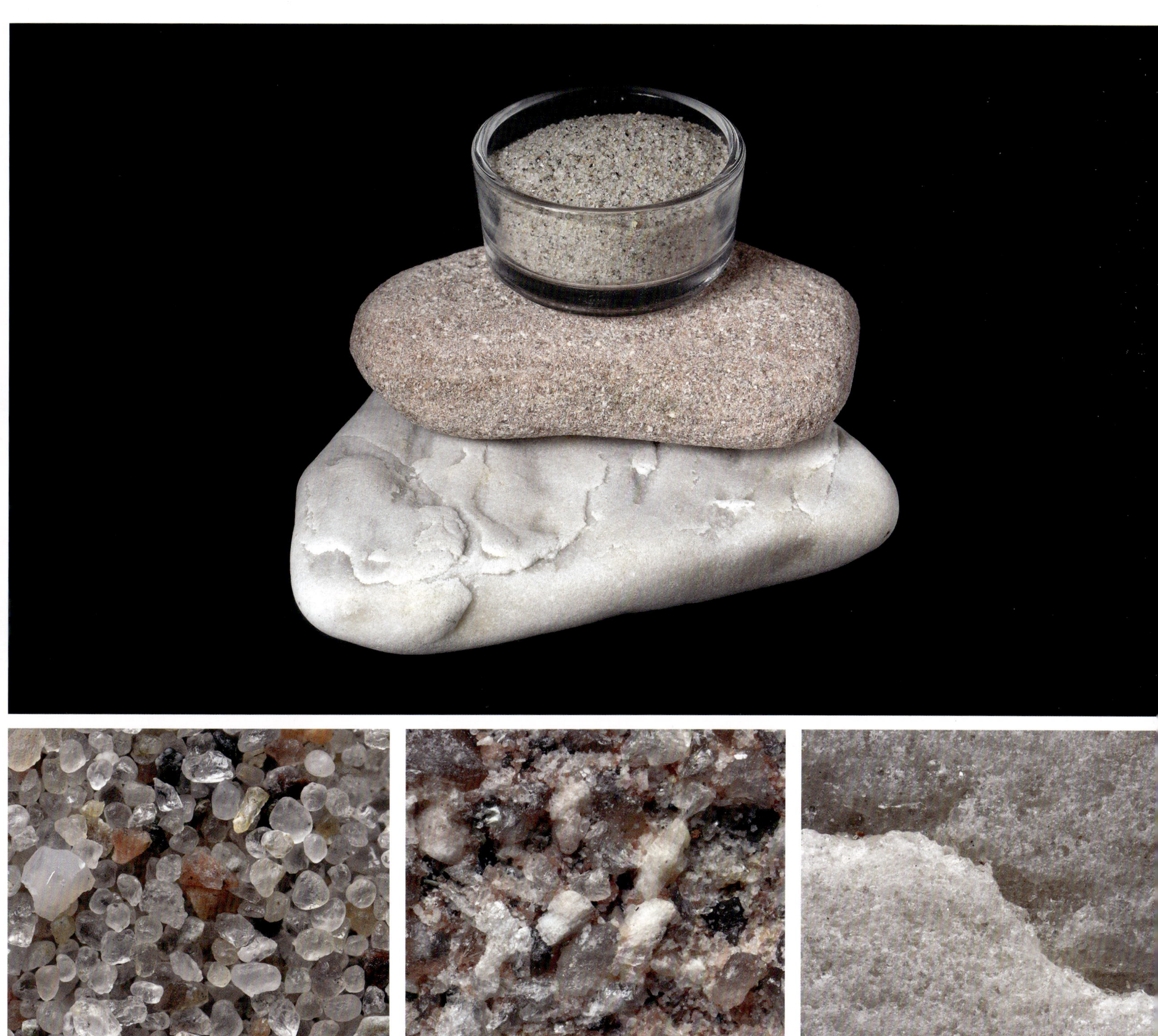

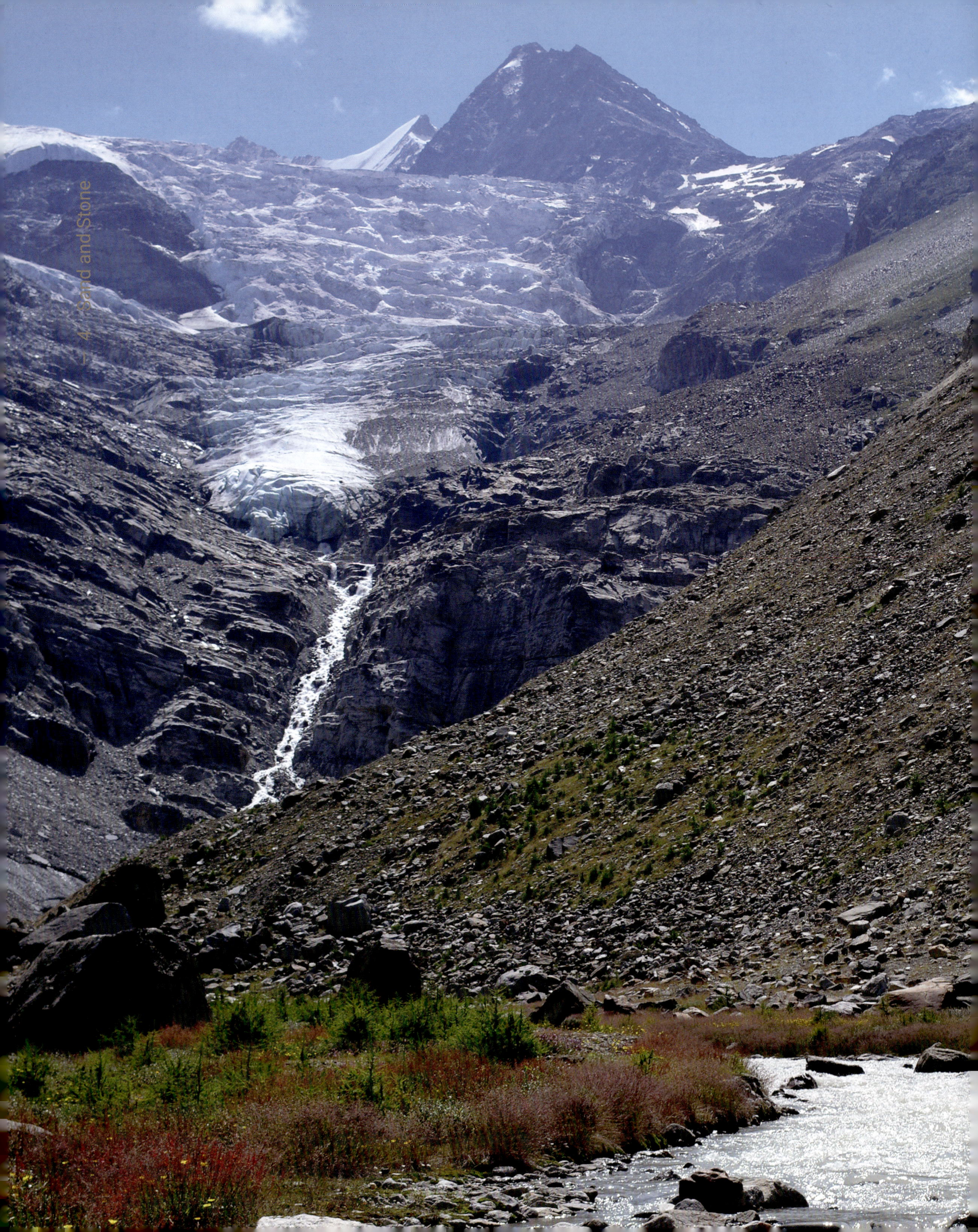

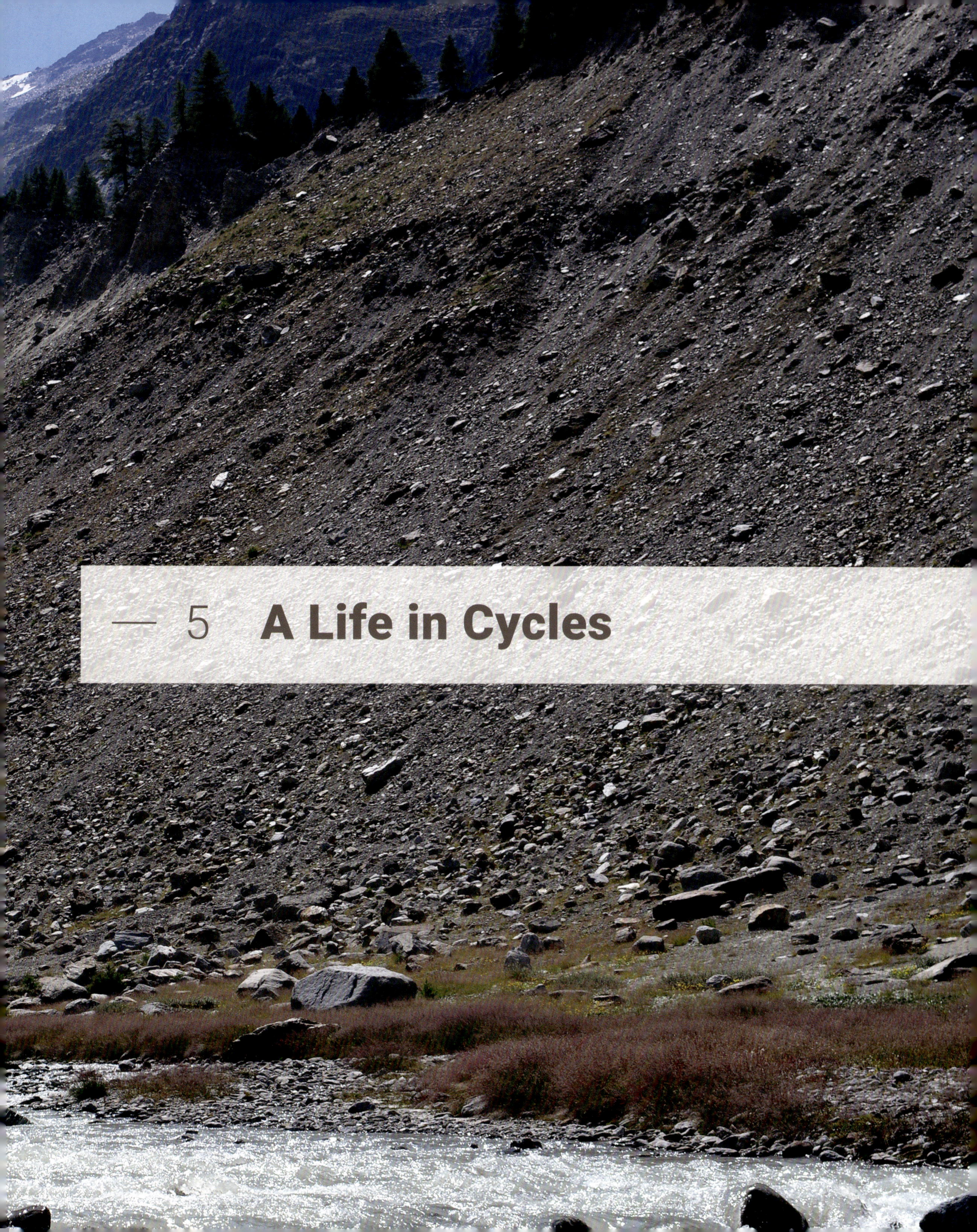

— 5 A Life in Cycles

The way up
and the way down
are one and the same.

—Heraclitus of Ephesus,
Fragment 60

Sand and rock have already been discussed. But how does rock become sand—what mechanisms come into play? What stories are hidden behind it, and how can they be discovered? All these questions will be addressed in this chapter, while we are trolling on the beach and experiencing amazing things behind the life of sand grains.

We began as minerals, petrified
And emerged as plants, we wilted
And grew into animals, we perished
And became human, forgetting everything before
Except in spring, when we almost
Remember the green,
Humanity wanders
On an unfolding path, and though we seem
To be asleep, there is a vigilance within
That guides the dream: to awaken us one day
Back to the truth of what we are

—Jalad ad-Din Rumi, ***Mathnawi****, around 1250, V 2211-20*

Previous double page:
Gasenried, Ried Glacier (2016), Switzerland

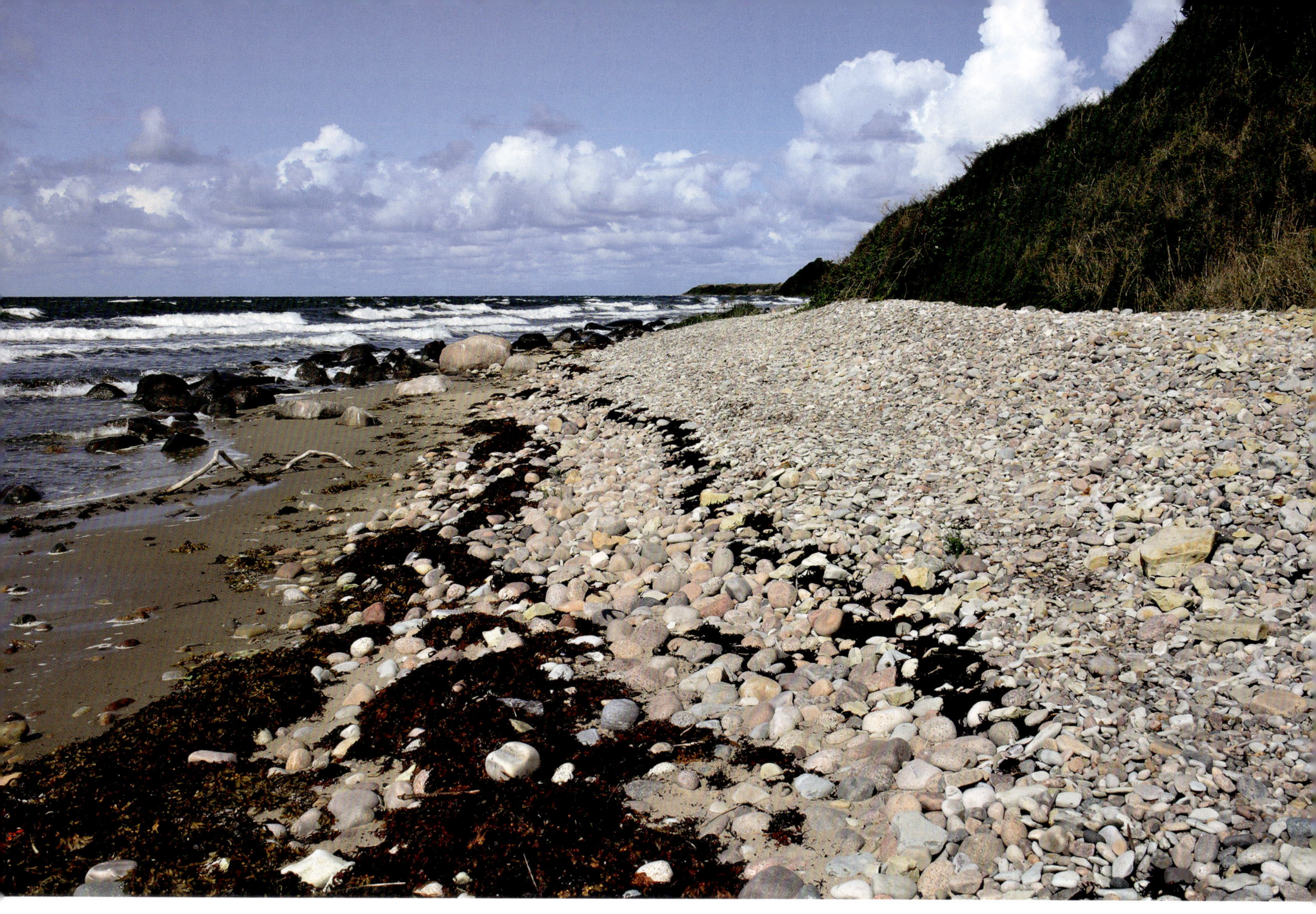

5.1 A Walk on the Beach

Let's imagine we're hiking on a rainy day on an island in the Baltic Sea, somewhere along the waterline. Rounded, colorful boulders lie shiny and wet in the sand, which is continually released by the receding waves. We had just decided to turn back when we noticed a cream-colored striped stone. It is a sandstone, as is usually found here, but it contains an unusual structure. Apparently, it's a trace of a living being. This makes us curious—it's not a fossil, but the trace of an animal, as we suspected. What if we could reconstruct the origin of the stone in our mind and its way here over the course of millions of years? So, let's collect it, look at it more closely, deal with it, and start the attempt to tell its story: the story of a beach stone. With this, we begin our chapter.

Figure 5.1 West coast of Bornholm, beach rocks (debris) originating from the Ice Age

In the region known today as southeastern Sweden, a small creature crawled onto the sandy soil of the sea approximately 540 million years ago. The sand, of course, was much older and came from the remains of a huge mountain range through the process of erosion, the Gondwanaland, which had long ceased to exist. At this time, the Lower Cambrian period, the sea already contained a number of life forms that had evolved since the inanimate Precambrian period. The land, on the other hand, was completely uninhabited, empty, and barren. No plants covered the ground, nor were there any animals. The continents as we know them today did not yet exist—they had different outlines and were located at different places on the globe during the course of continental drift. Africa and South America were not yet separated. The small Swedish animal, which might have resembled a lugworm of today, crawled into the sea near the Tropic of Capricorn. That's where Sweden was in the Lower Cambrian period.

Over the millennia, a number of sand layers had already formed, each with slight differences in color depending on which ocean currents were prevailing at the time, carrying with them and eventually depositing corresponding sand loads. According to its habit, the animal dug a small tube vertically into the ground, to stay there or to lie in wait for prey driven by the current. It may also have fed on the sand it picked up while digging and sifted through for food, perhaps seeking to stabilize the small tube with secretions. When the animal left its tube again—of course we do not know why—the surrounding loose sand flowed or seeped back into the empty tube and filled it up completely, losing the original stratification. The animal, however, did not return. It left two small squiggles on the seabed.

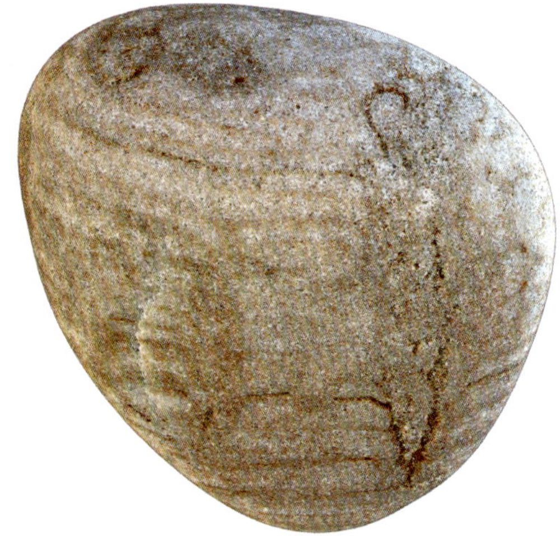

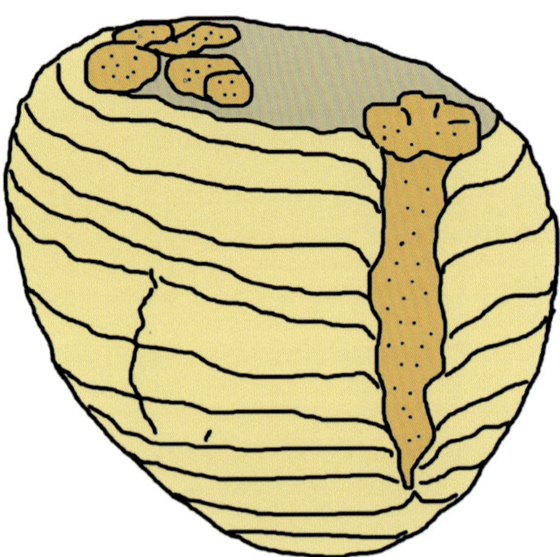

Figure 5.2 Beach drift from Bornholm, Denmark, with trace fossil *Monocraterion tentaculatum* (shown on the right as a diagram). Width 50 mm.

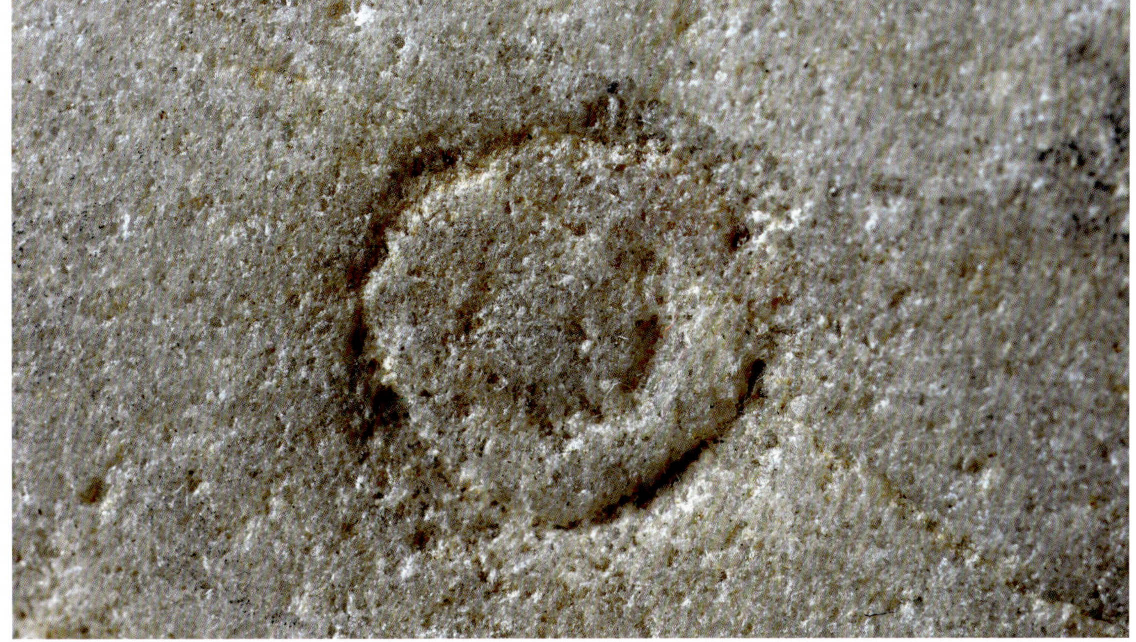

Time passed, and the sea poured more layers of sand over the scenery, millennium after millennium. Less durable mineral grains disintegrated and were transported away or went into solution; quartz remained. The sand sorted itself and increased its packing density like balls in a glass do when it is shaken slightly. The higher and thicker the sand layers piled up above the old piece of seafloor, the greater the weight bearing down on it. The pressure rose continually, almost all the water had meanwhile been displaced from the spaces between grains, the sand increasingly solidified, and only the water pressure in the pores still prevented the grains from breaking. During the following millions of years, the sand grains were finally pressed against each other so strongly at their points of contact and the process of a pressure dissolution reaction began. In the process, silica from the surface partially dissolved, circulated between the grains, and redeposited elsewhere as a siliceous binder. At the direct contact points, the crystal lattices grew together until pressure peaks equalized. Sandstone was formed from the formerly loose marine sand.

All these processes took place simultaneously over very long periods of time. While on the surface of the seafloor, new sediment layers are added on which other animals crawled, the layers far below solidified more and more. The Cambrian period was followed by the Ordovician period. Trilobites dominated the sea, volcanism the land. In the following Silurian period, the earth's equator ran over southern Scandinavia—the sea expanded considerably. In the Silurian period, the first real fish with jaws appeared, as well as tropical corals. Vascular plants conquered the mainland bit by bit, and the first arthropods ventured out of the water. They experienced a year that had four hundred days, because at that point in time the earth spun faster on its axis, so more days went by each year.

Figure 5.3 Baltic Sea, Hohwacht, Germany. Surface of a sandstone with the fossil trace *Monocraterion tentaculatum*. The inner circle corresponds to the diameter of the dwelling tube, and the outer circle represents the boundary of the trickling sand. Image width 20 mm.

The formation of the Caledonian Mountains reached its peak and folded today's Norway into a mile-high mountain range. In the Devonian period, Africa, South American, the Near East, Australia, and Antarctica still formed the vast continent of Gondwana.

More than one hundred million years have already passed since our little animal left its living tube. In the Devonian period, the development of fish then made tremendous progress, armored fish in particular. On land, the first lycopod-like plants appeared, later the first trees. In the following Carboniferous period, insects flew for the first time among the plants of the coastal marshes. There, huge forests formed, which contributed to the mighty coal deposits and became characteristic of the age. Scale trees and *Sigillaria* cf. *brardii* formed forests on land. The Variscan mountain building reached its main phase when the united North American and European plates collided with Gondwana and caused massive plate folding and mountain building. Even today, marine lime from the Devonian ocean can be found on the peaks of the Himalayas at altitudes of almost 30,000 feet (9,000 m), where they then disintegrate into sand.

Above our piece of seafloor, the Old Red mainland piled up and was slowly eroded away by water, wind, and rivers. Its sediments collected in offshore depressions and grew to thicknesses of several miles. In the Permian period, following the Carboniferous period, the first coniferous forests appeared, while on land, reptiles increasingly prevailed over amphibians as the climate became warmer and more desertlike. In the receding and evaporating Zechstein Sea, the 3,280-foot-thick (1,000 m) salt layers were formed, which today provide us with the valuable raw material. The subsequent Mesozoic era, which is made up of the Triassic, Jurassic, and Cretaceous periods, is considered the age of dinosaurs, which first appeared in Germany in the Keuper. They disappeared from the earth completely only with the impact of the Chicxulub asteroid sixty-six million years ago, at the end of the Cretaceous period and the beginning of the Paleogene period. Over 70% of life-forms at that time disappeared with them. What might have become of them without this cosmic catastrophe?

The sand grains of the piece of seafloor from the Lower Carboniferous period had in the meantime firmly cemented to stone. The trace of our animal, however, remained visible, although invisibly still hidden deep underground. While the continents shifted, formerly sediment-collecting depressions rose to mountain ranges, where they were exposed to erosion. Rain and rivers transported the debris again toward the sea. More than half a billion years passed in such a way until the Pleistocene epoch, 1.5 million years before our calculation of time, when large climatic fluctuations began. Warm periods and cold periods alternated—the average annual temperature dropped by up to 27°F (15°C). Large glacial layers developed at the North Pole and moved southward across Scandinavia. In some places, the thickness of the ice sheet was over 9,800 feet (3 km). The glaciers carried enormous masses of rocks, sand, and mud with them as products of erosion.

After the retreat of the ice during the warm periods, these debris masses then covered the land up to 328 feet (100 m). But what happened to the piece of seabed on which our animal once crawled?

Layer after layer of the uplifting rock was further eroded as the land rose, until one day, Cambrian layers came to light. Water penetrated cracks and crevices in the consolidated sandstone, frost burst the bonding, and the dwelling tube of our animal became part of a fist-sized fragment that was finally picked up by the approaching ice and transported farther south. The ice pushed the stone forward mile after mile until it came into contact with other stones and boulders, freezing here and there for a few thousand years, until one day it was finally released. This probably happened at the end of the last great glaciation period, the Vistula Ice Age. There has been no further advance of the ice since then. Its melting relieved the land masses in such a way that they have been rising almost 3.15 inches (8 cm) per year up to the present day. In the millennia that followed, waves and tidal changes tossed the stone around, further flattening and rounding its shape, as grains of it broke loose from the matrix or whole sections were detached during heavy impacts. Grains of the stone are sand grains of the primeval sea from the Cambrian period, which thus, after traveling through time over more than half a billion years, mixed with billions and billions of sand grains, forming today's beach. Grain by grain, the dwelling tube of our animal was exposed lengthwise and became visible, and with it the alternating cream-colored layers of the subsoil. Until finally, at the end of an afternoon walk, our gaze fell on this very structure. Today, the trace fossil is called *Monocraterion tentaculatum*. Much more is still not known about it; only that it is one of the oldest traces of complex life that can be found in Germany. How good that we became aware of this little treasure and witness to the earth's history.

The history of the sandstone we are looking at is at the same time a history of the sand grains from which it was formed. After the bonding material was dissolved through erosion, the grains—trapped almost like in a time capsule from five hundred million years ago—are released again. In principle, history could now start again from the beginning, but under the condition that the past time has already left its traces on the sand grains. These grains with all their past history would then, released from the time capsule of the sandstone, set out on a new journey that begins where the Baltic Sea picks them up today. A potential beachcomber on a continent in the distant future might then pick up one of the grains again, bound in completely different stone, and look at it with interest. History follows history. Provided, of course, that mankind and some attentive individuals in its midst may still exist in a few hundred million years.

Era		System / Period	Stage/Age	Start (Millions of Years Ago)
Phanerozoic	Cenozoic	Quaternary	Holocene	
			Pleistocene	2.6
		Neogene	Pliocene	
			Miocene	23.8
		Paleogene	Oligocene	
			Eocene	
			Paleocene	65
	Mesozoic	Cretaceous	Upper	
			Lower	142
		Jurassic	Upper	
			Middle	
			Lower	200
		Triassic	Upper	
			Middle	
			Lower	251
	Paleozoic	Permian	Lopingian	
			Guadalupian	296
		Carboniferous	Pennsylvanian	
			Mississippian	358
		Devonian	Upper	
			Middle	
			Lower	417.5
		Silurian	Pridoli	
			Ludlow	
			Wenlock	
			Llandovery	444
		Ordovician	Upper	
			Middle	
			Lower	488
		Cambrian	Upper	
			Middle	
			Lower	542
Precambrian	Proterozoic	Neoproterozoic	Ediacaran	630
			Cryogenian	850
			Tonian	1000
		Mesoproterozoic		1600
		Paleoproterozoic		2500
	Archean			4000
	Hadean			4600

Figure 5.4 An overview of the geological time periods. The depiction of the individual time periods is not to scale (according to Hann 2018).

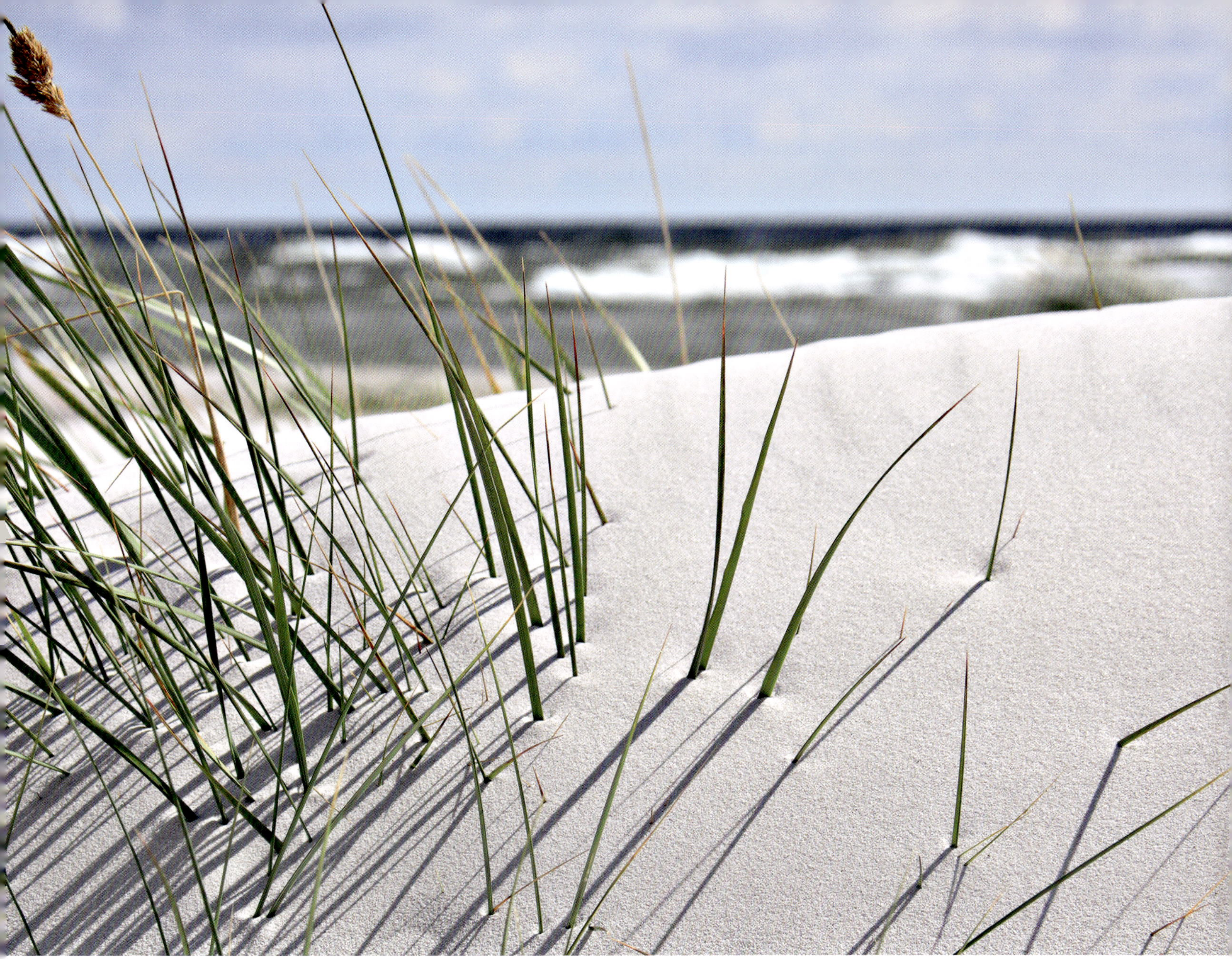

5.2 A Fossil Sand

But let's have our beach lover calmly remain in the present and walk on his island a good bit farther toward the southern tip, where the coastal strip opens up to an exceptionally beautiful, bright, and wide dune landscape. Unlike before, there are no boulder rocks, only sand, here in northern Europe, an almost unreal impression of southern flair. But the special thing is the sand itself. It is not the common beige-brownish or grayish color, but it shines and glitters in the sun: uniformly and brilliantly white.

Figure 5.5 Long, sandy beach with dunes near Dueodde. Southern coast of Bornholm, Denmark.

What makes this sand unique is its fineness, purity, and almost perfect sorting. The sand of Dueodde, which is the name of the nearby town at the southernmost tip of the island, was once world famous for these characteristics. Dueodde sand was used as blotting sand at a time when people still dipped their goose quill in an inkwell to write. In those days, inks were not as fast drying as the synthetic products available today, and the ink application was not as even as that of modern writing instruments. It was therefore customary to sprinkle sand from a special tin onto the paper being written on to absorb the excess ink; "blotting sand," in other words. The finer the sand that was at hand, the better it could absorb the ink components that were still liquid. To draw attention to the uniqueness of the Dueodde sand, an entire shipload of it was even sent to the World's Fair in London in 1938.[80]

So, what is the secret of this sand, and how does it get its properties? To answer these questions, our beach walker must turn his attention to the flat cliff plateaus along the coast between Snogebaek and Dueodde. Here, the subsoil consists of green sandy slate from the Cretaceous period. In the Upper Cretaceous, the southwestern Baltic Sea area was flooded by the sea, and sandy deposits formed in the shallow water near the coast of the Cretaceous Sea. During the Upper Cretaceous, a total of three of these green formations were formed. The older ones belong to the "Arnager Formation"; the youngest one, which is also considered to be the end of the Cretaceous, is counted as the "Bavnodde Greensand Formation." The age of the layers is about sixty-five to eighty million years. Green sands and greenschist alternate with limestone beds in the outcrops along the coast. The layers are particularly beautifully exposed near the villages of Arnager and Bavnodde, after which they were named.

The sediment of the greensand layers is only loosely bound, contains quartz sand as well as loamy and silty portions, and weathers very quickly to greenish sand. This gets its color from a remarkable mineral, glauconite, which is present mostly in the form of small, elliptical grains, about the size of sand grains. It is a complex green clay mineral, a phyllosilicate, which is accompanied by other phyllosilicates (illite, smectite, and chlorite) and other minerals in different proportions. When exposed to oxygen, the color changes from green to brown. Often, glauconite grains represent the spouts of tiny foraminifera, a very diverse phylum of shell-bearing protists that are classified as root feet. Glauconite is bound to sedimentary rock and is formed in shallow marine environments up to a maximum depth of 650 feet (200 m) in moderate to tropical climates.

Because glauconite does not hold up well when it comes to erosion, it was quickly washed out of the loose slates of the formation, after which the greenschist and the greensands eroded to fine quartz sand relatively quickly, whose color is a pure white because of the lack of further components of any significant quantity. Because of currents close to the coast and wind transportation, the sorting grade further increased, and the washed-out sand deposited around what is today Dueodde.

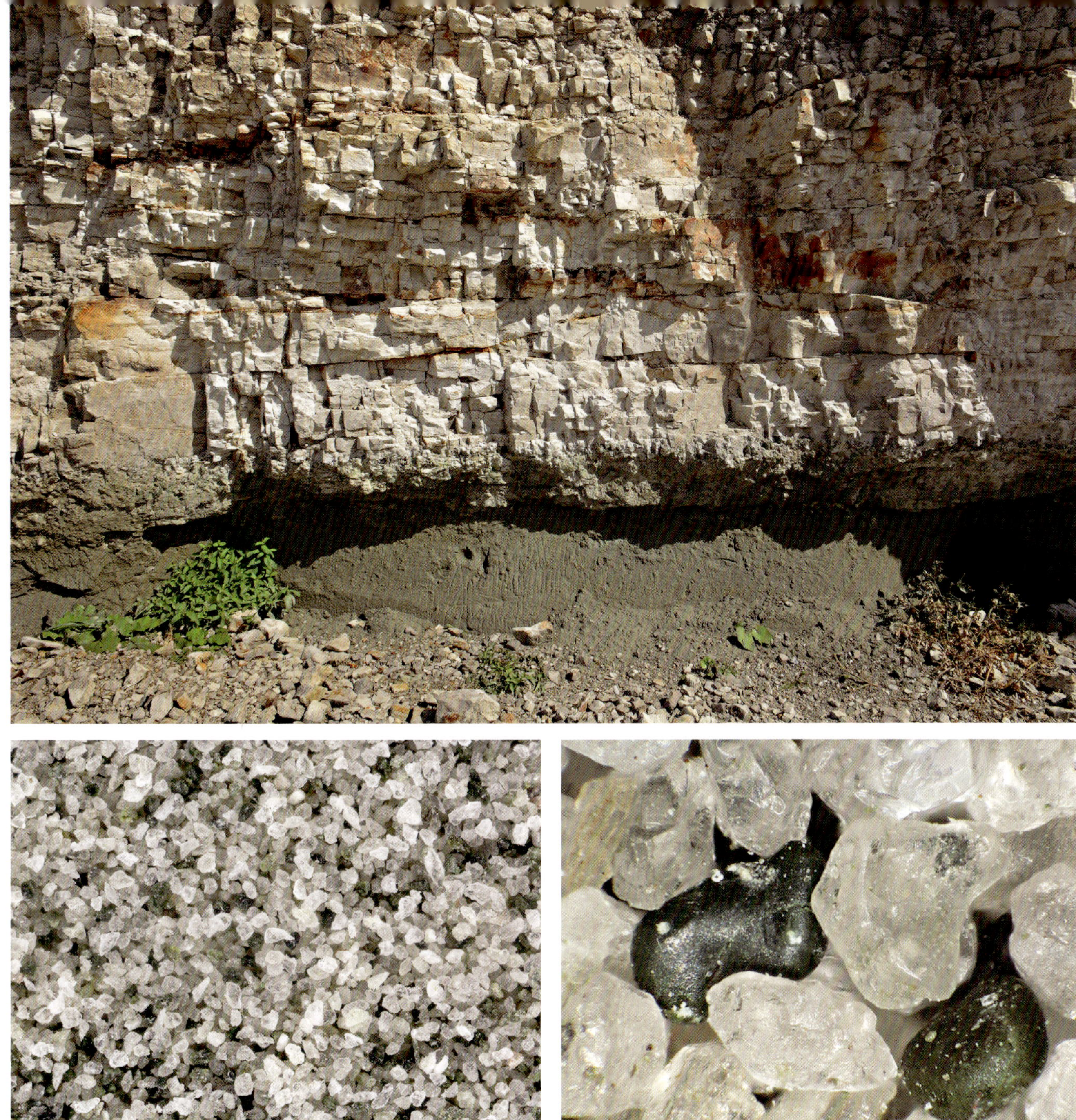

Figure 5.6 Bornholm, southwestern coast near Arnager, Denmark. The greensand formation at the bottom of the bedrock can be seen, which is overlain by limestone beds of the Upper Cretaceous period.

Figure 5.7 Arnager greensand. Very fine sand with grain sizes below 0.1 mm. Very good sorting. Reflected light. Image width 5.2 mm.

Figure 5.8 Bavnodde greensand. The relatively angular, only slightly rounded quartz grains are visible, as well as the green, elliptical, well-rounded, soft glauconite grains. Reflected light. Image width 1.0 mm.

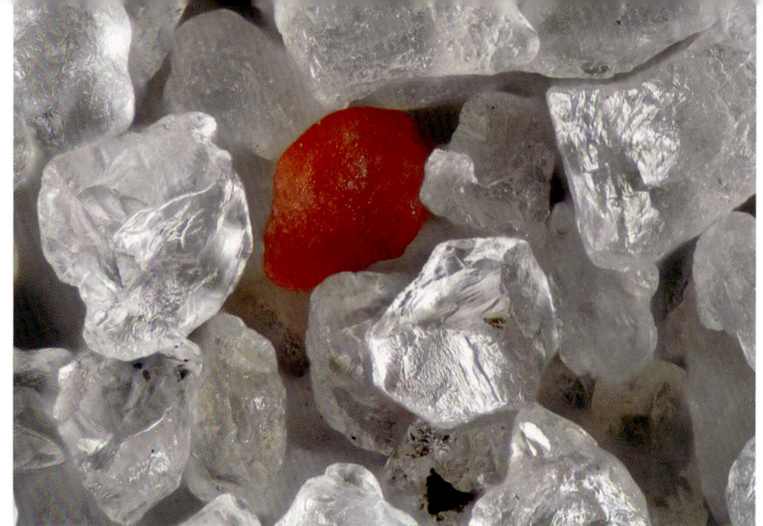

The short transportation distance was not enough to increase the sphericity, but the grains still show a recognizable rounding compared to the original stones. It can be assumed that the Cretaceous sand, which contributed to the formation of the greenschist, was already presorted by its origin and transport but was only insignificantly rounded. Dueodde sand is therefore a Cretaceous fossil sand that could probably one day be re-enriched and compacted in a marine depression. In this case, too, another story would then run its course.

Both stories are set on the island of Bornholm, and this is no coincidence. Bornholm is like a small showpiece of geology: the scene and also the site of many geological eras and their stratigraphic sequences, from Precambrian to ice-age relics. Bornholm's crystalline rocks are products of Danish Polish orogeny, which has not been known in research for very long.[81] Due to its location in the middle of the Baltic Sea, interesting and diverse sands can be found on Bornholm's coasts. For the geologist, the conditions are almost paradisiacal since the coastal sections are largely covered by glacial till from the Scandinavian countries. Almost all important rock types are found here within a small area. A challenge for collectors.

Both stories, moreover, have a common core: rocks become sand and sand becomes rocks again, a cycle that we have already seen in a similar way in the transformation of the types of rocks among themselves. But there is a special, sand-specific aspect that is added here. For a defined grain of sand, there is another process that is added to the end of a geological process that influences the same grain. The grain was changed but stays recognizable as an individual, so to say, in the following process. Rocks, on the other hand, change in texture and composition, are decomposed, are deindividualized, and become unrecognizable. There are, of course, exceptions in both cases.

Let's return to the question we posed in the beginning: Where does sand come from and what becomes if it, or, more exactly, what becomes of the individual grains?

Figure 5.9 Dueodde sand—composed almost exclusively of quartz. Occasionally, colorful quartz grains can be present, but they do not change the overall impression of white. The diameter of many grains is about 0.07 mm on the lower limit for the classification into the grain-size-fraction "sand." Reflected light. Image width 1.4 mm.

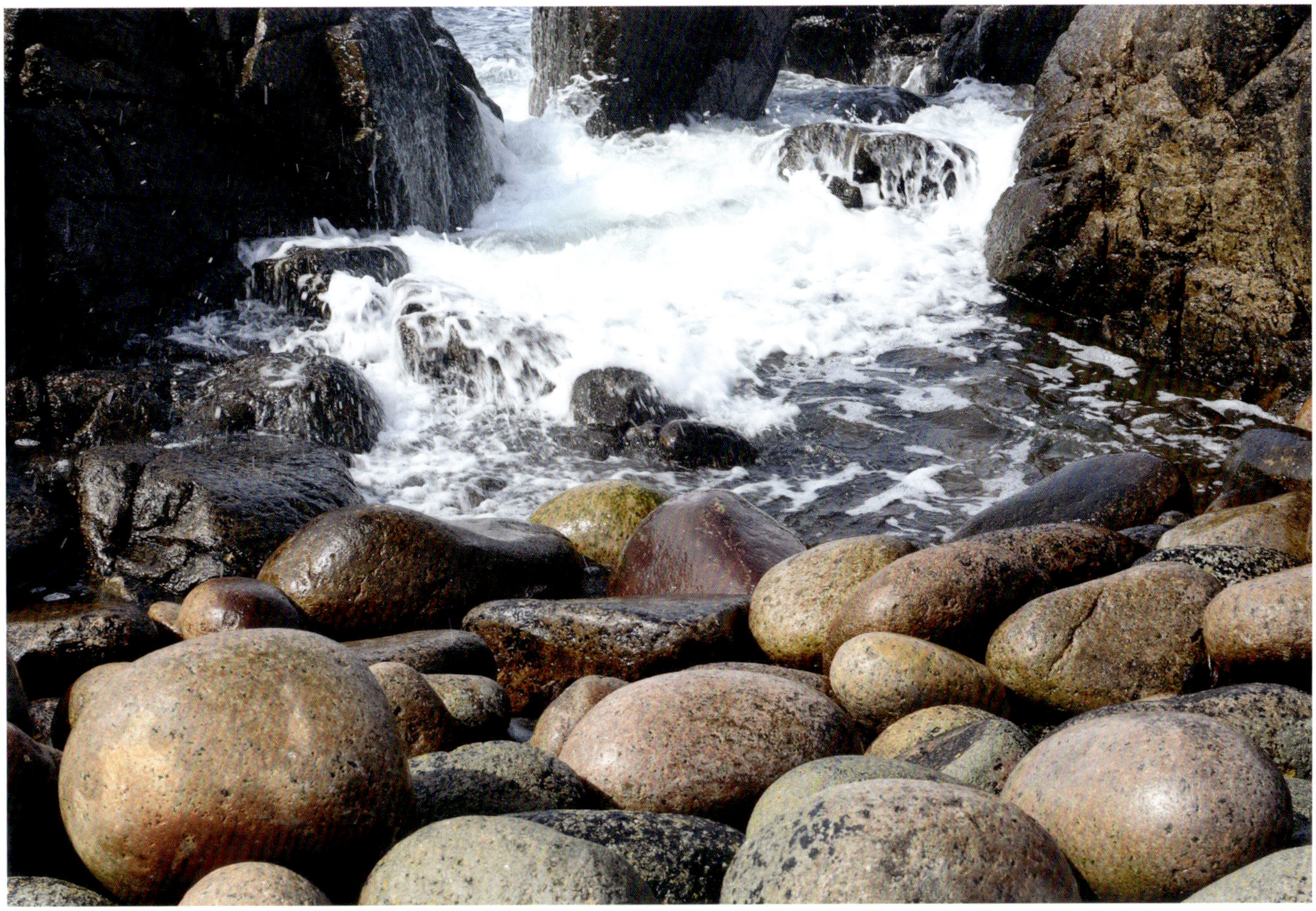

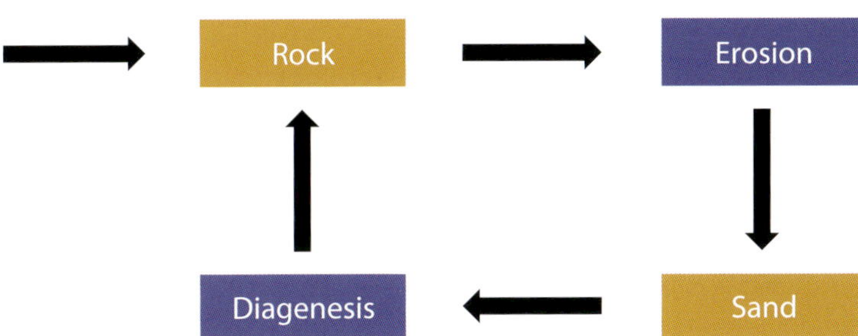

Figure 5.10 Glacial bedload and bedrock on Bornholm's northwestern coast at Salomons Kapell. One recognizes boulders in the foreground with diameters around 12 inches (30 cm), which consist of Middle Swedish porphyries, granites, and ignimbrites. The surrounding rock is 1.4-billion-year-old Hammer granite, named after the nearby fourteenth-century castle Hammershus.

Figure 5.11 Simplified depiction of a sand cycle

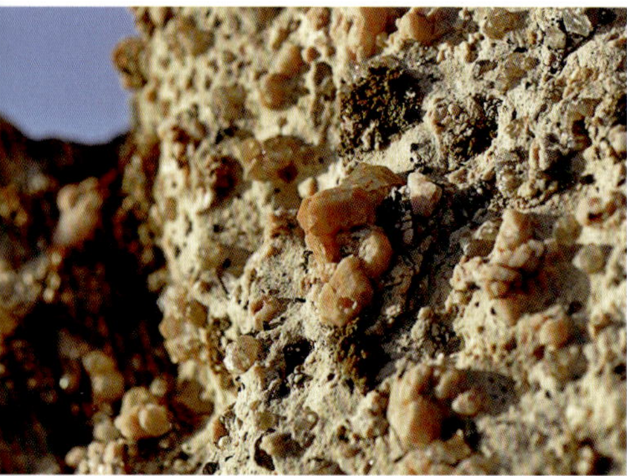

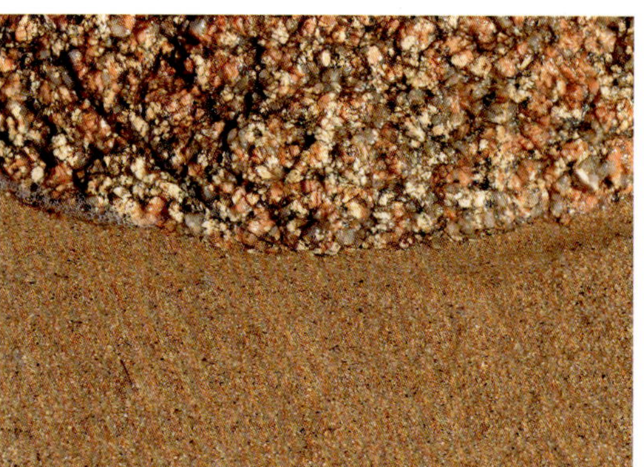

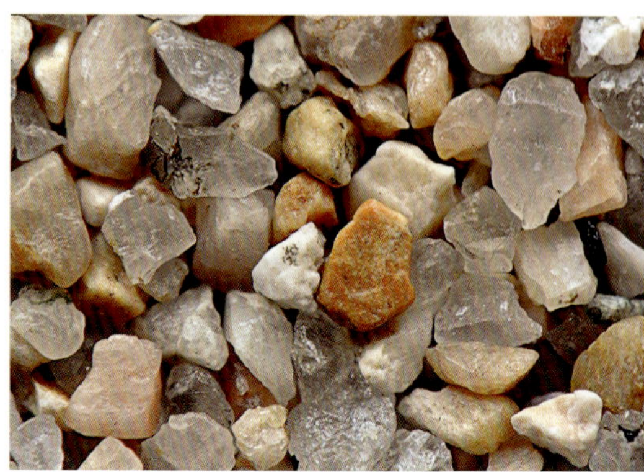

5.3 Life Cycle of a Sand Grain, Part 1

In chapter 3, the degree of maturity of clastic sediments was introduced as a characteristic. Increasing compositional maturity can be described as increasing quartz content and simultaneously decreasing content of more-erosion-prone minerals such as feldspar, olivine, and pyroxenes. At the very end, pure quartz remains as the most mechanically and chemically resistant material, which is also one of the most abundant. We therefore focus on quartz in the further discussions on origin, transport, and deposition.

Whereas it was previously assumed that quartz sand forms in rivers from the gradual abrasion of coarser pebbles, it was recognized as early as the beginning of the nineteenth century that it instead results quite predominantly from the decomposition of crystalline rocks. This realization is still valid today. Since granites are the most-common plutonic and thus crystalline rocks, they are the main source of all quartz sand grains in general. If metamorphically formed plutonic rocks are added, it can be concluded that quartz sand originates almost exclusively from granite, rocks similar to granite, and the gneiss of the continental masses. As a result, sand is very closely connected with the occurrence and erosion of crystalline rocks. The dimensions of sand grains therefore are closely related to the grain sizes of the minerals in granite, as we have already mentioned. We now are going to look at a typical life cycle of a grain of sand of continental origin a little more closely; "continental" because sand from oceanic islands and archipelagos is composed completely differently (group 4) and follow other rules in their creation.

Figure 5.12 Igneous rock on an exposed coastal cliff in the northeast of Sardinia, Italy. The eroded but still connected to the rock, partially idiomorphic quartz and feldspar crystals are clearly visible. The grain in the middle of the picture will soon detach, and a sand grain is created.

Figure 5.13 Creation of sand on the coastline directly from the nearby rock. You can see in the upper part of the picture the strongly eroded granite with red feldspars and the correspondingly colored sand below it, mostly with the same composition as the stone next to it. Image width around 50 cm.

Figure 5.14 Detailed picture of the coastal sand from the previous picture. The very young sediment is recognizable with its low rounding and the low compositional maturity of the mixture. Image width about 8 mm.

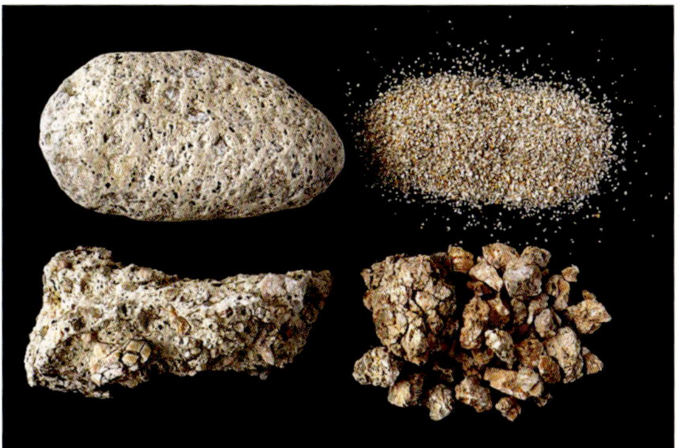

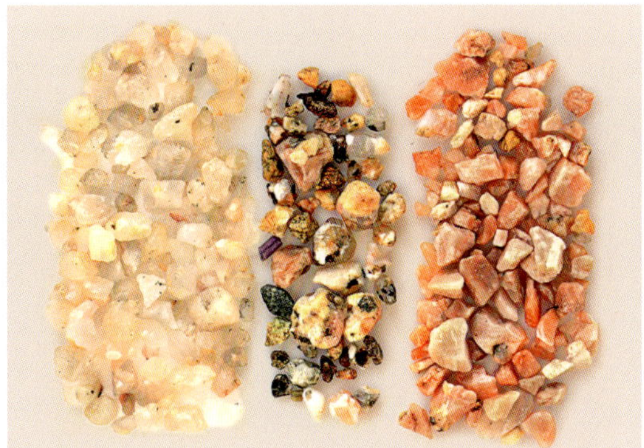

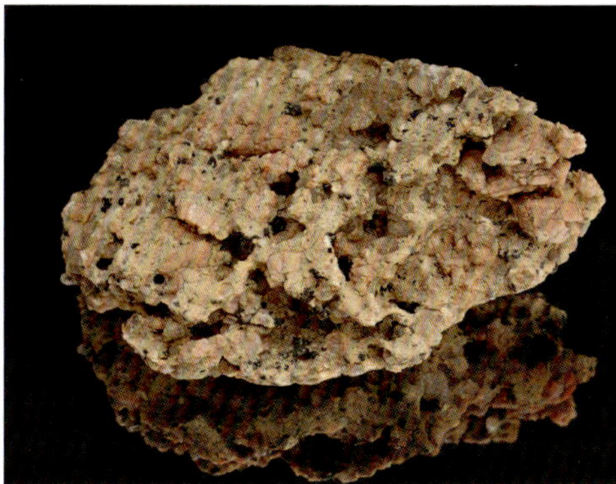

Continental sand is a product of gravity. From its formation, such as from the breaking off from the source rock in a mountain range, to its deposition in a marine basin, a grain of sand is exposed to the free forces of nature and follows gravity until it reaches the next-deepest point in its journey—as we will see—but it does not have to end there.

Next, let's look at the life of a single imagined grain of quartz sand in more detail and follow its journey toward the ocean. Of course, there is a myriad of possible courses, basically as many as there are grains of sand. Many of these life paths are similar in their basic pattern, and we will choose one of them that is quite typical. Crystalline rocks tend to be very massive in their structure. In early times, granite was considered a symbol of strength, indestructibility, and age. However, granite is by no means indestructible. Quite the contrary. Due to its coarse crystalline structure and its composition of very different resistant minerals, it tends to be sensitive when it comes to physical, chemical, and biological erosion.

Once pores, splits, or washouts have occurred through decomposition processes, temperature changes, or pressure stressors, plant roots can penetrate or infiltrated water can freeze and break out particles of various sizes out of the structure. The result is initially crumbly rock crust and, depending on how it continues, ultimately sand.

Figure 5.15 Erosion processes demonstrated on a coarse crystalline gangue from northeastern Sardinia, Italy. The samples come from proximity to each other. Rounded beach stone, eroded hand-sized specimen from a weather-exposed cliff, rock grit at the base of the cliff, and sand at the sample site.

Figure 5.16 Sand analysis from the previous picture. The sand is relatively young and, after only a short transport, compositionally immature, with only slightly rounded edges. Quartz (*to the left*), rock and shell fragments (*in the middle*), and alkali feldspars (*to the right*). Some of the rock fragments come from debris that is not of local origin. After a certain amount of time, only quartz and a portion of the fragments will remain. Cala Bitta, Sardinia, Italy.

Figure 5.17 Erosion process of a rhyolite. Rhyolite is the volcanic-ignimbritic equivalent to granite. The pores of the stone are recognizable—some components have been completely dissolved. So-called tafoni textures have formed. Sardinia. The width of the rock lying on the mirror is 80 mm.

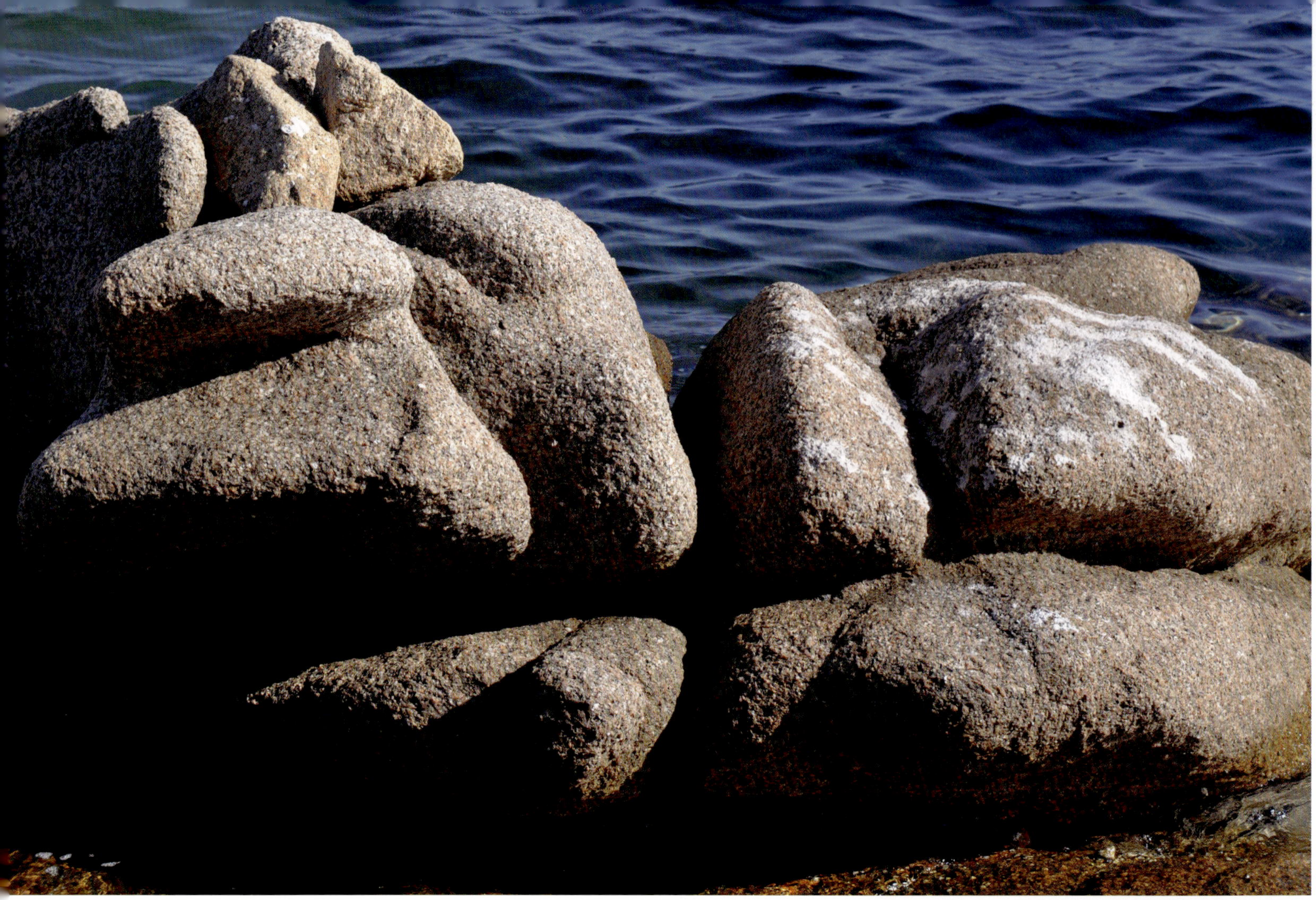

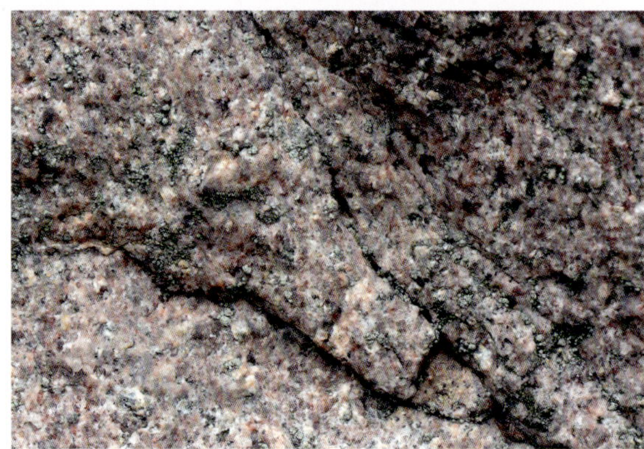

Figure 5.18 The erosion process progresses along cracks in the rock and leads to the rounded blocks ("woolsack texture"). Granite rock, northern Sardinia, Italy.

Figure 5.19 The weathering process of granite. Lichens promote the chemical erosion of the rock. The figure shows the surface of granite rock in a weather-exposed location in Bornholm, Denmark. Soon, the wedge-shaped area will be separated from the block of rock by frost heaving.

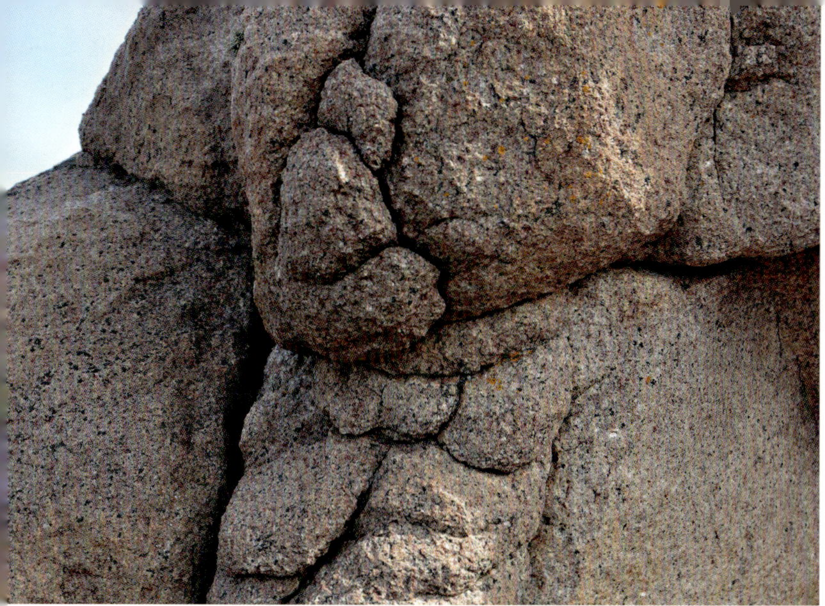

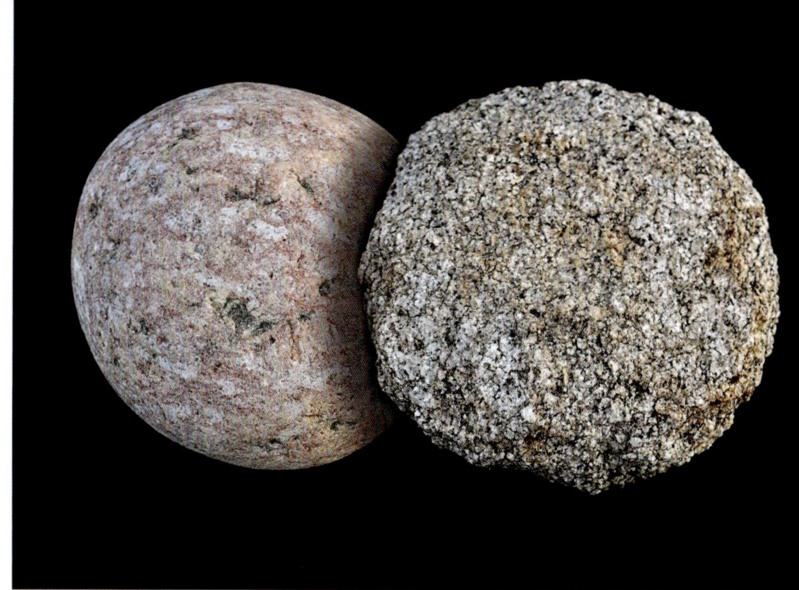

The amount of physical and chemical erosion is very dependent on the geological altitude and the exposure of the affected rock. While chemical erosion processes predominate in the lowlands and valleys, the higher a mountain range rises, the more exposed the rocks are to mechanical factors. Mountains are thus eroded and leveled piece by piece in geological periods. Wind, precipitation, and temperature changes wear away at the rock, which disintegrates inexorably and whose components accumulate in debris cones, the alluvial fans, around the mountain peaks and slide down to the valley. Rain showers wash the fragments and small components, including our freshly formed grain of sand—it is made of quartz—in rivulets to the nearest stream. A long journey toward the sea begins.

Let's follow it further. The mountain stream or the meltwater coming out of the glacier has enough energy to carry the grain of sand in a turbulent, swirling flow down the slopes into a river of the valley. Everything else now depends on its diameter. Very small grains can remain in suspension and are rapidly carried away as a suspended load in the current. Sand grains of typical dimensions, however, roll and bounce along only short distances across the bottom of the river, an often-intermittent process that can last a long time. This is because flows in channels, as with riverbeds, theoretically have zero velocity directly at the bottom. The flow processes near the bottom can be physically described with the help of the so-called boundary layer, which is formed by friction processes between the flow and the riverbed. Upward, toward the water surface, the boundary layer merges into the main flow of the river. The thickness of the boundary layer depends largely on the roughness of the bottom, which is greater for a gravel bed than for a rather smooth sand bed. This explains why, depending on the boundary conditions, small particles such as sand grains can sometimes stay on the bottom for a very long time if the sinking velocities are greater than the velocities of the upwelling suction or vortex forces.

Figure 5.20. Typical, so-called wool-sack-shaped weathering. Vang granite, northwestern Bornholm, Denmark.

Figure 5.21 *Left:* Unweathered boulder from the Danish coastline (diameter 2.56 inches [65mm]). *Right:* Heavily weathered granite from the Norwegian highlands. The shape of the stones has different causes in this case. Boulder stones are increasingly rounded by stone-to-stone collisions in the sea surf. The high mountain stone, on the other hand, takes on its shape through abrasive glacier transport and the subsequent cup-shaped weathering on the earth's surface, as is typical for granite.

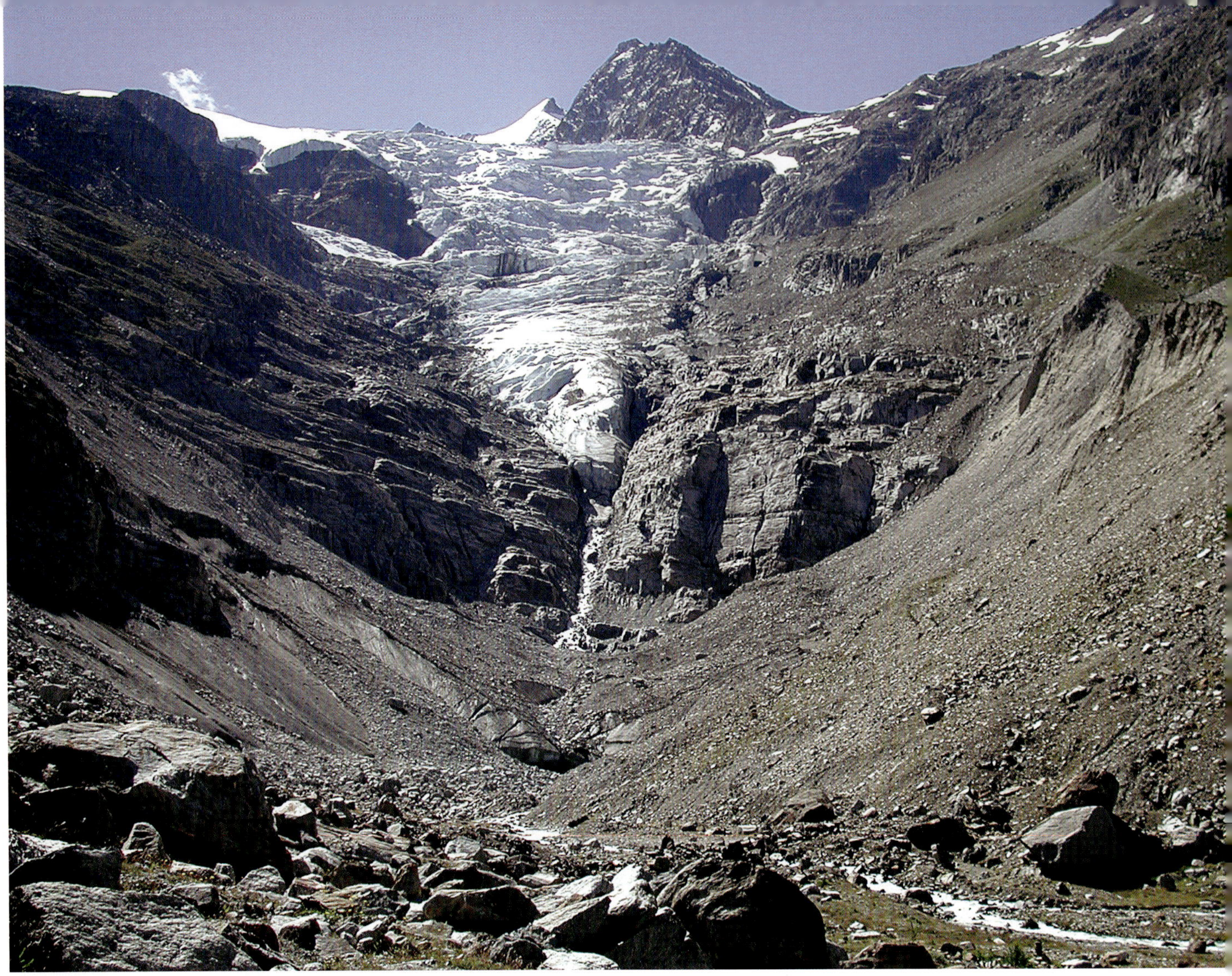

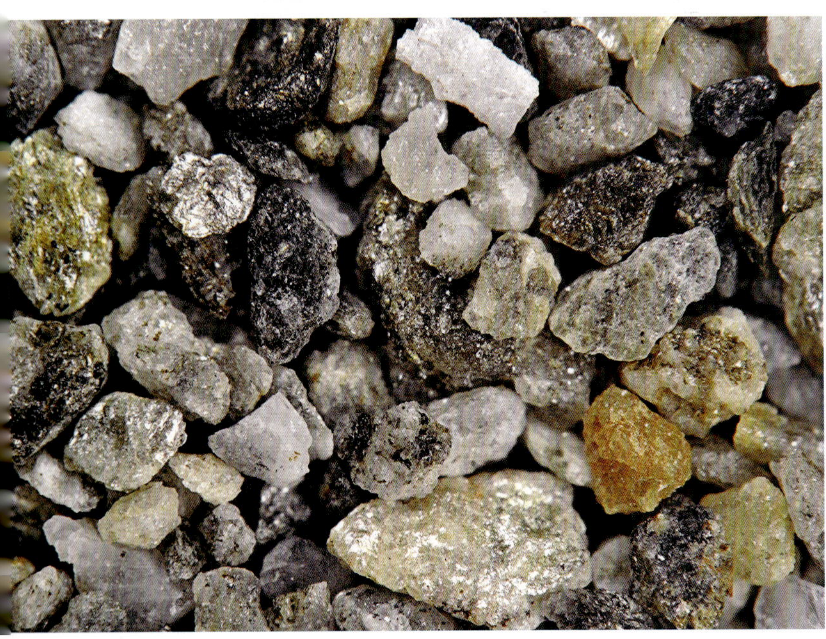

Figure 5.22 Glacial valley with stream, lateral moraines, and sandy alluvial fans. Birthplace of millions of new sand grains. Ried Glacier, Switzerland, 2010.

Figure 5.23 Sand sample from the valley of the Ried Glacier, Switzerland. The grains are edgy and fresh and are composed of fragments mostly of metamorphic and mica-rich rocks from the surrounding mountains. Image width 8 mm.

In addition to gravity, which causes the grain of sand to sink, other forces act on it: the frictional force of the current and the buoyancy force, the latter caused by the current flowing over the grain of sand. Buoyancy forces try to lift the grain of sand, just as air flowing over a wing pulls it upward due to the negative pressure created by its shape. This results in bouncing movements of the sand grains due to the interplay of buoyancy, propulsion in the current, and renewed sinking at another location. Rivers thus have, in addition to suspended loads in the main flow, rolling loads on the bottom and bounding loads in the boundary layer, which compose the billions of sand particles that are transported along the riverbed.

Along the further course of its journey, however, our grain is very likely to be deposited in the flow shadow of some bulge in the river or at the edge of one of its alluvial fans for an indefinite period and gradually covered by other sediments until it is released again one day—for example, by meltwater streams—and allowed to continue on its way. This intermittent process is a characteristic of fluviatile transport. Depending on the gradient and the type of river, and of course on the size and nature of the sand grain, a distance of 62 miles (100 km) can take between one thousand and one million years,[83] whereas small particles (such as silt or clay) may cover the same distance—in suspension—within a few hours. The particle diameter thus quite considerably decides the ratio of all affecting forces and thus has a significant impact on chronological events and the fate of the grain.

Through all this, the quartz grain we have been observing proves to be extraordinarily stable and resistant. In contrast to some other minerals (feldspar has already been mentioned in this context), it has only a vanishingly low solubility in most chemicals and is also not mechanically insensitive to the grain-to-grain collisions that constantly take place in river water. The water delays and dampens the movements, as anyone who tries to throw two balls against each other underwater can easily see.

It is to be presumed that the grain of sand will probably be spit out on the beach of a sea one day. About 80%–90% of coastal sands reached the sea via river transport; the sediment load of the Mississippi River alone accounts for 470 million tons annually.[84] If we now had the opportunity to put the grain under a microscope, we would make the astonishing observation that it looks largely undamaged. True, the corners and edges are broken, and the surface is littered with characteristic marks, but it has not taken on a distinctly round shape overall, as might have been expected. It looks almost exactly as it did when it weathered out of the rock of the high mountain region.

Even very long and powerful currents, such as the Mississippi, are demonstrably not in the position to significantly round sand grains made of quartz—the abrasive losses are only around 5%. Even recurring transport through such currents would not be enough to fully round the sand grain and approach the spherical shape.[85] How can it be that we frequently find all the partially perfectly rounded sand grains on beaches?

Experience has shown that coarser rock fragments and pebbles round off considerably after relatively short transport distances. If one follows a mountain stream downhill on foot, one can observe the effect of the increasing rounding of the edges of the stream pebbles. Theoretically, stream pebbles can even be marked and traced with some effort.

With sand, on the other hand, the situation is much more difficult. Individual sand grains cannot be marked and tracked in respect to their property changes. Sedimentologists are therefore depending on statistical measurements that allow statements to be made about the relative distribution of grain sizes as a function of the distance traveled. If samples are taken at different points in a long stream, there should be—according to the assumption—a clear increase in the average roundness. Such analyses were carried out in the 1940s by the then-leading US geologists William C. Krumbein, Haakon Wadell, and Francis Pettijohn. Krumbein, in particular, worked intensively on the characteristic quantities of shape and roundness for sand grains and introduced corresponding measurement methods and definitions, most of which are still valid today. The measurements on rivers, however, did not bring any clear correlations. As described, no valid function between transport distance and roundness could be found, especially on the Mississippi. The result was explained by the fact that the field measurements were falsified by sand loads introduced from tributaries, so instead investigations were carried out in the laboratory with centrifugal drums. In this way, transport paths of many thousands of miles could be simulated as if in fast motion by rotating drums filled with sediment.

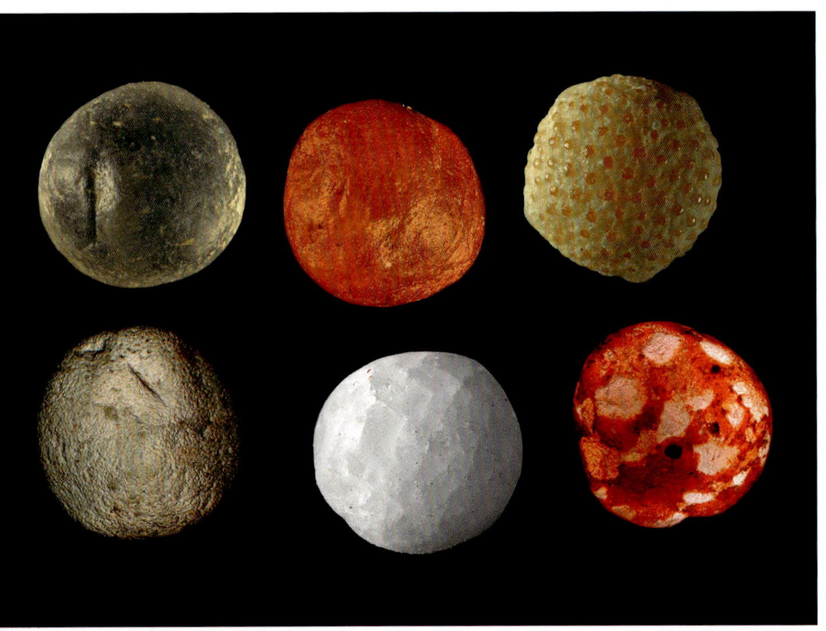

Figure 5.24 Examples of almost perfectly spherical sand grains. Depictions are scaled so they all have the same diameter in the picture. *Upper row*: Santa Cruz Lomas de Arena, Bolivia, diameter = 1,090 *μm*; Santa Cruz Lomas de Arena, Bolivia, diameter = 1,140 *μm*; Michelmas Cay, Australia, diameter = 1,240 *μm*. *Bottom row*: Murzuk, Libya, diameter = 750 *μm*; Is Arutas, Sardinia, diameter = 2,600 *μm*; Colorado sandstone, USA, diameter = 660 *μm*.

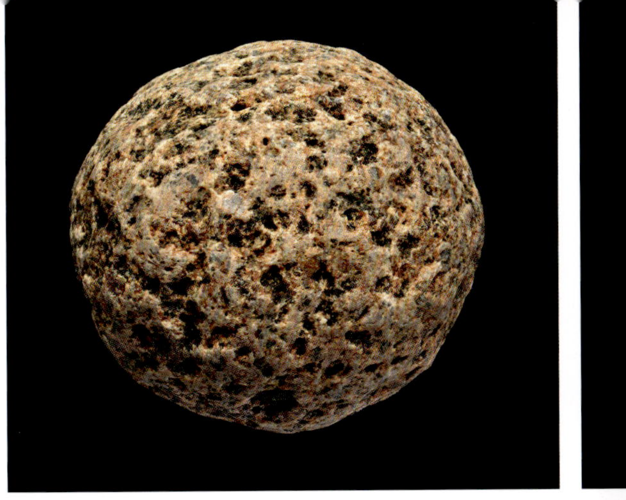

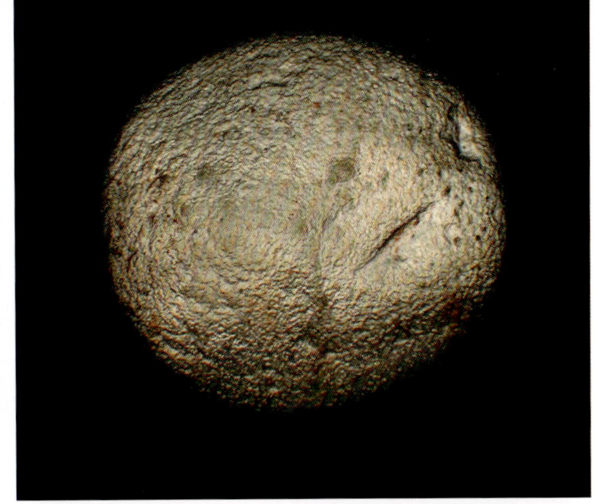

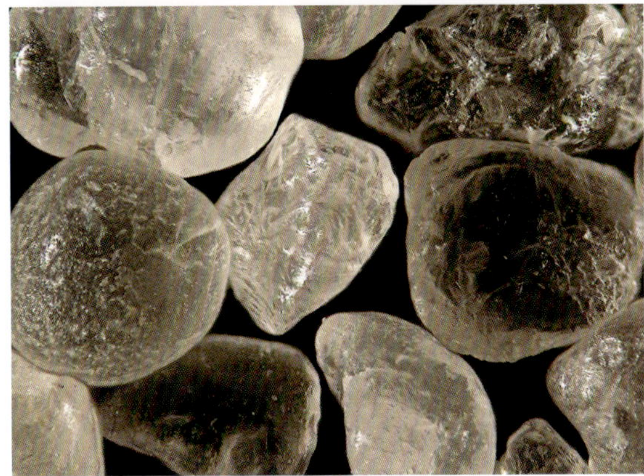

With particles that were gravel sized, reproducible values were obtained. However, it remained difficult to realistically measure the influence of different grain-sized particles on each other in the laboratory at a reasonable cost with a justifiable amount of time and effort. Krumbein used a flood event in 1938, which covered large areas around Los Angeles with debris and sediments, to confirm the findings obtained with pebbles in a large field experiment. In fact, he succeeded in describing the degree of rounding with the help of a formula.[86] Only the sand, surprisingly, did not seem to follow this formula.

The many almost perfectly spherical sand grains found on coasts and in sandstones and rivers did not seem to be sufficiently explained by experimental and theoretical approaches. Geologists eventually turned their focus to ocean beaches, which hurl sediments against each other with higher kinetic energy and move them repetitively rather than linearly due to the back and forth of the waves. While it could be demonstrated that seashores apparently round sand grains faster than rivers can, the apparent paradox of spherical grains was not yet explained.

Figure 5.25 Comparing a round stone with a spherical sand grain. *Left*: Granite sphere from Vanggranit, Bornholm, Denmark, beach, diameter 85 mm. Right: Sand grain from the Sahara, Murzuk, Libya, diameter 750 μm. The volume ratio between both round particles is around 1:1,500,000. The characteristics when it comes to erosion, behavior in currents, etc. are very clearly different among stones, pebbles, and sand grains.

Figure 5.26 Coastal sand from Rerik auf Rügen, Baltic Sea, Germany. Detailed shot. Monomineralic quartz. The grains are distinctly rounded, are partially spherically formed, and have a significantly textured surface. Reflected light. Diameter of the round grain in the left of the picture about 460 *μm*.

5.4 Life Cycle of a Sand Grain, Part 2

The mystery was revealed to researchers over the years, with a surprising solution: the journey of the sand grains was far from over once they reached the beach! Anyone who has ever held a piece of sandstone in their hands already held the solution. Billions of grains of sand are washed into the oceans minute by minute and accumulate in large sedimentation troughs over long geological periods. The grains are successively compacted by the pressure of the heaping masses and begin to bond together through crystallizing mineral solutions circulating in the pore space: sandstone is formed. Our quartz grain that we have been accompanying from the beginning of its journey in the mountains has thus become a component of a newly formed rock, a sedimentary rock. And one day in the distant future, it will most likely be part of the mountain-building process again. Over the course of many millions of years, the steady continental drift will lift above sea level the sandstone that had previously solidified in the depressions, possibly forming the ridge of a newly unfolded mountain range, whose rocks will now be worn down again by wind and weather and may release the grain of sand a second time, releasing it into a cone of rubble at the base of a mountain. The cycle beings again. The grain again makes its way toward the sea, a little more spherical, however, than millions of years before. The riddle has been solved: a life in cycles.

As long as quartz grains do not fall below the top 6 miles (10 km) of the earth's crust, they are basically indestructible. Erosion, transport, deposition, diagenesis, uplift, erosion, transport, and so on. With each of these cycles, the grains continue to round further but lose little volume. It is estimated that about half of all sand grains have undergone up to six of such cycles.[87] The cycles vary in duration from case to case, but their magnitude can be estimated at about two hundred million years (similarly to the Wilson cycle).[88] Thus, if we examine a Devonian sand grain, it would be possible, purely by calculation, for the grain to have undergone ten of these cycles after being released from a granite block of a long-degraded bedrock about 2.4 billion years ago. However, nobody can exactly reconstruct such a sequence through history.

The well-known and spectacular rock formations of the Colorado Plateau in the southwestern USA are a good example of this. When, during the formation of the giant continent of Pangea a quarter of a billion years ago, the seas of the Paleozoic era slowly receded, and the climate became increasingly dry and hot, a huge inland desert was formed in the area in what is now the states of New Mexico, Colorado, and Utah. In the process, sandstones of earlier formations again eroded and decomposed into sand and were deposited by winds (aeolian transport) in basin depressions and reconsolidated into sandstones.[89] The geologist McKee was able to conclusively prove the aeolian depositional conditions for this formation.

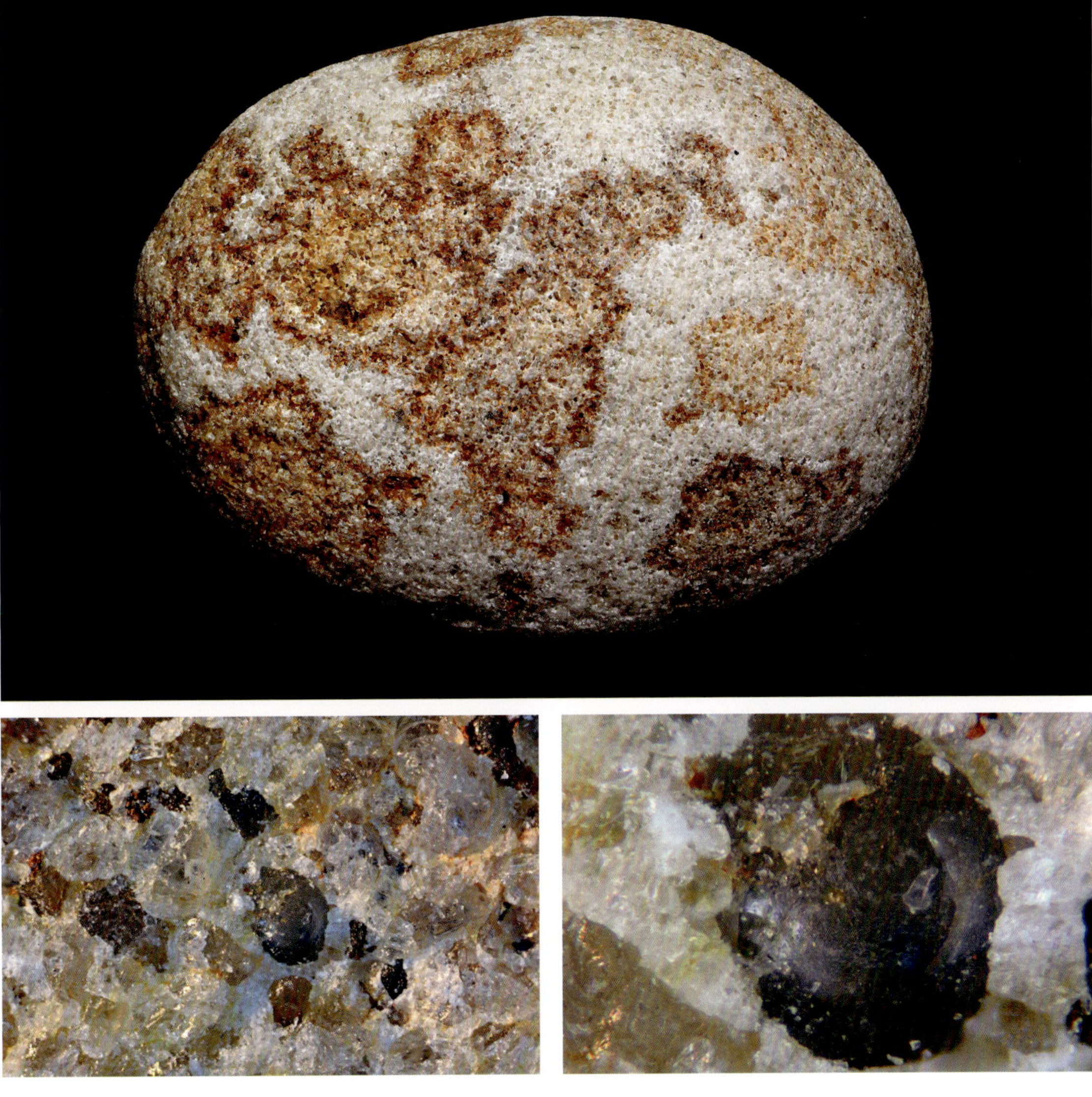

Figure 5.27 Sandstone (image width 48 mm) from the beach debris of N-Fehmarn, Germany. The brown coloration is due to secondary penetration of iron-containing solutions. *Bottom left*, there is an enlargement of the surface of the sandstone with an image width of 2.3 mm. *Bottom right*, the central grain of sand is enlarged once again. It is significantly rounded and would have found itself after weathering in at least its third cycle, if you consider a cycle being an embedding in sandstone. It is very probable to assume two to three previous cycles. The grain is made of quartz; clear quartz tends to appear gray to dark gray when with other grains. Image width 0.65 mm.

The sand grains of the Colorado Plateau today find themselves in at least their third erosion cycle.

It is worth noting that at the time the grain from fig. 5.28 might have been formed, the world was completely different and has been changing continuously ever since, just as the grain of sand, bordering on indestructibility, was exposed to the forces of nature over and over again, with only small adjustments to its shape, as phases of almost infinite rest under the surface of the continent alternated with phases of turbulence, abrasion, and decomposition. Nothing more fascinating can be imagined than looking through a magnifying glass or a microscope at a well-rounded grain of sand and imagining the history behind it, which, after all, can extend over times corresponding to half of the age of the planet.

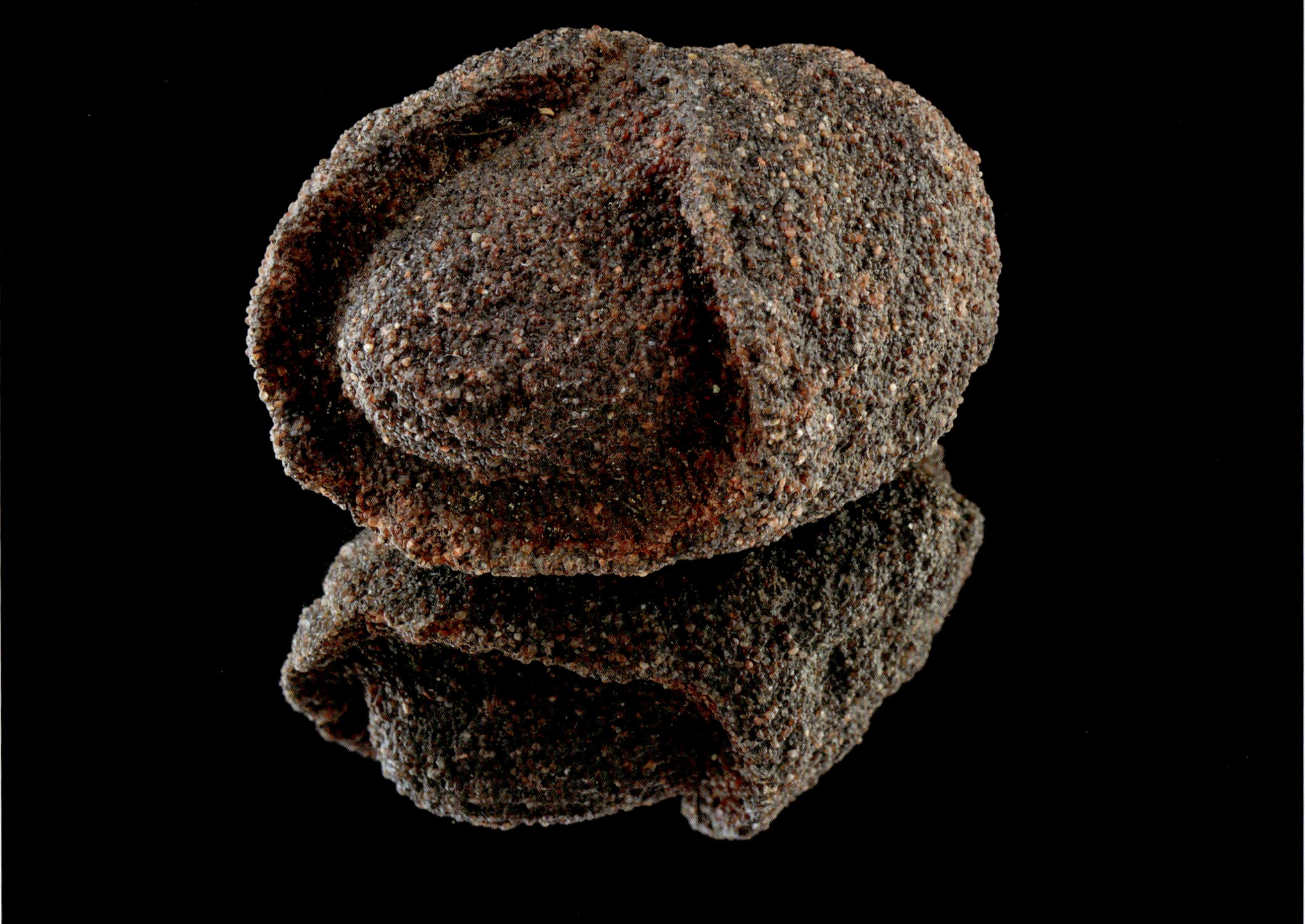

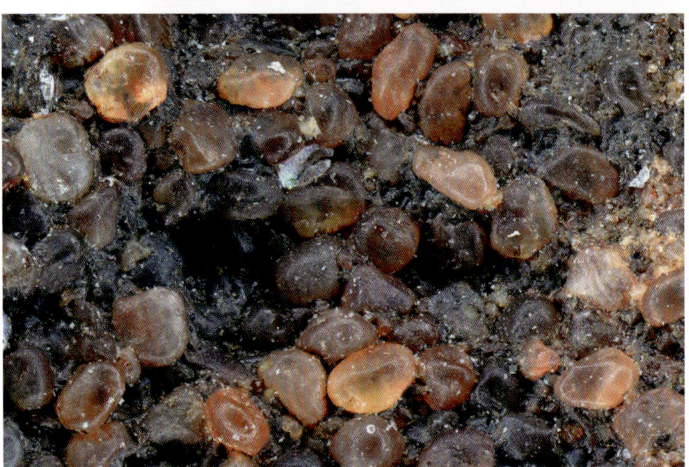

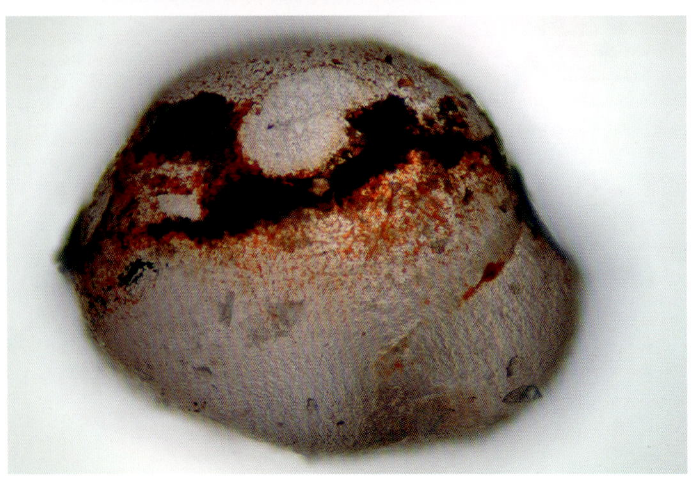

Figure 5.28 Weathered and well-rounded sandstone from the rocky dunes in Colorado, USA. The rock (43 × 32 mm) shows traces of bowl-shaped temperature weathering. Due to the extreme temperature changes from day to night and the resulting volume changes, concentrically arranged microcracks form, which ultimately lead to breaks.[90] *Center left*: Detailed shot of the sandstone surface, with well-rounded quartz grains in a dark matrix (image width 5 mm). The sand grains show in both form and texture traces of aeolian transport and depositional conditions. *Bottom left*: Weathered-out, single sand grain in bright-field transmitted light (diameter 600 *μm*). The typical red coloring of the grains in the dry desert is present only on the surface and is caused by iron oxide. Former contact areas to adjacent grains are visible.

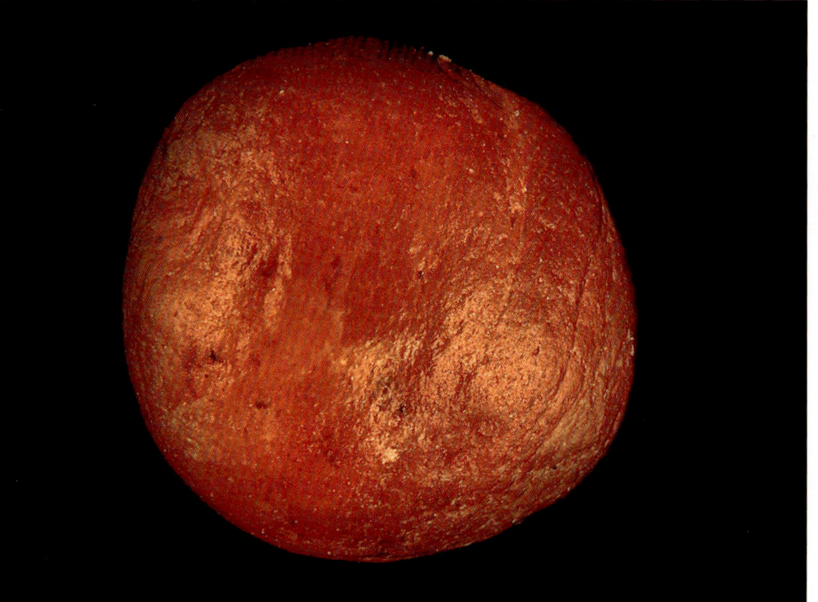

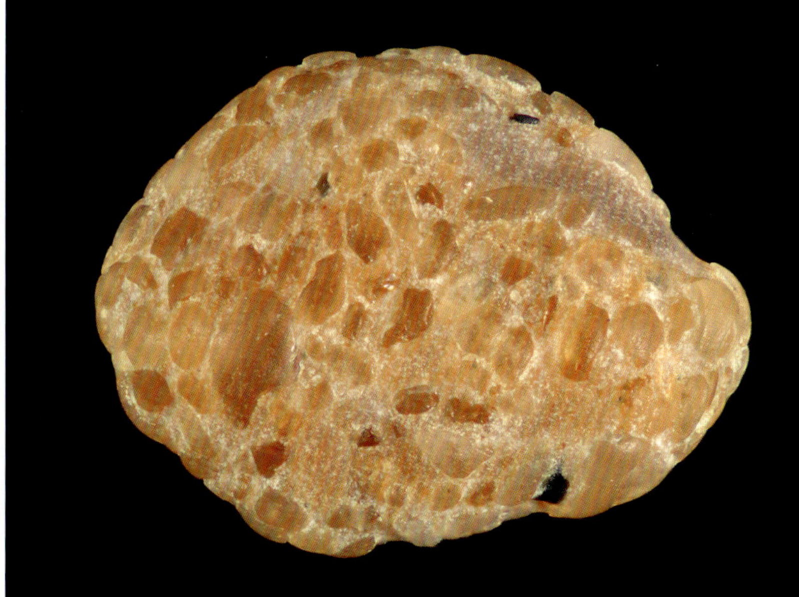

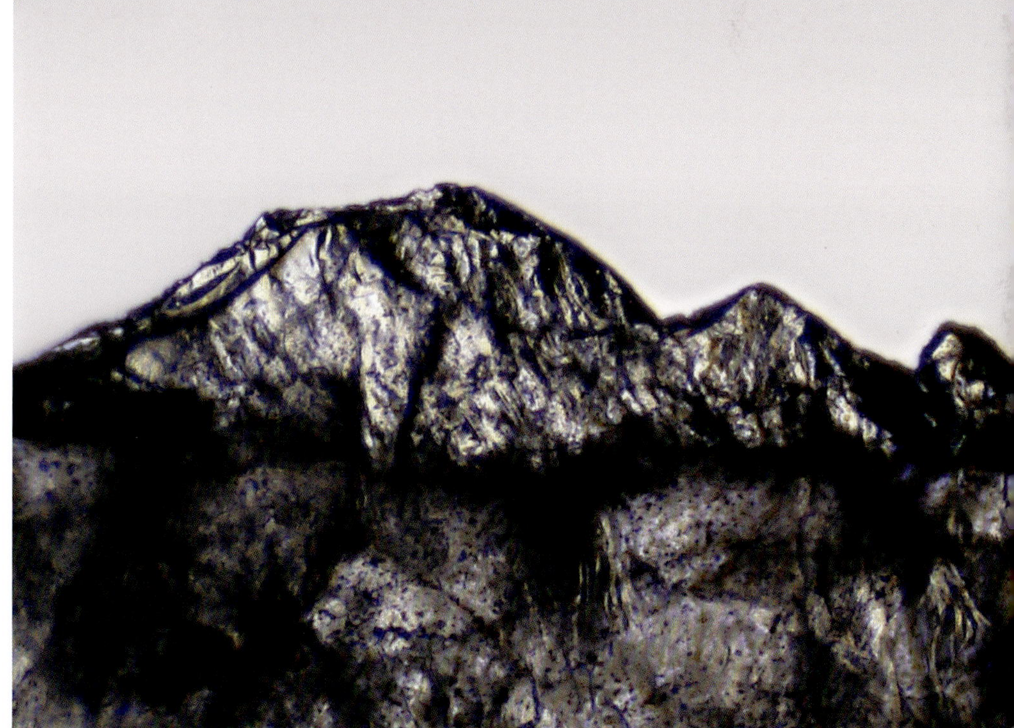

Figure 5.29 No one is in the position to fully reconstruct the history of such a spherically rounded grain of sand. Santa Cruz Lomas de Arena, interior desert, Bolivia. Reflected light. Diameter 1,140 *μm*.

Figure 5.30. Sand grain from Tel Aviv–Jaffa, Israel. The special thing about this grain is that it consists of quartz grains from a preceding cycle. These subgrains are well rounded and probably not in their first life cycle. See also fig. 11.36. Image width 3.8 mm.

Figure 5.31 Sand grain from Naheufer, Bad Kreuznach, Germany. The edge of the grain of sand pictured here looks like the contour of a mountain: the erosion product of which could be the start of any grain of sand. The sand grand therefore symbolizes the abstract process of the life cycle of sediment very objectively and self-referentially.

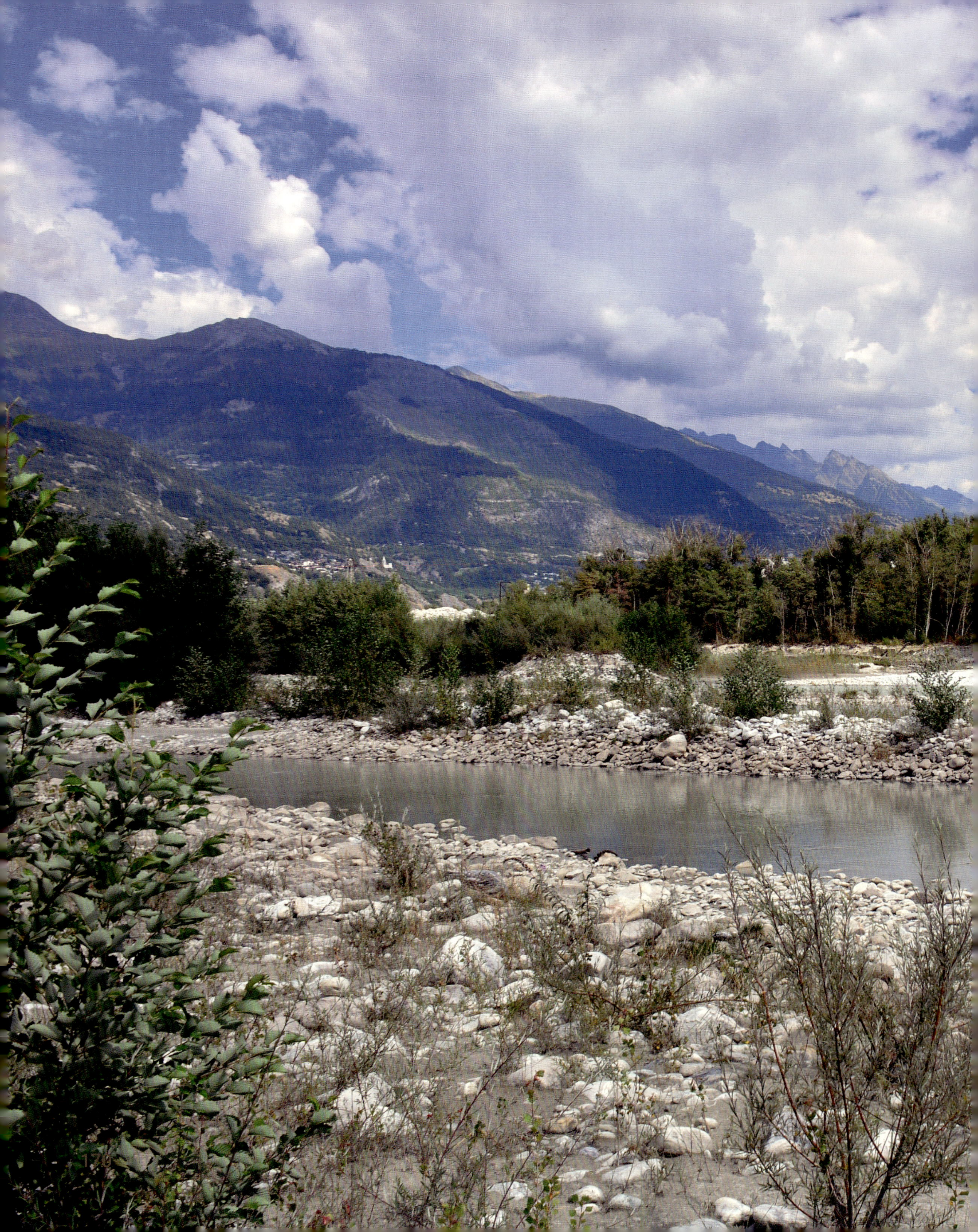

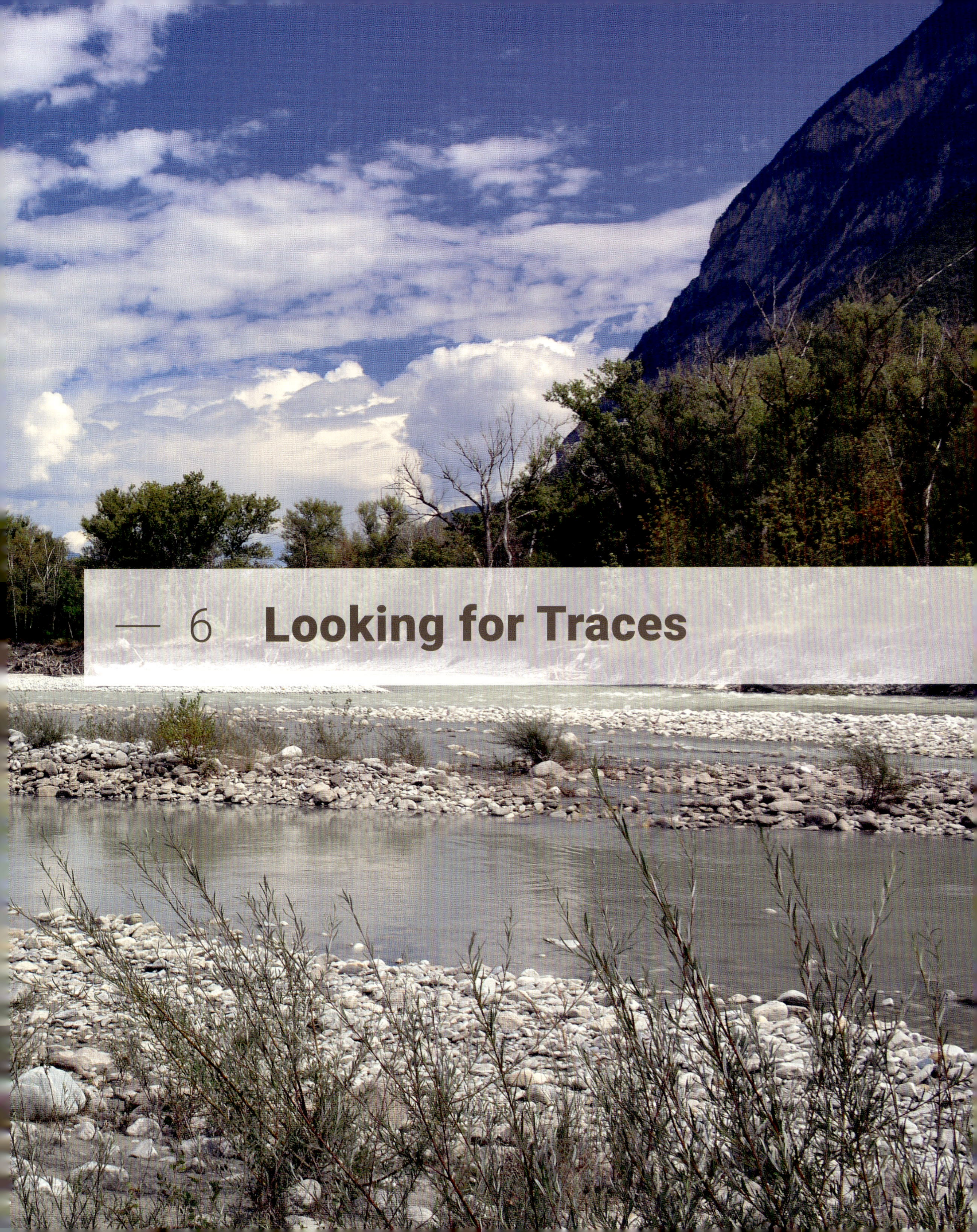

— 6 Looking for Traces

> Even mistakes are useful during considerations because they draw attention and give the sharp-eyed person an opportunity to practice.
>
> —*J. W. v. Goethe*, ***On Granite***

Sand grains, as we have seen, have very strange characteristics. Their creation is not necessarily followed by their decomposition. Many of them spend their existence in cycles of an almost unimaginable duration. But how do we know that? And can we perhaps learn even more about their lives full of change? On our search for traces, we will look at the inside of sand grains and delight in their surfaces. Both are chapters in a sand grain's diary. We now want to open it up and thoroughly page through.

> "Then I took a grain of coarse sand, of the kind used in this country for scrubbing pewter dishes and other household utensils: I brought this grain of sand before the magnifying glass with which I had observed the little animals, and I must say, as a result of my careful measurement by eye, that the axis of the grain of sand was more than a thousand times longer than the diameter of one of the little animals, which I saw in large quantities."[91]

This is what the man reported to the members of the Royal Society of London in a letter on September 16, 1692; namely, Antonie van Leeuwenhoek, one of the first to carry out extensive and systematic investigations with a microscope. Of course, Van Leeuwenhoek's letter was less about the sand and more about the discovery of "little animals," bacteria, whose—at least at the time—unimaginably small size he compared with the smallest relevant individual piece occurring in everyday life, the grain of sand, as a unit of measurement.

But Van Leeuwenhoek was definitely also intensively involved with sand. He was the first scientist to recognize and describe the individual and typical characteristics of each grain of sand:

Previous double page:
Rhone near Siders, Switzerland

> "You can't find two particles in any amount of sand that are completely alike, and even though at first glance they seem to be, they are completely different,"

he wrote much later, in 1703, to the Royal Society.[92] In addition, he was probably the first to consider and experiment with abrasion and the mechanisms for the formation of the shape of the same grains. He also added self-made illustrations of meticulously depicted individual grains, which he drew on the microscope. Van Leeuwenhoek, however, had a very distinct gift of imagination, and so in some sand portraits there are also pictures of tiny, humanlike beings, which he thought he recognized in the respective grain and which he also described in detail in his notes. Beings in the inner world of grains of sand.

Not much less imagination was needed in our last chapter, in which we told the hypothetical story of a very real grain of sand—the path it may have taken in the course of one of the cycles in its long existence. But we have examined only one of the many conceivable paths of the grain. It could just as easily have fallen onto a glacier after being released from the original rock, to remain frozen in perpetual ice for hundreds of thousands of years. Or it could have broken away from a sandstone formation in the mountainous region on the edge of the Sahara after millions of years and been carried away by the wind. Also, it could have taken the direct way into the sea, as a former component of a rocky cliff at the ocean's coast. Many more scenarios are imaginable. Each grain of sand has its own history.

At this point at the latest, we want to ask ourselves again whether these stories can also be told in reverse. Is it possible to extract a story from a grain of sand? Can its origin be revealed? We already had to answer that question with "no." But we do not want to leave it at that. Events, especially geological events, leave traces. We will follow these traces and see that at least certain conclusions are possible, which allow us to get a small insight into the individual past of single grains of sand. Thereby two basic questions can be pointed out, which will be treated separately and in detail in the following:

Question 6.1 Which rock did the grain originally come from? Was it released from a sedimentary rock after a preceding cycle, directly and primarily for an igneous rock, or was it subjected to metamorphic processes? Likewise, hydrothermal formation in veins or fissures would be possible. These questions can be answered with some accuracy mainly by examining the interior of the grain. The grain shape and other factors are also to be considered.

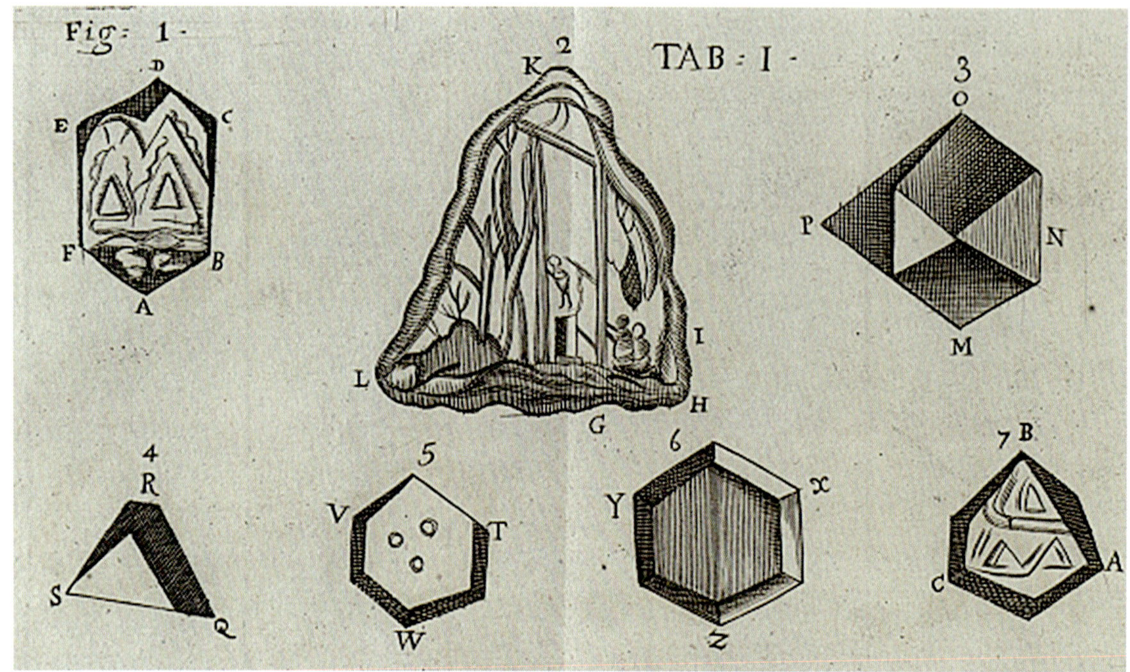

Question 6.2 To which milieu, to which environment can the grain be attributed? Was it once transported by wind, river, or glacier? Did It come from a desert, a lake, or the sea, and how was it finally deposited? For this second set of questions, we would instead have to look at the exterior of the grains with all its traces and structures, which as a result of the attacking physical forces and chemical effects have inscribed themselves on the surface as in a diary.

Figure 6.1 Depiction of sand grains according to Van Leeuwenhoek. Pay attention to the middle grain, which seems to contain a human form in the middle and in the bottom right corner. *Philosophical Transactions* 24 (1703): 1537. © The Royal Society.

6.1 Rock of Origin, or, The Inner World of Sand Grains

So let us first deal with the rock of origin and the clues by which it can still be recognized in the grain of sand. In the case of quartz grains, we are interested exclusively in the imperfections, all the small deviations from an ideal crystal. Let's dare to look inside the sand. To say it up front, we will unfortunately not find the small Leeuwenhoek creatures. In order to be able to make statements about the rock of origin, we have to look for characteristics that cannot be changed or overwritten by transport and abrasive influences, or only insignificantly. We restrict ourselves to such characteristics that can be determined without too much effort with instruments, with just the help of a microscope and its polarization device. We will examine four characteristics in turn, three of which deal with the interior of the grains.

- inclusions in the sand grains
- behavior under polarized light
- crystalline structure
- grain shape

6.1.1 Inclusions in Crystalline Sand Grains of Quartz

If we had a fully round, homogeneous, and monocrystal and grain in front of us, of course there would not be any clues we could trace back. But almost all quartz grains include foreign substances, so-called inclusions in their crystalline matrix, the witnesses that portray the creation history of the crystal. These inclusions make up somewhere between 1% and 5% of a host crystal.93 During the growth phase, crystals can take up and encompass foreign bodies, which are crystalline, gaseous, liquid, or glassy amorphous. Also, during recrystallization processes, foreign particles or minerals are formed in the host crystal.

Liquid and gaseous inclusions are particularly interesting. These usually form during the formation of the magmatic or volcanic rock deep inside the earth by trapping tiny amounts of liquids in the constantly growing crystal. They remain sealed there as a time capsule for millions of years. The dimensions of these fluid inclusions are on the order of a few thousandths of a millimeter. Many inclusions are even smaller, with dimensions of less than 1 micrometer, and can hardly be recognized even with a microscope.

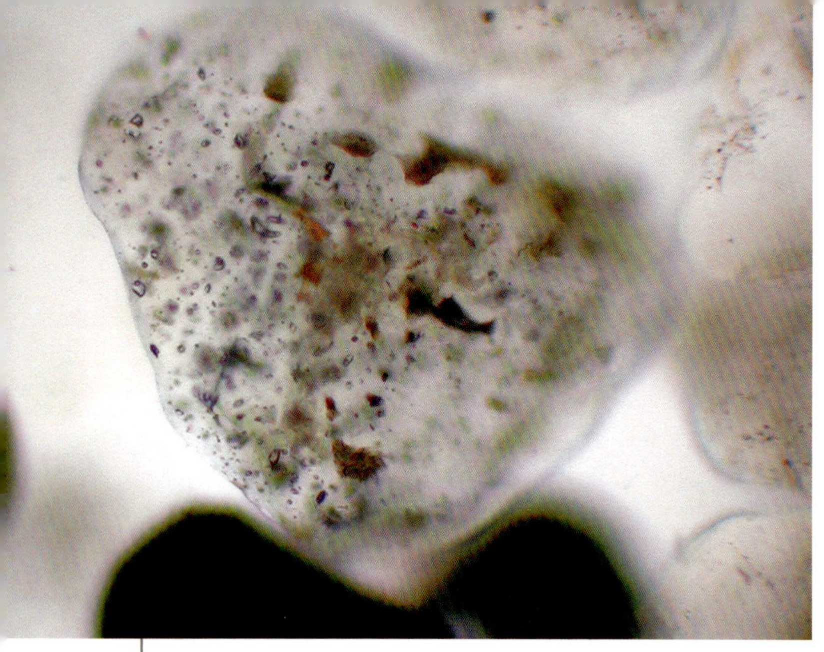

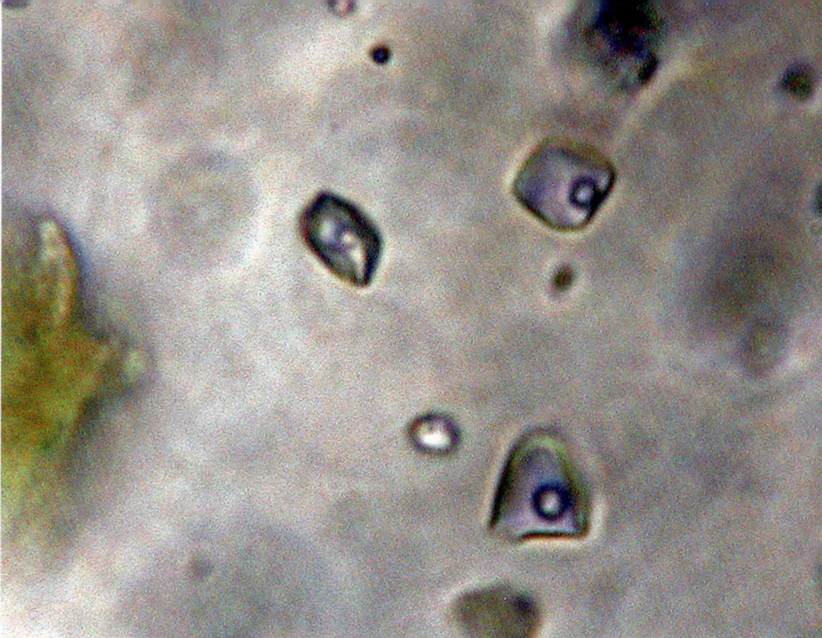

The opacity of milk quartz is due to millions of such microinclusions. According to Roedder's estimates, their density there is about one billion inclusions per cubic centimeter![94]

Solid inclusions are usually somewhat larger and in rare cases may have dimensions of up to a millimeter or more. More common are multiphase inclusions, which for example consist of a liquid droplet and a gas bubble. This type of fluid inclusion is sometimes called "libelle" (from Latin *libella* = water level).[95]

In the microscope we can observe with amazement that the fluid inclusions are in permanent motion. This is due to the molecular motion in very small fluid inclusion. It is incredibly fascinating and almost uncanny to imagine that the tiny and often-frantic movement of the small gas bubbles inside such crystals and sand grains has been going on for millions of years, unobserved, hidden in the deep darkness of the earth's interior and in massive rocks. Billions and billions of micromovements in billions and billions of sand grains and crystals above and below the earth's surface, without pause.

Van Leewenhoek's fantastic drawings of the inner worlds of sand grains were nothing more than representations of inclusions with their sometimes-bizarre shapes. Admittedly, we must give him credit for the rather simple optical equipment, which, however, precisely for this reason arouses our admiration and, at least in the field of biology, delivered amazing results. The first systematic research on mineral inclusions appeared after 1900.

The goal of all inclusion investigations is to obtain information from the substances within the inclusion. Suitable measurement methods can be used to determine physical data such as density and transformation temperatures. Depending on the method used, the chemical composition of the substances can also be measured directly or derived indirectly from optically obtained data.

Figure 6.2 *Left*: Sand grain with many inclusions from Bjerregard, Denmark. *Right*: Fluid inclusion in the sand grain pictured to the left. Highly detailed shot of a sharpness level in the center of the grain. The fluid inclusion to the top right in the picture has a diagonal of only 4.8 *μm*, the small bubble within it a diameter of about 1.5 *μm*. The small bubbles find themselves in constant motion!

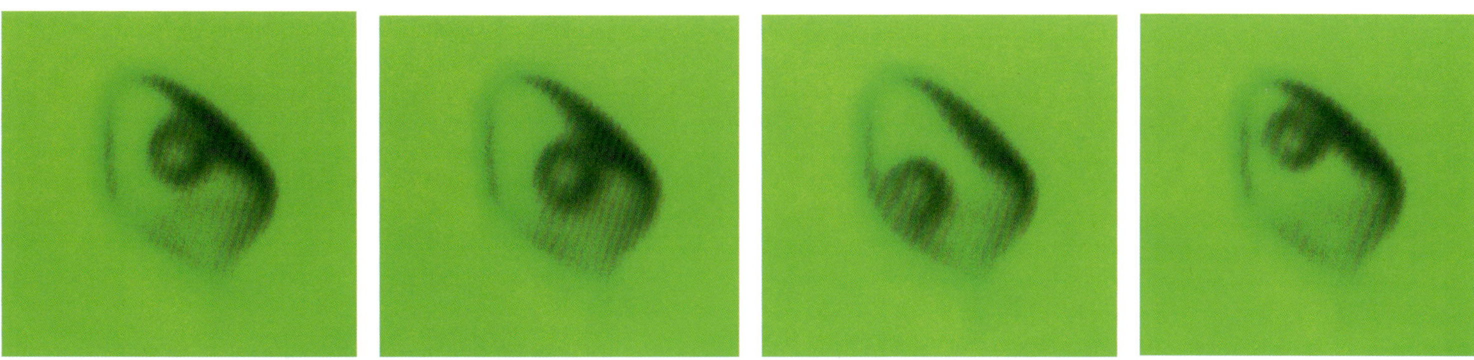

Fluid inclusions are multiphase inclusions that resulted from an originally hot and homogeneous fluid being trapped in the growing crystal during rock formation. During cooling in geologic time, the volume of the cavity remained largely the same, while the fluid physically contracted more, so that negative pressure created a gas phase, which is now visible as a bubble in the fluid. Assuming that the inclusion did not undergo any external changes in the further temporal sequence, an interesting and multifaceted piece of information is thus available. In addition to the chemical composition of the fluid, the formation temperature of the crystal or the inclusion time can be reconstructed relatively easily with the aid of so-called microthermometry. The process is reversed, as it were, by heating the sample together with the vial on a special heating table under constant observation through the microscope until the gas bubble disappears. The temperature determined in this way is called the homogenization temperature and, together with other data (e.g., degree of filling, salinity, isochoric course), is one of the most important pieces of information for inclusion research, since it allows conclusions to be drawn about the formation conditions of the inclusion. With such methods, even statements about the formation of sedimentation processes are possible. For example, on a Cretaceous sandstone, the homogenization temperature of accretionary seams on sand grains could be determined to be 122°F (50°C). This corresponds after conversion with the geothermal coefficient to a sinking depth of 3,281 feet (1,000 m) at the time of cementation (example according to Almon in Füchtbauer 1988[96]). It is almost unbelievable that the small bubbles tell the geologist that the sandstone was 3,281 feet (1,000 m) below the ground when they were formed.

If there are liquid inclusions, their salt content (salinity) can be measured. Pure water freezes at 32°F (0°C).

Figure 6.3 Fluid inclusion with gas bubble in a sand grain from the Ried Glacier stream, Switzerland. The inclusion was shot three times close together. The movement of the small gas bubble from picture to picture is easily recognizable. The inclusion has dimensions of only 5.5 × 7.3 *μm*. Transmitted light with contrast-increasing green filter.[97].

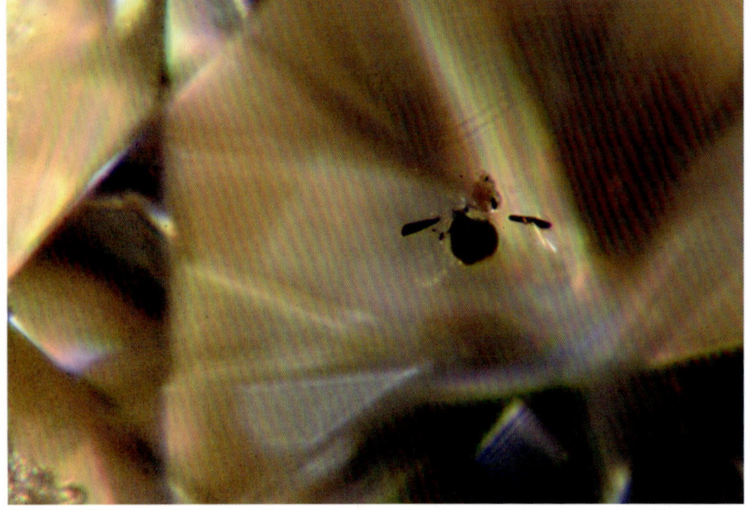

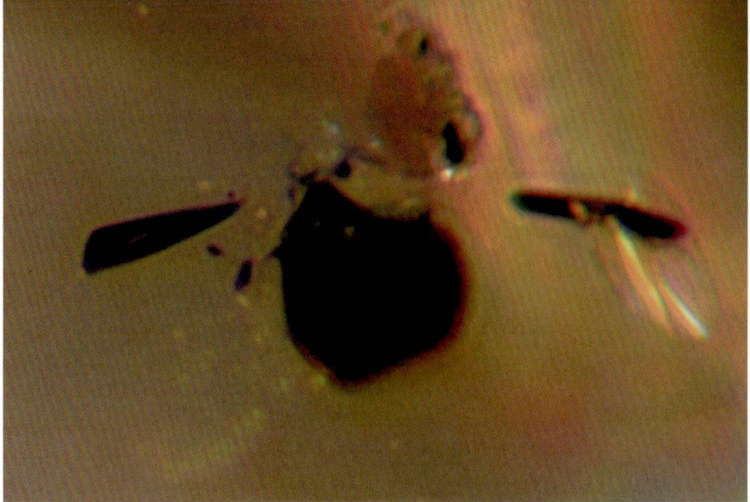

If the inclusion contains dissolved salts, these are known to lower the freezing point. If the freezing point of the liquid inclusion is known by measurement on the heating table (cryometry), the salt content can be calculated retrospectively and provide important information, for example, for deposit research. With suitable methods, larger inclusions can be opened to analyze the contained gases and liquids directly. Spectroscopic methods (Raman spectroscopy) allow statements about the chemical composition to be made through direct measurement down to a size of 2 *μm*.

The scientific preoccupation with mineral inclusions found and finds use in the research of warehouses of rare and important minerals such as wolfram and fluorite or copper. Sometimes even extraterrestrial material can be researched for its formation conditions through inclusions. Inclusion research is also still used in paleoclimatological research for tracking oil stockpiles.

In the production of glass, especially special glass, pure quartz sands are needed, which are extracted in open-pit sand mines. Any kind of additives, such as inclusions, change the optical properties of the glass. Suitable "glass sands" are therefore determined by inclusion analyses, and their composition is monitored.

Gemstones, especially diamonds, are considered particularly valuable if they have as few inclusions as possible or, in the best case, none at all. Diamonds are called "flawless" if they show no inclusions when viewed through a 10× magnifying glass. They fetch the highest prices and are rarer the larger they are. However, it should be noted that typical inclusions, with their characteristic compositions, distributions, and derivable origin, can be a guarantee for the authenticity of diamonds and other gemstones. Although it is already possible to specifically produce fakes with equally fake inclusions, these can usually still be identified by experts.[98] For some gemstone types, among them corundum and diamond, specific inclusion typologies are known, which can provide a fingerprint of the associated deposit (i.e., the place of discovery).

Figure 6.4 Inclusion in a diamond. Due to the extremely high pressure and temperature conditions in the formation of diamonds in the earth, there are no liquid inclusions. In the picture, there is a multipart inclusion probably made of graphite or a graphite liner. Jewelry diamond Montblanc special edition. Image width: *left*, 640 *μm*; *right*, 160 *μm*.

A completely inclusion-free ideal diamond, on the other hand, is virtually indistinguishable from an equally inclusion-free fake.

For the enthusiast, small inclusions give the stone an almost personal touch, which individually and unmistakably preserves the geological history of formation in the stone. This is especially true for diamonds, with their very unusual and interesting history of formation. This can be divided into three chronologically separate sections. In a Precambrian early phase, magmas rose into subcrustal regions and formed expansive ovens in which olivine aggregates accumulated. In a much-later phase, probably in the Carboniferous period, the ascent of the magma masses under high pressure occurred in fissures of the earth's crust, with differentiation of the magmas and transformation of the olivine nodules and peridotites to eclogite rocks. At pressures up to over 100,000 bar and temperatures around 2,372°F (1,300°C), diamonds formed in periodic spurts. Apparently, the diamonds crystallized from fluid carbon dioxide. After many millions of years passed again, the diamonds were finally ejected along with the original rock in events resulting from volcanic eruptions. Kimberlite developed, a brecciated debris rock with remnants of eclogite as the parent rock. This third phase took place in the South African diamond deposits during the Cretaceous period.

The diamond originates from inaccessible depths of the earth. Its inclusions allow a unique insight into its exotic formation conditions and therefore contribute implicitly, but nevertheless quite substantially, to its almost mythical fascination. They were formed during all three phases of diamond genesis and are never liquid but always solid, and mostly crystalline in nature.

Different generations of inclusions can be distinguished in crystals. A distinction is made between primary (syngeneic), secondary (epigenetic), and pseudo-secondary inclusions. Primary inclusions are formed during the growth of the crystal and are therefore often located on its growth zones. Secondary inclusions, on the other hand, form only after crystallization of the host crystal has occurred. They are usually bound to forms resulting from a later transformation of the crystal. These include cracks from mechanical destruction due to metamorphic processes or chemical remodeling due to corrosive processes. Secondary inclusions often occur in large numbers at mended crack surfaces. Pseudo-secondary inclusions are formed during growth but show characteristics of secondary formations.[99]

In order to gain information about the formation of a grain of sand, it is necessary, or at least helpful, to be able to distinguish between primary and secondary inclusions. In this case, one is in the position to make statements about the crystallization itself and the time afterward, which portrays the further course of its development history.

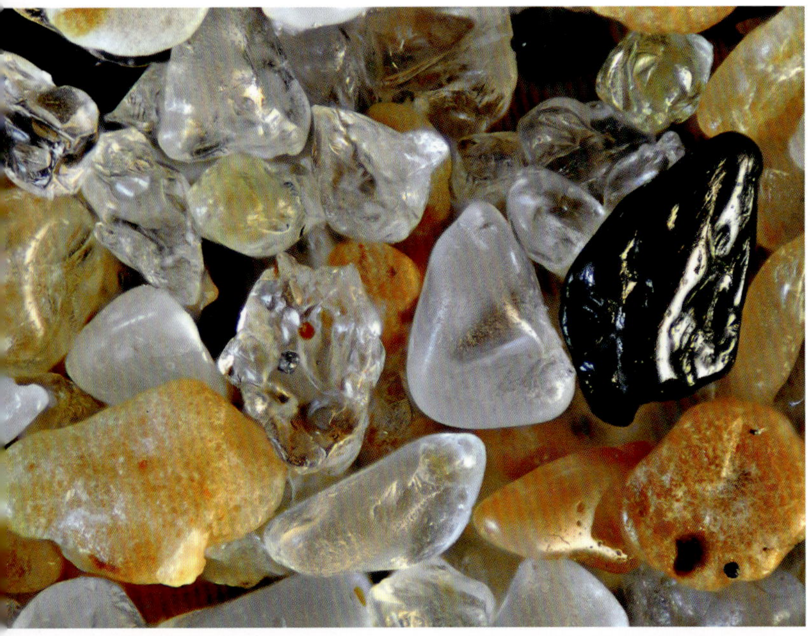

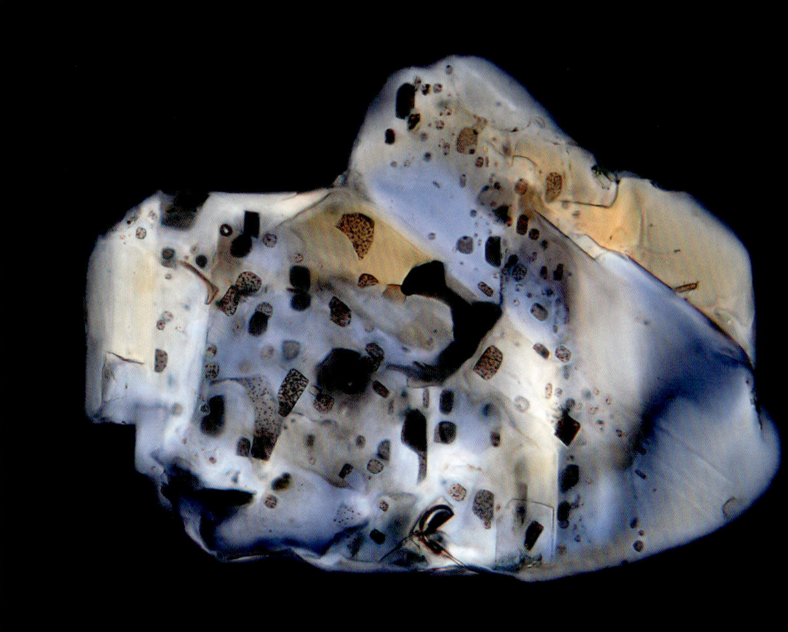

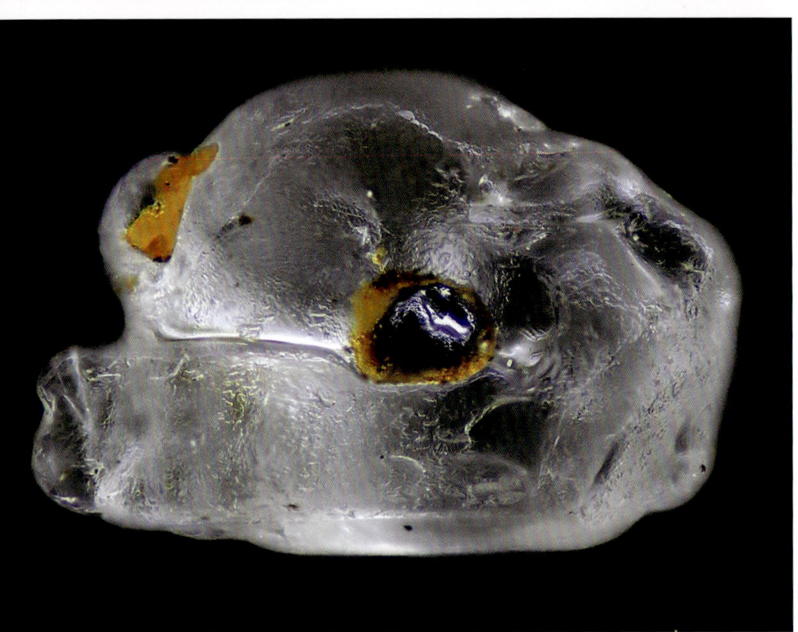

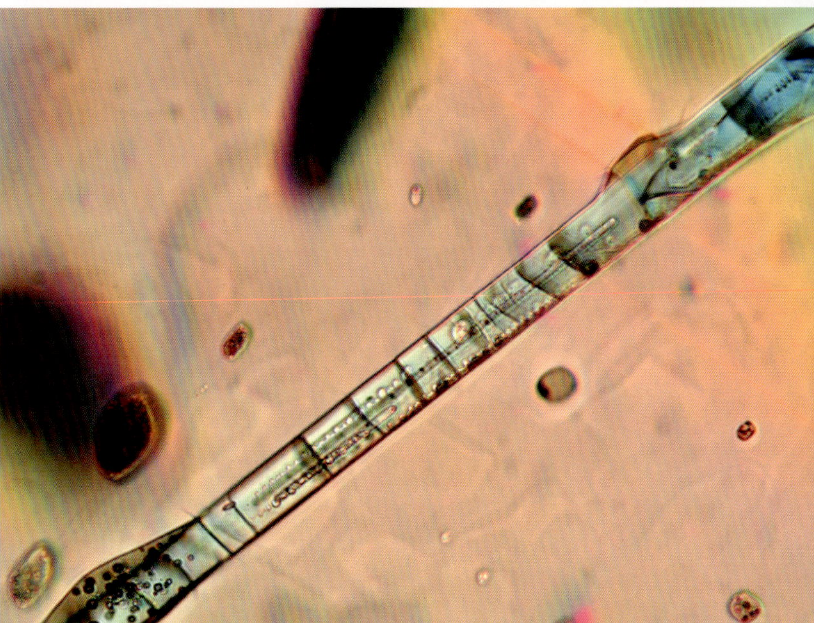

Figure 6.5 Marine sand from Tangalle, Sri Lanka. The middle grain of sand contains two significant inclusions, and the bottom one is of metallic nature. These are primary inclusions that developed during the crystallization of the grain of sand. The grain has a width of 400 *µm*.

Figure 6.6 Sand grain from Whatipu Beach, Auckland, New Zealand. The grain is made of quartz and shows characteristics of chemical erosion. The strong and typical rounding points to a history in an aeolian environment. The central and primary formed inclusion, from an ore mineral, is easily visible, as is its yellow oxidization product. Transmitted light, grain size 500 × 710 *µm*.

Figure 6.7 Sand grain from Lovina, Bali, Indonesia. The large number of multiphasic, solid inclusions is clearly recognizable. In the middle, there is an opaque, dark, melt inclusion. Grain size 315 × 420 *µm*. Transmitted light, crossed polarizers.

Figure 6.8 Detailed shot of a complex, multiphasic inclusion in a sand grain of clinozoisite in transmitted light under crossed polarizers. Diameter of the tube-shaped inclusion 12.6 *µm*. Lovina, Bali, Indonesia.

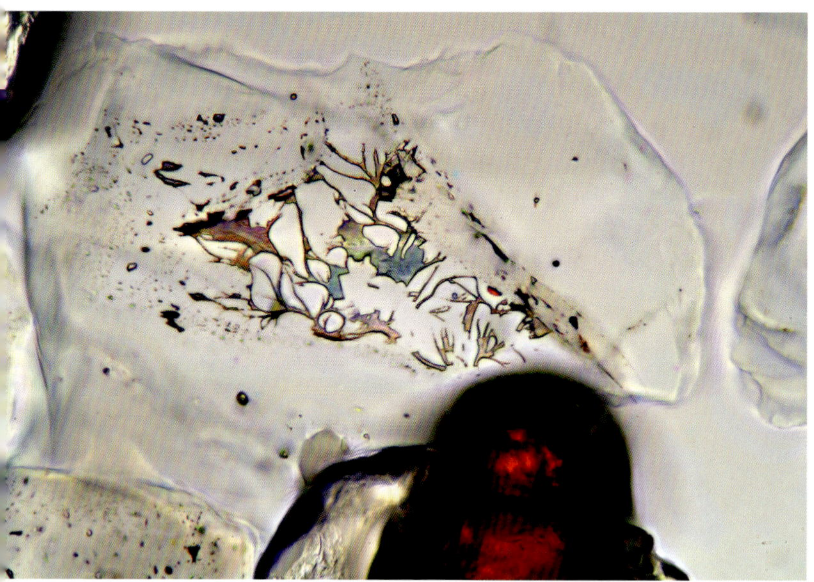

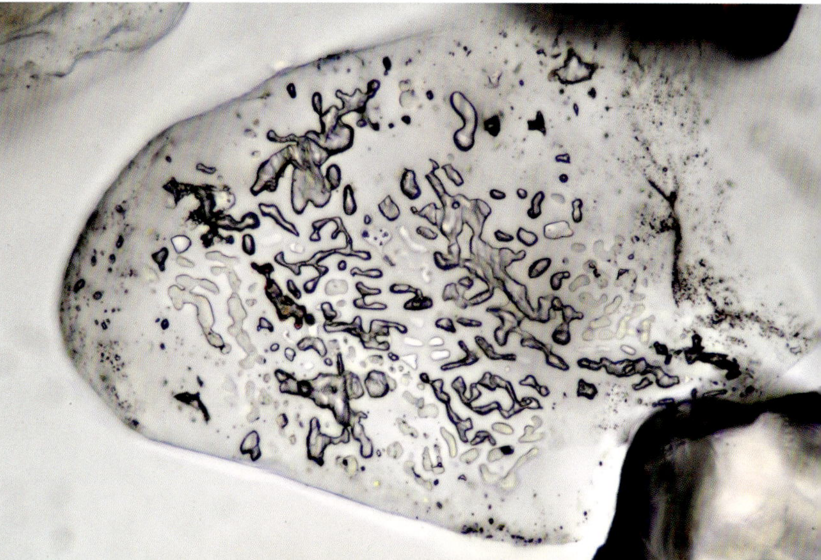

This becomes possible only when both inclusions are present in one sand grain. To be thorough, it's worth mentioning that there are many other minerals along with quartz that contain inclusions that can be diagnostically beneficial.

Here are a few criteria by which primary inclusions can be recognized.[100]

— presence of single inclusions, no groups, in an otherwise inclusion-free grain
— large dimension (about 10% of the crystal)
— groupings with random distribution of few inclusions in the space
— straight-line arrangement associated with growth zones of the crystal

Criteria for secondary formed inclusions include the following:

— very thin or shallow inclusions
— Massive accumulation of inclusions on the surfaces unrelated to crystal orientation. They reflect the healing processes of the cracks.
— curved or bowed surfaces that do not correlate with crystal directions

We should insert a particularly impressive example of insights that investigations of sand grains and their inclusions can provide at this point. It concerns the history of the zircons from the Australian Jack Hills: we already briefly dealt with it in chapter 3. The Jack Hills of Western Australia are located about 500 miles (800 km) north of Perth and belonged to the Yilgarn Craton.

Figure 6.9 Sand grain from a mineral placer on the beach near Bjerregard, Denmark. The inclusions are of secondary nature and probably came to be alongside the healing of a crack. Grain size 170 × 320 *μm* (a brown rutile on the bottom edge of the picture).

Figure 6.10 Sand grain from Bjerregard, Denmark, with a flat (planar) arranged group of secondary, monophase inclusions. Image size 220 × 320 *μm*.

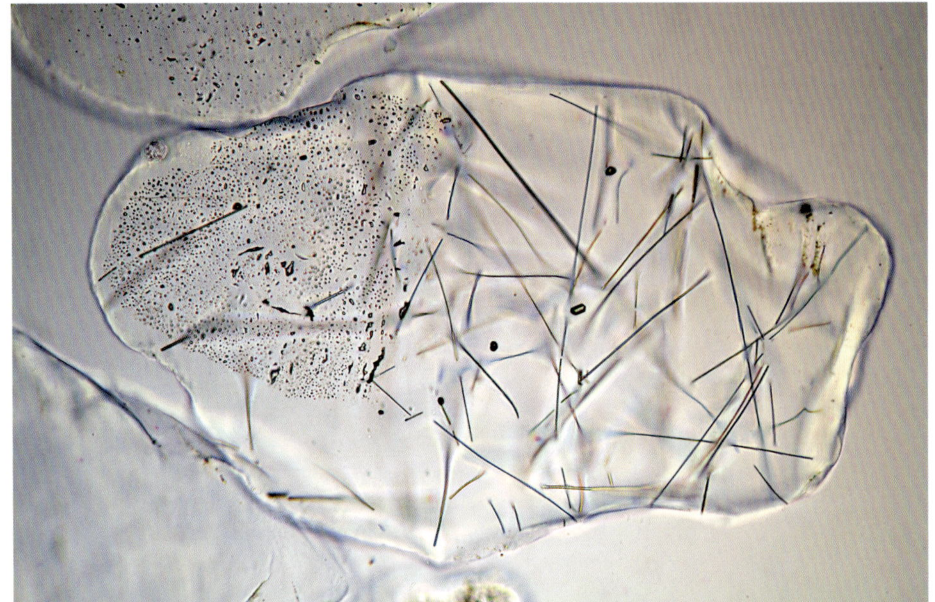

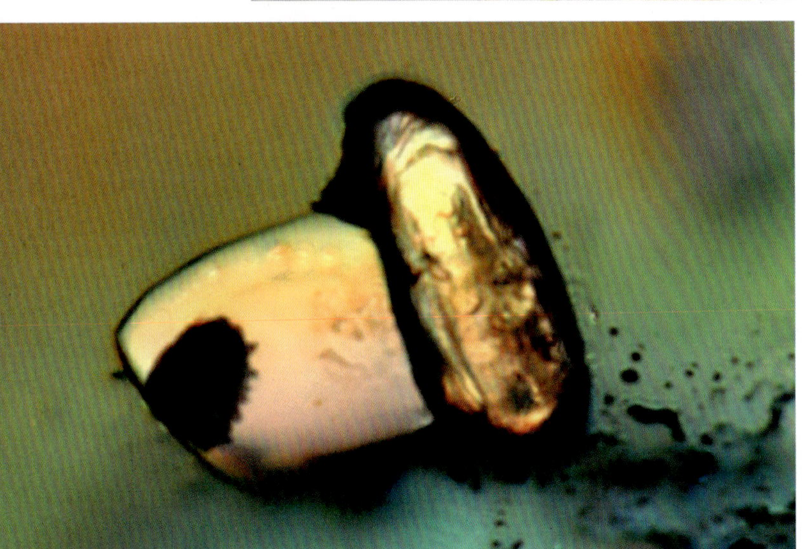

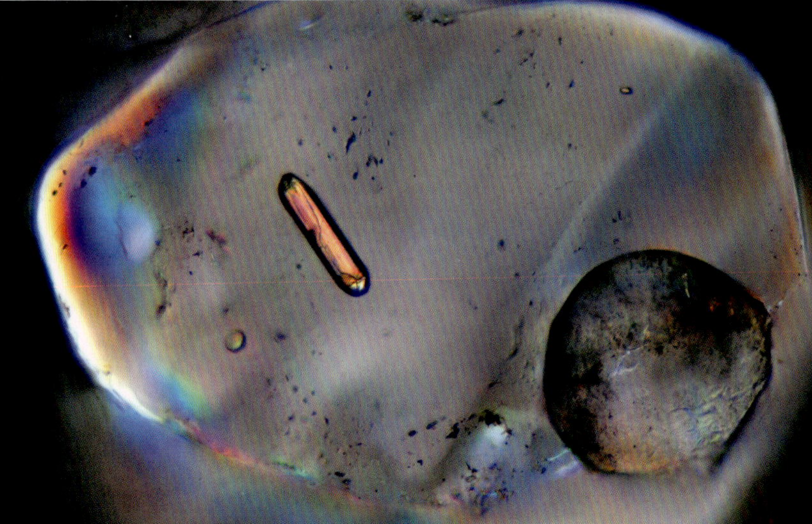

They were formed about three billion years ago and are now largely eroded. Erect, nearly vertical layers of red and orange metasediments consisting of sandstones and conglomerates are found there. Alternating layer packages with poor sorting suggest that they are sediments of former alluvial fans and fluvial deposits. The sediments are also metaphorically altered, as inferred from the presence of minerals such as disthene. The sandstones are predominately detrital quartz. What makes the formation so outstandingly important is the fact that much-older zircon crystals are found in the metasediments, which are about three billion years old.

Figure 6.11 Quartz grain from Bjerregard, Denmark. On the left side, a field of secondary inclusions can be seen, which probably formed in the healing of a crack. To the right side, rutile nails of primary nature. Grain size 280 × 475 *µm*. Transmitted-light bright field.

Figure 6.12 Crystalline inclusions grown into each other in quartz grain from Fehmarn, Germany. Image width only about 80 *µm*. Transmitted light, crossed polarizers.

Figure 6.13 *Right*: Quartz sand grain from Bjerregard, Denmark, with crystalline primary inclusion. It is an idiomorphic apatite crystal (length 60 *µm*). At bottom right in the grain, an unusually large inclusion. Transmitted light, crossed polarizers.

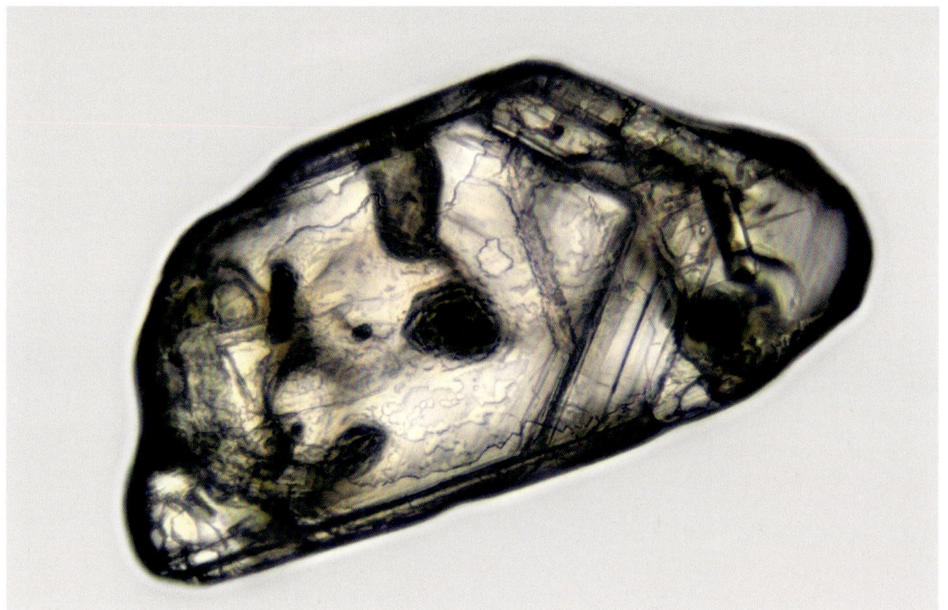

As already mentioned, zircon is ultrastable against erosion factors of any kind. Some zircon crystals from Jack Hills have the additional special feature that they hide small diamonds as inclusions.

Researchers crushed large quantities of parent rock to extract the crystals in sufficient numbers for study. The zircons are more than four billion years old and thus even older than the 3.6-billion-year-old gneisses in which the sediments are embedded. Some of them, with an age determined by uranium decay of 4.404 billion years (at Erawandoo Hill), are the oldest pieces of earth matter known. Since then, they have been the subject of intensive research by international teams of scientists. But the zircons hold even more amazing information. Measurements of the isotope ratio of oxygen (^{16}O to ^{18}O) of gas inclusions showed that the zircon-bearing rocks must have encountered water during their formation. Since the zircons themselves are detrital, they were therefore eroded from even-older parent rocks through weathering processes. Further investigations showed that the parent rock of the zircons had granitic character. This is astonishing because granite can be formed volcanically only in the context of continental crusts. Until now, however, it was assumed that continental crusts existed much later. Moreover, diamonds form only very deep within the solid subsurface of continents under very high pressure. Thus, at a time when the parent rock of zircons solidified from the melt, the earth already possessed continental plates or their fragments, and there was water on a significant scale. If the hypotheses from measurements and observations of the zircons are to be confirmed, the earth was already covered at this early time by a primeval sea and at least partly by continents.[102] The earth must have experienced, therefore, an appropriate cooling to hold liquid water. This possibly came from carbonaceous chondrites, not from comets as had been previously assumed.

Figure 6.14 Sand grain formed from a detrital zircon crystal rounded at the edges. A central foreign crystal inclusion is easily visible in the center of the grain, as well as further, seemingly primary inclusions. In addition, the growing areas of the crystal formation phase are easily recognizable on the right side of the crystal. Heavy mineral placer near Hohwacht, Germany. Dimensions of the crystal are 260 × 150 *μm*. This length/width ratio, under 3, points to an igneous parent rock.[101]

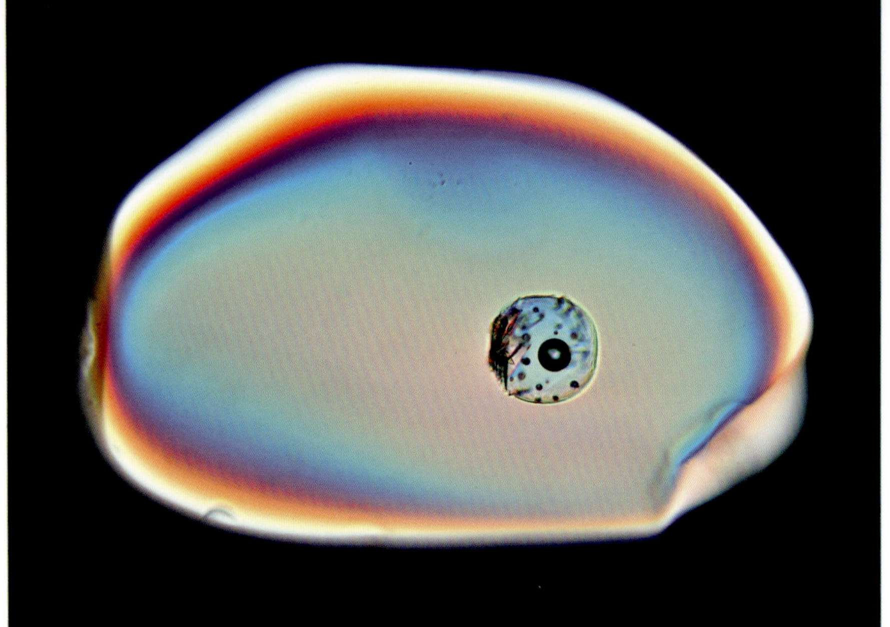

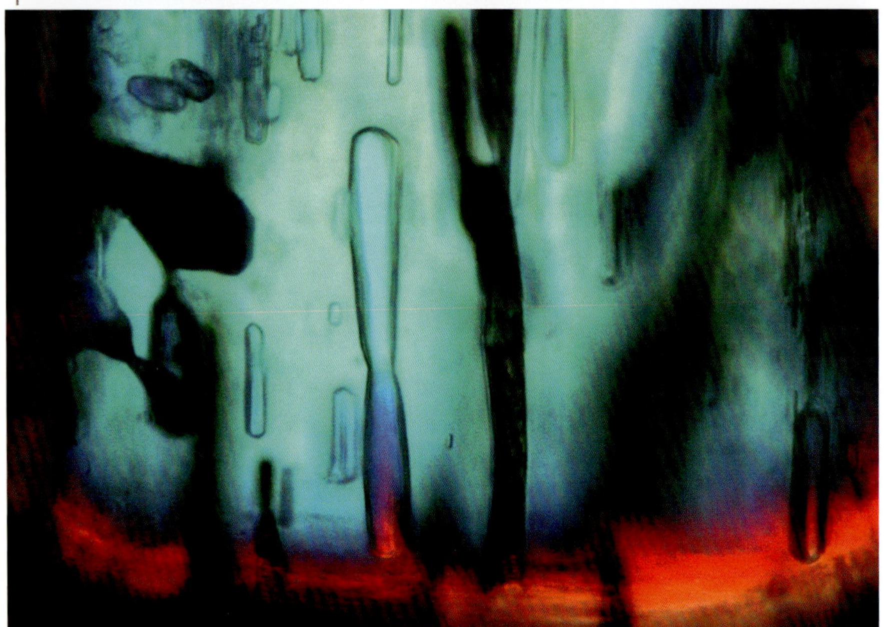

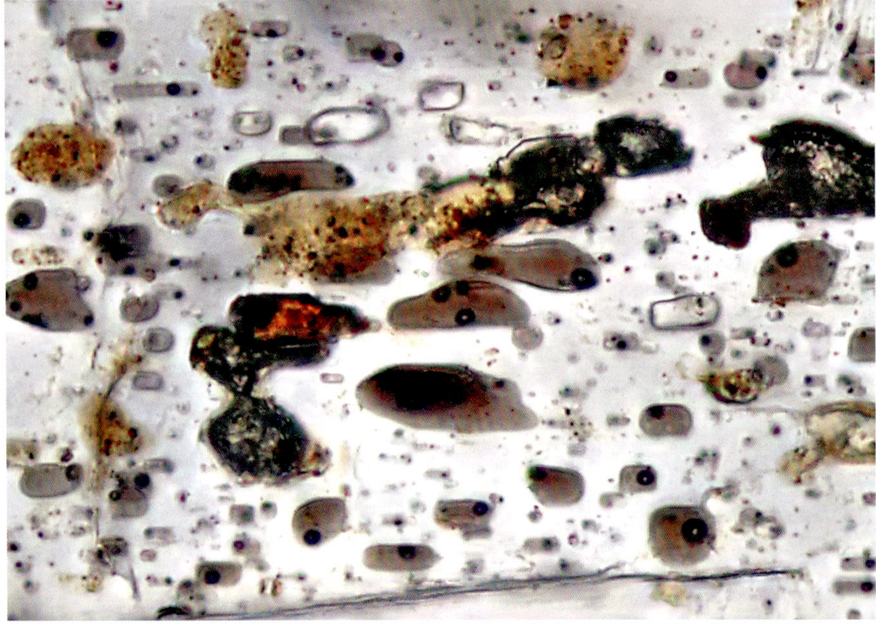

Figure 6.15 Well-rounded sand grain of quartz with primary three-phase inclusion. Crossed polarizers. Whapitu Beach, Auckland, New Zealand. Transmitted light. Grain size 245 × 365 *μm*.

Figure 6.16 Primary, growth-synchronous, tubular inclusions in a sand grain of disthene, Kerala, India. Crossed polarizers. Transmitted light. Image width about 200 *μm*.

Figure 6.17 The inside of a clinozoisite. Sand grain from Bali, Indonesia. Oriented melt inclusions with brown coloring and solid gas bubbles (type 2), pure crystalline inclusions, and melt inclusions of amorphic form (*top right*) and further inclusions of unknown composition, but amazing beauty. Transmitted light. Image width 160 *μm*.

In the year 2015, researchers discovered graphite crystals in the younger zircon of Jack Hills (4.1 billion years old), whose isotope ratio of carbon ^{12}C to ^{14}C corresponded to that of life-forms. That would be a new record for the presence of life on the young earth.[103] What remains is the realization that a few tiny grains of heavy mineral sand from inclusionary zircon are capable of telling quite astonishing stories. They provide insight into interrelationships that contribute significantly to an understanding of the earliest period of Earth's history and perhaps lead to correcting or even partially revising assumptions made since then.

Let's go back to our topic and, along with the temporally determined inclusion genesis, consider the most commonly occurring types of inclusions (types 1–5), which can be observed on sand grains with the help of a microscope.

Type 1: Watery solution with gas bubble. This is the most common type, which is only absent in volcanic sand grains. Identification is easy, especially when the gas bubble moves ("fluid inclusion"). Only one gas bubble occurs at a time, provided the cavity is not too fractured. The liquid is mostly clear and colorless.

Type 2: Silicate melt with gas bubble. If a two-phase inclusion is present, where the gas bubble does not move, it most likely is a melt inclusion when of plutonic origin. Normally, multiple gas bubbles are present. Gas bubbles from inclusion type 2 do not move and often have asymmetric forms. The homogenization temperatures for glass/melt inclusions tend to be over 1,832°F (1,000°C), and with that, over those from type 1.

Type 3: Fluid carbon dioxide with gas bubble. Gases and liquids change into the fluid phase when a sufficiently high pressure is present above a specific temperature. In this state, they possess properties of both gases and liquids. This means that, on the one hand, such so-called fluid mixed phases expand like gases when the pressure drops, but on the other hand, they have the properties of liquids and can dissolve salts.[104] If a fluid mixed phase is trapped during crystal formation, its components separate when it cools below the critical temperature. This is 705°F (374°C) for water and only 88.3°F (31.3°C) for carbon dioxide. The latter is often reached during microscopic examination due to the heating effect of the light source. Above 88.3°F (31.3°C), two-phase carbon dioxide inclusions therefore also return to a fluid mixed phase, where the gas bubble disappears. With selective temperature control, a CO_2 inclusion can therefore be easily identified by measuring its homogenization temperature, which then corresponds to the aforementioned critical temperature. Renewed cooling brings the bubble to the surface again.

Type 4: Watery solution with liquid CO_2 and gas bubble. A three-phase inclusion of wall-covering water, within which the CO_2 is found, within which another gas bubble can be observed. As with type 3, the inclusion can be identified when the two CO_2 phases homogenize at temperatures over 88.3°F (31.3°C).

Type 5: Watery solution with solid crystal and gas bubble. If an isotropic cube is present, it is a salt crystal (halite). This can be checked with crossed polarizers below a turning platform. Other types of crystals can be proved optically with certain effort.

Two quotes may follow our excursion into the world of inclusions, the inner world of sand grains. The inclusion researcher Edwin Roedder called for patience and warned against hasty conclusions from only apparently consistent data. The second quotation in particular recalls Goethe's admonition that it is in the transition from experience (i.e., measurement) to judgment (i.e., conclusion) that the greatest dangers lurk for the researcher. So, from Roedder,[105]

> "One is tempted to conclude that the price of success in fluid inclusion studies is eternal vigilance."

> "It is surprisingly easy to get beautiful, consistent, reproducible, but incorrect numbers."

Apart from all the systematics and the possibility of making inclusions scientifically useful, one should always keep in mind what beauty is hidden in some inclusion scenarios. Floating clouds of miniature bubbles, crystals within crystals, bubbles that have been dancing for millions of years, and hair-thin geometric structures can be found. Inclusions are what makes sand grains unique, despite their unimaginable variety and number.

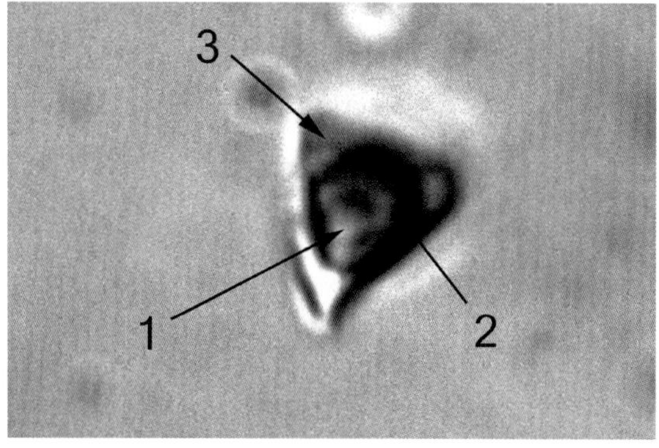

Figure 6.18 Sand grain from Bjerregard, Denmark. A three-phase inclusion can be seen here with a moving fluid inclusion (smallest bubble 1 in the center). The small bubble has a diameter of 1.2 *μm*, the one surrounding it a diameter of 3.8 *μm*. The inclusion is located in the middle of a quartz grain, which explains the bad picture quality. Oil immersion, bright field.

Next, we will leave the world of inclusions and come to the next characteristic that can give us clues to the source rock of the sand grains.

6.1.2 Behavior under Polarized Light

We have already mentioned "crossed polarizers" several times in the captions. What is this all about? Polarizing filters are well known to photographers, and they are also indispensable to the closer examination of sand. Two of them are needed for our little experiment. To do this, place a few clear and larger grains of sand on one of the two filters, which you hold over a white sheet of paper or over a flashlight. Then we place the second filter over it and turn it until the background turns black. We now notice that some of the grains of sand suddenly appear in bright colors, the interference colors. This is the arrangement when talking about crossed polarizers. The physical background is easily explained. We learned in chapter 3 about sunlight simplified as a bundle of light waves. These waves oscillate in all possible directions. A polarizing filter lets through waves that oscillate in only one direction, the passing direction. If you put a second filter on top of it, rotated 90 degrees, the pass direction of the first filter is blocked as a result and the background remains black. But why do you still see the sand grains? The sand grains in the center have a property called birefringence, which causes the wave paths to split up and interfere behind the second filter, which then causes colorful appearances. The more detailed, rather complex physical background may not be of interest here. But it important to know that the emerging color sequences are not only very beautiful, but crucial for the determination of the minerals, an effect that is used in polarizing microscopes.

If we were not to turn the sand grains around their own axis, we would be able to observe that their brightness systematically changes during a 360-degree turn, whereby they appear dark in four positions exactly 90 degrees apart (extinction positions). At a distance of 45-degree from these extinction positions, the maximum brightness is reached. Why are we interested in this? Simply because sand grains that were under strong tension, such as a metamorphosis, don't follow this rule.

They have a so-called undulous or wavy extinction, so they don't become dark homogeneously, because the extinction position wanders over the grain like a shadow. This means that pressure stress of crystals can be detected. Externally applied stress of sufficient magnitude thus leads to the deformation of crystals, which can become optically noticeable in undulous extinction. At even-higher stress levels, groups of very small inclusions are increasingly deposited in the crystal and lead to the formation of deformation lamellae, also called Boehm lamellae. These are found mainly in quartz grains of metamorphic overprinted rocks.

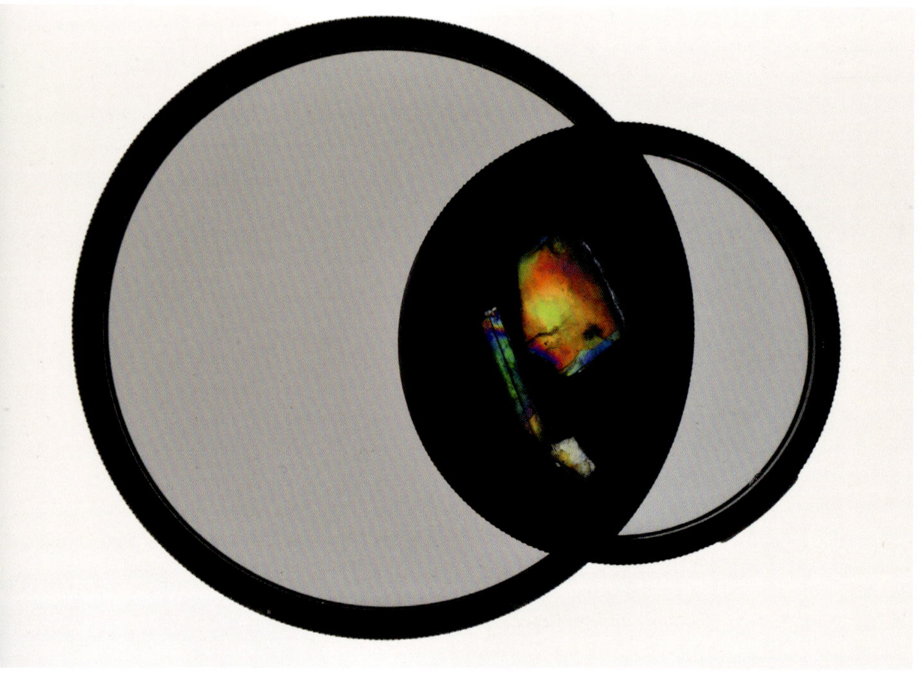

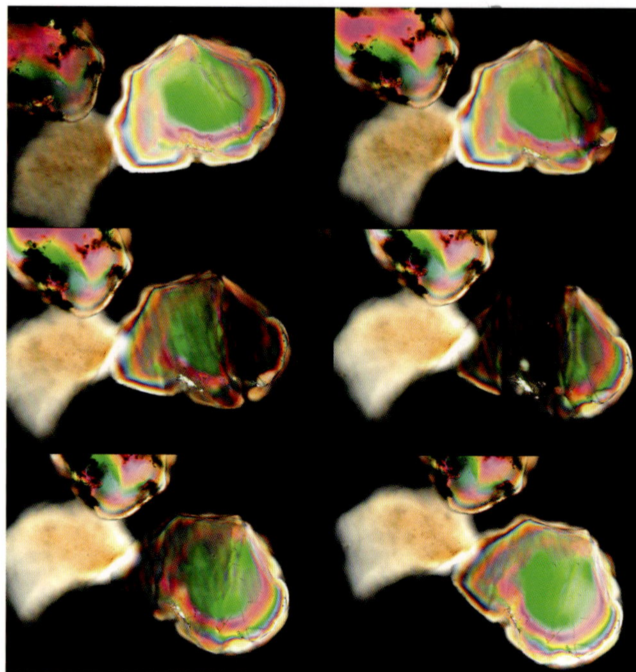

They can also be observed under the microscope in polarized light. If the stress increases even further, cracks and fractures appear in the grain, until finally polycrystalline aggregates are formed by recrystallization processes.[106] We can thus infer by appropriate observation of sand grains how high the rock pressure was that they were once under. A very astonishing phenomenon in itself.

Figure 6.19 Two common polarizing filters turned 90 degrees and laid over each other, so they don't let any more light through (black middle). If you insert a birefringent and transparent crystal between both filters, this will appear bright and glow with colorful interference colors. Muscovite platelets are brought in in the picture.

Figure 6.20. Sand grain from Bjerregard, Denmark. Six phases of a significantly distinct undulous extinction. Order: from the top right to the bottom left in steps of a 10-degree grain turn clockwise. It is clear to see that the extinction moves over the grain partially as a shadow. Grain of quartz. Crossed polarizers.

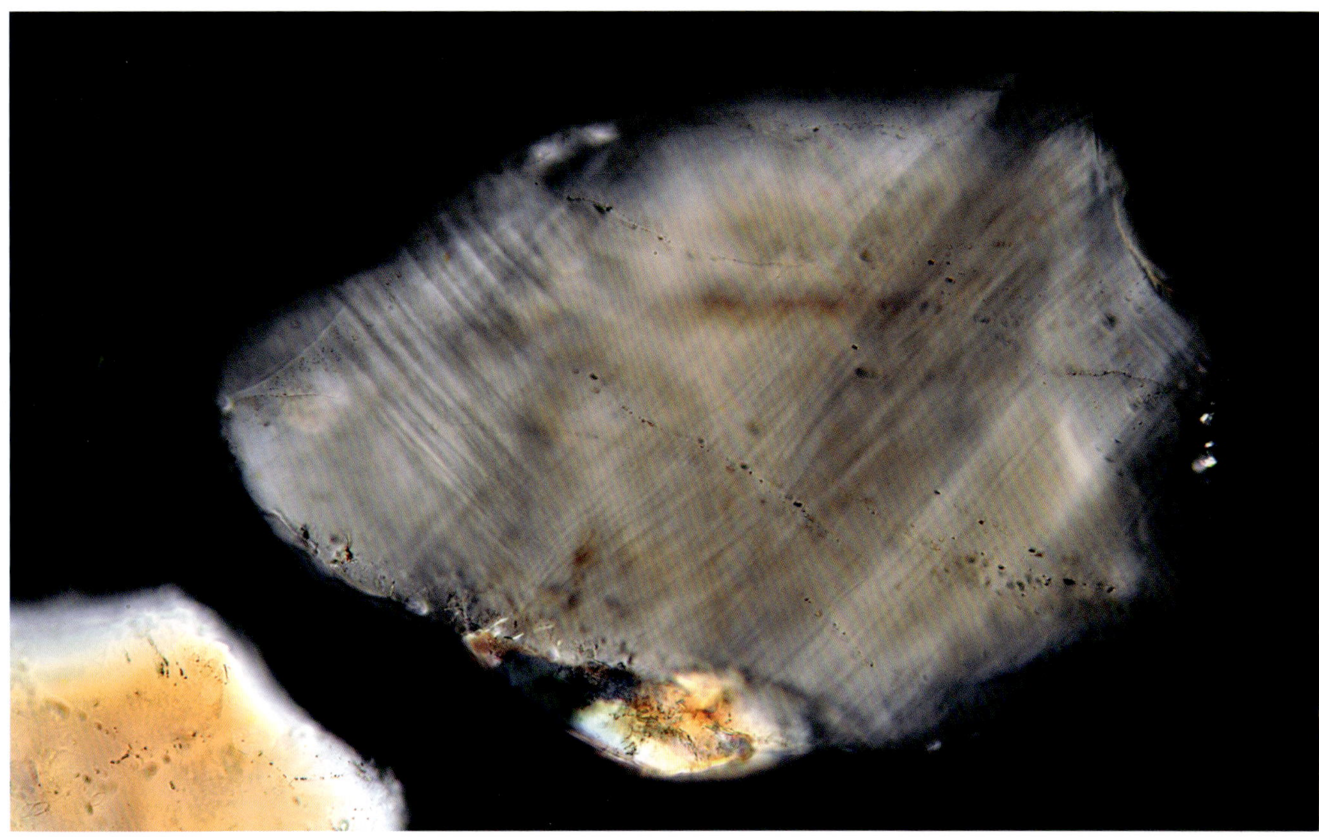

The stress-induced structures can be completely reversed by recrystallization of the sand grain. Similar processes from everyday life can be recognized. Steel, for example, can be hardened by stressing the microstructure through heating and subsequent cooling. Similarly, the microstructure softens again by releasing the lattice tensions through annealing. Accordingly, quartz grains in the stress-free state and also in the state of very high stresses do not exhibit undulosity but homogeneous extinction. This seems strange at first, but it can be explained by the recrystallization processes at very high pressure and temperatures during extreme metamorphism.[107]

Figure 6.21 Sand grain of quartz from Sehlendorfer Beach, Baltic Sea, Germany. The striped structure has to do with Boehm lamellae, which are oriented here in at least two directions. According to this, the grain has undergone a significant metamorphism. Grain size 310 × 465 µm.

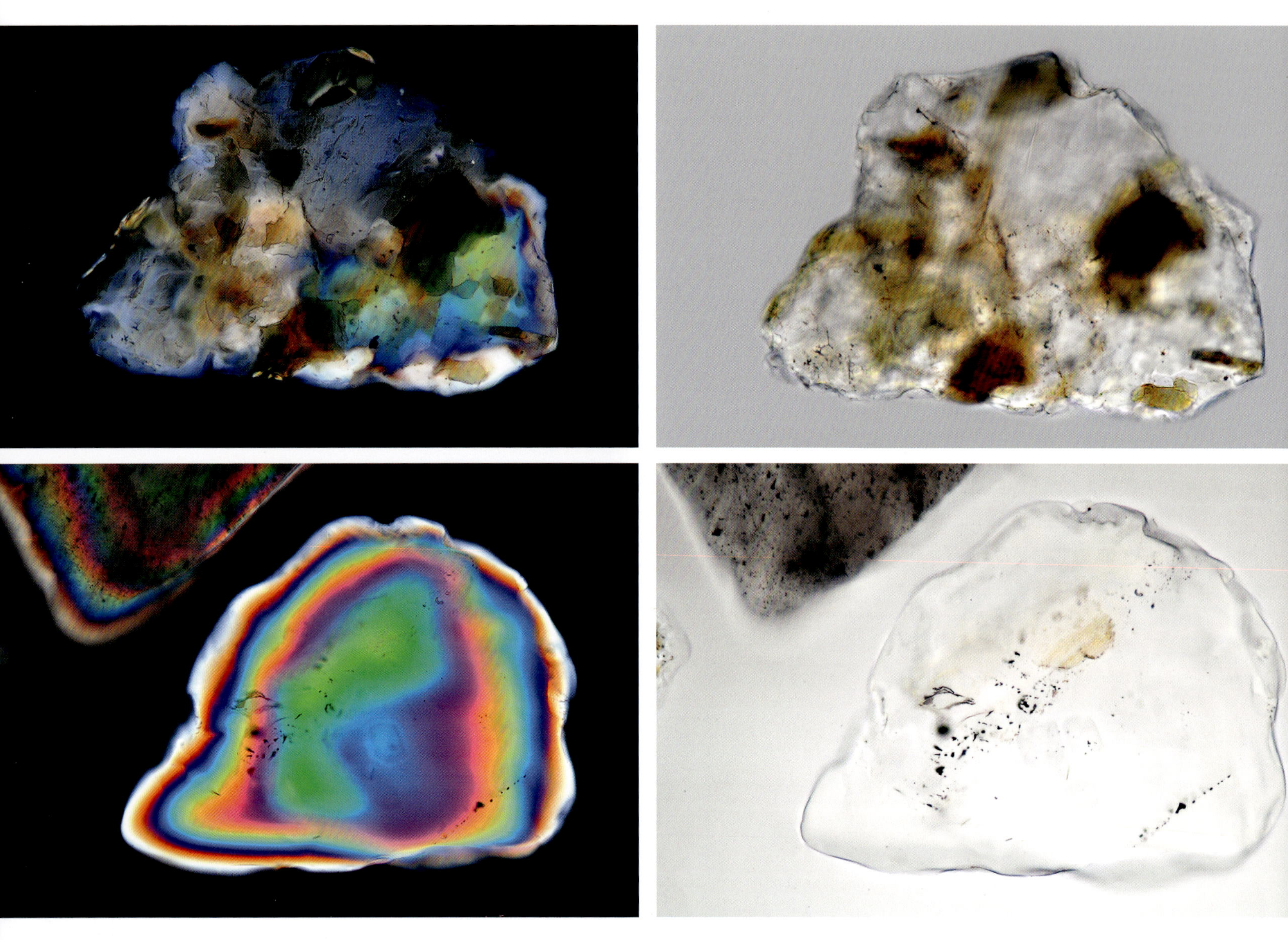

Figure 6.22 *Upper row*: Polycrystalline sand grain, Ried Glacier valley, Switzerland. Grain size 220 × 380 *μm*. *Left*, respectively: Pictures with crossed polarizers; *right*, respectively: pictures in a bright field.

6.1.3 Crystalline Structure

If we observe midsized sand grains, between 200 and 600 *μm*, we can more or less recognize two types of internal structures. One type of sand grain is homogeneous and, other than possible inclusions, doesn't have any further substructures. They are monocrystalline. In contrast, there are also sand grains that seem to be composed of multiple smaller grains. These sand grains are called polycrystalline; however, they are not composed of diagenetic sand grains that grew together in sandstone, but rather from crystallites of different formation genesis. Upon closer research, it is additionally important to differentiate what the borders of the subgrains are like. Smooth, polygonal, and saturated (irregular-toothed) borders are found here. In addition to the nature of these borders, the shape of the individual subgrains can be described as tending to be equal in size and rather elongated.

6.1.4 Grain Shape

In addition to the listed characteristics of the grain interior, the shape or form of the sand grains can also be used diagnostically. Grain shape and degree of rounding have already been discussed in chapter 3. In our current context, other characteristics are of interest, which are referred to by technical terms in literature. Idiomorphic grains are bounded by their crystal surfaces (contrast: xenomorphic). Idiomorphic crystals show their characteristic crystal shape in the rock, while xenomorphic crystals have not developed a characteristic shape of their own; for example, because their crystallization took place at a later time between other, already crystallized minerals.

With this prior knowledge, we now want to approach the question of possible source rocks of the sand grains. Here we use the following classification, always related to quartz sand grains, and summarize characteristics in brief profiles:

— igneous source rock, plutonic
— igneous source rock, volcanic
— hydrothermally formed source rock, gangue rocks
— metamorphic source rock, metamorphites

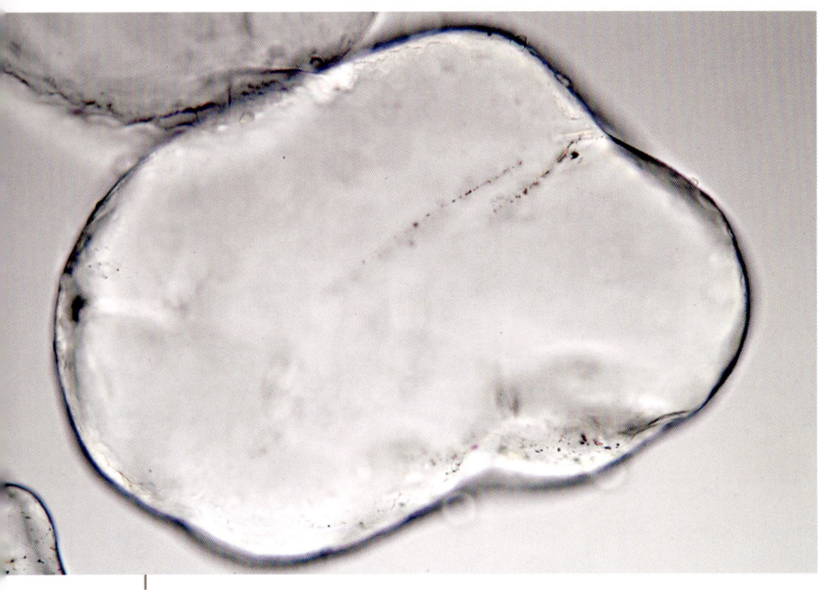

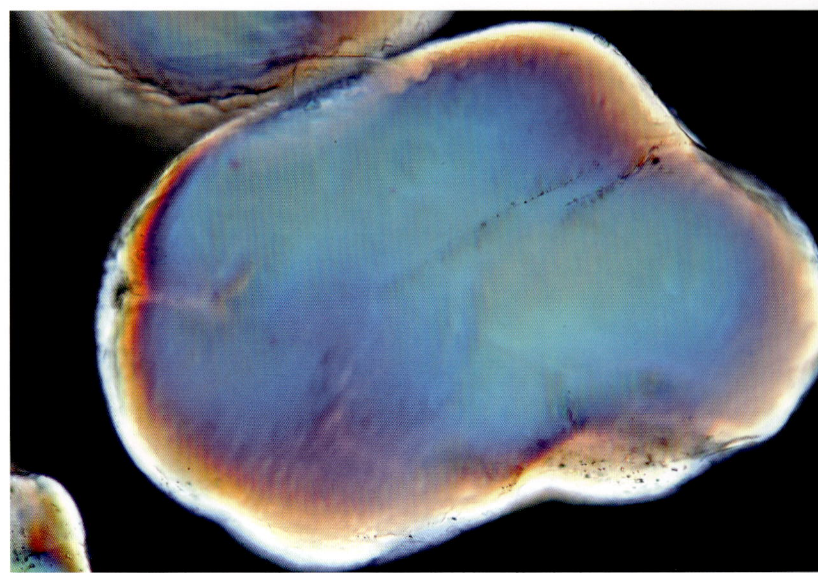

6.1.5 Profile: Igneous Source Rock, Plutonic

(a) Inclusions

- single and multiphase inclusions
- > 50% of multiphase inclusions, often in larger numbers
- bead-like inclusions
- no glass inclusions
- mineral inclusions irregularly distributed, not common
- mineral cracks tend to be colorless

(b) Behavior under polarized light

- homogeneous or slightly undulous extinction

c) crystalline texture

- monogranular, monocrystalline

(d) Grain shape

- irregular shape
- mostly equidimensional
- xenomorphic (without significant crystal surfaces)

Figure 6.23 Quartz grain from Pellworm, North Sea. The grain is monocrystalline, as the right picture reveals with crossed polarizers, and shows extinction. It contains two small rows of bead-like inclusions but otherwise lacks features. It most likely originates from granite rock. Grain size 230 × 235 *μm*.

6.1.6 Profile: Igneous Origin Rock, Volcanic

(a) Inclusions

- only solid inclusions, no liquid phases due to the high formation temperatures
- Glassy inclusions, melt inclusions
- > 50% single phase
- inclusions partially irregular and amoeba shaped

(b) Behavior under polarized light

- homogeneous extinction (crystallization happens mostly without external pressure stress)
- (c) Crystalline structure
- monocrystalline

(d) Grain shape

- idiomorphic, often bipyramidal (high quartz)
- crystals often with straight edges and rounded corners
- corroded grain borders, partially larger corroded areas
- grains mostly crystal clear and appear polished

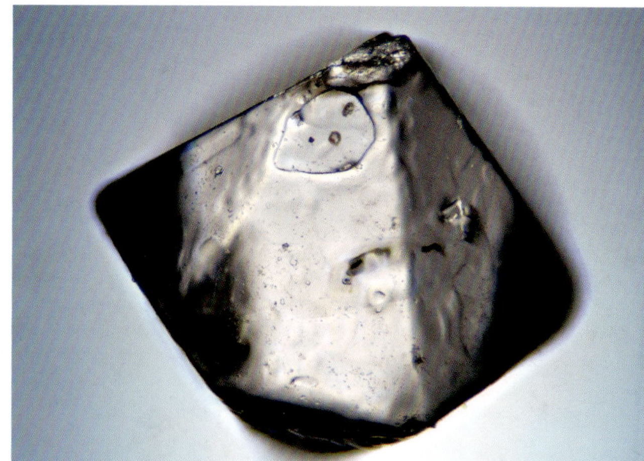

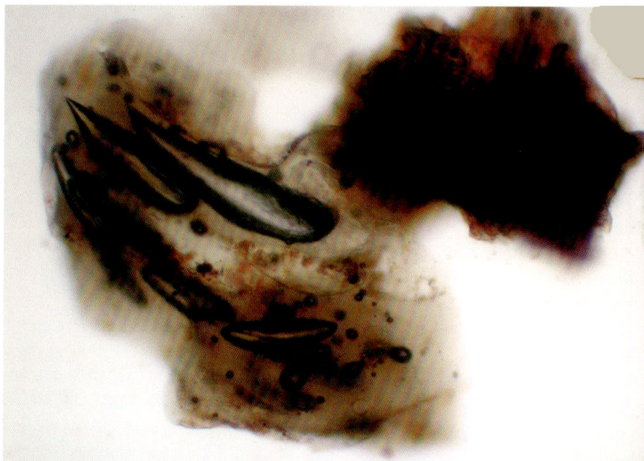

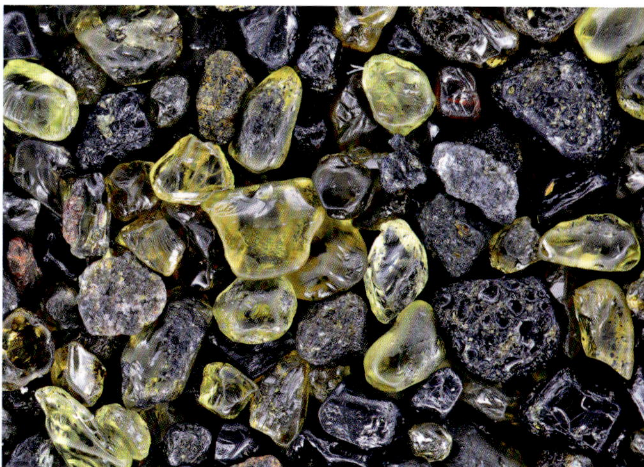

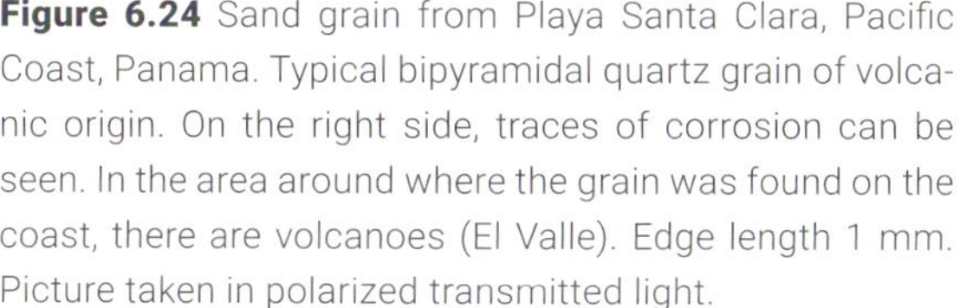

Figure 6.24 Sand grain from Playa Santa Clara, Pacific Coast, Panama. Typical bipyramidal quartz grain of volcanic origin. On the right side, traces of corrosion can be seen. In the area around where the grain was found on the coast, there are volcanoes (El Valle). Edge length 1 mm. Picture taken in polarized transmitted light.

Figure 6.25 Sand grain from the banks of Lago Conguillo, Chile. The lake from whose banks the sample originates lies at the foot of the Sierra Nevada volcano. The grain is made of a microcrystalline to glassy matrix with large tapered-shaped gas bubbles. These hold a gas/water steam mix that was captured at the time of the eruption. Transmitted light.

Figure 6.26 Sand sample from the Azorean island Failal, Faia Praia do Norte, Portugal. The picture shows olivine and volcanic glasses. Typical fracture shapes can be seen as well as the partially polished, partially matte, and partially bubbly structure of the surface. Image width 5.5 mm.

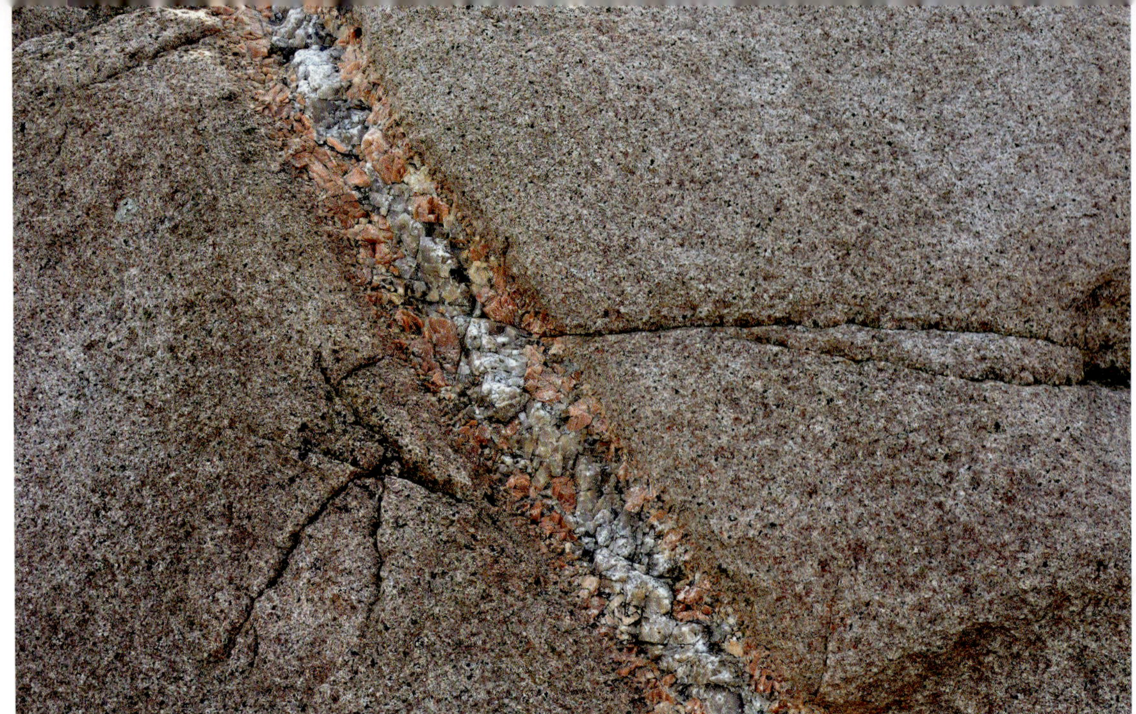

6.1.7 Profile: Hydrothermally Formed Source Rock, Gangue Rocks

Hydrothermal formation of minerals and quartz (e.g., "gangue quartz") occurs in fissures, veins, and cavities by precipitation from hot, mineral-rich, aqueous solutions. Temperatures can reach up to 752°F (200°C) and overlap with processes of low-grade metamorphism. Gangue rocks and pegmatites developed in the top of granite plutons. As the melt rose from the earth's interior, cracks formed in the surrounding rock, into which the rock melt penetrated. After slow cooling, larger cracks and veins yielded rock bodies with coarse-grained textures that crystallized into pegmatites. In the process, mostly white quartz grains developed, flanked by reddish feldspar bands. Aplites occur in narrower crack systems, often outlined by tectonics. While these aplitic gangue rocks are light colored and extremely fine grained due to rapid cooling, and individual minerals cannot be distinguished with the naked eye, pegmatites represent the exact opposite. The defining characteristic of pegmatites is their coarseness. They are minerals dominated by quartz and potassium feldspar, and often also by rare minerals, which can be found in exceptionally large crystals even up to a meter in size. Pegmatites are generally striking and colorful.

During further magma rising, mineral-rich, watery, and superheated solutions penetrated the neighboring rock in crustal levels above the pluton tops and crystallized there in the so-called hydrothermal phase, which followed the previously described (pneumatolytic) stage. Temperatures during this hydrothermal formation were between 752°F (400*°C) and 932°F (500°C). Water was always present in a fluid and evaporated state, which led to the characteristic inclusion* patterns of the hydrothermal quartz.

Figure 6.27 Pegmatite vein in a granite pluton near Hammeren, Bornholm, Denmark. The light quartz grain area of the gangue is easy to recognize, which is lined with reddish coarse crystals of potassium feldspar. The width of the gangue is about 15 cm.

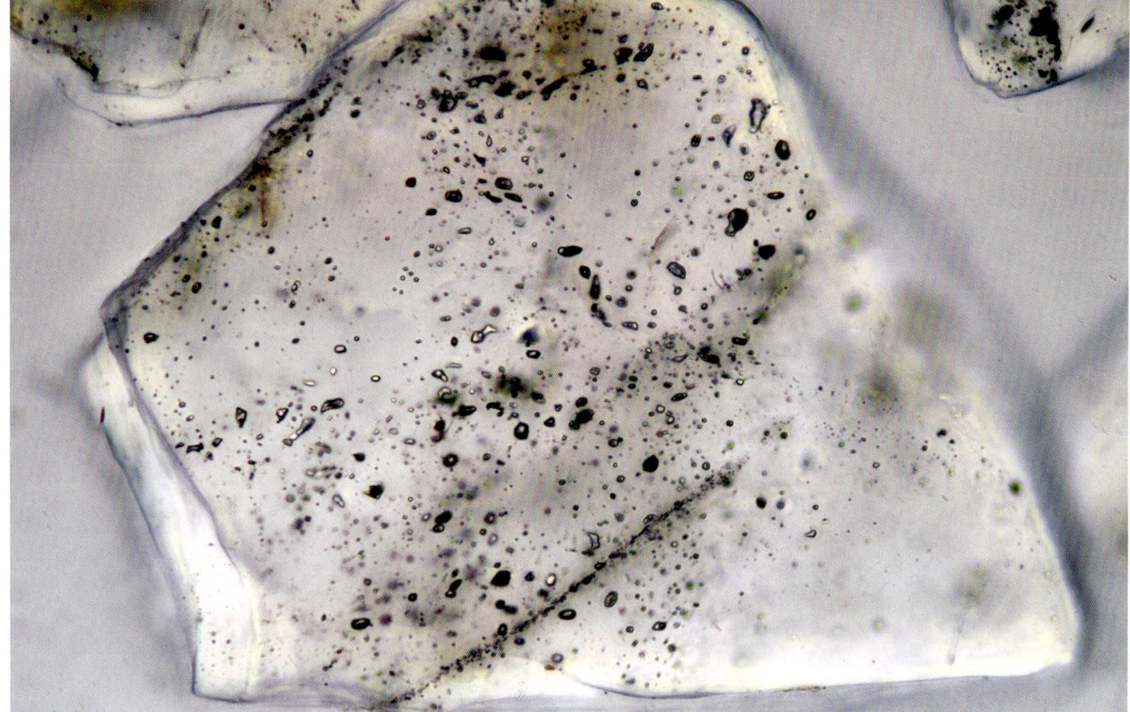

All three summarized mineral formations have the presence of hot mineral solutions as a common feature. Depending on the formation conditions in the veins, local pressure increases could also occur. This is helpful for understanding the features listed in the profile.

a) Inclusions in sand grains

— many liquid inclusions and linear inclusions ("bubble trains")
— often-milky appearance due to the many inclusions; for example, milk quartz
— no or few solid crystal inclusions
— if solid crystal inclusions are present, then pegmatitic formation
— often-wormlike-looking chlorite inclusions

b) Behavior under polarized light

— Extinction tends to occur mostly homogeneously, but slightly undulous extinction possible. Shadows travel not without fractures over the subgrain borders.

(c), (d) Crystalline texture and grain shape

— no typical characteristics

Figure 6.28 Hydrothermally formed quartz grain from Bjerregard, Denmark. Irregularly formed. Many inclusions, many of which are two-phased with liquid and gas bubbles (fluid inclusions). No solid crystal or gas inclusions. Homogeneous extinction. Image width 360 *μm*.

6.1.8 Profile: Metamorphic Source Rock, Metamorphite

In the case of sand grains from metamorphic source rocks, somewhat-more-complicated conditions are present, since different degrees of metamorphism can be distinguished, which also leave different traces. Even the type of compressive stress is important—only asymmetric stress leads to a deformation of the material and to features that can be identified on the sand grain if necessary.

a) Inclusions in sand grains

Needlelike inclusions of minerals associated with metamorphic formations predominate; for example, garnet, disthene, and rutile. Fluid inclusions are very rare, while CO_2 inclusions increase, especially at higher metamorphic grades. Glass inclusions, as we know them from volcanic sand grains, are not found. Cracks and fissures from mechanical stress are clearly marked with secondary inclusions.

b) Behavior under polarized light

Starting at about 1,112°F (600°C) to 1,472°F (800°C), quartz grains recrystallize and, with it, also lose for the most part their optical characteristics that could point to a previous pressure stress. At lower temperatures, the aforementioned undulous extinction comes into play. If it is very distinct over the whole grain, metamorphic gneiss is probably the source rock; if it is only weakly pronounced, granite is also possible. If, however, the sand grain consists of several subgrains with smooth borders, the undulous extinction associated with slate or recrystallized metamorphic rocks is to be interpreted as the source rock. In contrast, strong metamorphism—for example, in gneisses—leads to interlocked grain margins (suturated) and elongated grain shapes.

c) Crystalline texture

— compound grains composed of more than ten subgrains (reliable characteristic)
— suturated (interlocked) or granulated grain borders
— Boehm lamellae

d) Grain shape

— elongated
— xenomorphic

With all the stories we imagined behind the grains of sand on the beach, which may have spanned hundreds of millions of years, we previously went to the interior of the grains to learn what kind of rock it may have come from. Let us now leave the diagnostics of the interior and turn to the shell, the dermatology as it were, the outside of the grains. Thus, we turn to the second question from the beginning of this chapter.

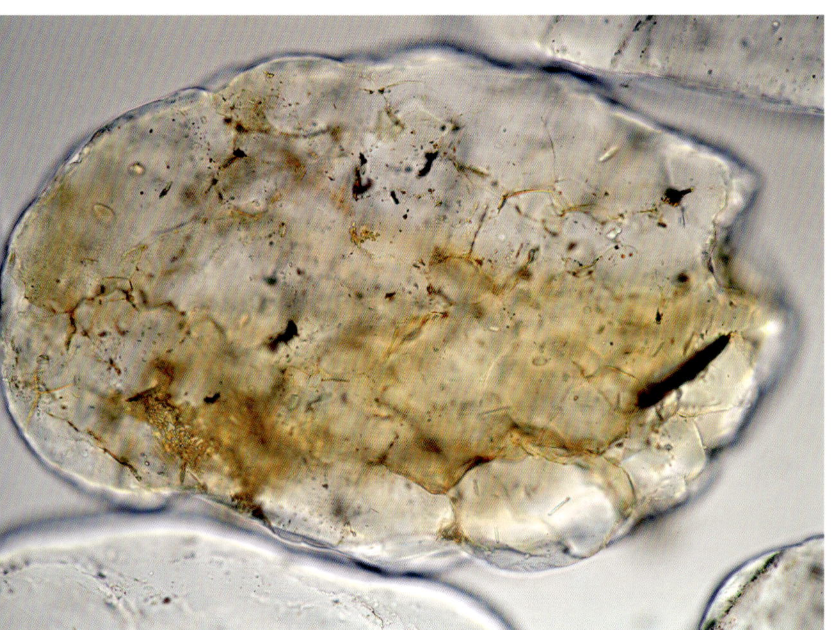

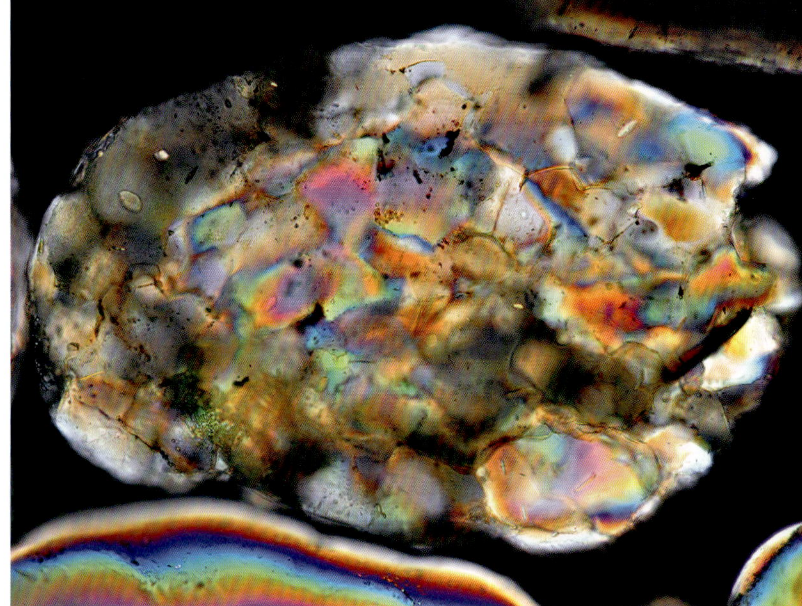

Figure 6.29 Polycrystalline and slightly elongated sand grain from Fehmarn with partially elongated subgrains. Clear undulous extinction that differs from subgrain to subgrain. In the subgrain in the bottom right and in other places, there are additional small rutile needles that are recognizable. With it, there are multiple characteristics of a metamorphic source rock present. Medium to strong metamorphism is to be inferred. *Left*, transmitted light in bright field; *right*, crossed polarizers. Grain size 310 × 475 *μm*.

6.2 The Transport and Deposit Milieu, or, "It's the Surface That Counts"

The source rock, or the place of crystallization of the sand grain, determines the mineral content and primary inclusions from the start. Both remain unchanged over the life of a grain of sand. In addition, there is another group of characteristics that, in contrast to the previous one, can be found on the grain surface and are influenced by external factors. These influences result from the different environmental milieus in which sand grains reside during their long cyclic existence. All these environments leave traces. We will try to decipher at least some of them.

As before, we are limiting ourselves to crystalline sand grains of quartz. Starting from the supply area, the place of origin of a grain, erosion processes decompose the respective source rock to sand and further erosion debris and accompany the grain from its formation to its respective end. The migration of the grains is summarized under the term *transport regimes* and can last very long and include completely different external influences, until finally in equally different *deposition regimes* the sediments enter a resting phase. Although transport is dormant during this time, chemical changes can take place. This is especially true of subsidence diagenesis, in which forming sediment is removed from the influence of the earth's surface through tectonic processes. This may be followed, as we had seen, after further very long periods of time, by an uplift phase, which may culminate in a second release. Once again, a supply area is created, and the next cycle begins. At each state of this geological cycle, new mechanical or chemical traces are produced on the surface of the grains, but older traces are overprinted and obliterated; the process is dynamic.

How are all these influences connected? Let us imagine a sphere of fine-grained quartzite and place it in a riverbed. We had shown that what happens next then depends disproportionately on the size of the object, here the sphere. Let's choose it to be about the size of a fist and leave it to its fate. Provided that the river flows leisurely along in a stony bed, it seems plausible that the rock goes through a rolling motion and receives collision marks by collisions against other stones over time. These are quite obviously dependent on factors such as flow velocity, size, and material of the particles in the riverbed, and so on. Quite crucial is the material of the observed object. A quartzite ball has the same properties in all directions, while a layered material, such as shale, would wear quite differently, layer by layer. The quartzite sphere would remain spherical and rolling, while slate would wear flat and would advance by pushing rather than rolling in the riverbed.

It is easy to see that the shape of the resulting surface marks will also differ significantly. The various parameters therefore influence each other. The thought experiment may illustrate how difficult it is to obtain statements about even significantly smaller particles, the sand grains, which can assume all forms of motion in water and air in almost infinite quantities and are exposed to forces that have a shaping effect on them for an indefinite amount time in an indefinite environment. In addition, there are further possibilities of being exposed to trace-forming influences. Let us think here of movement under loads, as occurs in glaciers, or of chemical attack by solution or deposition in warm, humid environments. To describe all this would take us much too far. Let us therefore concentrate on five typical environments.

- desert milieu, aeolian transport
- river milieu, fluviatile transport
- glacial milieu, glacial transport
- ocean and coastal milieu, marine or littoral deposition/transport
- lake milieu, limnic deposits

In the following, we will go through them one after the other and look for traces and textures on the surface of the sand grains found there that could give us clues as to whether they belong to one of these environments. For this purpose, the investigation method must be adapted to the now-changed task. We present it briefly here for better understanding.[108] With a few exceptions, the surface textures we are looking for are in the size range of 1 to a good 50 *μm* and nowadays are usually examined and documented for research purposes with the aid of scanning electron microscopes (SEM). In scanning electron microscopes, samples are placed in a vacuum chamber and scanned line by line with an electron beam. Electrons reflected or scattered from the sample are collected by detectors and processed to produce a black-and-white image. This method has much-higher resolution and magnification capability than light microscopes because the wavelengths of the electron beam are much smaller than those of light. The "scanning tools" are finer, so to speak. Scanning electron microscopes are not part of the standard equipment of the interested sand observer and probably exceed his possibilities considerably. We therefore want to show how sufficiently good images of quartz grain surfaces can be obtained with quite simple means, using normal light microscopes and bright-field transmitted-light illumination. Usually, the grains are completely embedded in a medium on a glass slide as a scatter preparation and are provided with a thin cover slip for transmitted-light examination. The method used here is based on embedding the grains to be observed not completely, but only up to their center, in a medium that has a refractive index as similar as possible to that of the grains themselves.

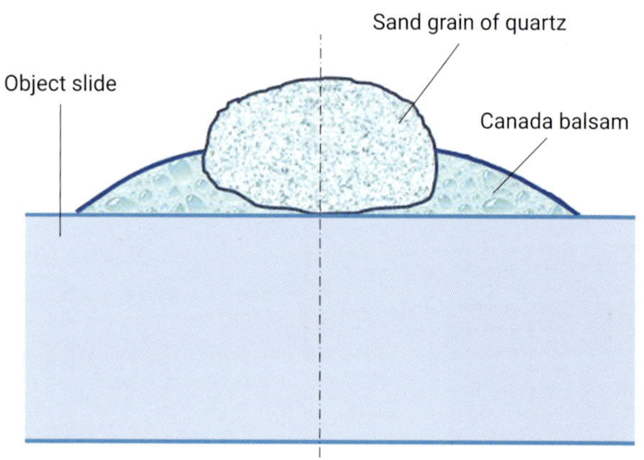

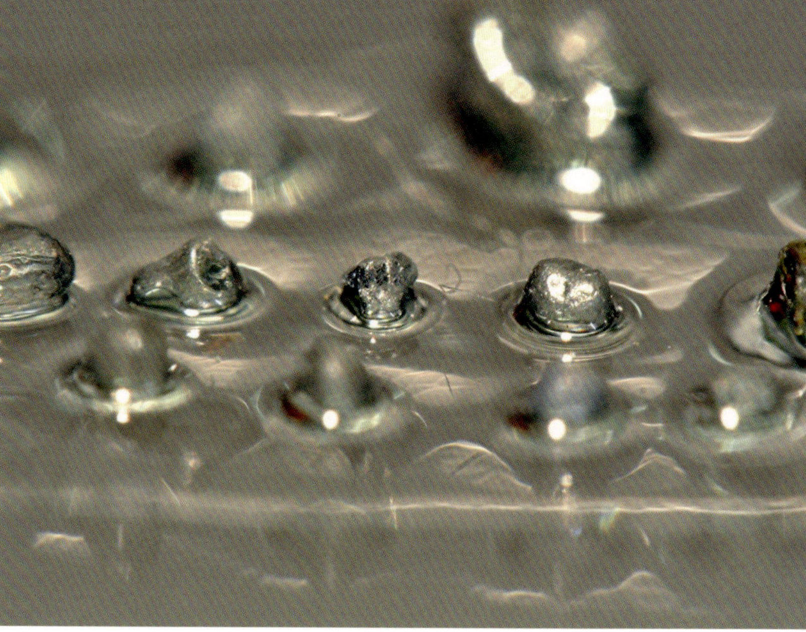

For quartz grains, the time-honored Canada balsam, a natural tree resin, produces good results. Thus, the grains peek out about halfway from the mass of embedding material and are transilluminated from below. Cover glasses are not used.

Structural features such as small cracks or indentations on the underside of the grains fill with the liquid embedding medium; when it is heated and better optically suppressed, the smaller the difference is in refractive indices between the grain and the embedding medium. This is necessary because only the upper surface of the grain is to be examined. Ideally, the light beam does not "notice" the difference between the embedding medium and the grain surface, and the structure of the lower surface disappears. What remains visible is the free, nonwetted grain surface against air, whose structures stand out clearly and in high contrast. However, the method is suitable only for grains that are largely transparent and as free as possible from inclusions.

In this chapter, this *transmitted-light texture method* is used very regularly. The verification on test grains showed that the surface information in the magnification range is valid and thus useful for our purposes. Notwithstanding this, the scanning-electronic-imaging method remains, of course, more powerful when it comes to order of magnitude. Since only the surface of the grains is to be considered, imaging by reflected-light illumination is obvious. A comparison with the transmitted-light texture method is difficult since different image impressions are produced. This is important to know, especially when asking how the grain "really" looks. Each imaging technique has its own specific contribution to answering this question. For our purposes, the transmitted-light texture process usually turns out to be the more suitable one. In order to achieve a sufficiently high depth of field, meaning to correctly represent the third dimension, fifteen to a hundred images were exposed for each photo in just as many levels with distances between 1 and 10 μm and then processed via stacking software to form a complete photo.

Figure 6.30. Embedding of the sand grains in Canada balsam as preparation for viewing the surface texture. The light comes from below, respectively. *Left*, schematic view; *right*, photograph of completed sample, with twelve sand grains of differing sizes.

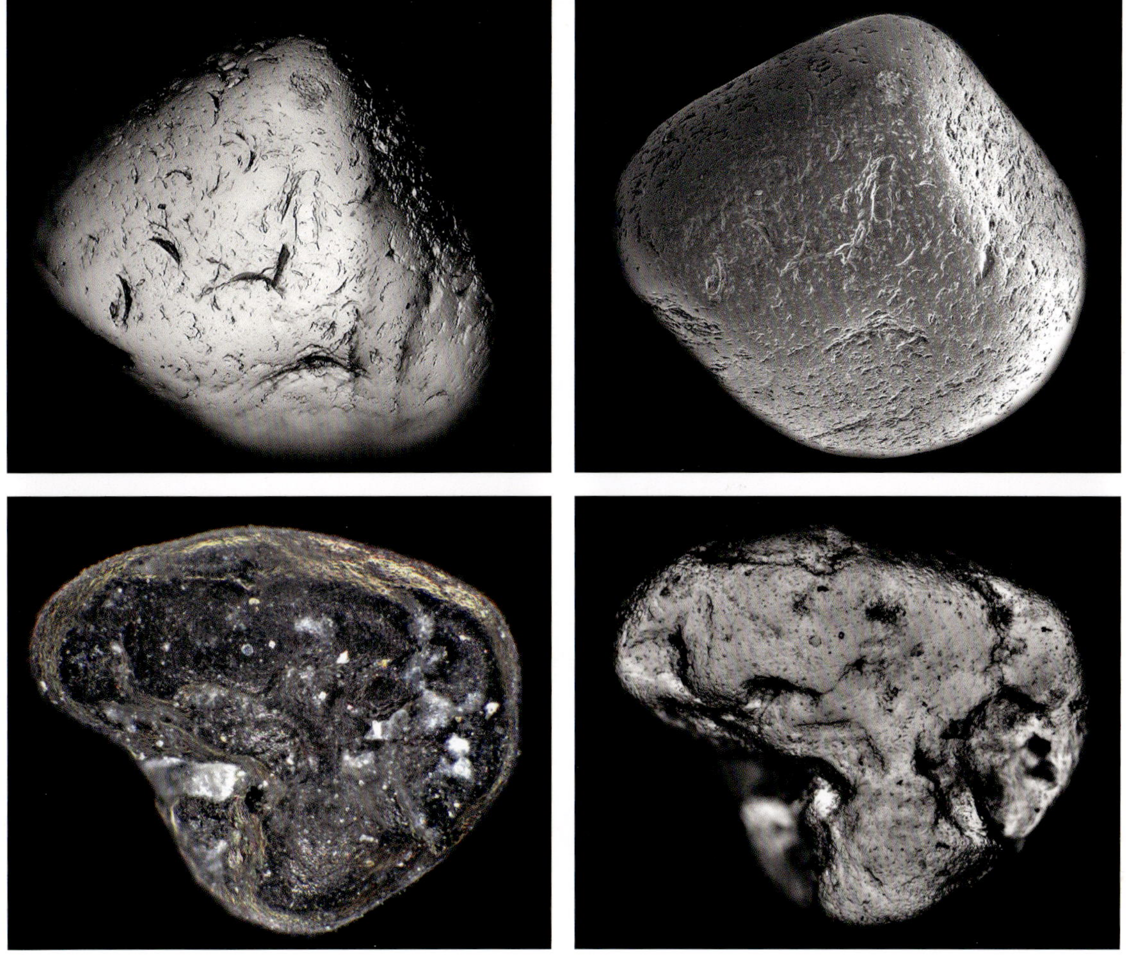

This creates a comprehensive, often-three-dimensional impression of sharpness. The goal was to achieve a balance between an aesthetic image impression and the most realistic reproduction possible.

Figure 6.31 Sample grain from the sands of the Upper Crustaceous period (Santon) near Haltern, Germany. *To the left*, a light microscope image using the described texture method ("transmitted-light texture method"), with ten layers with distances of 10 *μm*. Image width 600 *μm*. *To the right*, another image for comparison of the same grain at a similar angle with the scanning electron microscope—the sample was previously covered with a thin gold layer.

Figure 6.32 *Left*: Reflected light; *right*: Transmitted-light texture method. In a direct comparison it can be seen that the transmitted-light texture method in this example shows a finer pictorial gradation of the surface characteristics compared to the reflected-light depiction. Artifacts and misinformation are not recognizable. If the sand grain has opaque areas, the method is not realistic, because these show up as dark spots, which can lead to false interpretations. In an ideal case, multiple lighting methods should be used to get the most information possible. Murzuk, Libya. Grain size 500 × 600 *μm*.

For the analysis and interpretation of surface textures on quartz grains, extensive literature can be referred to. The research of the pioneers of this field, such as David H. Krinsley of City University in New York, John C. Doornkamp of the University of Nottingham, and William C. Mahaney from Oxford University, is most cited.

In their basic investigations, the abovementioned authors analyzed and cataloged sand grains of known origin. The grain surfaces were searched for patterns and features, with the aim of being able to specifically recognize the location when of unknown origin. Mahaney (2002) lists in his catalog forty-one different characteristics that can be found on the surfaces of sand grains.[109]

It should be noted in advance that a clear classification does not fundamentally exist. As with the inclusion investigations, valid conclusions of the origin can be drawn only from the combination of several pieces of information, even if in some cases with individual grains a diagnosis can be made without a doubt. After all, cases are known from literature in which the individual history of a single grain of sand can actually be traced back and reconstructed six hundred million years.[110]

The process chosen by Bull and Morgen (2006) is exceptionally interesting. The authors took on the task of developing a reproducible method by which one could determine if two sand samples were of identical origin or not. This is of high importance when it comes to forensic evidence, especially when ruling out a match in the case of the exoneration of a suspect in a crime investigation. For this purpose, the researchers chose a descriptive method that allows each grain of sand to be individually assigned a set of characteristics comprising up to five levels. The characteristics of a discovered sand sample are then defined through the frequencies of the individual characteristic combinations. By comparing these to the known reference samples, collected materials from crime scenes can be analyzed for matches. The large amount of data that was collected from the authors from research of over seven hundred cases was processed with the help of appropriate databases. Such investigations must be very secure and court-approved because they could decide the fate of human beings.

With these preliminary considerations, we will now look at the four previously defined milieus in succession (deserts, rivers, glaciers, oceans) and try to find out how the respective geophysical behaviors are depicted on the surface of the sand grains and to prove this by looking at a few examples. In addition to their informational content, the illustrations of the grains come out clearly and with high contrast, simply and beautifully in their respective individuality, and support the assumption that there may not be two completely identical grains of sand in the world.

6.2.1 Desert Milieu, Aeolian Transport

In dry deserts, the absence of a dampening water layer causes grains of certain sizes to round off more quickly during collisions than in the case of fluviatile or even glacial transport. The wind takes over the transport function. In laboratory tests from the 1960s, abrasion values were found that were a hundred to a thousand times higher compared to fluvial transport.111 From this, it was concluded for a long time that sand grains in deserts should have a significantly higher degree of rounding on average. More-recent investigations, however, do not completely confirm this assumption.

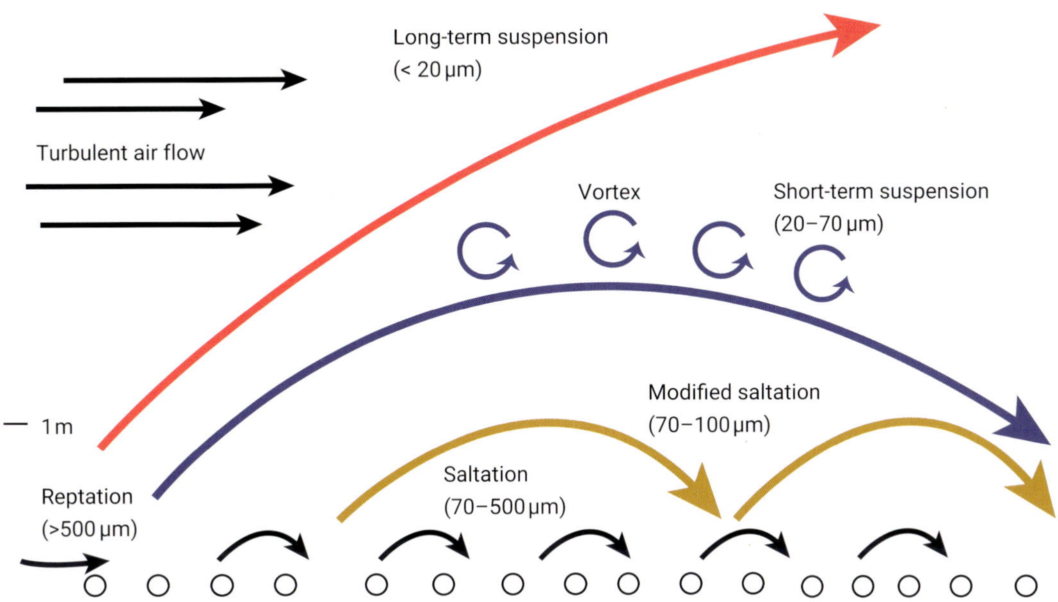

Comprehensive studies show that well-rounded grains tend to be rare. In dune sands, subangular grains dominate—only about 8% of the sand samples could be characterized as "well rounded."[112] Elongated, strongly rounded grains are very characteristic of dry, hot deserts, which still show their shaping from the parent rock.

However, when it comes to the degree of rounding of desert sand, there have been differing results. While the Tunisian deserts tend to have a large proportion (50%) of well-grounded grains, the samples in Bahrain have only 12%, and in the Kalahari merely 2%.[114] It can be generally said that coarser grains (phi ≈ 0) are well rounded, while finer grains (phi ≈3) tend to have been created subangular. A high degree of rounding can be observed in areas that have recycled sediments as the source of their sand, which means only that the grains have already been through at least one previous life cycle.

Edge wear due to direct grain-to-grain impact occurs preferentially with sand grains of a defined size class. Very small grains with diameters under 20 *µm*,[115] which aren't attributed to sand, tend to be in long-term suspension, meaning they are carried directly by the wind.

Figure 6.33 Mechanisms of wind transportation. Reptation refers to creeping processes on the ground, saltation to the occasional lifting from the ground and the flight along trajectory curves. Suspension refers to more or less remaining in a state of limbo long term. All processes are directly dependent on the grain size. Graphic according to Blümel 2013.

There, they can remain in a kind of suspended state in which they reach great heights of over 3 miles (5 km), depending on the weather conditions, and are deposited again only after longer distances of sometimes over 621 miles (1,000 km). Up to 100 million tons of dust loads are transported from the Sahara northward across the Mediterranean Sea every year. Mostly, this dust is then washed out by precipitation and forms recognizable yellowish coatings without a fertilizing effect. Annually, the worldwide approximately 1.5 billion tons of desert dust are introduced into the atmosphere and deposited elsewhere. Of this, an estimated 0.9 billion tons originate from the Sahara.[116]

When looking at short-term suspensions (20–70 *μm* grain diameter), the transport distances are comparatively shorter. Grains with a diameter of over 500 *μm* roll across the ground without flying up and do not have the ability to build up large amounts of kinetic energy. They tend to go through rather crawling movements, which are caused by collisions or also by wind. The sand grains that have a diameter between 70 and 500 *μm* demonstrate a phenomenon that is known as saltation.[117] The wind drives the grains upward, but due to their size and weight, they cannot be held in the current and fall back to the bottom in a typical curve, where they meet the grains already lying there, from which they then bounce back elastically like balls and are again carried away by the wind. During this permanently continuing bouncing, enough relative velocity is generated between the grains, and thus kinetic energy, to cause abrasive wear.

More-precise diameter limits cannot be declared without further information, since these in turn depend on density, wind speed, and additional parameters. Also, static electricity, which builds up in the grain-air mixture, must also be counted among the influencing parameters. Even at higher wind speeds, the layer of jumping grains ("jumping freight") does not reach above knee height. For the beginning of saltation (Latin *saltare* = to jump), wind speeds between 7 and 12 mph (12–20 km/h) are required. The ratio between the jump height to the jump length of the grains is between 1:6 and 1:15, with the maximum saltation heights between 3 and 10 feet (1–3 m) also occuring.[118] The percentage of sand grains that are moved through high-energy saltation processes is between 75% and 80%. The energy-rich collisions lead to compaction waves that run back and forth inside the sand grains. With these processes, material fatigue on the surface can occur, as well as defects in the crystal lattice, to which clusters of nanoparticles about 0.003 *μm* in diameter can deposit. These surfaces are highly reactive and are sensitive to chemical and physical attacks.

Many desert sand grains show a surface texture that can be looked at with simple magnifying glasses. The grains appear frosted or matte, like sandblasted glass. Therefore, it was originally thought that this effect was also due to the mechanical blasting action of the sand.

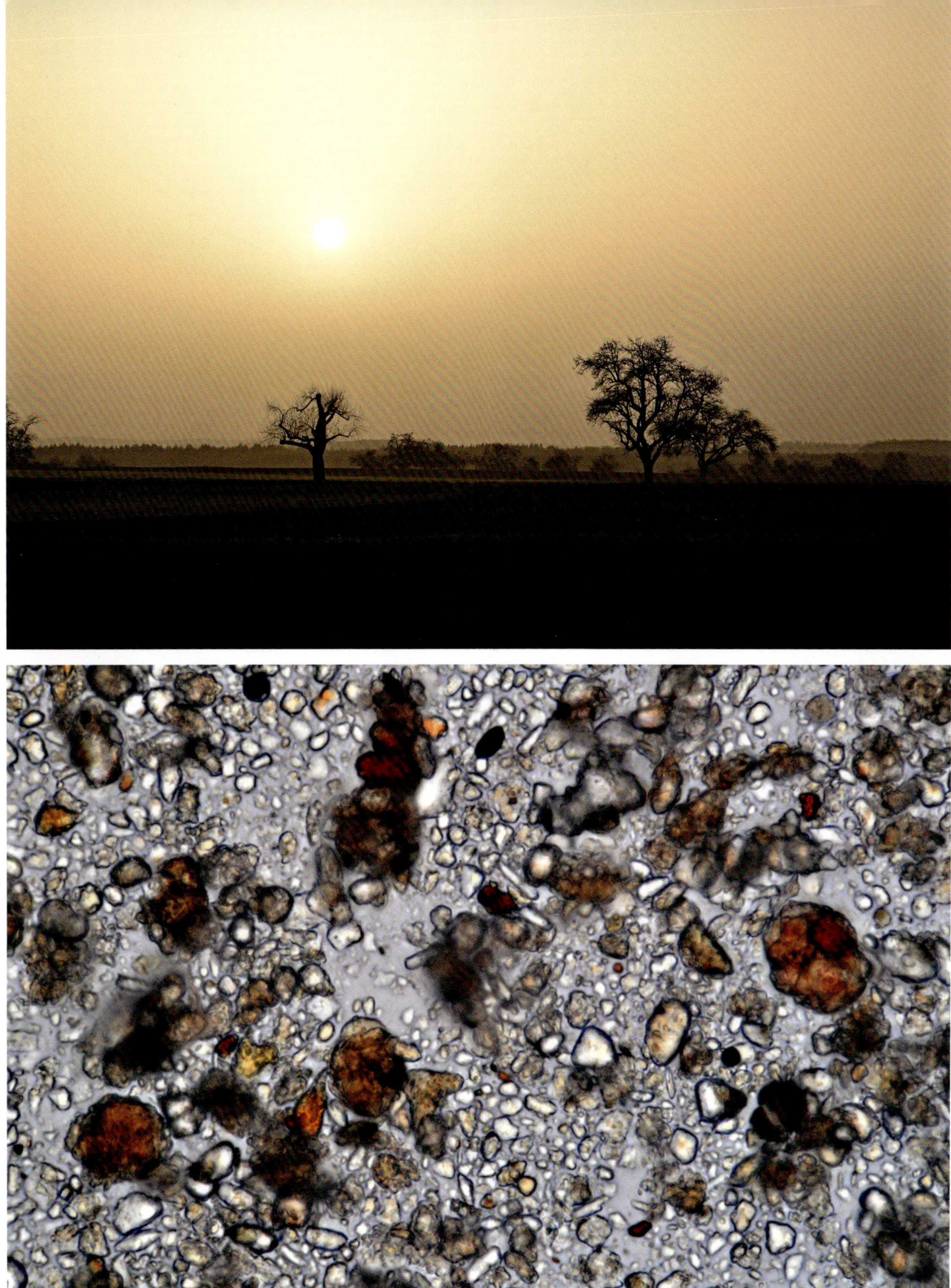

Figure 6.34 Through southern winds, dust from the Sahara can be transported in large amounts far over the Alps. This yellowish lighting (dust turbulence) was captured on February 24, 2021, north of Stuttgart around 5:30 p.m. in an otherwise clear sky. The picture (*below*) shows a sample of the accompanying Sahara dust under a microscope. The larger, red-brown grains in the right half of the picture have a diameter of 19 *µm*. Image width 200 *µm*.

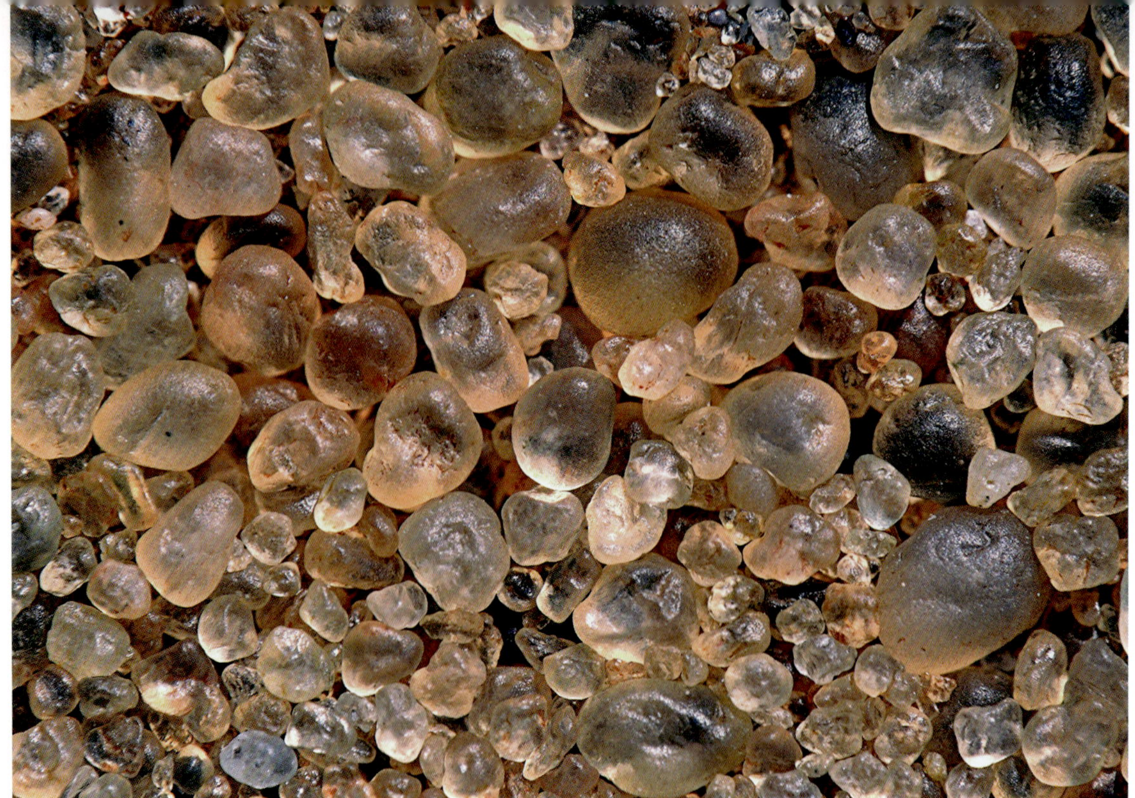

In contrast, more-recent investigations rather support the hypothesis put forward by Kuenen (1959), according to which mainly chemical processes cause the matting. The microscopically thin film of condensation water (dew), which forms around the grains in the dry deserts as a result of large temperature differences between day and night, is sufficient to produce tiny etching spots on the surface, preactivated by the mechanical impacts described, which thus dull it. In contrast, pure impact marks caused by collisions tend to be larger on average. Both effects work together. Similarly, the surface of the occasional frosted grains found in fluvial or even marine environments is due to numerous and oriented etch spots, as shown by Margolis and Krinsley (1971). Under suitable conditions, the balance between abrasion and dissolution or deposition processes can also lead to frosted grains in natural environments, primarily in extremely turbulent underwater areas.

Also, the opposite process, the polishing of originally matte grains, can be explained through chemical processes. In this case, silicates in solutions are left behind in layers after the film of water is evaporated off the surface of the grain, which leads to a type of healing and smoothing of the scratches and imperfections. Over time, the grains appear smoother, and previously sharp broken edges are rounded.[120] Margolis and Krisley (1971) cite lab experiments in which researchers rotated plastic bottles filled with quartz spheres and water for 490 days. In the end, they could prove that a silicate solution with 300 ppm was created with this procedure.

Figure 6.35 Desert sand from the Sahara (Murzuk, Libya). Monomineralic, well-rounded quartz grains with well-recognizable matte surfaces. Image width about 7.3 mm.

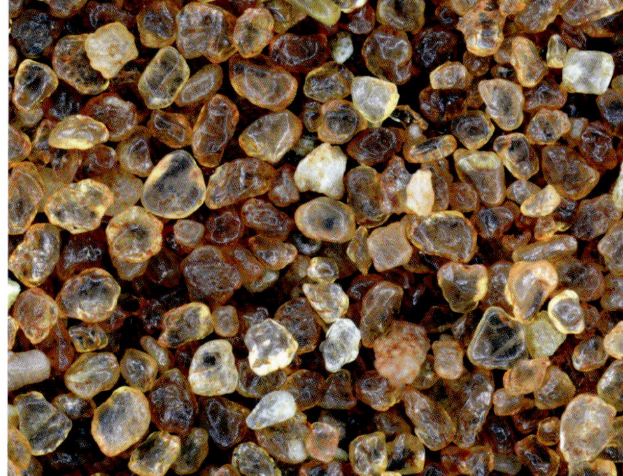

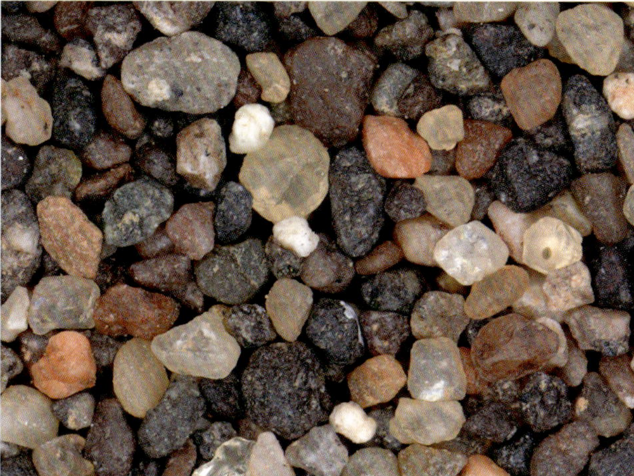

If comparable conditions occur in natural environments, it can be assumed that silicate accumulation effects could lead to a smoothing and polished appearance of grains (defrosting). The exact mechanism of defrosting is still unknown. High-energy collisions that occur at very specific points may cause temperature increases above the melting point in the region of the contact points, as is suspected. This effect could also lead to shiny surface areas that are occasionally observed.[121] Polished grains resist subsequent abrasion much better. Polished-looking sand grains are also found on seashores with strong surf; however, the polishing effect here is more comparable to the action of polishing drums as used in the jewelry industry.

Now, what surface structure is significant for desert sand grains? Generally, the classification listed here according to Mahaney (2002) describes three characteristics of microtextures that lead to a good prognosis of the potential aeolian origin of quartz grains. We will consider modern and hot deserts, which are different from colder dune areas close to the coast or fossil sediments.

1. The first characteristic is shown in the so-called *upturned plates*. These plates appear on the surface as more or less parallel steps that are the result of microfractures along cleavage planes of the crystal lattice. They can be artificially produced and reproduced in suitable experiments in the laboratory, but under natural conditions they are usually rounded by chemical or abrasive action. These modified plates, in addition to the mechanisms already described, account for the matte appearance of many desert sand grains.[122] Some authors point out that the type, frequency, and spacing of the erect plates depend on the grain size and wind speed. Erected plates are to be understood as the result of impacts under high energy at correspondingly high wind speeds.[123]

2. A further characteristic consists of elongated or equidimensional indentations, which are preferably found on grains with sizes above 450 μm. The elongated shape could, as mentioned, indicate the rock of origin.

Figure 6.36 Desert sand from a sand dune in Sossusvlei, Namibia. Grains have round edges; are shiny, only partly matte, and well sorted; and have pretty low sphericity. Almost all grains are covered with a red ferrous oxide layer. Image width 5.1 mm.

Figure 6.37 Desert sand from Valle de la Luna near San Pedro de Atacama, Chile. The Atacama Desert is the driest desert in the world (excluding polar regions) and has existed for fifteen million years. The sand sample is well sorted, with rounded grains of low sphericity. The very low compositional maturity stands out. Image width 5.25 mm.

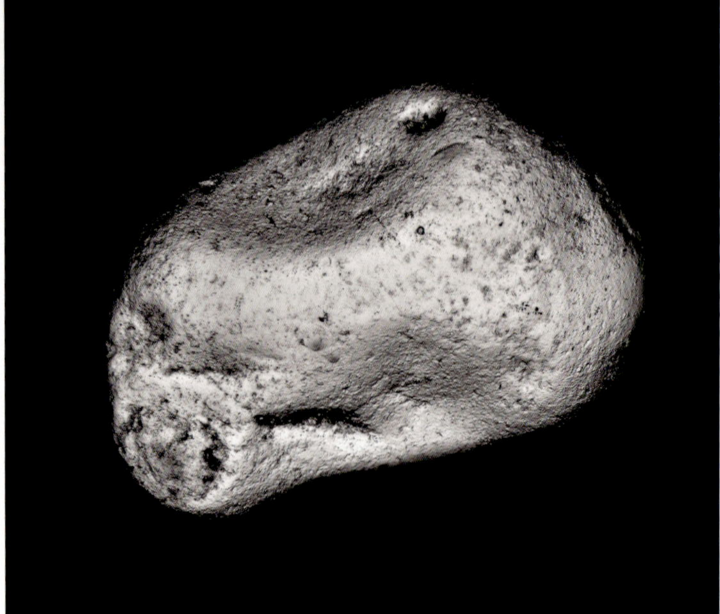

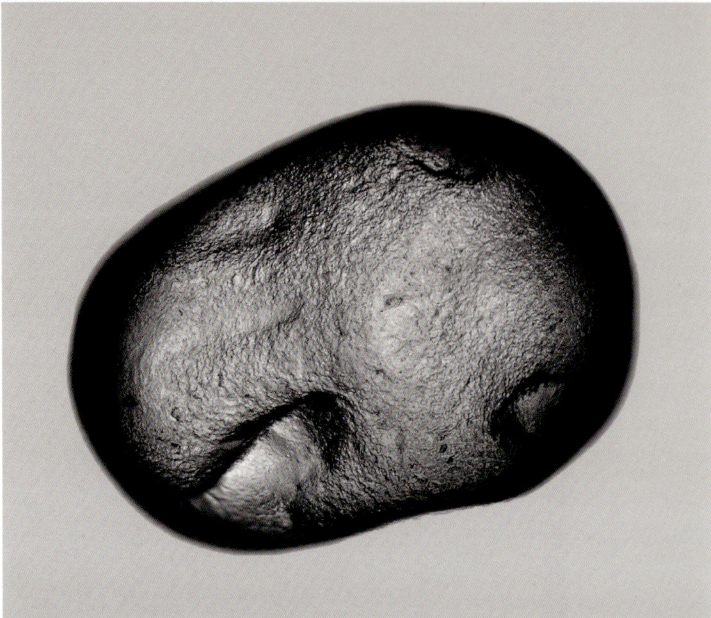

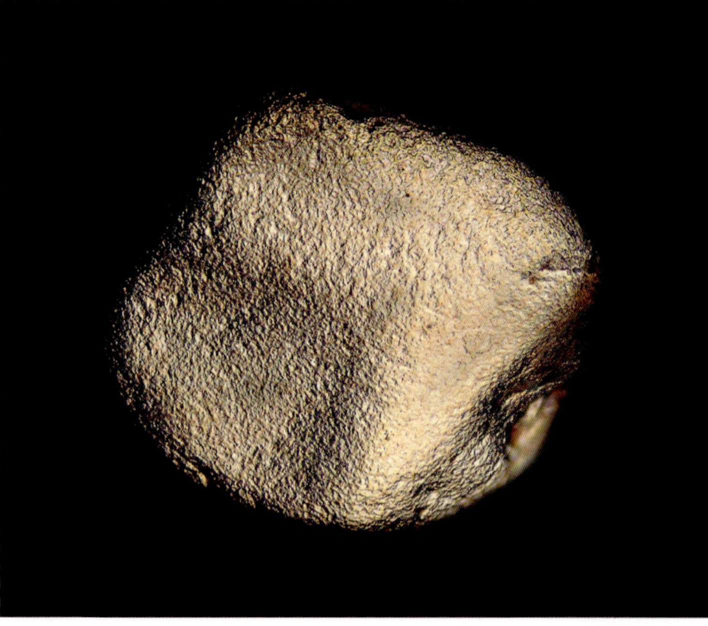

Figure 6.38 Desert sand grains from Waterberg, Namibia. Elongated, well-rounded shape with a relatively smooth surface. Transmitted texture lighting. Grain size 425 × 560 *μm*.

Figure 6.39 Desert sand grain from the Sahara near Murzuk, Libya. Characteristic, elongated, rounded sand grain with matte surface. The so-called *upturned plates* are recognizable. There are strong indentations on all four corners. The matting continues partially in the indentations. Transmitted texture lighting. Grain size 380 × 540 *μm*.

Figure 6.40. Sand grain from the Sahara near Murzuk, Libya. The rounded form with the structure of upturned plates on the surface is characteristic of a dry, interior desert. Macroscopically, the grain seems frosted, which is no longer recognizable in the enlargement shown here. Grain diameter around 900 *μm*. Transmitted texture lighting, natural colors.

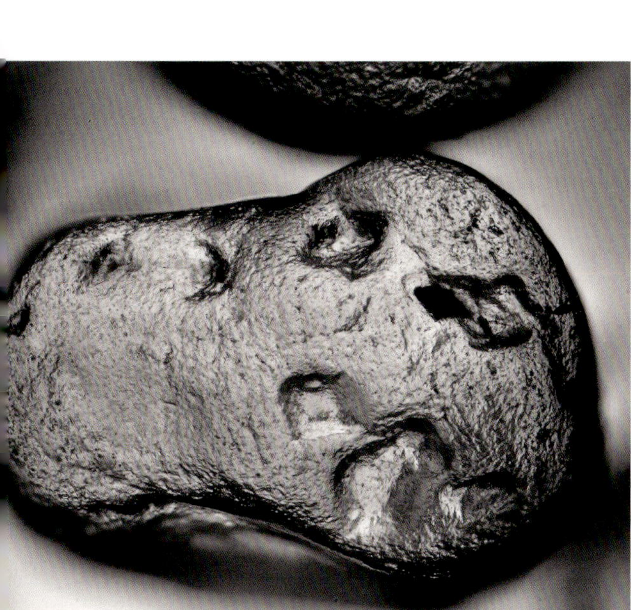

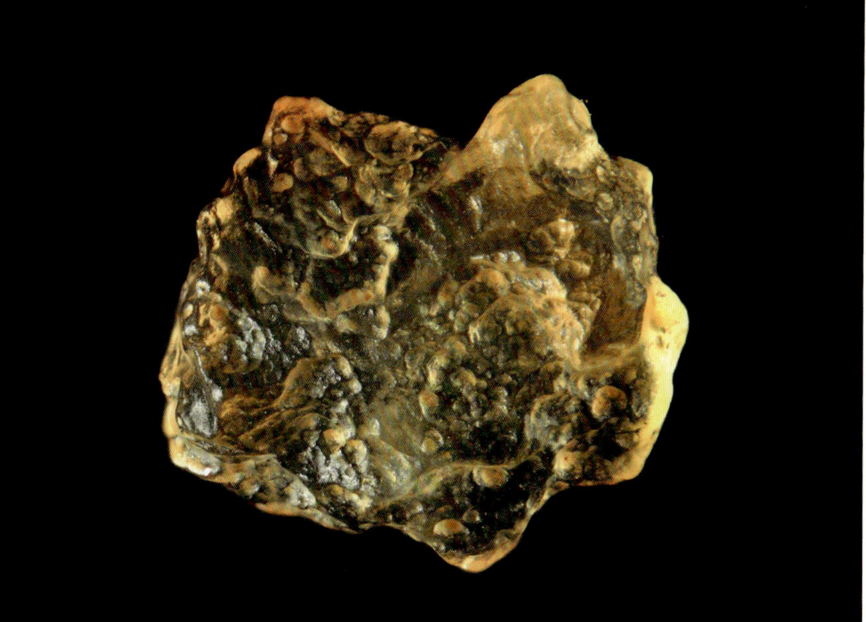

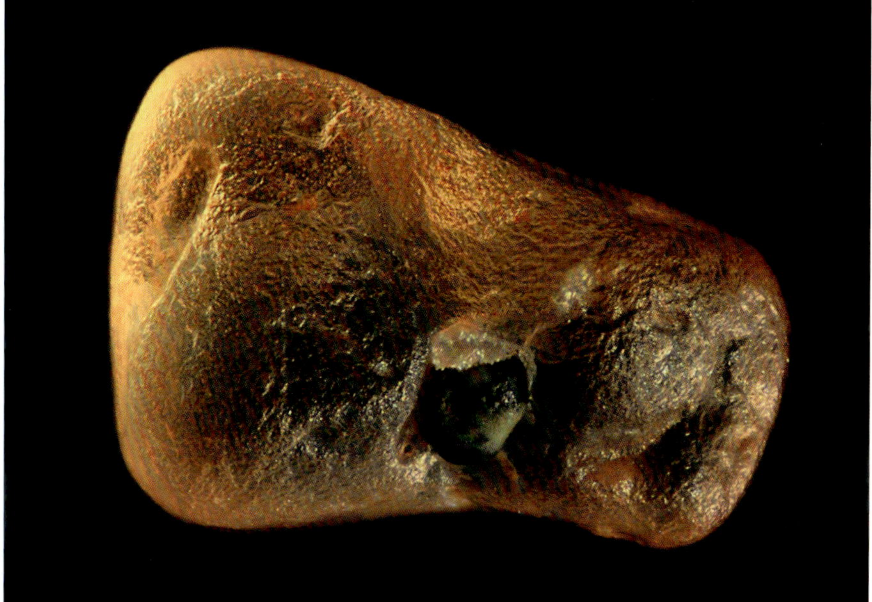

Figure 6.41 Sand grain from the Sahara, Murzuk, Libya. The grain shows typical characteristics of aeolian transport: elongated form, rounded bulbous corners, equidimensional indentations due to high-energy collisions, and upturned plates. The latter leads to a matte appearance of the grain, and the interiors of the larger indentations are comparatively mostly free of plates. Transmitted texture lighting. Grain size 470 × 670 *µm*.

Figure 6.42 Sand grain from the interior desert Lomas de Arena near Santa Cruz, Bolivia. The grain is not typical of this desert but shows noticeable mineral precipitation, meaning chemically dominated surface characteristics. Reflected light, diameter about 1.4 mm.

Figure 6.43 Sand grain from the desert near Abu Dhabi, United Arab Emirates. The grain shows the typical form with elongated indentations, bulbous corners, and a reddish surface, which comes from the deposits of hematite. The small, dark "grain" in the middle is probably an exposed mineral inclusion. Reflected light, grain size 360 × 520 *µm*.

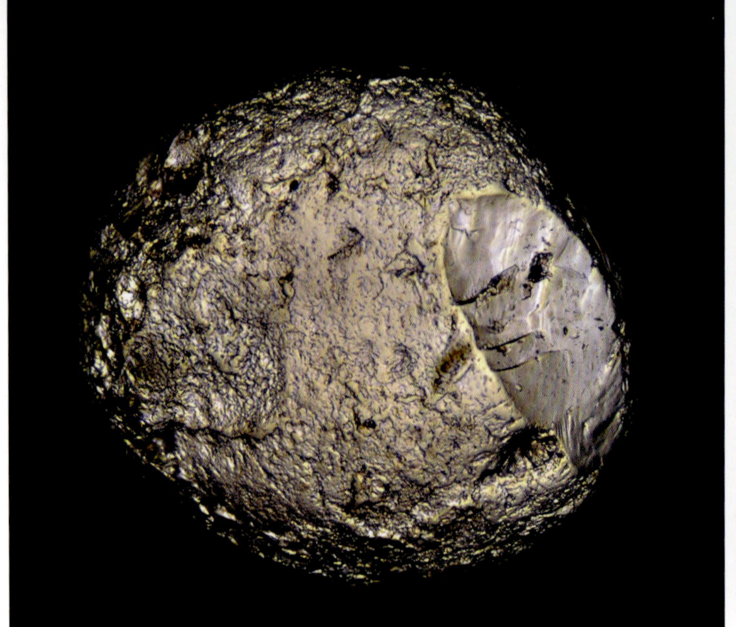

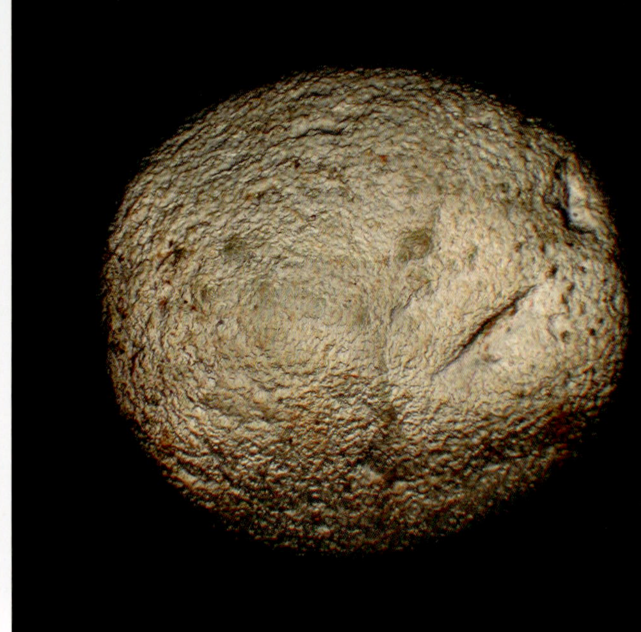

Larger indentations on rounded grains are a very strong indication of aeolian environment. In the case of grain shape, "bulbous corners" are of additional diagnostic importance in this context. The indentations apparently originate from direct, hard collisions of medium to larger grains, each with each other or with the grains creeping and rolling in the soil area, during the saltation phases.

3. Arc-shaped or circular fracture structures are found on many aeolian grains. Other than direct impacts, physical and chemical erosion processes are suspected as causes.

4. Smooth surfaces on rather medium to smaller grain fractions resulting from solution and deposition of silicates. The influence of chemical processes on the surface and also the shape of the sand grains depends very much on the type and duration of weathering exposure. Due to large temperature differences, condensation processes occur at night even in hot desert areas on the ground and in areas close to the ground. In this process, oxides are deposited in addition to silicates, which additionally round structural elements and thus glaze the surfaces. The often-observed reddish color of the desert grains results from the deposition of thin iron oxide films.

And finally, we will introduce a sand grain from Kühlungsborn as another interesting example. The grain comes from an area of sand of the Baltic Sea near Rostock and should have marine/littoral characteristics. But you can't find much of it. Instead, the grain has almost all characteristics that we have described for the aeolian environment, and looks from the exterior to be very similar to the already presented desert sand grain from the Libyan Sahara near Murzuk, which we use to compare. The sand grain from Kühlungsborn is almost perfectly rounded, so it most likely has already run through multiple cycles. The surface has a few sickle-formed collision marks, which can be attributed to the marine but also the aeolian milieu.

Figure 6.44 *Left*: Sand grain from Kühlungsborn, Baltic Sea, Germany. The grain has the typical characteristics of aeolian origin. *To the right*, a fresh break with *conchoidal fractures*, so horn- or shell-shaped breaks. Dimensions 440 × 510 *μm*. Transmitted texture picture while keeping the colors. *Right*: Quartz grain from Murzuk, Libya. Diameter 750 *μm*, transmitted texture picture.

Significantly aeolian, however, are the rounded, upturned plates that largely cover the surface as well as "meandering ridges" in some places. In addition, many small "pits" can be seen, small impact marks that the Sahara grain lacks. Of note is the large and fresh fracture on the right side. This results from a high-energy collision that must have occurred temporally after rounding in aeolian environments, since the fracture surface is not altered or worn, and only the edges of the fracture are slightly rounded.

How the sand grain found its way to the coast is unknown. At any rate, it has spent a long time in the desert or in a landscape similar to a desert in the past and found itself in an environment with high wind energy. It is possible that a combination of coastal dunes and time in a high-energy surf zone near the ocean is the cause. A long history full of movement is behind it, and we can unlock only a small part.

6.2.2 River Milieu, Fluviatile Transport

The fluviatile milieu is very inconsistently expressed with respect to its depositional conditions. Many sand grains of continental origin, coming from a crystalline rock, will have taken their first steps in an alluvial fan. In such alluvial fans, landslide debris flows with many boulders merge with the fine material suspended in the slope water. Their slope usually decreases faster than in rivers.124 Once the sediment loads from the alluvial fans have been released into the river, the sediments begin a long journey toward the sea, following the path of gravity.

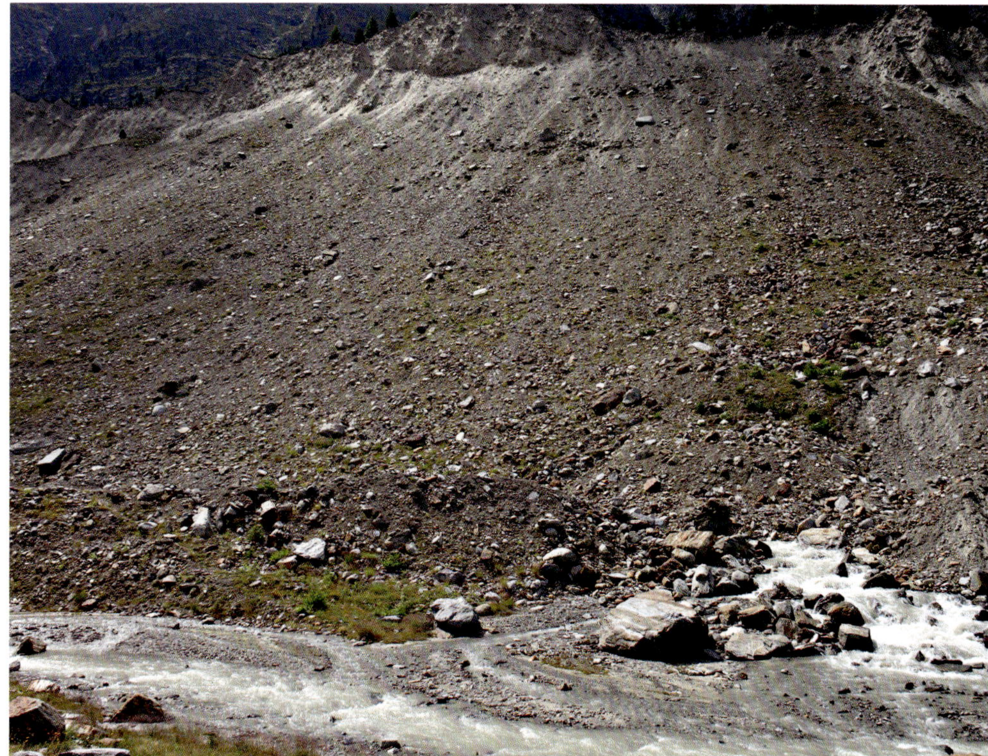

Figure 6.45 Alluvial fan on a glacial moraine slope. The sediment sliding down is caught by a river created by melting water, whose dull color hints at the large suspended load it's carrying. Glacial and fluvial milieus overlap here. Riedbach, Wallis, Switzerland.

Our task of assigning the attribute "fluviatile transport" on the basis of surface observations from the sand grains is connected with many difficulties. For this, it would have to be possible to identify physical and chemical properties in fluvial environments that leave specific traces on the quartz surfaces. Unfortunately, this is not the case. The range of conditions to be expected is very wide: from slow-flowing streams meandering in broad valleys to torrential mountain streams carrying tons of rocks and mud down to the valley, loaded with meltwater. Many sand grains, whose antecedents are known to be fluviatile, have additional glacial traces, which is also not surprising for mountainous regions. Fluvial environments cause abrasions to sand grains through free turbulent particle flows. In addition, looping effects can be detected in rather rare cases at surcharging contacts. Direct grain-to-grain collisions under high energy leave small V-shaped impact marks or V-shaped fractures. These occur rather infrequently in fluviatile environments. As described elsewhere, fluviatile transported grains round very slowly and therefore have angular to subangular shapes. Kuenen (1959) investigated degrees of rounding in fluviatile transport and found that transport distances equivalent to twelve times the equatorial circumference of the earth would be required to achieve good rounding. However, depending on the life cycle of the grain, no more than 620 miles (1,000 km) of transport by rivers can be expected on average.[125] It follows that, in summary, mechanical abrasion through fluvial transport is almost negligible. Even with transport distances of around 6,200 miles (10,000 km), which would correspond to about ten average life cycles, the abrasion value would only be around 2% according to these studies. Elsewhere, siliciclastic precipitates on surfaces are described as characteristic of fluviatile sedimentation environments.[126] Krinsley and Donahue examined sand grains from exclusively fluviatile environments worldwide, which did not show any glacial history. They also concluded that these types of grains do not develop any typical or special surface features other than minor wear.[127] In describing the surfaces of sand grains, the term "relief" is used, with a division into high, medium, and low.[128] In low relief, the surface has almost no topographic irregularities, whereas in high relief, it is distinctly furrowed and fissured and has bumps greater than 10 *μm*. In this sense, fluviatile grains have medium and sometimes strong relief, while marine and aeolian grains are more in the range of weak relief.

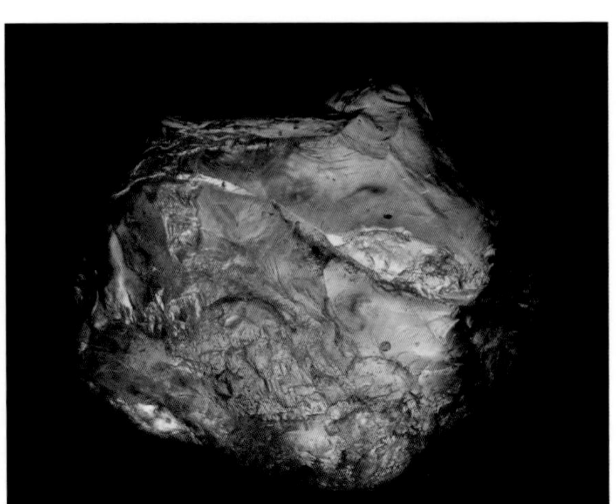

Figure 6.46 Quartz grain from the Po River near Casei, Italy. The grain is atypically well rounded but shows fresh breaks and lines. It has probably already been through multiple life cycles. Image size 625 × 730 *μm*, transmitted-light bright field.

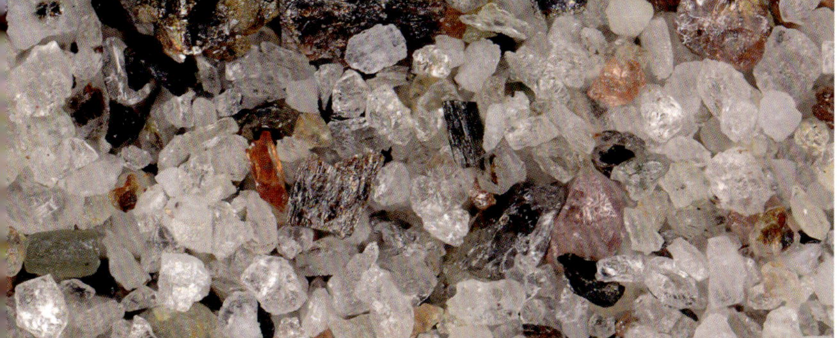

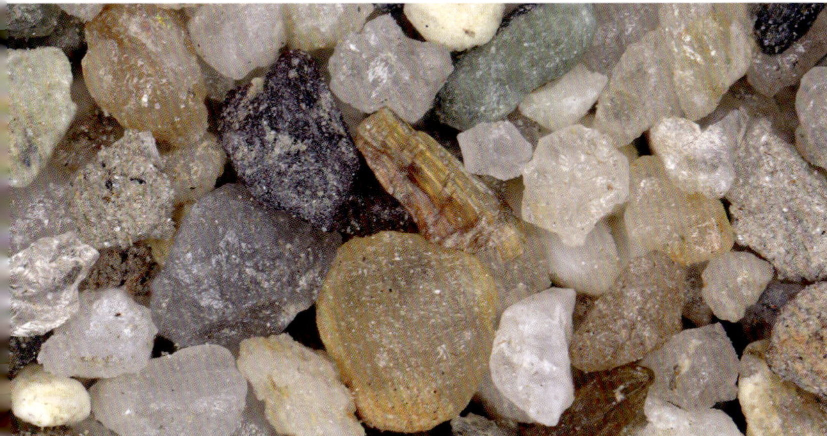

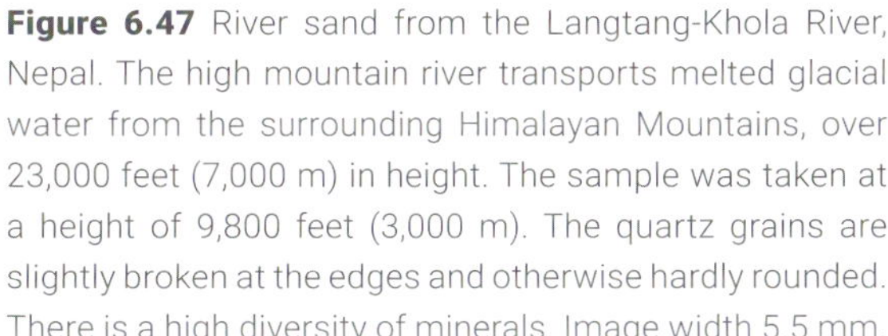

Figure 6.47 River sand from the Langtang-Khola River, Nepal. The high mountain river transports melted glacial water from the surrounding Himalayan Mountains, over 23,000 feet (7,000 m) in height. The sample was taken at a height of 9,800 feet (3,000 m). The quartz grains are slightly broken at the edges and otherwise hardly rounded. There is a high diversity of minerals. Image width 5.5 mm.

Figure 6.48 River sand from the Po River near Casei, Italy. The Po drains the southern Alpine region. The wide range of minerals of the sand grains reflects the large geographical area they come from. The grains are angular to subangular—their compositional maturity is low. Image width 5.9 mm.

Figure 6.49 River sand from the Vorderrhein near Versam, Switzerland. In this location, the Rhein breaks through the sediment masses of the Flims rockslide, which took place ca. 9,450 years ago. The grains are only slightly rounded with their high relief. *To the top left*, a transparent sand grain of muscovite. Image width 3.8 mm.

Figure 6.50. River sand (with attached silt) from the Elbe River near Lenzen, Germany. The grains are well to very well rounded, are partially spherical, and have a high compositional maturity. The sand is made of ice-age sediment and recycled sands from the Elbe Sandstone Mountains. The sample was taken from under the waterline. Image width 3.8 mm.

Figure 6.51 River sand from Rio Fitz Roy, Argentina. The river drains a source area rich in glaciers, so the grains reflect both glacial and fluviatile characteristics. Image width 5.25 mm.

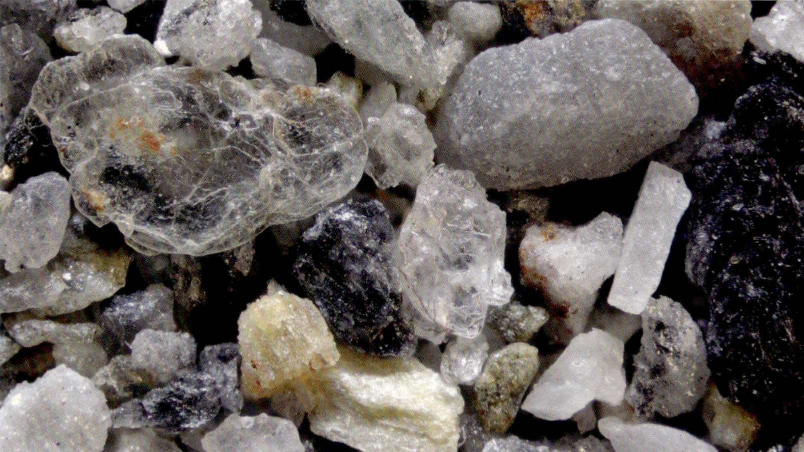

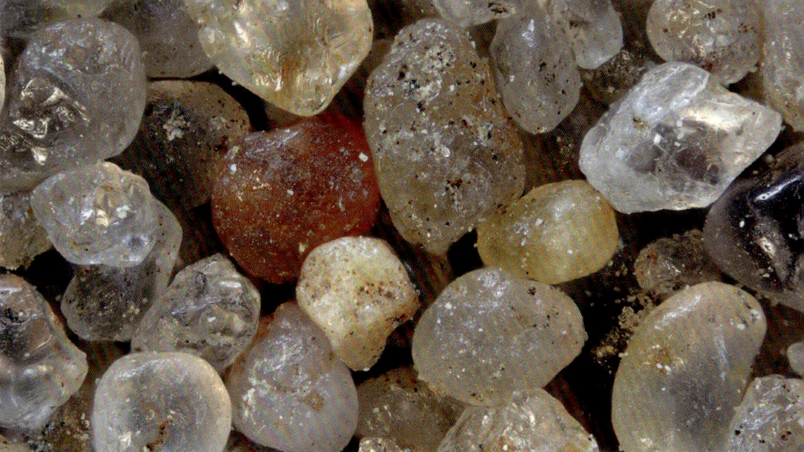

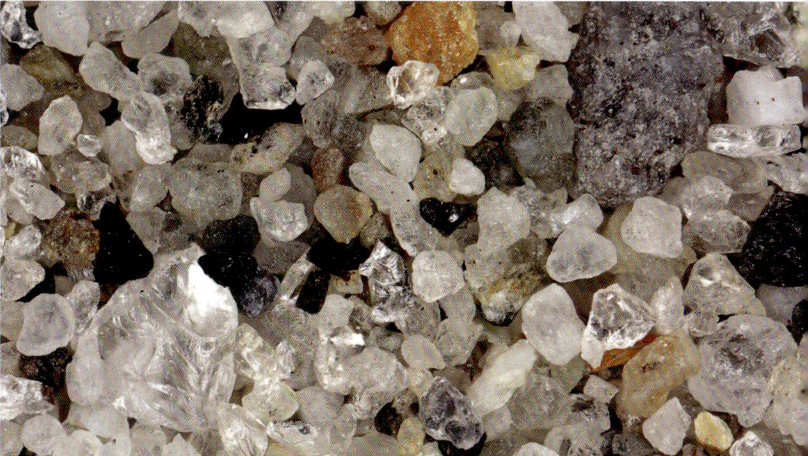

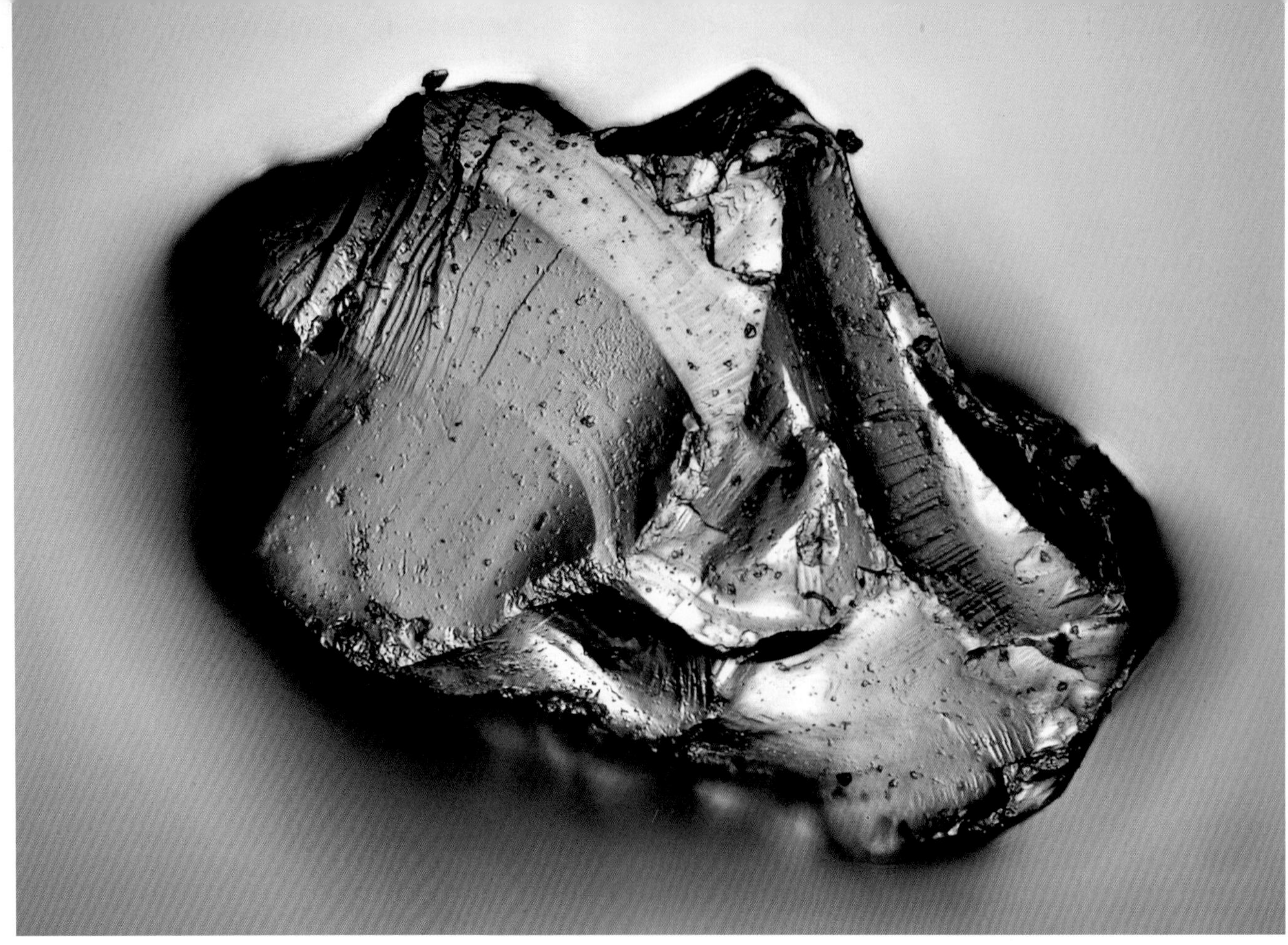

6.2.3 Glacial Milieu, Glacial Transport

As with all other environments, it must be remembered that sand grains not only may be of recent origin but may also bear traces of much-earlier cycles that have been preserved, at least in part. This is especially true of glacial transport. The idea is remarkable because with a grain of sand found close to a beach in a warm climate, one would not think of a past with a glacial background. However, if the grain originates from a coastal sandstone, the prehistory is characterized by the age of the sandstone and the place on Earth where the sedimentation of the rock once started in prehistoric times.

Glaciations can be detected by geomorphological clues, such as round humps and U-shaped valleys, or by traces left by glaciers on the rock surface through corresponding grinding marks. In particular, typical sediments and sedimentary sequences indicate previous glacial environments. These include deposits of moraines, boulder clay, fluviatile sands, and belt clays, but also erratics and dropstones.

Figure 6.52 Sand grain from Rio Fitz Roy, Argentina. The quartz grain has an angular contour without significant rounding. A glacial origin can be inferred. Abrasive traces on the ridges can be seen, as well as fragmentation with stepped break marks and silicate deposits below the middle of the grain. Grain dimensions 450 × 330 *μm*. Transmitted texture lighting.

Figure 6.53 *Right side*: Sand grain of quartz from Rio Fitz Roy, Argentina. The grain is angular, without significant edge rounding. Even the ridges are consistently sharp edged. The breaks are significant, with stepped break marks and smooth break surfaces being recognizable. To the top right of the grain, there is a shell-shaped fracture. A glacially influenced history is recognizable and geographically plausible. Grain size 585 × 995 μm. Transmitted texture lighting.

Even today, many landscapes are clearly marked by former glaciations. In northern Germany, as already described, glacial debris is of particular importance; it can provide evidence both of glaciation and of the former Scandinavian surface rocks that were transported southward by the ice. With careful observation, many bedloads can be discovered that carry clear traces of glacial transport.

For us, the question arises of what traces the glacial transport could leave on the surface of individual sand grains that are still recognizable later. This is not quite trivial, since the localization of glacial/nonglacial is less clear. In the environment of a glacier, very different environments exist, which produce correspondingly different or, in extreme cases, no traces at all on the grain surfaces. In "warm glaciers" the ice glides on a film of water (pressure solution), and in "cold glaciers" the ice is frozen to the ground and carries rocks and sediments with it during transport. Sediments of the carried or pushed ground moraine are exposed to very strong abrasive forces. Slow movement under high loading pressure of the ice leads to high normal stresses with simultaneous high shear stresses in the affected rocks.

Completely different conditions are found when sediments are transported frozen on the surface of a glacier or in its interior, where they are largely shielded from mechanical influences. The ice in the center of glaciers moves faster than at the margin or the surface; sediments therefore mix during transport. Despite these very different conditions, common features can be found in glacial sand grains. Even more than grains from fluviatile environments, they show the sharpest fracture edges and are clearly angular in nature. In this they surpass even the grains directly eroded from the source rock, since these are already chemically altered by eroding processes. Sand grains crushed and transported by glaciers show almost fresh, uneroded surfaces.

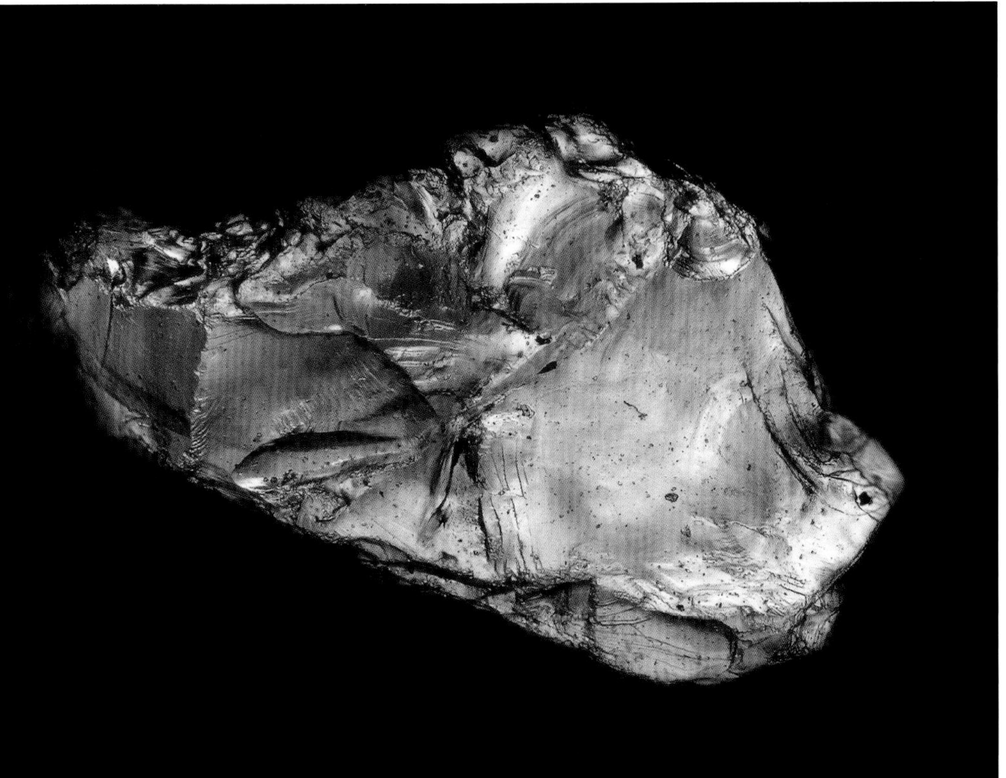

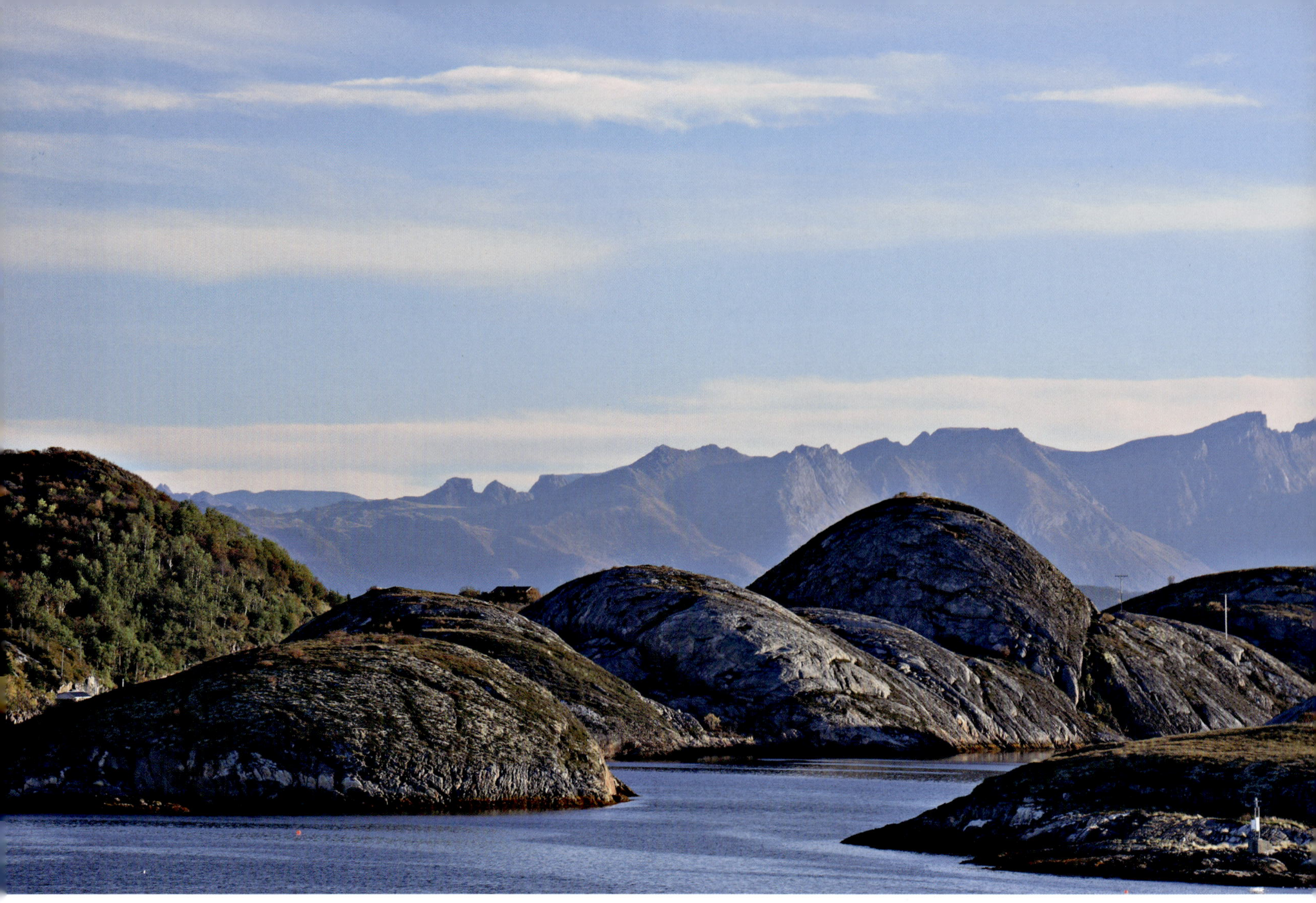

This leads to the question of how a fresh fracture occurs and how it can be identified. Generally speaking, a grain of sand that has a fractured surface has undergone a material failure. Here we refer to the "material" quartz, which, like glass or ceramics, belongs to the brittle materials. Brittle materials, as we know from everyday life, are susceptible to impact or even point loads. If we throw a ceramic cup against the wall, it will break; if we use a cup made of aluminum, it will come away with a dent. Technically speaking, the ceramic cup suffers a brittle fracture, while the aluminum cup suffers a plastic deformation failure. In either case, the material failure "force fracture" leaves a very characteristic structure on the fracture surface. Glass, some ceramics, and quartz show a conchoidal fracture in this process. The crack fronts are arranged concentrically and altogether look like a shell. Also, vice versa, we can conclude that in the presence of a conchoidal fracture on a grain of sand of quartz, either an impact-like event or a slow load increase over the point of material failure occurred in prehistory.

Figure 6.54 Glacially formed landscape in Norway in the polar region near Brønnøysund

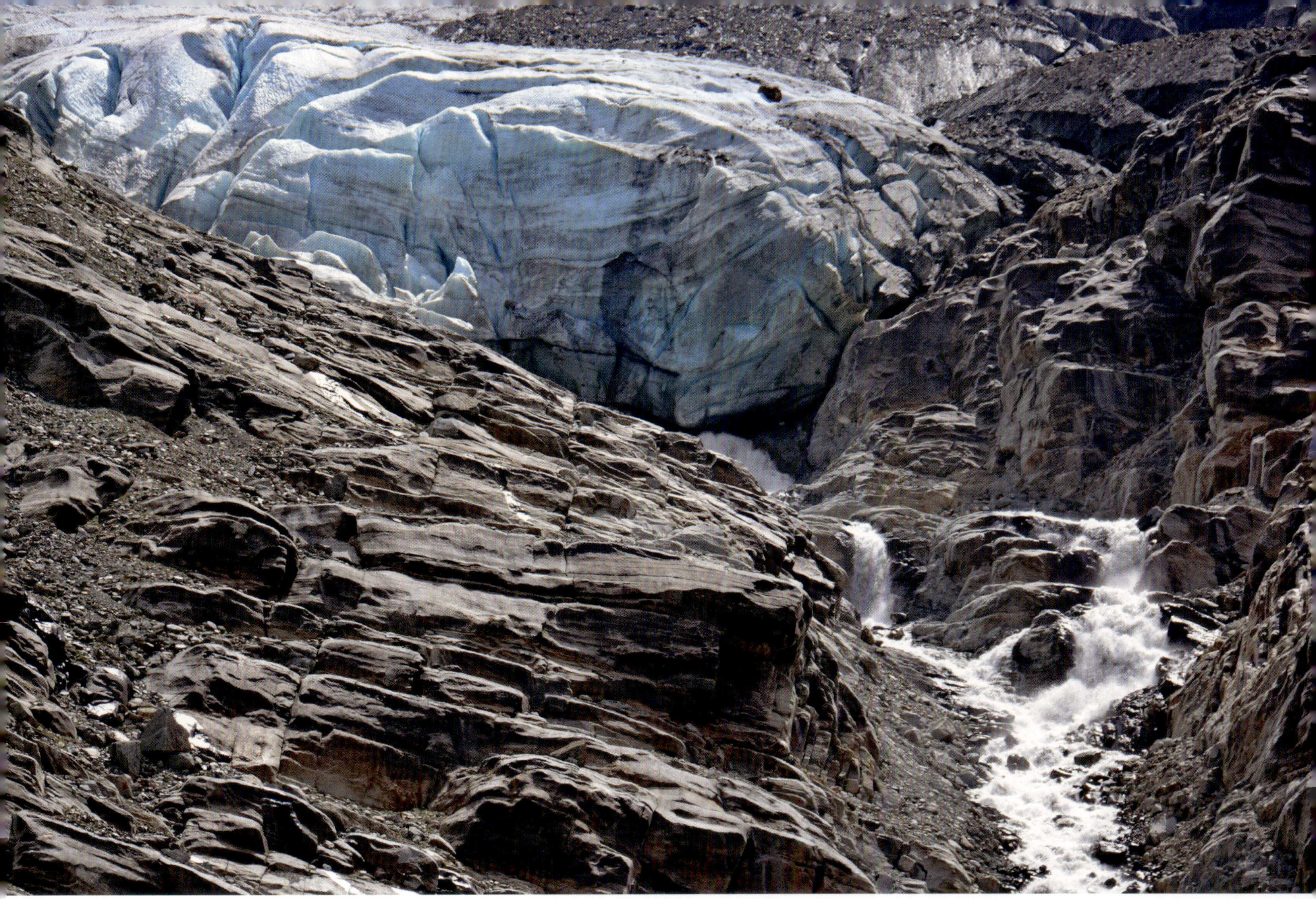

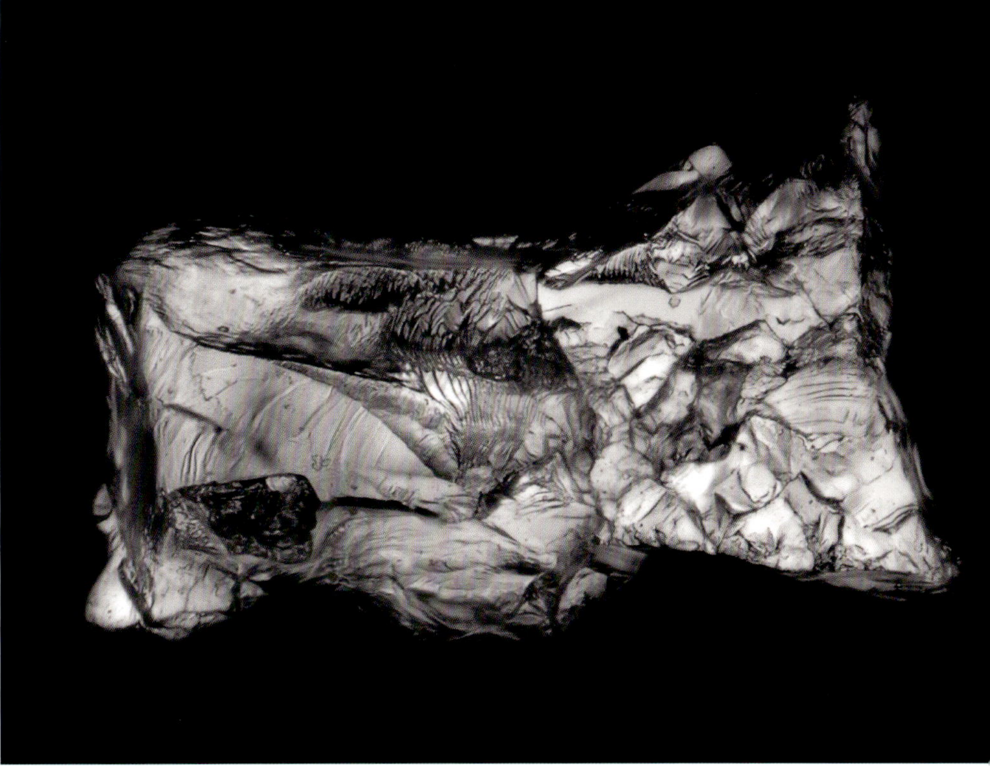

Figure 6.55 Ried Glacier near St. Niklaus, Switzerland. Glacial and fluviatile milieu cannot be easily separated in this area. The different transport situations to which sand grains can be exposed are recognizable.

Figure 6.56 Sand grain from the Ried Glacier, Switzerland, which was taken directly from the glacial water flowing out so that influences from fluvial transport could be ruled out. The edged contours with very low edge rounding, as well as the surface characteristics that are present, confirm this. Ridges from the breaks (*middle left*) don't show any abrasion. Far to the right is a recognizable shell-shaped break structure. Additionally, to the left side, there are multiple series of arch-formed steps that can be seen that similarly have signs of fresh breakage. Grain size 260 × 520 *μm*. Transmitted texture lighting.

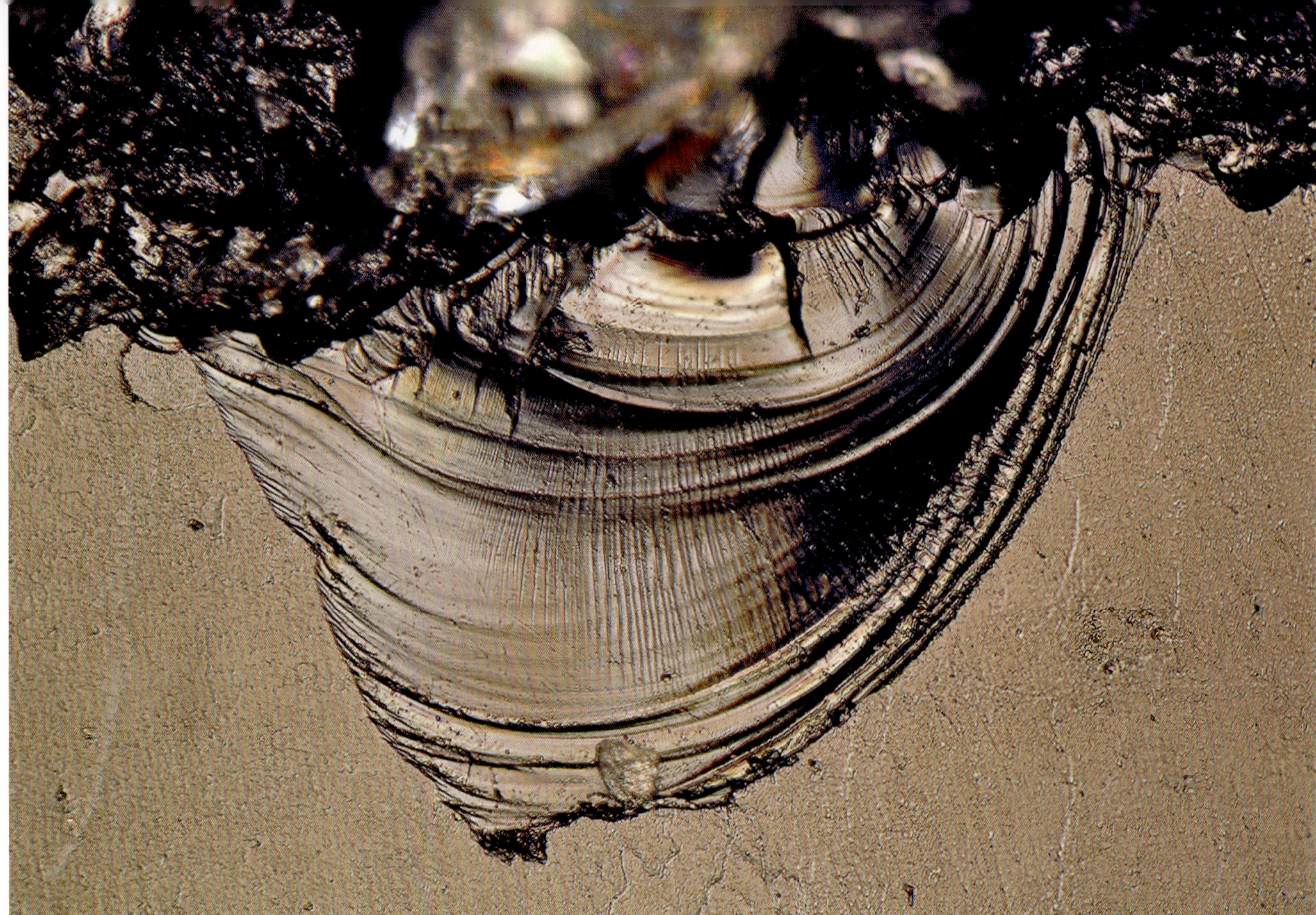

This aspect is essential, since shock load occurs very often in high-energy submarine areas on coasts, while slow load increases are more typical for glacial areas. In general, however, glacial transport events are expected to show a strong heterogeneity in terms of the stresses on the sediments involved.

With these considerations in mind, we can now turn to the specific surface characteristics of glacial sand grains. The following six primary characteristics are sorted, starting with the most important. If multiple characteristics can be observed on the sand grains, there is most likely a glacial history.[129]

1. High variability in the dimensions of shell-shaped break structures
2. Very large/high relief compared to aeolian or marine-influenced grains
3. Semiparallel steps. Deep cracks or breaks in the grain surface with distances over 5 μm, which were most likely caused by fractures from shear stress.
4. Arch-shaped steps, probably from impact stress. Deep cracks or breaks in the surface that were caused by impacts.
5. Parallel stripes of different lengths. The stripes can be evaluated as scoring marks that are caused by movement relative to sharp structures. Indentations caused by a sharp-edged counterpart ("scratch marks").
6. Fracture structures in an arrangement similar to roof tiles

Figure 6.57 Shell-formed break (brittle fracture) on the edge of grown quartz crystal with Wallner lines. The surfaces to the left and right in the picture are naturally grown crystal surfaces. Width of the break is 480 *μm*. Differential reflected-light inference contrast.

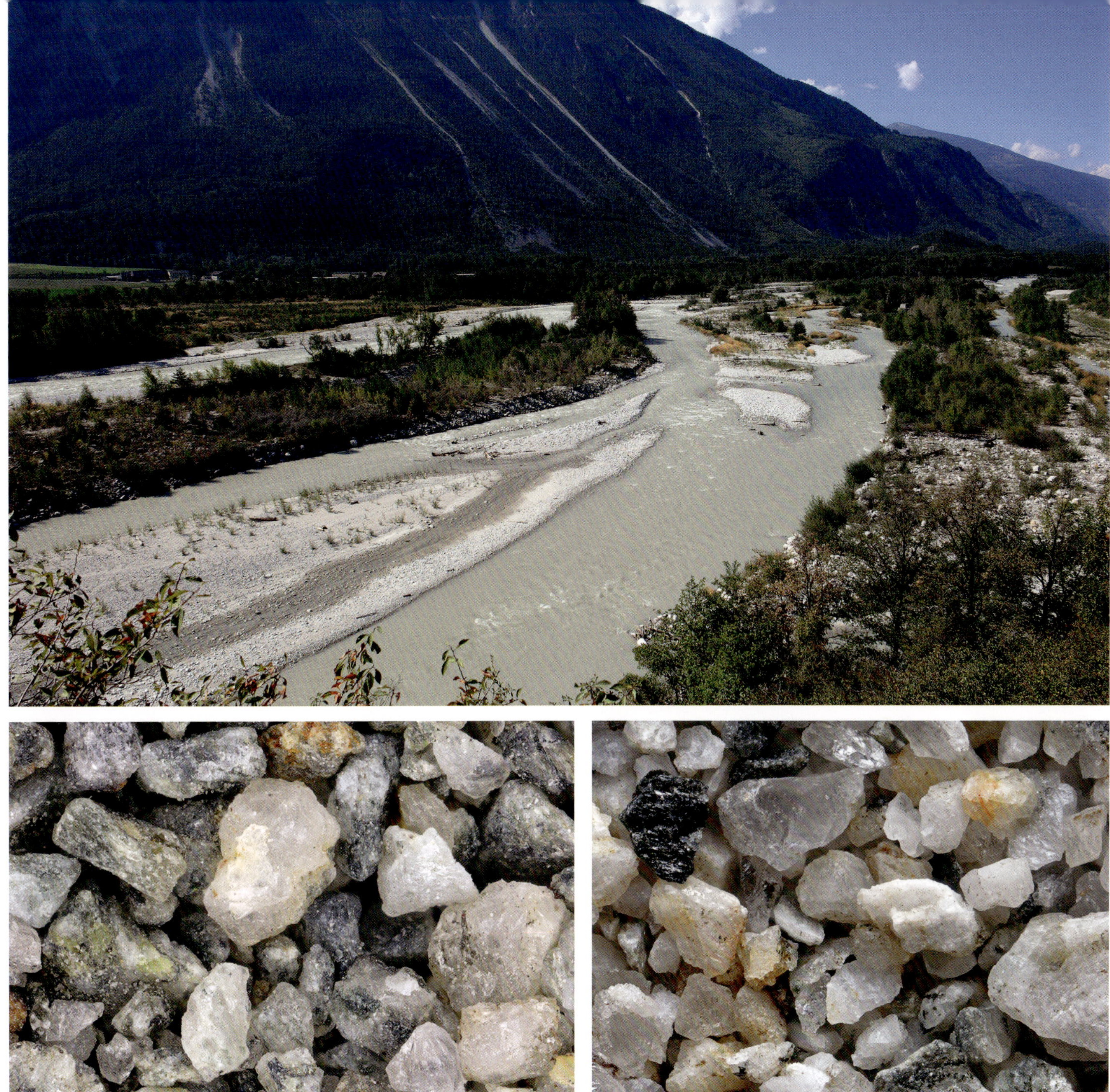

Figure 6.58 Meandering, young Rhone between Leuk and Siders, Switzerland. Glacial and fluviatile sediments mix here and are deposited by the river on the constantly newly forming gravel and sandbanks.

Figure 6.59 Sand from the Jungtal, Switzerland. The sample was taken less than 330 ft. (100 m) below the tongue of the Jung Glacier from the glacial stream. Macroscopically, there is no recognizable rounding. Image width 5.5 mm.

Figure 6.60. Sand taken from the fresh meltwater stream of the Rhone Glacier in the high valley of Gletsch, Switzerland. Typical edgy, unsorted, and fresh sediment with the lowest compositional maturity. Image width 7.3 mm.

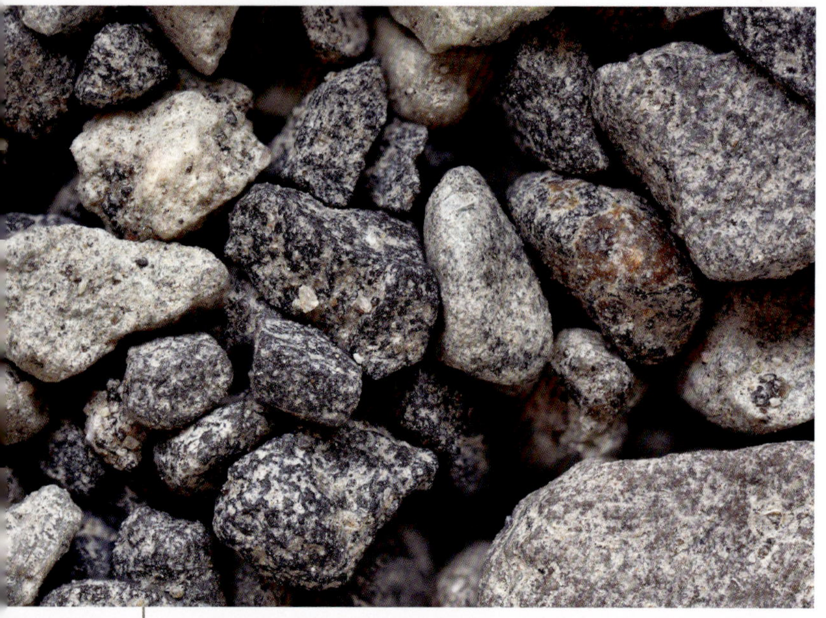

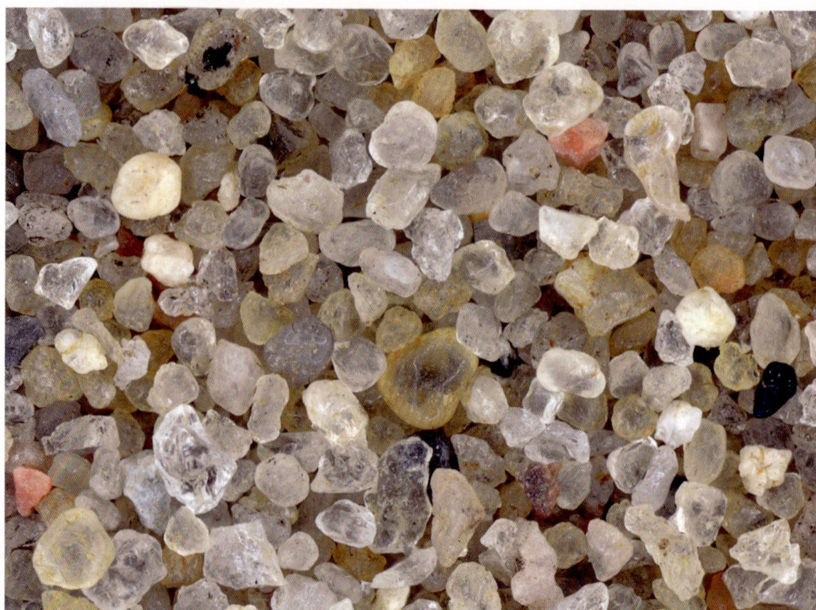

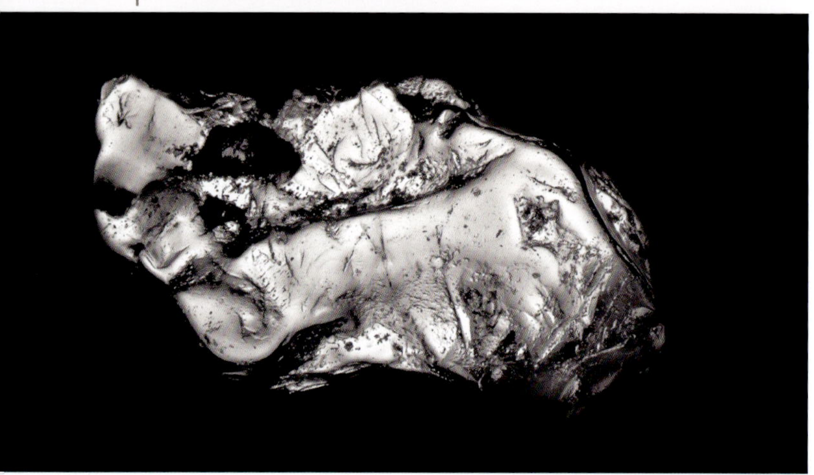

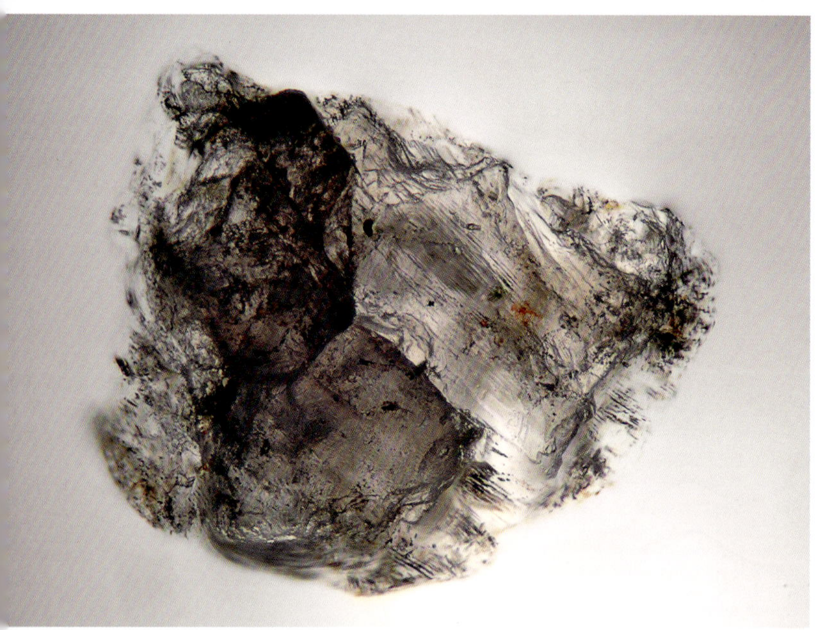

Figure 6.61 *Top left*: Sand taken from the glacial gate of Vatnajökull, Iceland. The grains are well rounded at the edges, which points to a turbulent transport environment in the meltwater area. Image width 9.1 mm.

Figure 6.62 *Top right*: Sand from Fürstenberg/Havel, Germany, so-called Mark Brandenburg sand. The landscape is characterized by Weichselian young moraine plates and sediments. The grains are partially very well rounded, which can point to aeolian phases. Many grains show their ice-age origins with additional fluviatile characteristics. The sample was cleaned with water before being captured and is therefore free of fine particles. Image width 5.5 mm.

Figure 6.63 *Middle*: Sand grain from Fürstenberg/Havel, Germany, from the area characterized by Weichselian young moraine plates and sands. The grain has a complex morphology and shows traces of abrasive wear (rounded corners). Additionally, score marks (*top right and lower middle*) are recognizable. Traces of erosion are spread over the whole grain. Upon further enlargement, impact marks can also be seen. The mix of fluvial transport and glacial characteristics tells of a history in the ice-age glacial and river systems. Grain size 380 × 860 *μm*. Transmitted texture lighting.

Figure 6.64 *Below*: Quartz grain from the Rothera Station, Antarctica. The grain is very typically triangularly faceted and shows linear structs to the top right. Glacial transport is very possible. Transmitted texture lighting in original colors. Edge length around 1,000 *μm*.

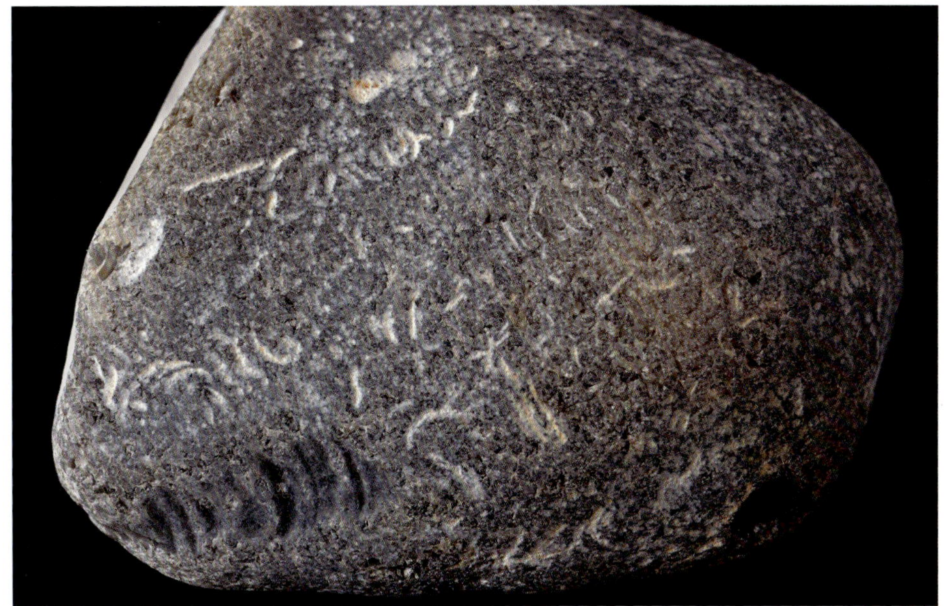

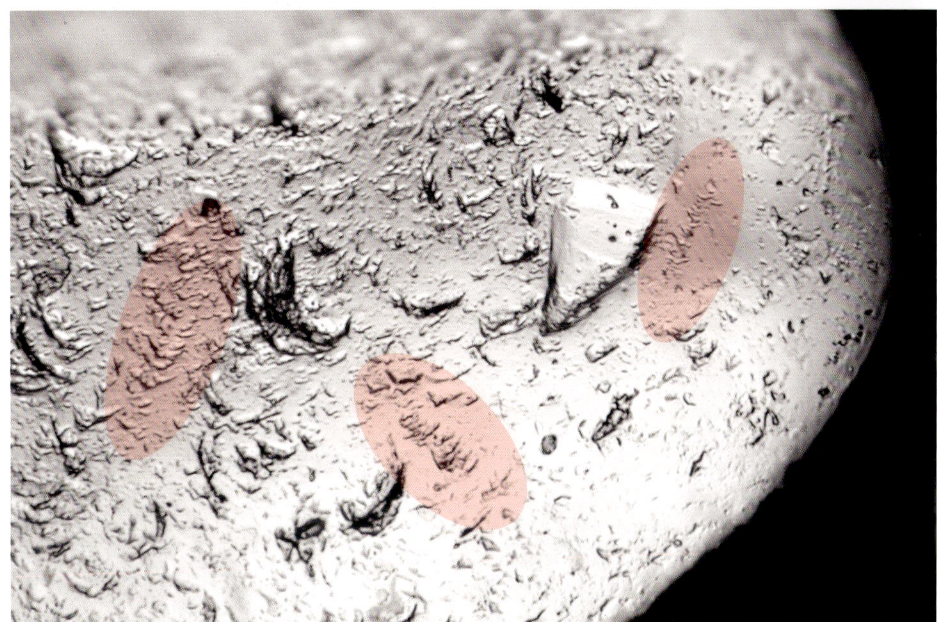

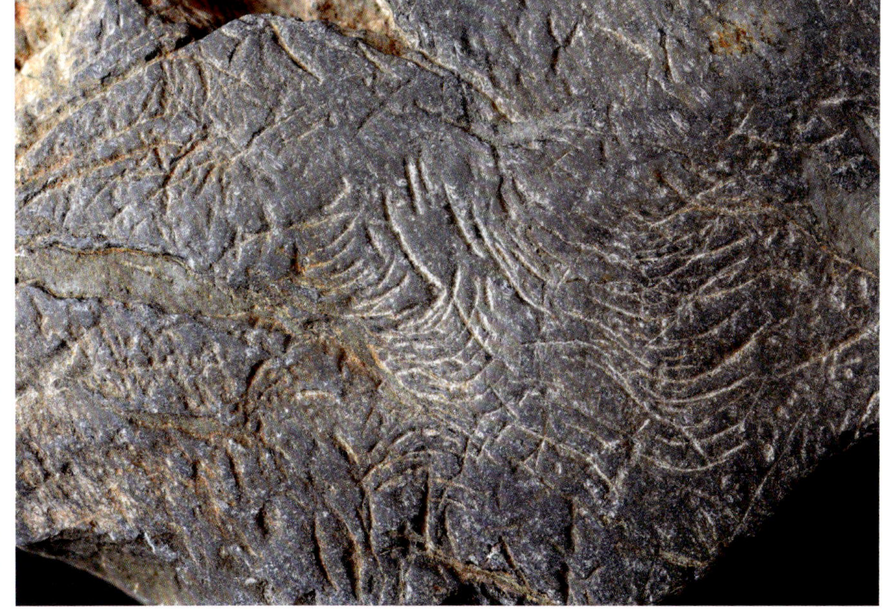

Figure 6.65 Flint from Fehmarn, Germany, with many whitish pressure marks. To the bottom left of the stone, a coarse structure can be seen, which with its size of a sand grain is characterized as a microstructure with "chatter marks." A sharp-edged counterpart impacted the firestone under high pressure and relative movement and left behind these deep indentations as a result of a stick-slip process. Image width 80 mm.

Figure 6.66 Grain from the coastal area of Kerala, India. The structures highlighted in the red ovals are "chatter marks" (width about 10 *μm*). Formation mechanism as with comparable macrostructures (e.g., on flints). Image with 280 *μm*. Transmitted texture lighting.

Figure 6.67 Flint from Hochwacht, Baltic Sea, Germany, with series of successive parabolic cracks (sickle marks). The Kerala sand grain from earlier is comparable here, with similar markings but much-smaller dimensions. Image width 75 mm.

Now, at the end of the chapter, we will take a look at glacial structures with a common and interesting phenomenon that simultaneously forms a segue into the marine or littorale milieu. It has to do with glacial sickle marks, impact marks, and parabolic cracks.

A "parabolic crack" is understood to be a smaller and single structure that can occur in large numbers but, unlike "sickle marks," is not usually arranged in steps one after the other. Sickle marks also have larger dimensions. Both forms undoubtedly can be traced back to glacial activity.[130] The sickle-shaped form of indentations actually occurs over four orders of magnitude (20 *µm* to over 30 cm) in different contexts, which are not yet completely clarified, but we now want to talk a bit about the reasons why.

Each type and expression of this phenomenon reveals that an interaction must have taken place between ice and rock, or between two bodies consisting of rock, with or without water. Relative movement is always necessary. On the one hand, this relative motion can take place slowly and uniformly, as is the case with the slow creep of glacial ice over bedrock, in which case parabolic cracks or sickle marks are formed. The movement can also be undirected and jerky, as expected in grain-to-grain collisions with or without water, in which case impact marks are produced. There is also a third possibility. If the continuous creeping movement is temporarily inhibited and then very suddenly released, an effect can occur that is called stick-slip. The involved bodies move jerkily in relation to each other, where phases of sticking and phases of slipping follow one another in immediate succession. Oscillations are induced, and chatter marks are formed on the involved bodies.

Earthquakes follow this stick-slip principle, just like the squeaking chalk on the chalkboard or the squeaking door hinge as examples out of daily life.

The flowing glacial ice layer causes, in addition to the relative movement above the bedrock, a very high contact force of the transported debris on the bedrock. The higher the contact forces due to the ice layers, some of which are many hundreds of meters thick, the greater the frictional forces between the sediments moved with the ice and the bedrock. This resulted in wear marks with abrasion and the well-known smooth grinding of glacially formed mountains. For sand, Mahaney (2002) was able to show that the thickest ice layers also result in the most intense destruction of the grains. Furthermore, the intensity of the grain eruptions is correlated with the temperature of the ground moraine in the glacier.

Figure 6.68 Sickle marks on a mound of granite near Hammeren, Bornholm, Denmark. Width of the marks is around 20 cm. The steep fracture edges lie on the top left side, respectively. It has to do with convex sickles with glacial movement from right to left.

Figure 6.69 Flint from the coastal strip near Vang, Bornholm, Denmark. The rock, with a length of 30 mm, shows two very significant complex sickle marks, which find themselves exposed on the tip side and are structurally similar to a glaciated mound. In addition, the surface of the stone is covered in many randomly distributed, sickle-formed impressions (impact marks).

Figure 6.70. Schematic depiction of a few surface characteristics and their names

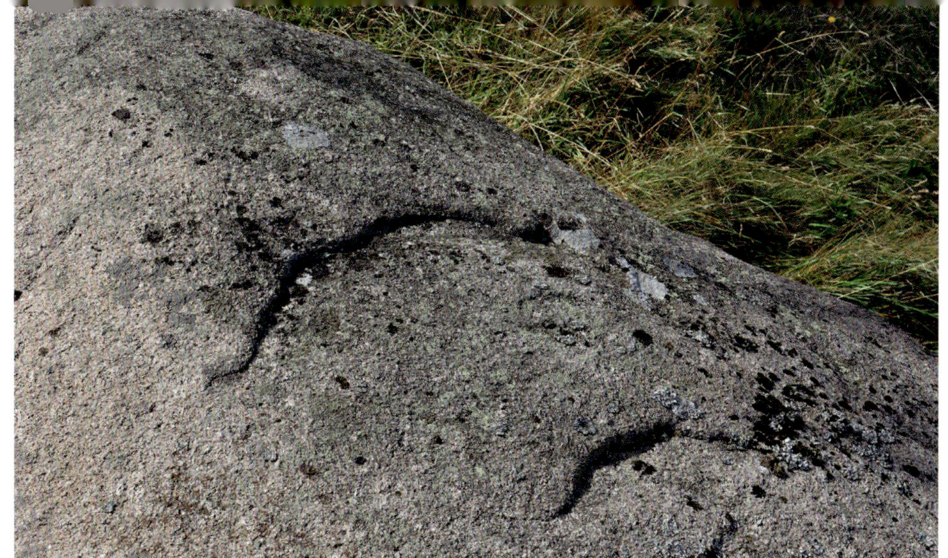

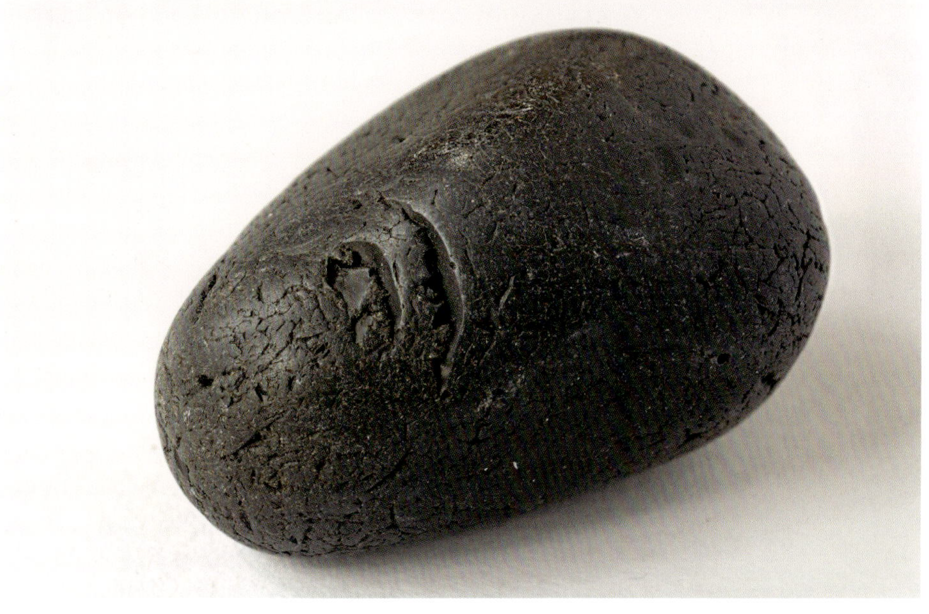

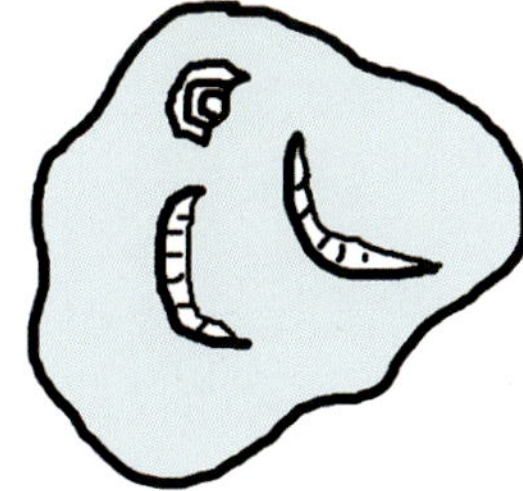

Convex sickle marks (below)
Graded arc (above)

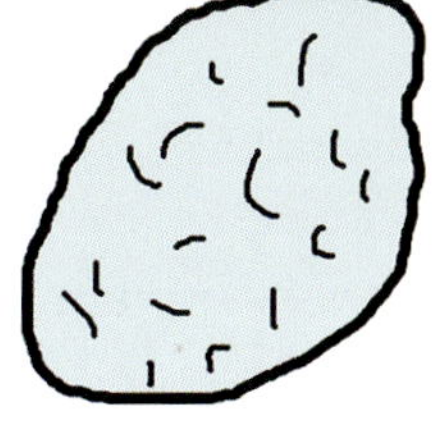

Crescentic fractures
Percussion/impression marks

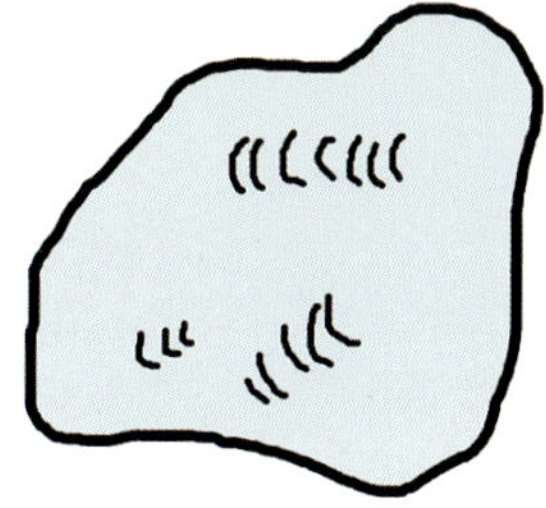

Chatter mark trails

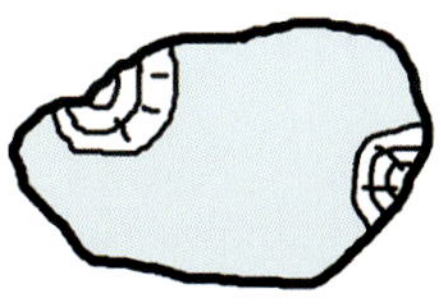

Conchoidal fractures

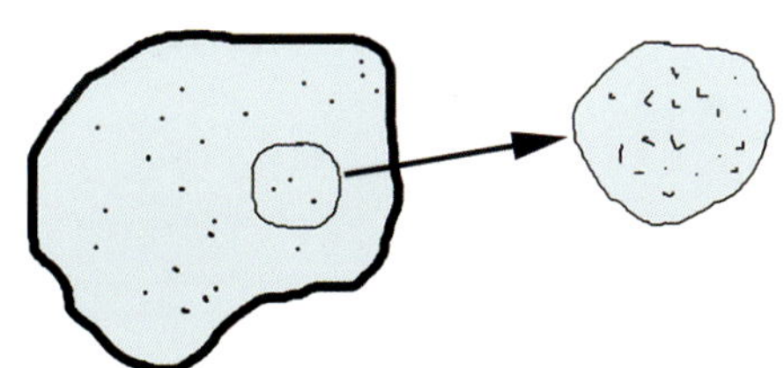

Pits and V-shaped percussion cracks

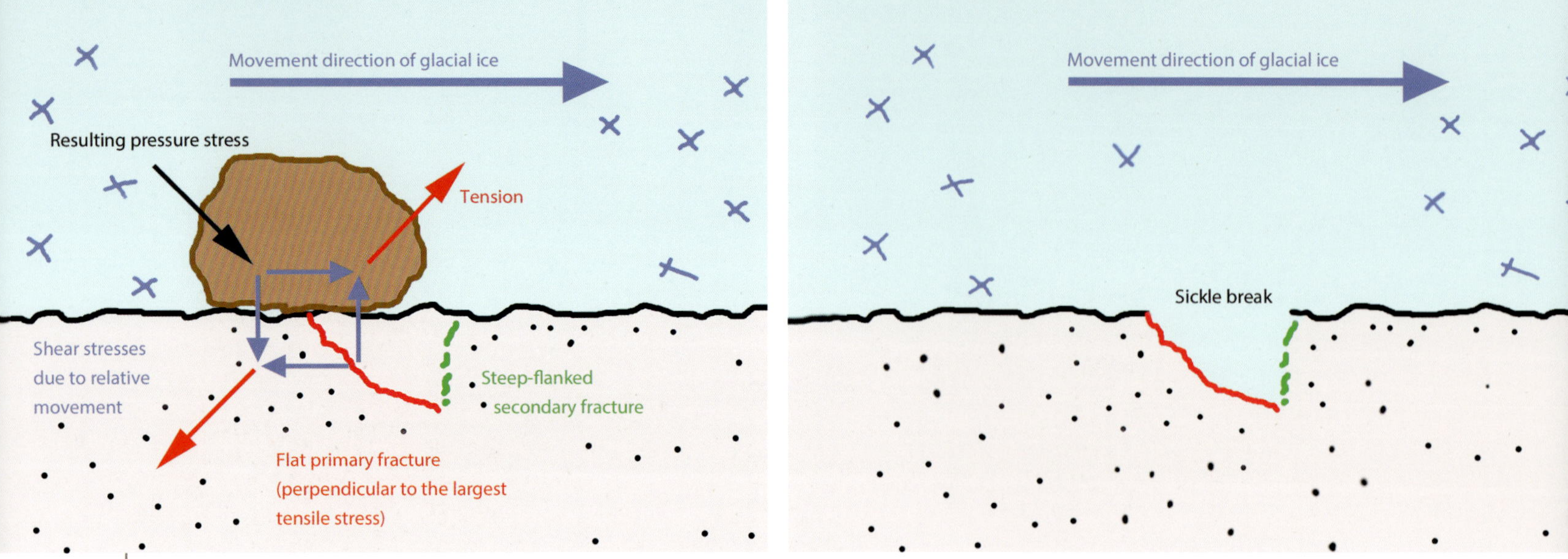

In very cold glaciers, the sediment itself is frozen and the frequency of grain-to-grain impact is reduced.[131] At higher temperatures, there are flowing water masses that strongly increase the frequency of contact and can sometimes lead to a complete destruction of the grains. The grains then show triangularly faceted basic patterns independent of their original form, with the original structure and surface mostly destroyed. In lab tests it could be shown that there are no other geological processes that can so fundamentally change the surface of quartz grains like glacial processes.

After the material breaks, sickle marks always have a flatter and steeper flank in relation to the cross section. Due to the force equilibrium at the friction point, it can be demonstrated that the movement of the ice always takes place from the flatter to the steeper flank (*see figure*). This is even independent of the shape of the sickle in the top view. In fact, two shapes can be distinguished here. So-called convex sickles have a steep flank that is convexly bent in the direction of the ice flow, whereas with "concave sickles" it is exactly the opposite.

Taking into account these considerations, sickle marks are oriented to the direction of the flowing ice current. Conversely, the flow direction of the former glaciers can be deduced from the arrangement of the marks in the bedrock. This also applies to glaciation periods long in the past, which is why sickle marks are a useful tool for reconstructing past climatic periods.

If we now mentally reduce the size scale, you can imagine that the described process can also be applied to smaller rocks. Indented sickle marks can also be found on boulder-sized blocks, on debris, and even on small flint nodules, especially the wallstones (German: *Wallsteine*). The wallstones from farther up (fig. 6.69) show exactly the expected sickle pattern, and it can be assumed that the rock frozen on ice came into contact with another object, and then the sickle form of the fractures formed according to the described mechanisms.

Figure 6.71 *Left*: Forces and fractures occurring in the contact area of boulder bedrock. *Right*: Cross section through the resulting sickle fracture after the boulder was transported away by the ice..

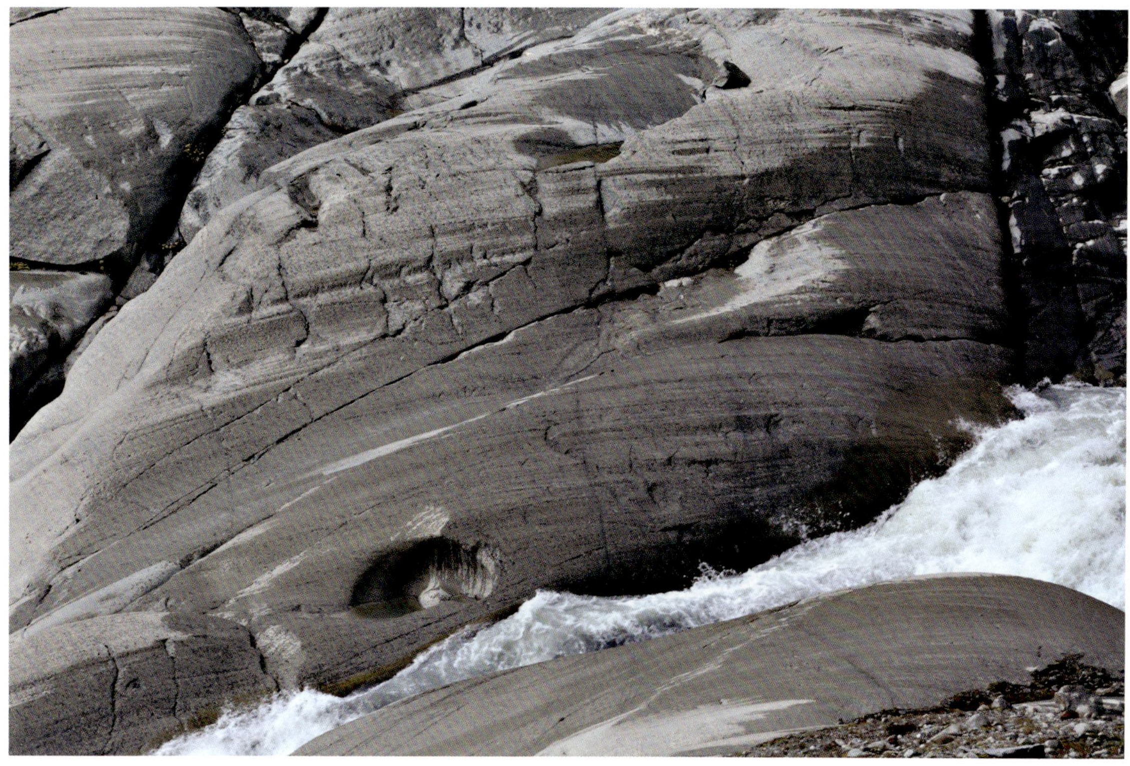

It is easy to see that in the case of boulders, there is not necessarily an orientation to the marks, since the boulders can change their position during transport in the ice.

The wallstones already mentioned represent a small peculiarity. They are flints in spherical or elongated-ellipsoidal form, which occur in various colorations from black to light brown. A commonality between them is a very smooth surface, almost polished, which makes it challenging to find them among the beach debris on the eastern North Sea and the Baltic Sea. They also sporadically occur inland in gravel pits that were loaded by the debris of the ice ages. If you break a wallstone, you can see that the surface is always somewhat lighter than the center and thus has a kind of patina.

Figure 6.72 Source of the Rhone, Furka Pass, Switzerland. *To the right*, beyond the picture, is the glacier with glacial lake. Many sickle marks on the polished granite. Former direction of movement from the Rhone Glacier, which caused it, was from right to left.

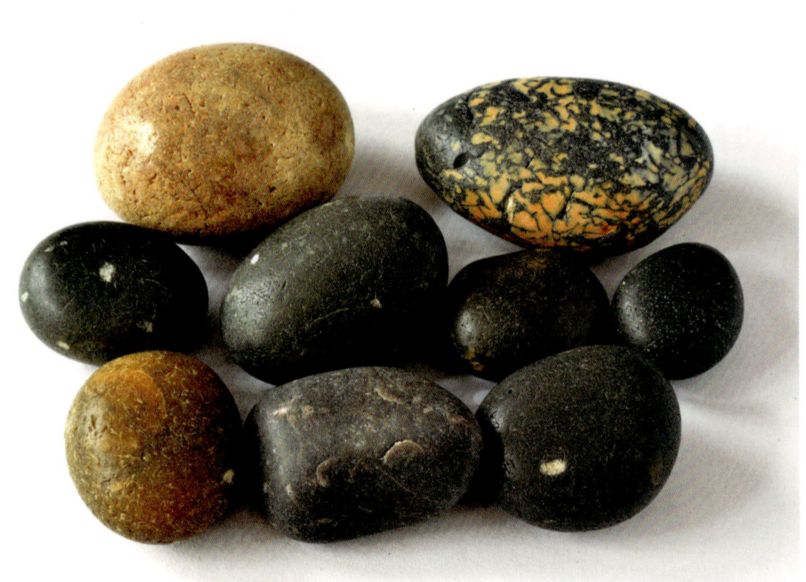

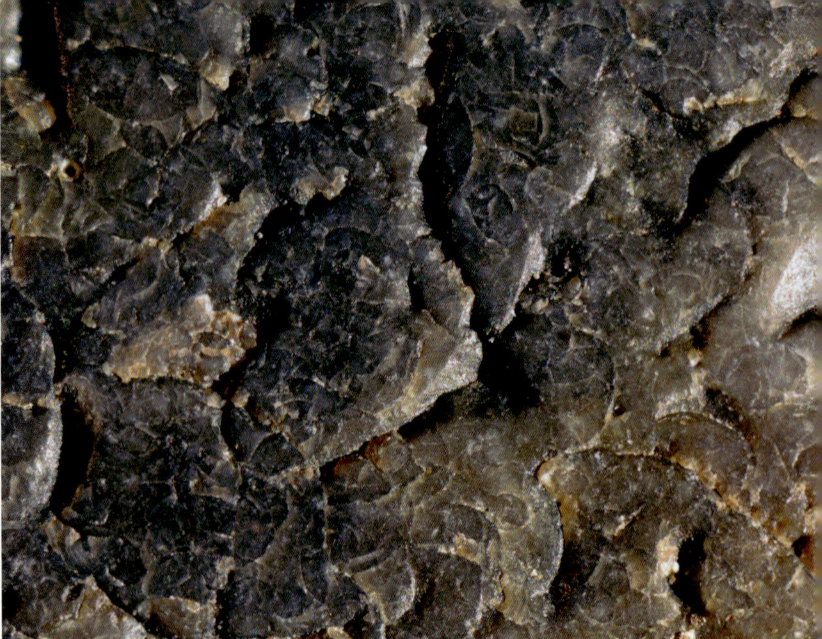

Of particular interest for our investigation is the fact that surfaces of the wallstones are covered with sickle marks. These are predominantly sickles, which, in contrast to those described above, are very shallow and therefore have more cracks. The term “impact mark” has become established in literature for this term. It can be assumed that a hard and randomly oriented stone-to-stone impact can leave such a mark on the surface of a piece of flint. The enlarged image of the wallstone surface shows innumerable microscopic cracks of the type described, which are then apparently further deepened to some extent by weathering or renewed impact. Particularly interesting are the areas where two crescents cross. It can be clearly seen that V-shaped splinters have erupted there and have accordingly left V-shaped marks. The comparison to the V-shaped impact marks on sand grains is obvious. Probably both have an analogous pattern of causes.

But where do the wallstones come from? Their history is not yet completely clear. The most likely explanation is a fascinating one: Wallstones were beach debris from a prehistoric sea. Cretaceous flints were unrolled hundreds of thousands of years ago in the surf of the ancient Tertiary sea, shaped into their present form, and probably finally piled up to form beach embankments. Millions of years later, the glaciers of the ice ages transported them to the south and distributed the wallstones over a very large area,
where we can still find them today, even far inland.

Figure 6.73 *Left*: Collection of wallstones from the coast of Bornholm, Denmark. All the stones have smooth, almost polished-looking surface. The whole surface of the stone is covered in impact marks. On the stone below the center, the large, sickle-formed marked are especially recognizable, with their broken edges appearing whitish. Diameter of the rocks is between 20 and 55 mm.

Figure 6.74 *Right*: Surface detail of a wallstone. In the enlargement, the amount of the small, sickle-shaped cracks (impact marks) becomes clear. The larger sickle marks have significant depressions and V-shaped marks formed in the overlapping area, which can also be seen macroscopically. Vang, Bornhom, Denmark. Image width 6.8 mm.

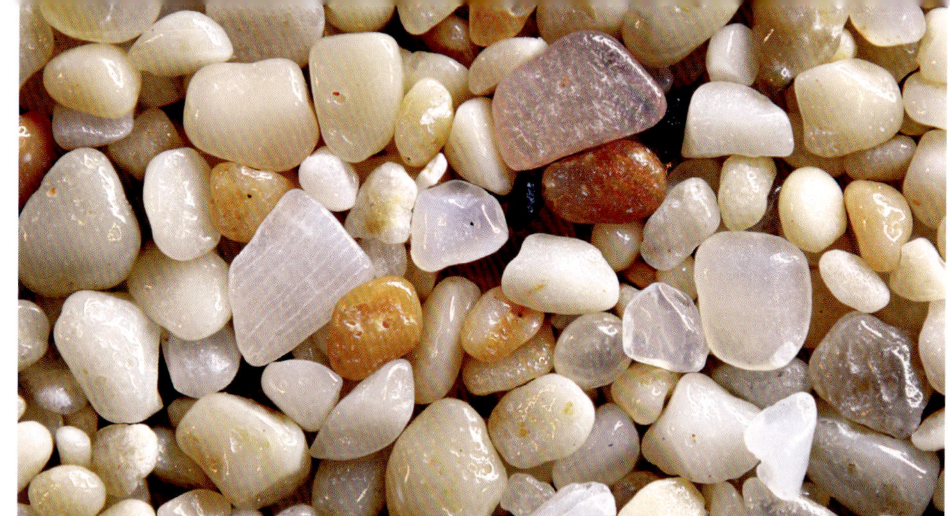

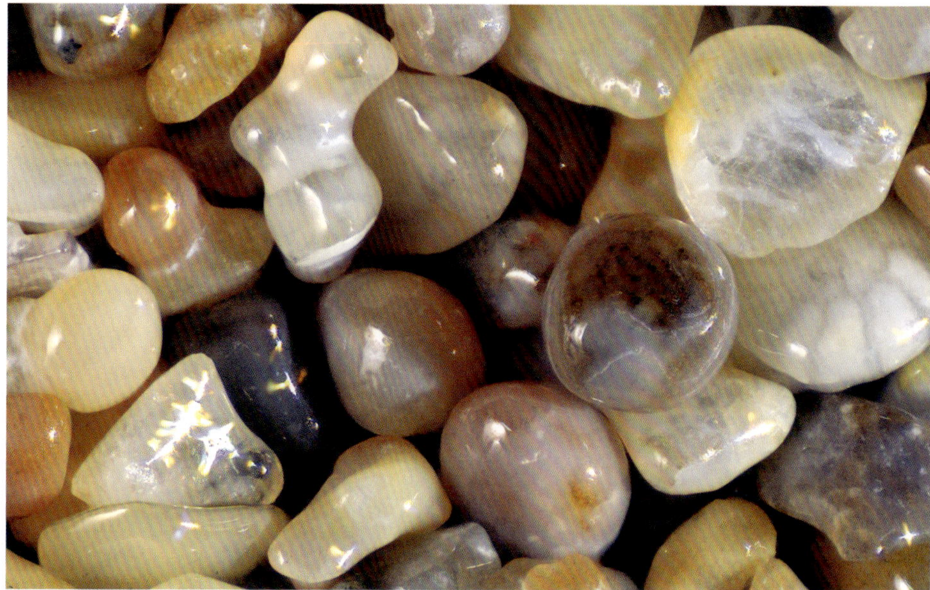

6.2.4 Ocean Milieu, Marine or Littoral Deposition

Can these findings be applied to sand? Are there sickle marks or impact marks on sand grains, and, if so, are they formed by the same principles and do they provide clues to the origin of the grains?

Let us start with the crescent-shaped impact marks, which we have conceptually distinguished from the sickle marks. Sickle marks are more deeply notched and show a flat and a steep flank, while impact marks form fine cracks only on the structure of the surface. Impact marks can also be found on individual grains of sand where they are, just like on wallstones and erratic blocks, irregularly distributed and randomly oriented. It is to be assumed that their origins can be traced back to free grain-to-grain collisions. The impact marks are most often found on marine sand grains.

Figure 6.75 Sand from the island of Boa Vista, Cabo Verde. The sand is made almost completely of calcareous residues of marine life. The crushed-shell debris has been highly polished by water and waves. Image width 5.1 mm.

Figure 6.76 Sand from Golden Beach, Cyprus. The partially calcareous, partially silicate sand grains with portions of ooids are almost all extremely polished, with almost no exceptions. Even nonspherical grains are polished on the surface. Image width 2.2 mm.

In particular, intermittent and reversing surf, which leads to colliding movements of the entrained sand and gravel being carried, causes collisions both between the grains themselves and between the rock debris and sand grains.

In comparison, fluviatile sand grains show only few or no impact marks. This is explained by the already mentioned low relative velocity, since the grains in a river flow largely in the same direction, excluding locally confined turbulent whitewater areas. Also, the combination "grain between two boulders" takes place permanently in the coastal zone. Higher contact times (the river length is limited; the time at the coast is principally not), opposite flow directions, and the mixture of grain sizes can be used as explanations for the accumulation of impact marks on marine sand grains. Scratch marks and incisions are also found on marine grains with high energy input. However, it should be noted and explained that, by far, not all marine grains show impact marks. Moreover, these are distributed very differently in grain samples. Some grains are littered with marks, while the typically larger portion does not have any impact marks at all. Perhaps this can be explained by the fact that not every surf zone has boulders, and thus in these areas, high-energy collisions are omitted or are significantly rarer.

Collisions that take place in water between grains are generally muted. If we throw a single grain of sand with a diameter of 0.5 mm through the water, it would reach a maximal and constant sedimentation velocity of about 225 mm/second if it fell freely and unobstructed. In the air, the grain would reach that same speed after falling only 2.6 mm! That is why aeolian environments lead to much-higher collision speeds among grains. It would then be expected that sand grains from the desert also have many impact marks. That is the case; however, the resulting marks are continually re-formed through abrasion. In addition, deserts sands are subject to chemical factors. Both lead to the already described "frosted" look of the surface through standing plates.

Similar grains are not found on coasts; quite the opposite. Many coasts act almost as large polishing machines. The energy of the wind constantly blowing over the sea is transferred to the surface of the water and leads to rhythmic wave motion, which is transformed into a never-resting, reversing motion at the coastline. In addition, there is a current parallel to the coast. Both form the drive of this polishing machine, which also constantly brings the sand particles into sanding contact with each other as well as with gravel and rubble. Each individual wave reorders and causes the sand grains to rub against each other. And this over periods of thousands of years. This is why perfectly polished sand grains from surf-exposed coastal sands are found much more frequently than in other milieus, as long as the appropriate mixture of sediments is available. The polishing mechanism works even with grains with relatively lower sphericity.

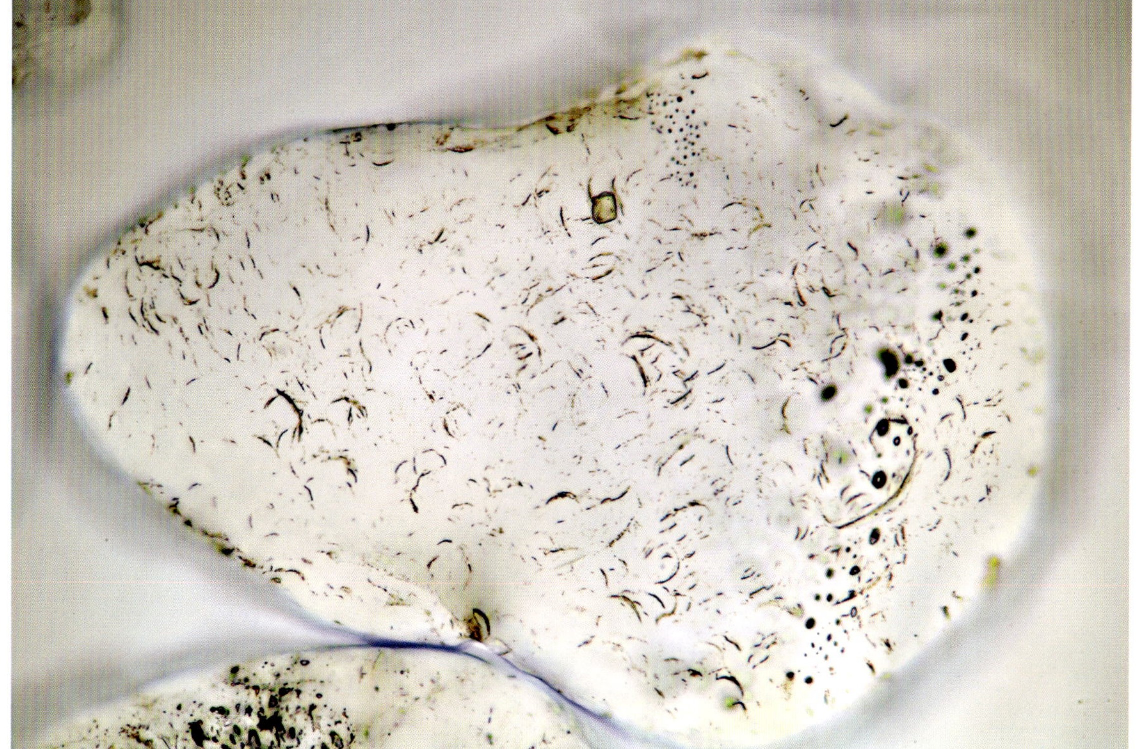

Let us return to sickle marks, which we had found in a pronounced form on rounded and glacially reshaped rock ridges. They occur together with the impact marks and are distributed, just like them, over all different size scales—from a grain of sand to a rock. Since both types always occur together, even on the same grain or boulder, the greater depth of the sickle marks could be caused by stronger impacts or by random impact accumulation at a particular site, or they could be the result of progressive erosion and dissolution processes of an initial impact mark, with subsequent chipping at the margins. If the latter were true, then the sickle marks would arise from the impact marks.

The case study of a sand grain from the beach of the island of Fehmarn (figs. 6.78 and 6.79) exemplifies the diverse and variously interpretable structures that can be found on sand grain surfaces, as well as the history of the grain that can be developed from the clues.

What other characteristics are typical of the coastal and surf area? For the sake of completeness, it should be mentioned again that "marine" naturally encompasses a wide range of sedimentation sites: from deep-sea to turbidite flows and their debris, to storm-swept cliffs, and also quiet shallow-water areas in wide bays. The literature usually distinguishes only between "low energy" and "high energy" areas but does not arrive at clear characterizations or definitions. We consider here the coastal and surf zone and group them under "littoral." The two microtextures listed below are assigned to the marine/littoral zone.

Figure 6.77 Sand grain from Fehmarn, Germany. The grain is covered with impact marks that are randomly oriented. The impact marks have dimensions between 20 and 80 *µm*. The black round points on the right side of the grain are inclusions. Bright-field transmitted light. Grain size 355 × 500 *µm*.

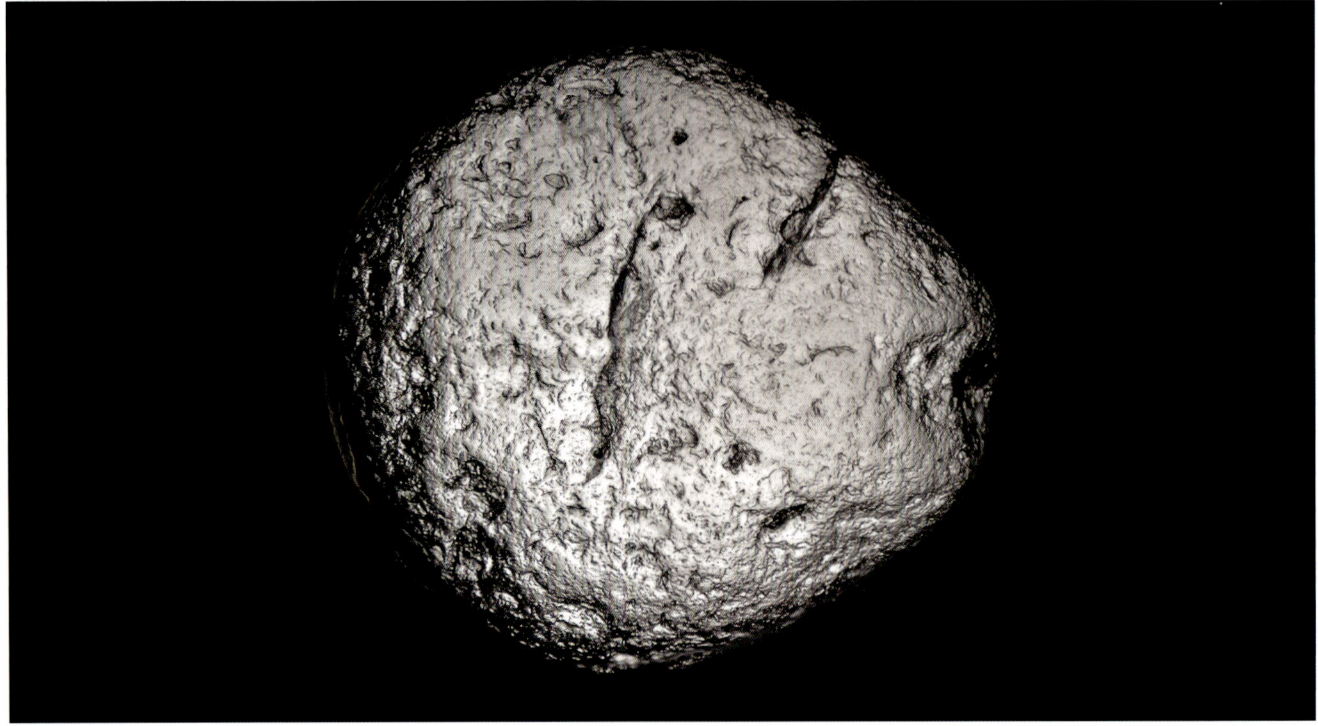

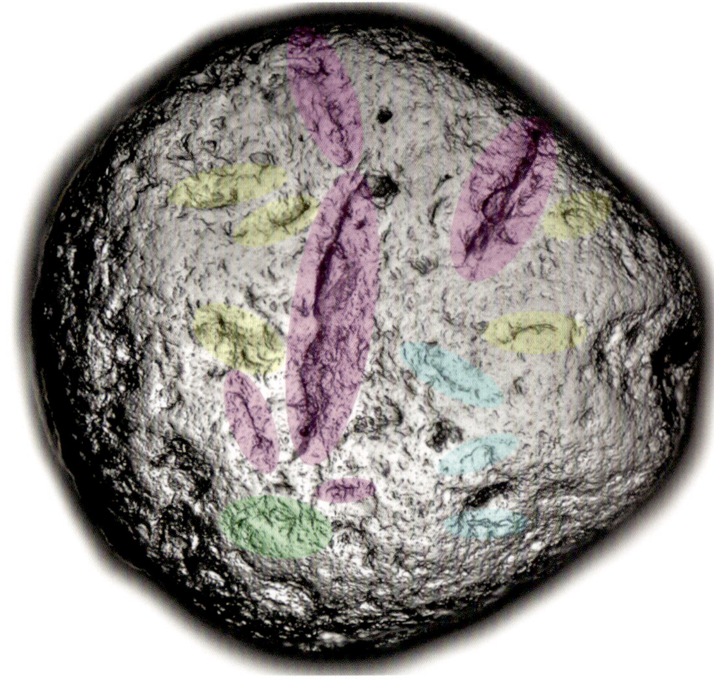

Figure 6.78 *Left*: Localization of a surface texture of a sand grain from Marienleuchte, Germany (fig. 6.79): violet: straight grooves; yellow: sickle marks, impact marks; light blue: wavy ridges; green: V-shaped impact marks

Figure 6.79 *Above*: Sand grain from Marienleuchte, Fehmarn, Germany. The grain is very well rounded and therefore does not find itself in its first cycle. The many large and craterlike depressions caused by impact are easy to see. In addition, the impact marks and the sickle marks as well as the many dotted marks on the surface are easily recognizable. Another characteristic that sticks out are the two large indentations that can be seen in the center of the grain and on the top-right edge. The size of the grooves points to the fact that they were not formed by collisions of the grain in its current size. It is more likely that they were formed by stresses or during a previous cycle of the grain (e.g., glacial). Along with the characteristics already mentioned, there are some aeolian structures that are recognizable (standing plates, impressions, meandering ridges). One could infer that the grain was located in a coastal dune region for a certain amount of time. From this, one could develop the following history: glacial transport with high stress and scoring, a stay in a dune area with aeolian stressors and significant rounding, then transition into the marine environment. Diameter 580 *μm*. Transmitted-light texture illumination.

Characteristic	Dimensions	Description
V-shape percussion cracks	< 5 μm	Caused by grain-to-grain collisions. The triangular form is crystallographically based. Typical for high-energy underwater areas such as littoral zones and whitewater.
Crescentic percussion marks	1 to 30 (max. 50) μm	Most common microtexture on quartz grains. Caused by hard, strong grain-to-grain collisions. Precursor to smaller conchoidal fractures. Typical for aeolian saltation and creeping loads. The most common grain diameter in dune areas is medium to coarse. Also after underwater collisions between sand and pebbles.
Conchoidal fractures	10 to > 100 μm	Caused by power impact or surface pressure. For minerals without pronounced cleavage surfaces. Dots on the breakage area are interpreted as traces of inclusions. Mostly on grains over 150 μm.
Graded arcs	3 to 400 μm	Concentrated series of fan-shaped arcs. Caused through very heavy, shock-like impacts or very high pressure.

1. V-shaped patterns, V-shaped impressions

These structures are mentioned by all authors. Especially by Krinsley and Doornkamp, the dimensions for the Vs are labeled as "very small." V densities are mentioned up to 6 V/μm^2, which can be found by using a light microscope. Other authors mention flank lengths of the Vs up to a maximum of 5 *μm*. Vs can also be the result of etching processes, which tend to occur in calmer water zones and differ from impact marks in the way that they are oriented and appear aligned with the crystal lattice (*en echelon*), with flank lengths up to 50 *μm*. In summary it can be said that the V marks created mechanically from impacts appear randomly on the grain and point to a moderate- to high-energy environment. For lower-energy milieus, oriented Vs can be expected that are caused through chemical etching processes. V structures appear on far over 50% of the littoral sand grains of high-energy areas, while the percentage lies under 50% for fluviatile transport (Madhavaraju 2009).

2. Straight or slightly curved furrows

Under high energy (so, strong surf zones), mechanical scratching processes underwater can lead to this type of surface characteristics. According to Krinsley and Dornkamp, the lengths can go to a maximum of 25 *μm*, while other authors subscribe to larger dimensions.

Figure 6.80. Summarized list and description of the surface textures that can be caused by free grain-to-grain collisions in different environments. The required impact energy for the creation of the characteristics through collisions increases from top to bottom. The specifications are according to Vos (2014).

Deeper furrows with diameters < 400 *μm* are excluded from grain-to-grain contact and can be assigned more to the glacial milieu or rocky surf zones carrying stones. In the littoral area, mechanically formed V marks as well as furrows/striae can be found and are thus characteristic.

The rounding of the sand grains takes place much more quickly in the coastal areas than in river transport and probably is topped only by aeolian abrasion. The rounding process itself is strongly dependent on the grain size, just as in other environments..

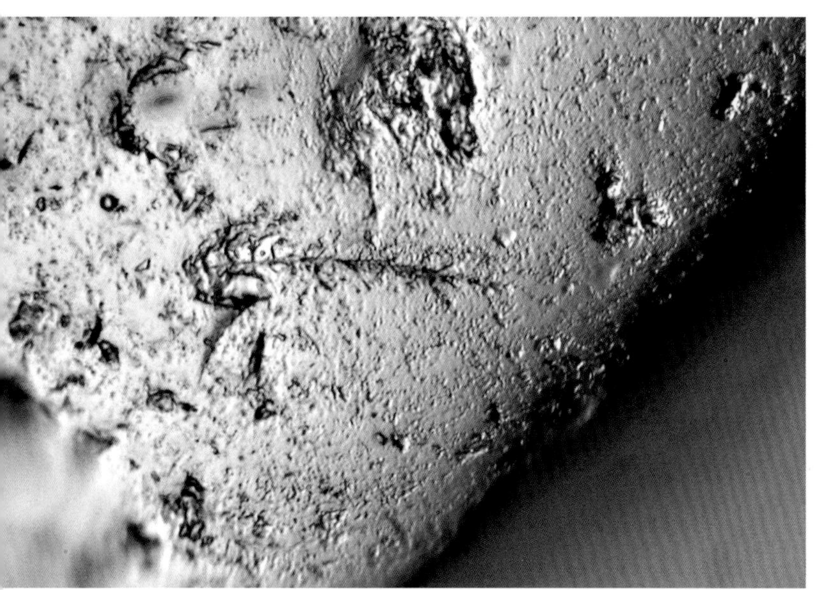

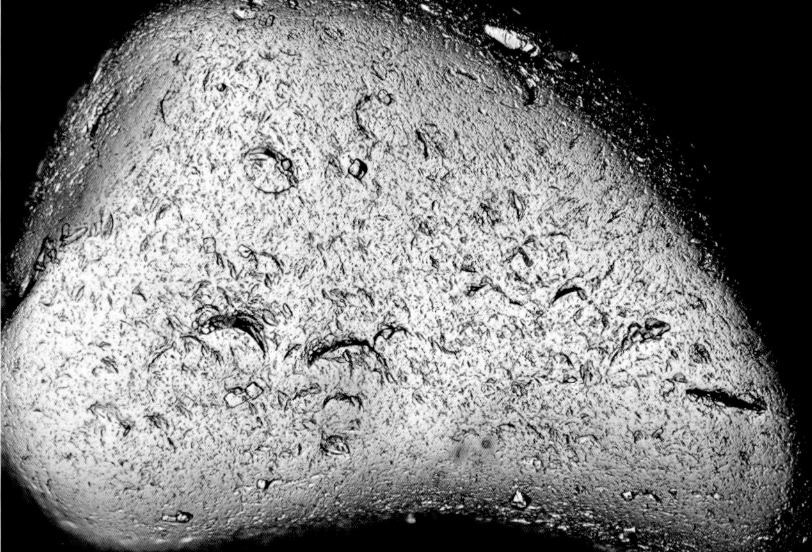

Figure 6.81 Detail of a sand grain from Kühlungsborn, Germany. The scratch mark in the middle of the picture has a length of 75 *μm* and is significant for high-energy surf zones. On the left end of the mark, there are a few larger Vs. The small, dot-like impact marks are clearly recognizable that cover the surface. Transmitted texture lighting.

Figure 6.82 Quartz grain from the Copacabana, Rio de Janeiro, Brazil. It shows a few sickle-shaped impact marks, but otherwise a high density of small V marks, characteristic of high-energy areas; *to the right*, a deep scratch mark. Image width 940 *μm*. Transmitted texture lighting.

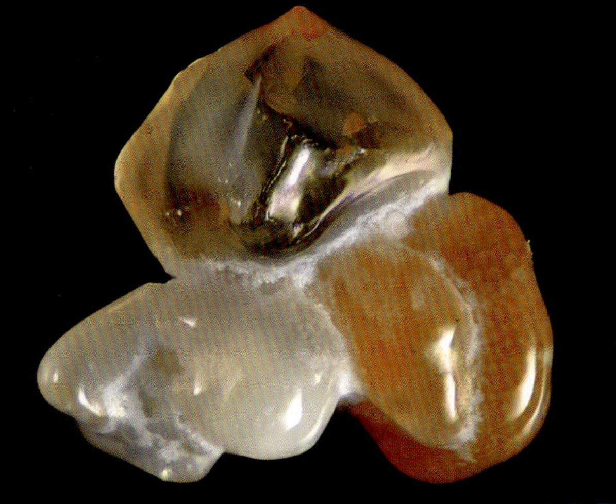

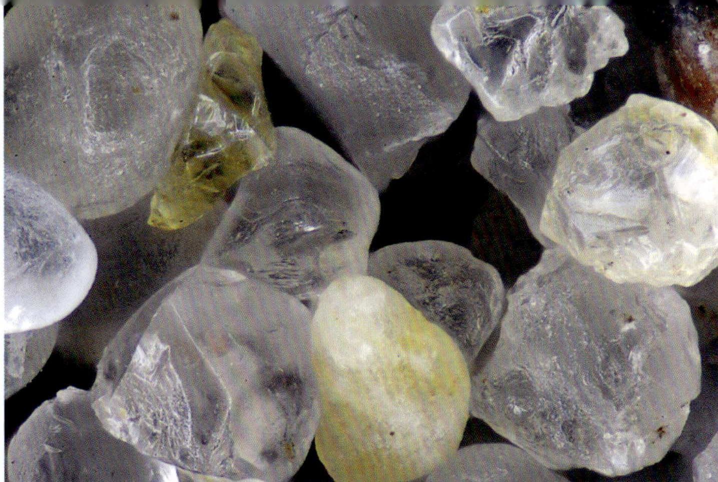

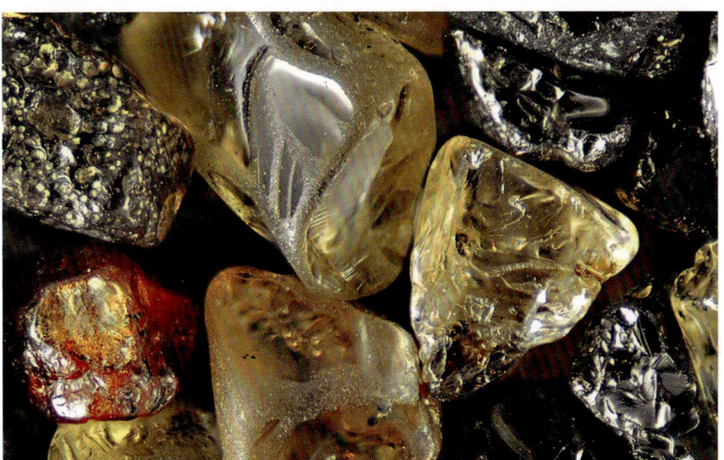

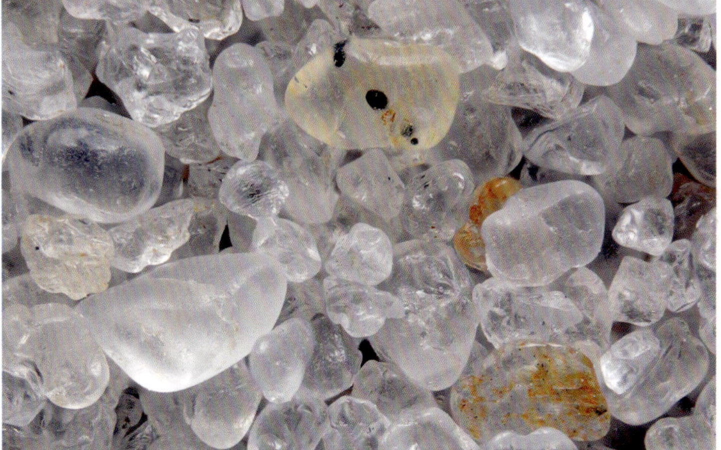

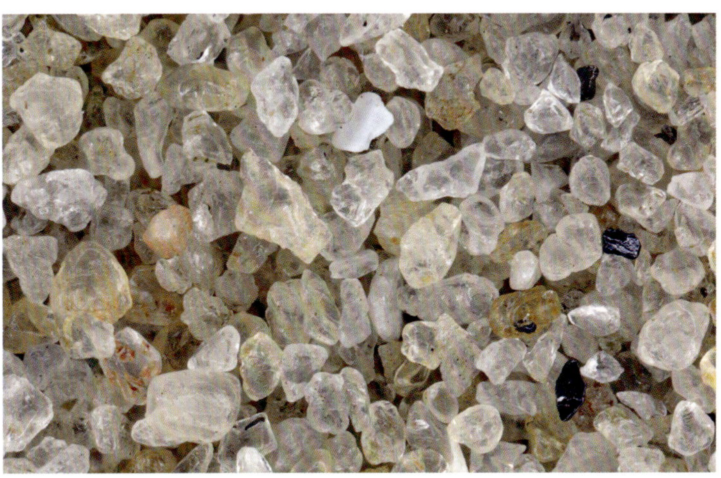

Figure 6.83 Sand grain from the Golden Beach, Cyprus. The sand grain, composed of three subgrains, is completely polished despite its jagged form; impact marks are not present. 540 × 620 μm. Reflected light.

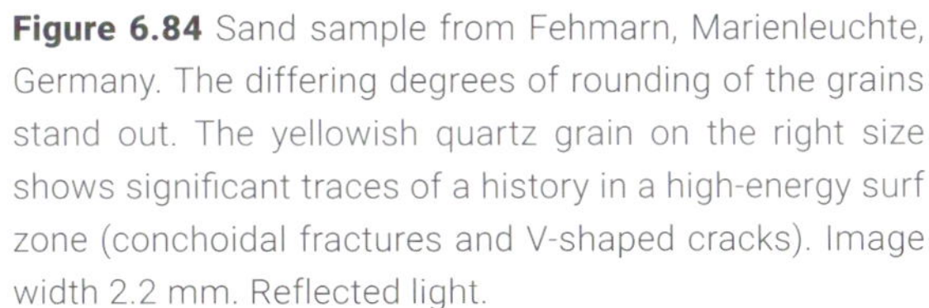

Figure 6.84 Sand sample from Fehmarn, Marienleuchte, Germany. The differing degrees of rounding of the grains stand out. The yellowish quartz grain on the right size shows significant traces of a history in a high-energy surf zone (conchoidal fractures and V-shaped cracks). Image width 2.2 mm. Reflected light.

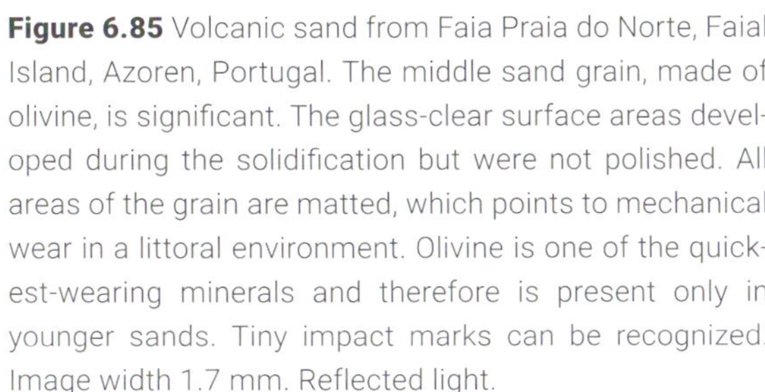

Figure 6.85 Volcanic sand from Faia Praia do Norte, Faial Island, Azoren, Portugal. The middle sand grain, made of olivine, is significant. The glass-clear surface areas developed during the solidification but were not polished. All areas of the grain are matted, which points to mechanical wear in a littoral environment. Olivine is one of the quickest-wearing minerals and therefore is present only in younger sands. Tiny impact marks can be recognized. Image width 1.7 mm. Reflected light.

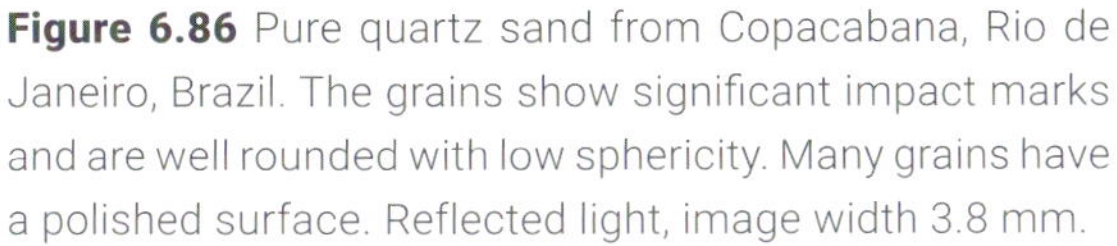

Figure 6.86 Pure quartz sand from Copacabana, Rio de Janeiro, Brazil. The grains show significant impact marks and are well rounded with low sphericity. Many grains have a polished surface. Reflected light, image width 3.8 mm.

Figure 6.87 The sand from Abad Turtle Beach, Kerala, India, is only partially made of quartz grains and has proportions of light heavy minerals such as kyanite and sillimanite. Many grains are rounded only at the edges, and there are only a few spherical or polished grains. Reflected light, image width 3.8 mm.

6.2.5 Lake Milieu, Limnic Deposits

Lakes differ from oceans, on the one hand, by their smaller size, and, on the other hand, through their low salt level. Also, there are no noticeable tidal influences in lakes. In lakes, sediments from the surrounding land are deposited, transported only small distances, and moved only moderately. Neither do large forces act on limnetic sediments, as we know from glaciers, nor do high current velocities of water occur, as has been described for rivers or seacoasts. From the point of view of sand, lakes are more or less large continental sedimentary basins. According to their formation, lakes can be divided into river lakes, glacial lakes, lagoons, epicontinental depressions, and rift lakes, to name the most important.132 River lakes originated from cutoff oxbow lakes and receive only periodic inflow; typical glacial lakes are found in the Alps, especially in the Alpine foothills, whose large lakes are of glacial origin. Lake Victoria in Uganda is an example of a shallow epicontinental lake formation, which is quite different from the usually very deep and permanent lakes in continental rift faults such as Lake Tanganyika.

For the reasons mentioned above, sand grains from limnic deposition do not show specific surface characteristics by which they could be identified. At least this is true for individual sand grains. Limnic sands as a whole are on average not very well rounded; they are also very poorly sorted and are compositionally immature, often with high proportions of silt. This usually makes it possible to distinguish them from fluvial sediments, especially since traces of chemical and biological weathering are very often present on the grains.

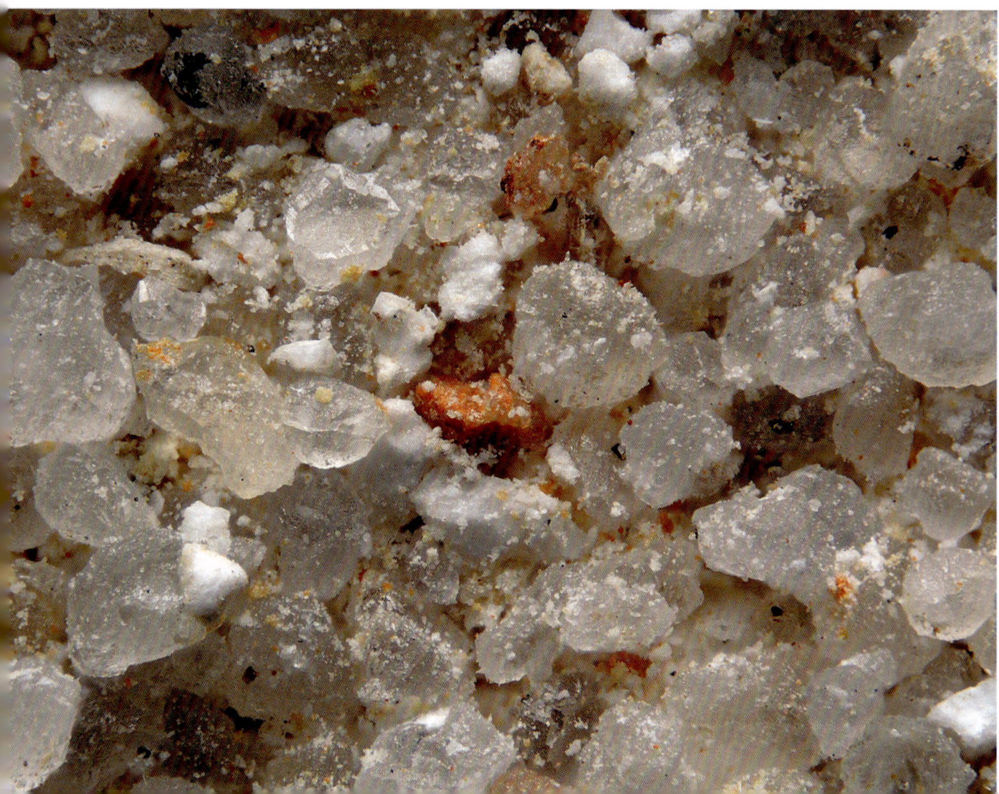

Figure 6.88 Lake Titicaca is the largest freshwater lake in South America and is located at an altitude of 12,533 feet (3,820 m). The sand shown comes from the Isla del Sol, Bolivia, and is a very fresh and immature sediment with a high proportion of silt. Reflected light, image width 3.45 mm.

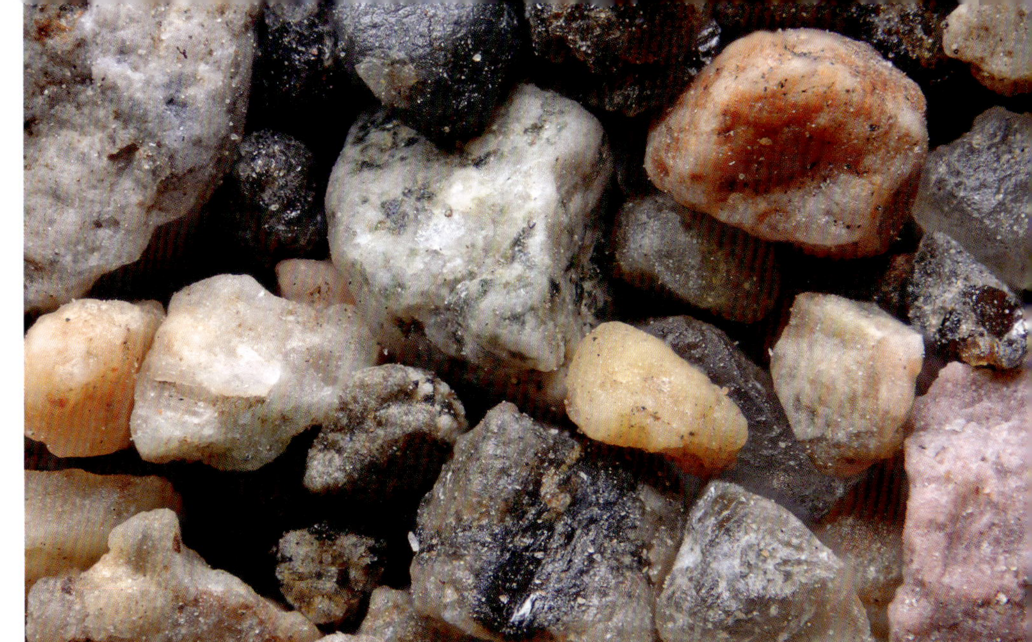

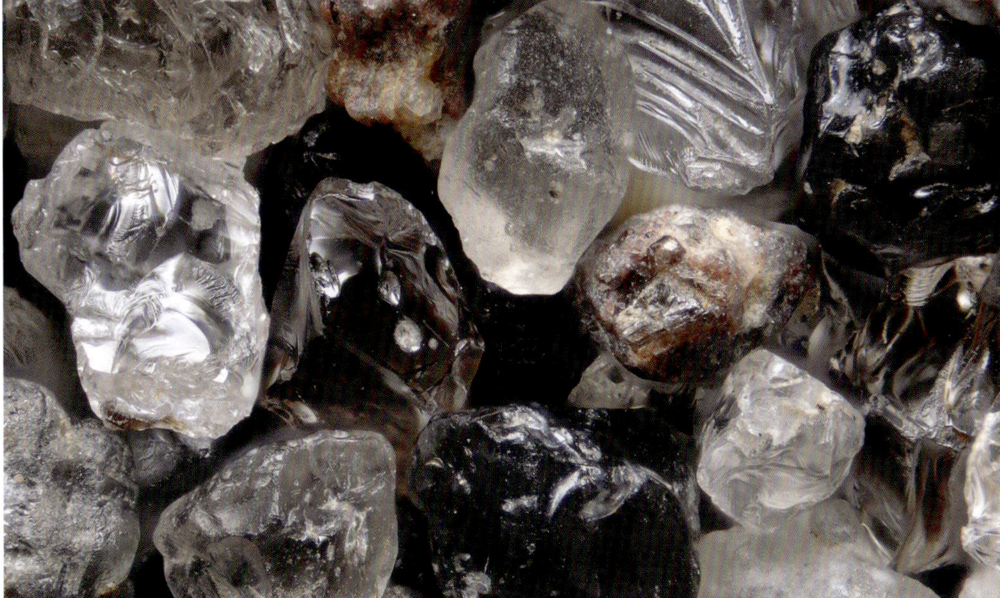

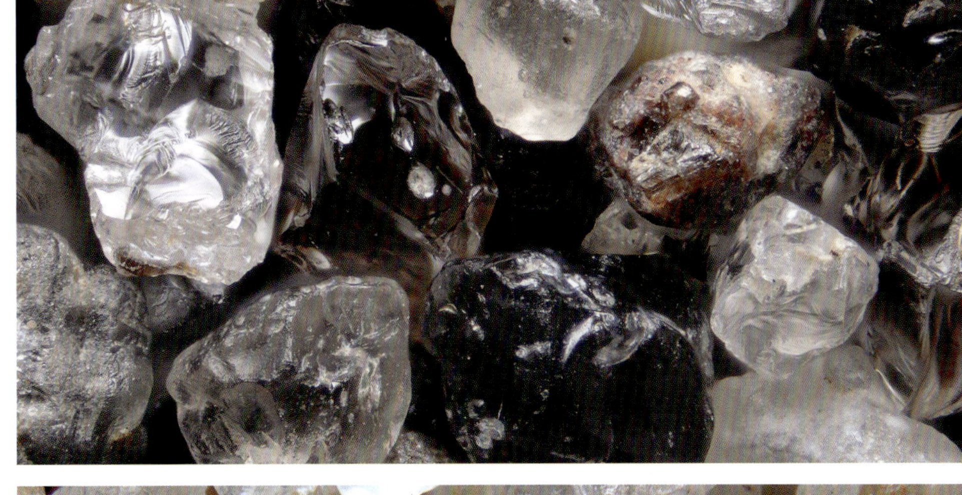

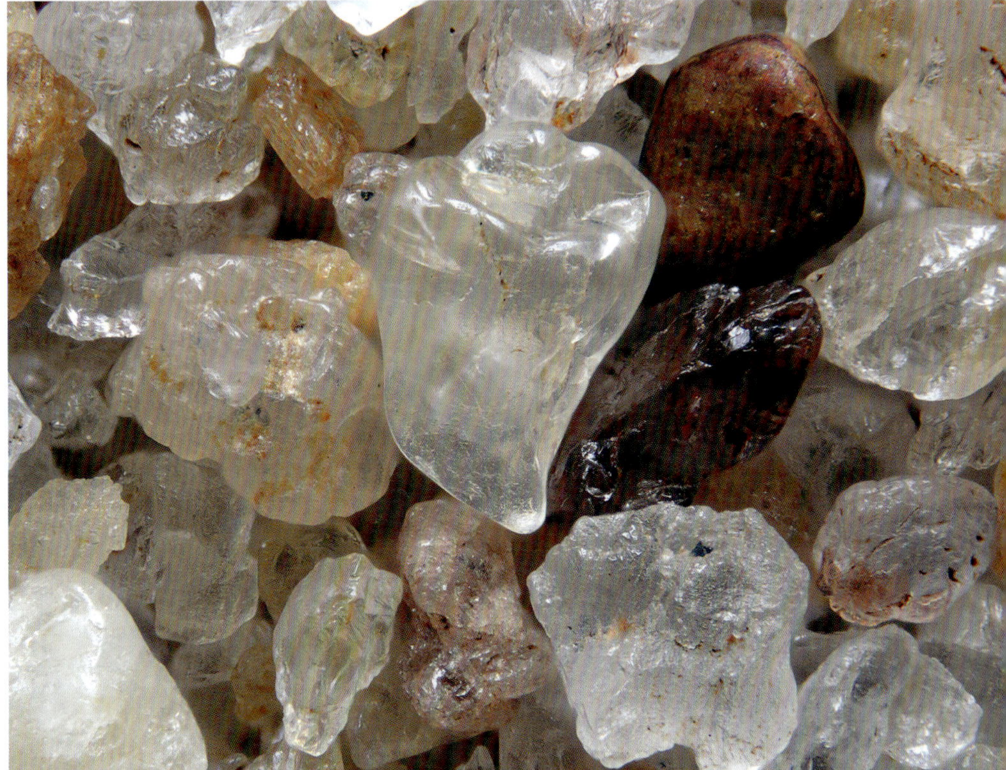

Figure 6.89 Lago Buenos Aires, in the mountain region of Argentina, is located only 663 feet (202 m) above sea level. The fresh and very immature sediment is composed mostly of small rocks broken off from the surrounding mountain region, as well as proportions of silt. Reflected light, image width 3.6 mm.

Figure 6.90. Yellowstone Lake, at an altitude of 7,743 feet (2,360 m) in Wyoming (USA), is the largest mountain lake in North America. The sand is composed of remnants of volcanic activity (last eruption was seven hundred thousand years ago). Clear quartz grains show conchoidal fractures and glacial traces with only minimal rounding at the edges. Reflected light, image width 3.45 mm.

Figure 6.91 Lake Victoria is the largest lake in Africa, and with a depth of only 279 feet (85 m), it is an example of a lake formation in an epicontinental depression (3,724 feet [1,135 m] above sea level). The sand comes from Entebbe, Uganda, and is composed mostly of moderately rounded quartz, which can often be found with an iron-containing coating. Reflected light, image width 3.45 mm.

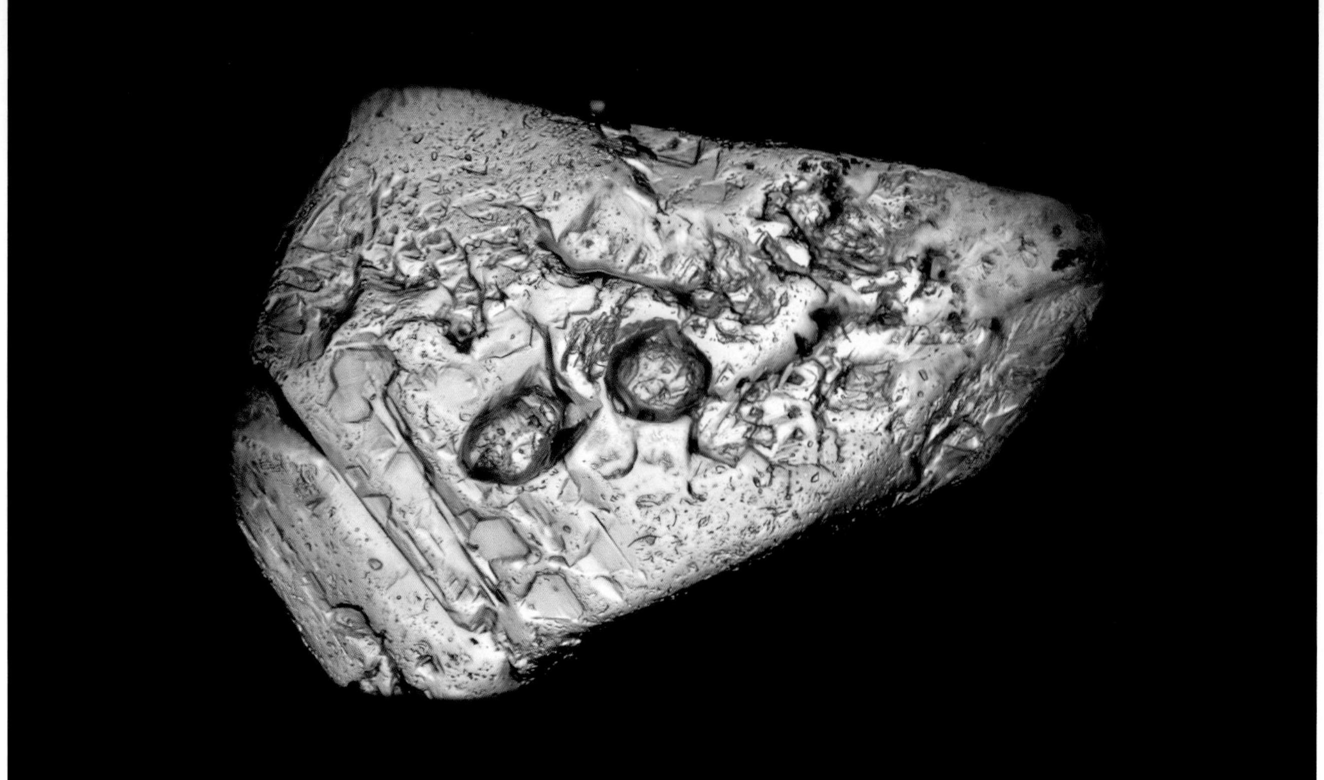

We have not reached the end of our search for traces. At the beginning, we had asked the question of whether and to what extent sand grain surfaces contain information that enables us to reconstruct their history. Indeed, it has been seen that a careful examination of the surface can give us clues to the life of a grain. There is something detective-like about this research: the more pieces of the puzzle we have, the better. Material, inclusions, structure, shape, and surface—in the best case, everything comes together to form a coherent picture. Most of the time, however, the examination will not come to a clear conclusion if only a single grain is available. A sufficiently large sample is needed, and statistics are called for.

The surface structure analysis of quartz grains still represents an important instrument in sedimentology today, with which valuable statements for the reconstruction of sedimentary deposition processes and cycles can be worked out with the help of further analysis methods.[133] The dedicated sand lover succeeds even with the help of a simple light microscope in obtaining a first and nevertheless fascinating impression of the topography of the sand grains, which is already suitable for drawing conclusions about their long life path through the history of the earth and thus deciphering signals from prehistoric times.

Figure 6.92 Sand grain from the eastern Australian coast near Byron Beach. The unusual grain is made of an optically single-axis mineral, but probably not quartz. *To the left*, polygonal etched figures can be seen. In the top left part, there is a collection of oriented V marks of chemical origin. The recognizable sickle-shaped impact marks on the bottom part of the grain, as well as the rounded edges, point to a littoral phase in the life of the grain. The two large holes in the middle have a diameter of 40 *μm* and were clearly formed chemically. The height of the grain over the point is 250 *μm*. Transmitted texture lighting. The grain is a good example for an observation by Pye (1990): "In the case where coastal sand dunes have experienced a long period of weathering, involving several stages of aeolian reworking and redepositioning, as on the coast of eastern Australia, the grain surface textures are dominated by chemical features, which include oriented etch pits, deep arcuate etch lines, and angular breakage features."[134] It almost seems as though Pye was looking at our grain..

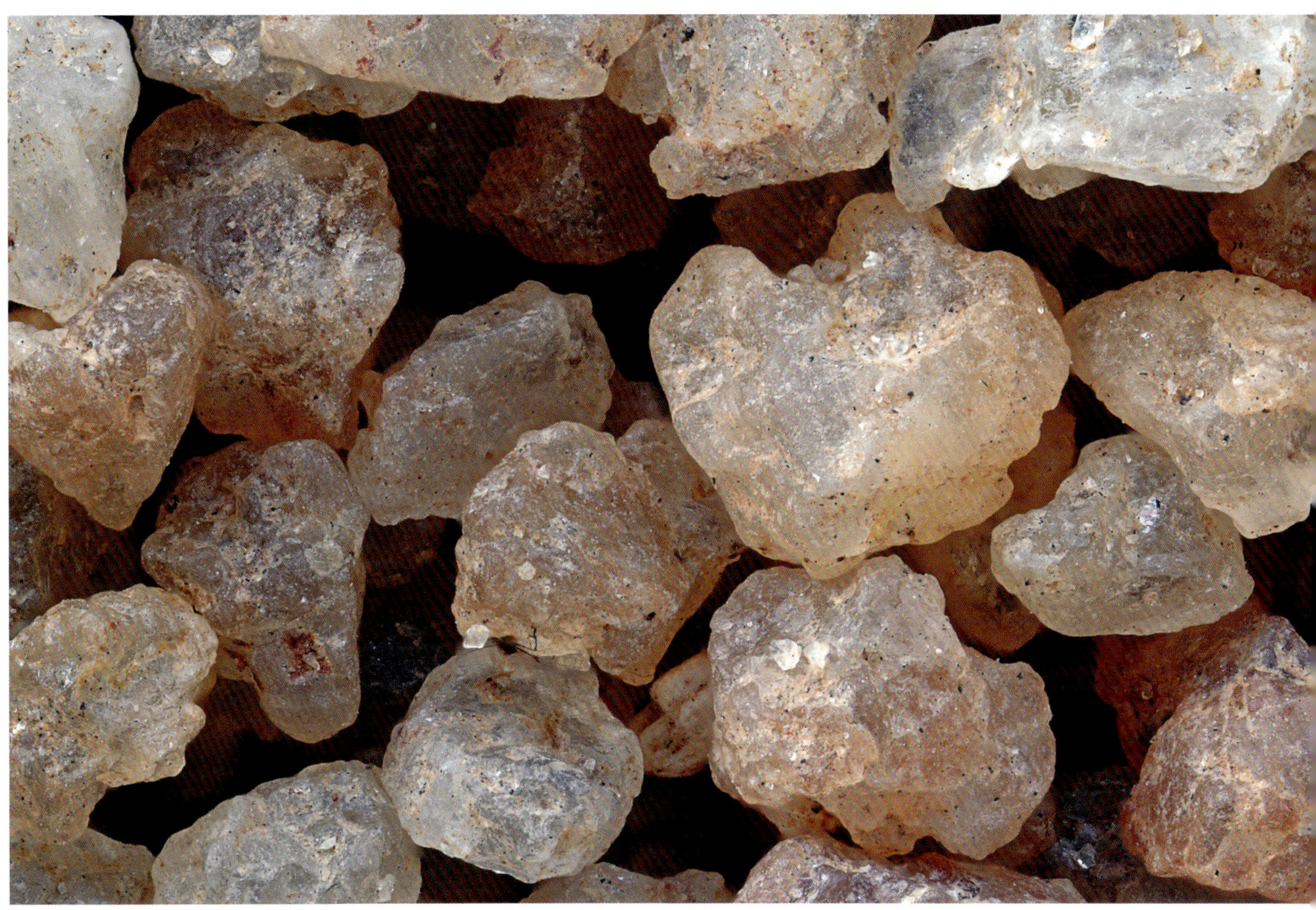

Figure 6.93 Sand sample from Binao, Ivory Coast. In contrast to all the other examples in this chapter, the sand sample displayed here can not be clearly assigned to any of the abovementioned milieus. The sand comes from a small inland town of the Ivory Coast, which lies in the middle of an expanse of rainforest 230 feet (70 m) above sea level. It is a land sediment. Neither ocean nor river, desert, or glacier is anywhere nearby. It is probable that the sediment was formed at its location a long time ago, so a product of erosion of a rock that has been transported only very short distances. The grains are only slightly rounded. Under these conditions, the erosion of a plutonic rock would result in a compositionally immature sediment with a high percentage of feldspar. The sand we see here is composed mostly of quartz. For this reason, it is much more probable that the sand is a decomposition product of sandstone that was deposited in its original sedimentation cycle possibly through fluvial transport, which would explain the low rounding of the edges. This consideration is supported by the growing lines present in the sample, which is typical for diagenetically changed sands. Image width 5.9 mm.

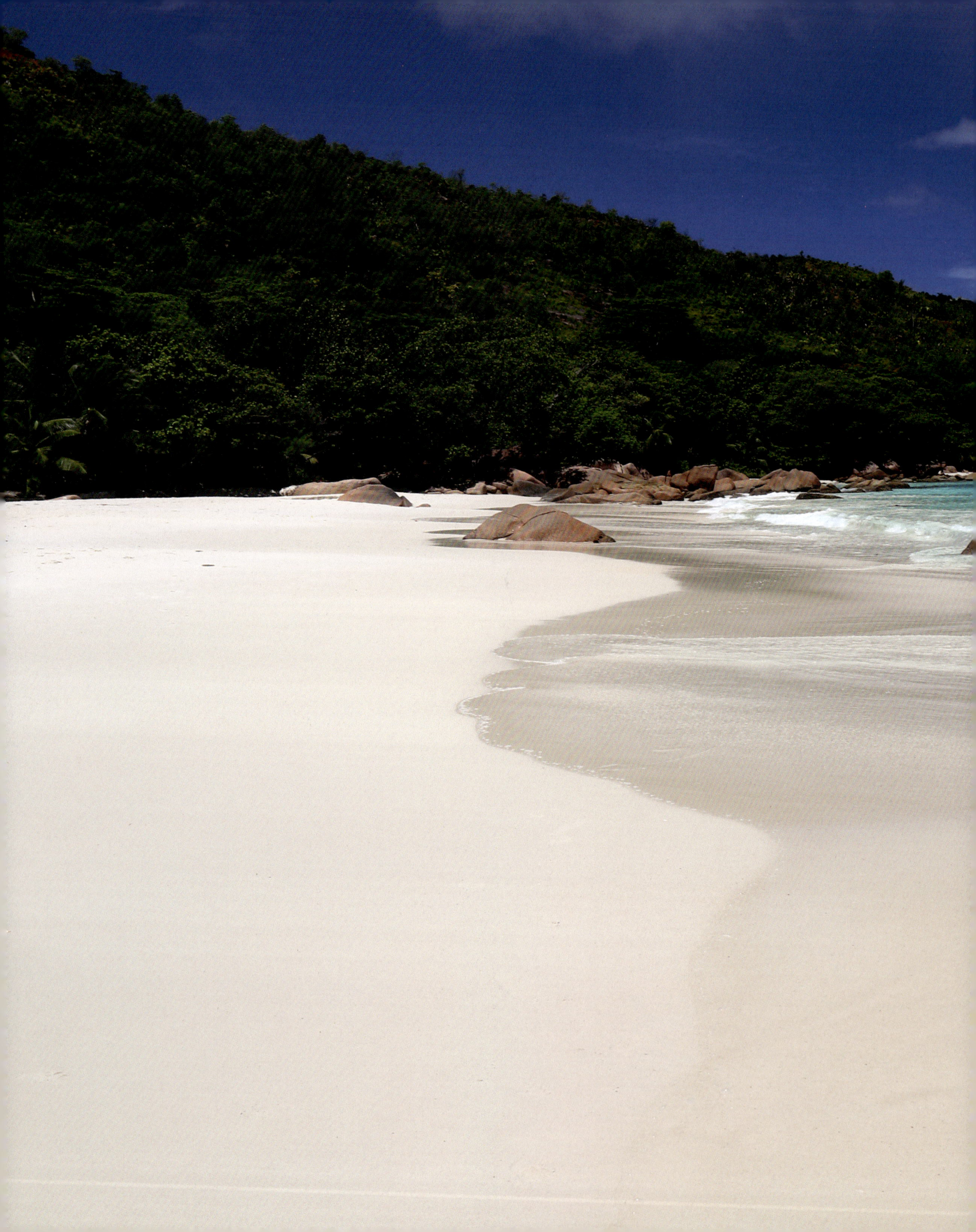

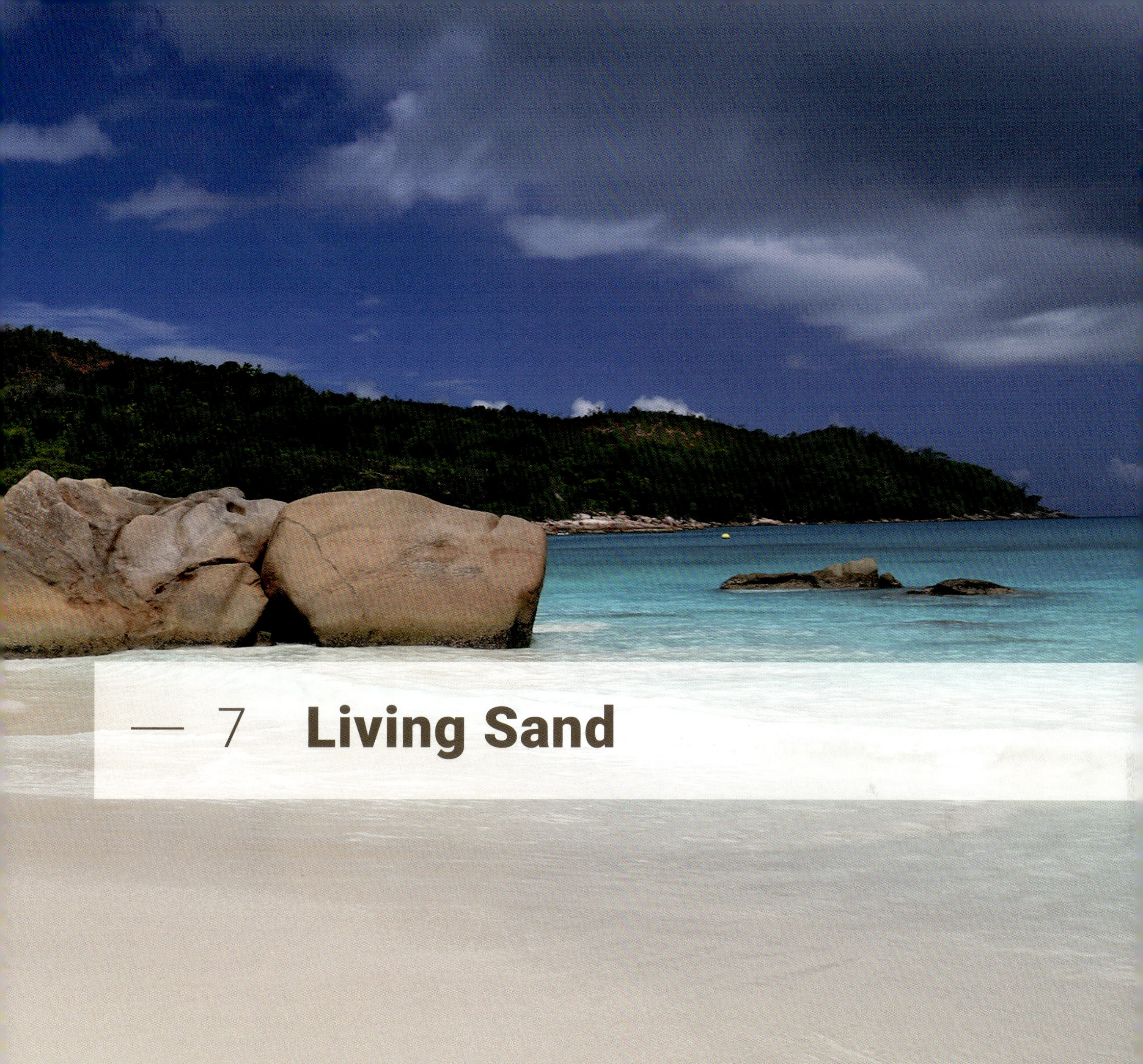

— 7 Living Sand

The first thing we notice about a living object is the form as a whole, then the parts of this form, their configuration, and connection.

—*J. W. v. Goethe,* ***Consideration of Morphology***

Sand and life. Can these two concepts be combined? A classifiable, granulated material as a product of erosion is how we have gotten to know sand. But just as before, we want to look closer once more and discover that sand can be created by life, grain by grain, and can also be home to life. Yes, even individual sand grains can hide life. Let's now discover sand as a colorful mixture of life and traces of life.

"When then in this way the organic body has been more or less destroyed in a way that a form is annulled and its parts can be considered as matter, then sooner or later chemistry enters and gives us new and beautiful conclusions about the last parts and their mixture."[135]

So far, we have exclusively devoted ourselves to the clastic components of sand. Sand as a result of the erosion of rocks and whole mountains. This is quite a one-sided view; a view also from the experience of people who, like me, live on large continents and their associated coastlines. Sand here is composed of different types of minerals, the abundance of which goes hand in hand with their resistance to weathering. Quartz is at the front and has been our most important object of investigation. But what if we turn our attention to completely different areas of the world and turn our thoughts to the tropical islands and archipelagos in the midst of great oceans; that is, to the beaches of the many vacation regions that we so highly value? Sand there is of a different nature. Indeed, we can venture the assumption that a shovelful of North Sea sand and a shovelful of sand from Mauritius probably do not have two even remotely similar grains. Wherein lies the difference? Let's go find it.

Previous double page:
Anse Lazio, Praslin, Seychelles

Figure 7.1 *Left*: A serpentinite block in Binntal weighing multiple tons near Fäld, Switzerland. This block has a very moving history behind it. The original material was a rock called peridotite, which is composed almost exclusively of olivine. Both can be formed only in depths of over 30 miles (50 km), so in the top of the earth's mantle. If peridotite reaches the earth's surface through tectonic processes in the presence of water, it changes to serpentinite. Serpentinites are grazed earth mantle arenas that were unfolded and highly metamorphosed (550°C and 8,000 bar)[136] during mountain building and finally were exposed by the erosion of the mountains. The block here comes from the 8,000-foot (2,500 m) Schwarzhorn. During the last ice age, glaciers moved it toward the valley and left it in its current position after melting, where it can be marveled at, among others. Original mantle rocks, the famous Finero peridotites, can be found in Italy in Valle Canobina, in the small town of Finero.

Figure 7.2 *Right*: Serpentinite surface on a debris block next to the Lengenbach Quarry, Binntal, Switzerland. The slickensides, which look almost polished, can be easily recognized, as well as the greenish color of the serpentinite with its light to rusty-brown erosion zones. *To the bottom right*, a yellow/black map lichen that shows silicate minerals. Image width about 80 mm.

7.1 Tropical and Continental Coasts and Beaches

Many of the tropical coasts are far from the continental crusts, whose main component is granitic and thus silicic rocks. As we have seen, continental crusts of silicon-rich rocks are lighter than the basaltic-gabbroid oceanic crusts and float like corks on the upper mantle. Also, oceanic crusts with 3–5-mile thickness (5–8 km) are much thinner than their continental counterparts, with thicknesses around 19–37 miles (30–60 km) of "crystalline bedrock." Since about two thirds of the earth's surface is covered by oceans, the largest part of the earth's crust also consists of the oceanic type, which represents the "normal case," so to speak, but remains largely hidden from observation. But there are exceptions.

The lithospheric plates of the earth are in constant motion; among them there are plates that move away from each other (divergent plate boundaries) and those that move toward each other (convergent plate boundaries). When an oceanic plate meets a continental one, the former pushes under the latter because it is heavier. In this process, which extends over geological time, parts of the oceanic plate or crust are usually scraped off during subduction as if by a plane. The scraped-off plate parts escape there by melting completely in deeper zones. During such processes and also during the similar but much more dramatic collision of two continental plates with accompanying complex mountain building, parts of the ocean floor crust are folded up and finally lifted up. If they come to light once by erosion, one can actually study oceanic crusts of the prehistoric times in airy mountainous regions. In geology, these separated ocean crust parts are called "ophiolites" and are described on the bases of their typical rock sequences (pillow lavas, gabbro, serpentinites, greenschists, etc.).

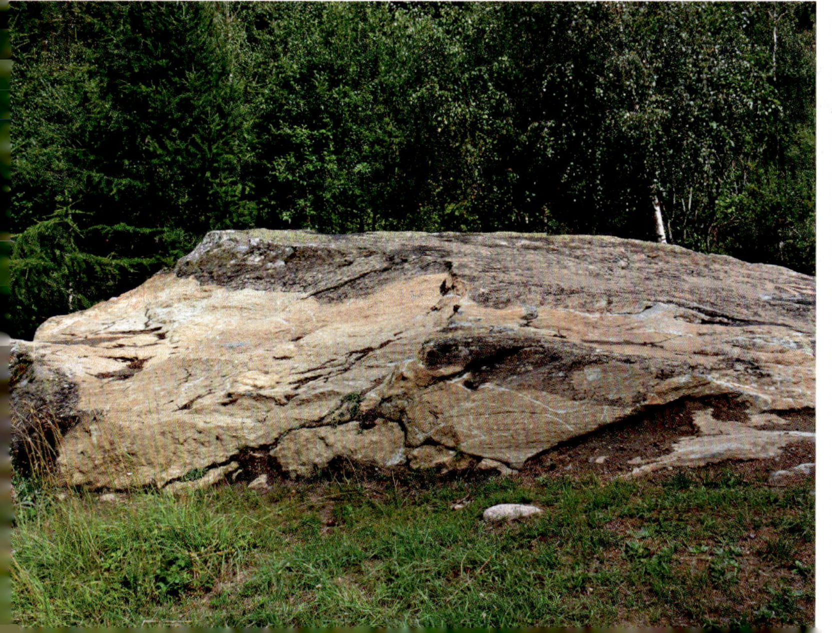

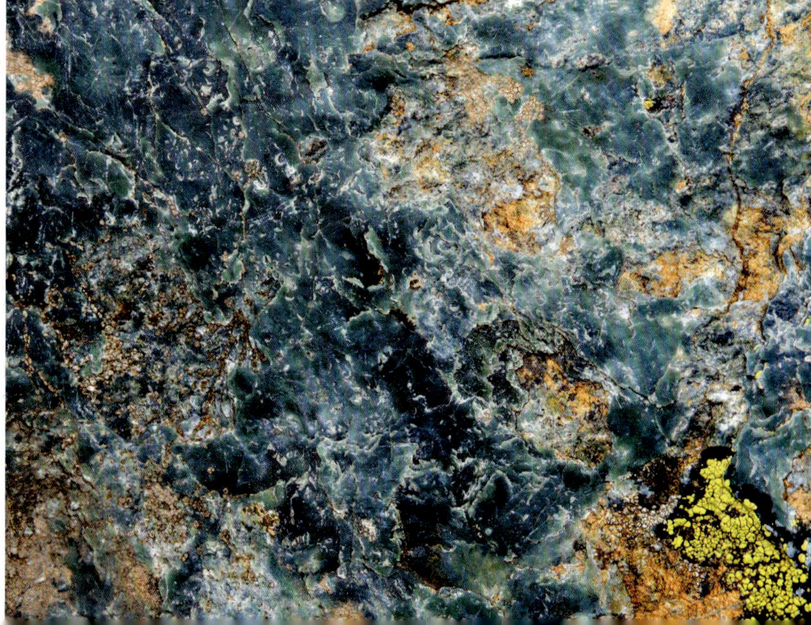

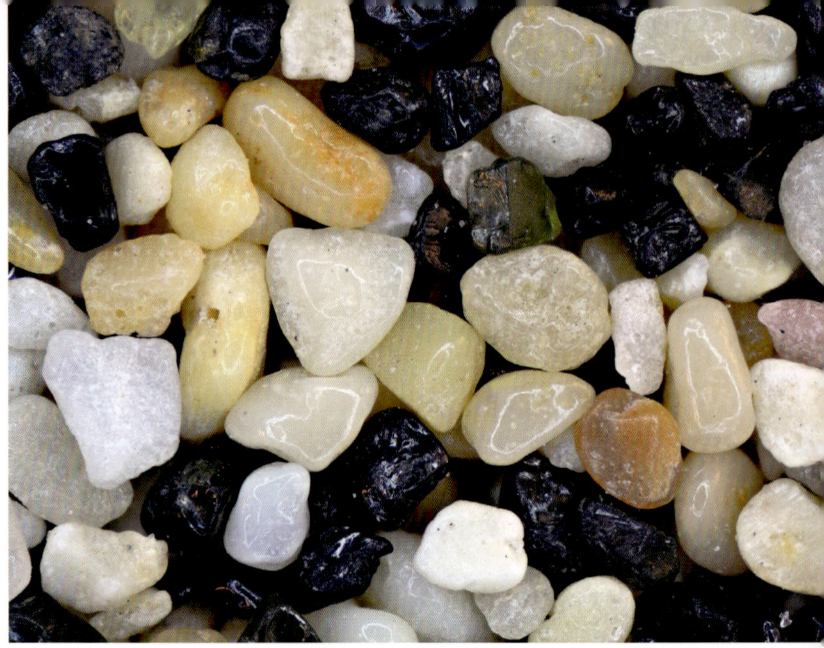

Almost all tectonic plate processes entail volcanic activity in their vicinity; this applies to convergent as well as divergent plate margins. If the magmas escape underwater, submarine volcanic mountains are formed, which are called seamounts. Their lava production is, depending on their location relative to the plate boundary, rather small, but sufficient to form islands of very different sizes. Seamounts are the most common type of volcano on Earth. In the Pacific alone, their number is estimated at more than one million.[137] Seamounts also exist in other oceans, such as on the Canary Islands off the coast of West Africa, which can be counted among the seamounts. In most cases, such volcanically formed islands and island arcs yield relatively little sediment mass when compared to continental environments. Pelagic, clayey sediments and volcanic ashes predominate in mineral components and are rather transient or easily washed out.

However, now we come to the actual topic of the chapter—the rich underwater life-forms around these tropical volcanic islands. Coral reefs surround the islands and provide a unique habitat for immensely rich flora and fauna. Billions upon billions of shell-forming organisms live, reproduce, and die again and again in the reefs, leaving behind shells, spines, teeth, skeletons, carapaces, and other components of their existence, mostly calcareous in nature, which eventually wash up on the shores one day. Beaches of fragments of life intermixed with particles directly from the center of the earth. Of course, the sands not only of tropical islands and coasts contain traces of life of this kind, but here their forms are particularly rich, and also only here there are so many beaches consisting exclusively of past living beings (group 4 sand; see classification from chapter 4). Basically, the sediments of tropical islands are almost exclusively of local origin. Quartz is always missing on these beaches because it is of continental formation.

The sands of continental islands are, in contrast, composed differently. Besides consisting of the remains of marine organisms, they are composed mostly of erosion products from mountains and continental land masses.

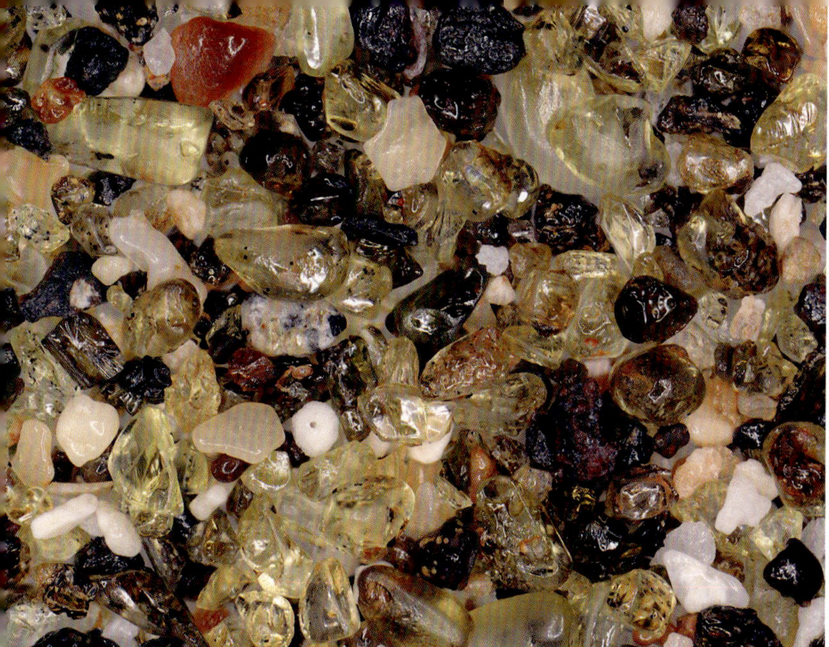

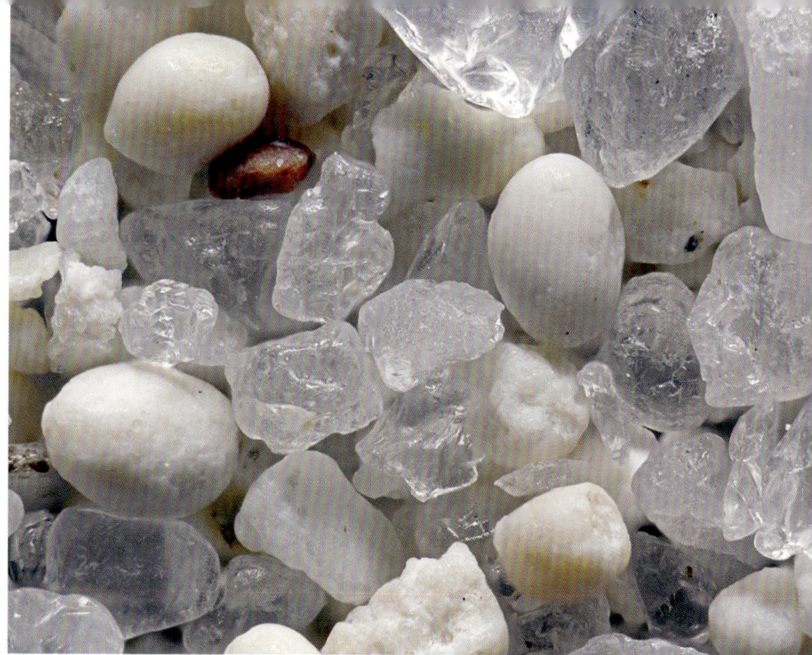

The erosive overburden is transported as sediment load via river paths into the sea and is deposited on the seafloor, sometimes in huge sand heaps in a fan shape. Under suitable conditions, offshore island chains known as barrier islands also form. One of the world's largest deep-sea fans of this kind is the Bengal fan of the Ganges and the Brahmaputra Rivers off the coast of Bangladesh, which also consists entirely of alluvial sediments. Its expanse stretches for a considerable 1,864 miles (3,000 km) to the south. The Bengal fan consists of the debris of the Himalayan Mountains and has an estimated volume of 3 billion cubic meters. This is more than the volume of the entire crust of the Tibetan Plateau lying above sea level.[138] The stratification has reversed. The sediments that once formed the long-since-eroded summit regions of the Himalayas now lie at the bottom on the alluvial fan and vice versa. In addition to the previously mentioned Ganges delta, the Mississippi River delta is also one of the largest in the world. The Mississippi discharges its sediment load of 470 million tons annually from the Rocky Mountains, just as the more westerly Rio Grande discharges its sediments originating from the Mexican highlands into the Gulf of Mexico. Due to the prevailing ocean currents, the Rio Grande forms shoreline-parallel barrier islands, but the Mississippi does not.

The famous Florida Keys, which border the east side of the Gulf of Mexico, are a group of tropical islands that are composed of old coral reefs and sands of different origins. They owe their formation to sea level fluctuations, which can be traced back to different ice ages. As an ice phase began over a hundred thousand years ago, the sea level started to sink dramatically and erode the old coral reefs. Gypsum and lime sands were created.

Figure 7.3 *Left side*: Sand from the Insel Sal, Cabo Verde. Dark, volcanic grains are mixed with white, biogenic grains, giving off a "salt and pepper" appearance. Cape Verde is an archipelago of volcanic origin. In addition to the islands, a row of seamounts also belong to the archipelago. Image width 3.8 mm.

Figure 7.4 Sand from Papakolea, Green Sand Beach, Big Island, Hawaii. The islands are, as the sand sample clearly shows, of volcanic origin. Green olivine, which was eroded out of the lava of the Mauna Loa cinder cone, dominates, but light biogenic grains and dark grains out of volcanic glass are also recognizable. Image width 5.9 mm.

Figure 7.5 Fine sand beach from Key West, Florida, USA. The pure quartz grains are only rounded but with low sphericity. Feldspars are completely missing. The individual ooids of lime (aragonite) can also be seen. The image width is only 2.2 mm.

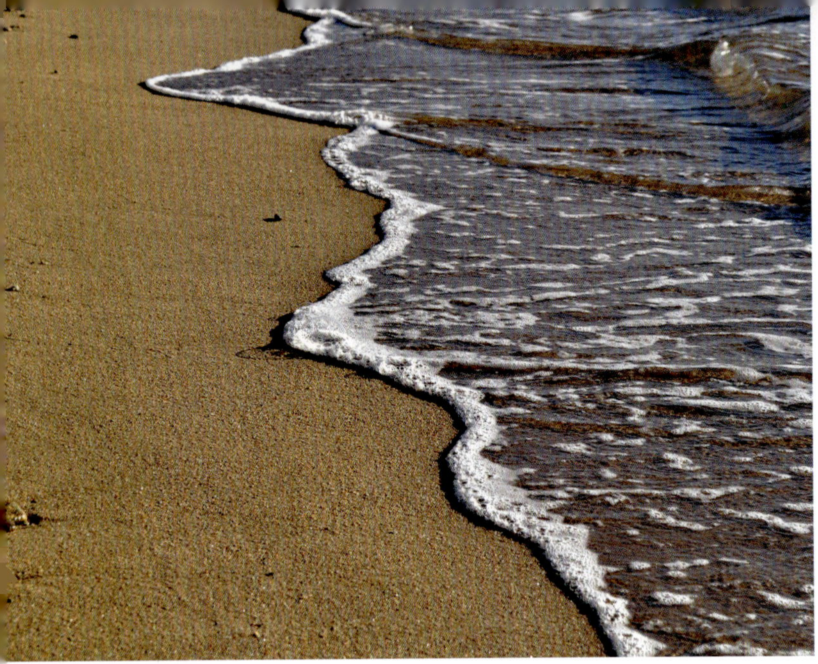
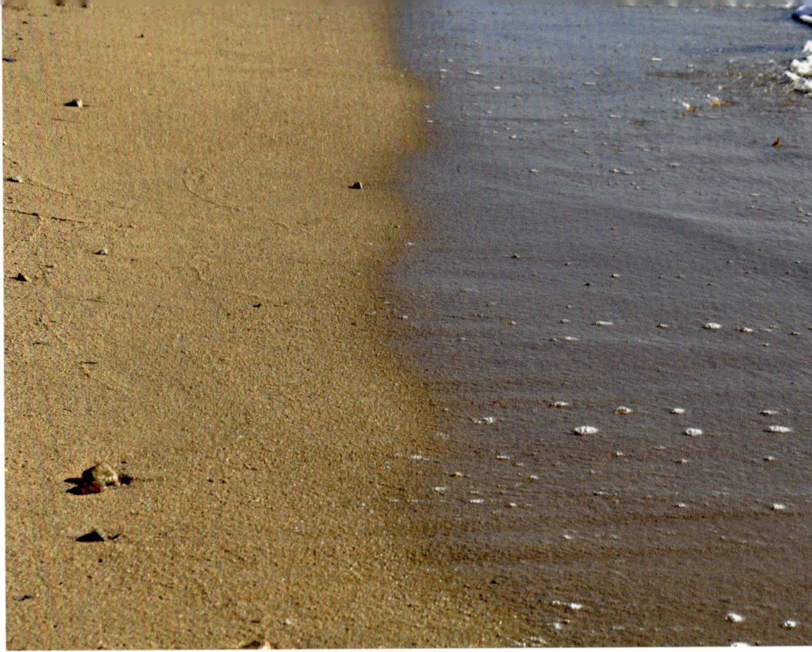

In addition, significant ooid sands of the aragonitic type formed in the shallow marine basin (see also chapter 4 as well as fig. 7.5). The current islands consist of calcareous sands, ooid sands, and very pure quartz sands, each intermixed with shell remains of the rich marine fauna. The origin of the quartz sands is unclear. They probably originated from eroded interbeds of the Key Largo limestone, and sediments of the surrounding mountains brought into the Gulf of Mexico by rivers or glacial transport.

Regardless of the composition of the coastal sands, the so-called intertidal zone is a very diverse and varied but also a very challenging habitat. It extends between the lowest low-tide line and the highest high-tide line. Microclimatic conditions here range from moderate temperatures and seawater overflow to sweltering heat during midday hours, with dried-out sand.

Small basins cut off from the sea often form, which are filled with new water only at high tide and otherwise have a high salinity due to evaporation. Despite, or perhaps because of, these conditions, intertidal zones are inhabited by a wide variety of creatures. Mollusks, crustaceans, starfish, sea urchins, worms, insects, and many more inhabit the narrow habitat that separates sea and land or, better, that connects sea and land. In the course of evolution, living creatures have adapted to the environmental conditions prevailing there and have known how to use them skillfully. The waves, which roll in eternal cycles, pose dangers but also bring food from the deeper areas of the sea, the subtidal zones, to the coastline. After their death, the organisms leave their shells and skeletal remains on the beach, where they are gradually crushed by chemical and physical forces and by the wave movement of the water. Sand is formed.

Figure 7.6 Part of the intertidal zone near Cala Bitta, Sardinia, Italy. The different sand areas saturated with water are recognizable. As a sand/water mixture, they have very different features that underground animals especially need to adapt to.

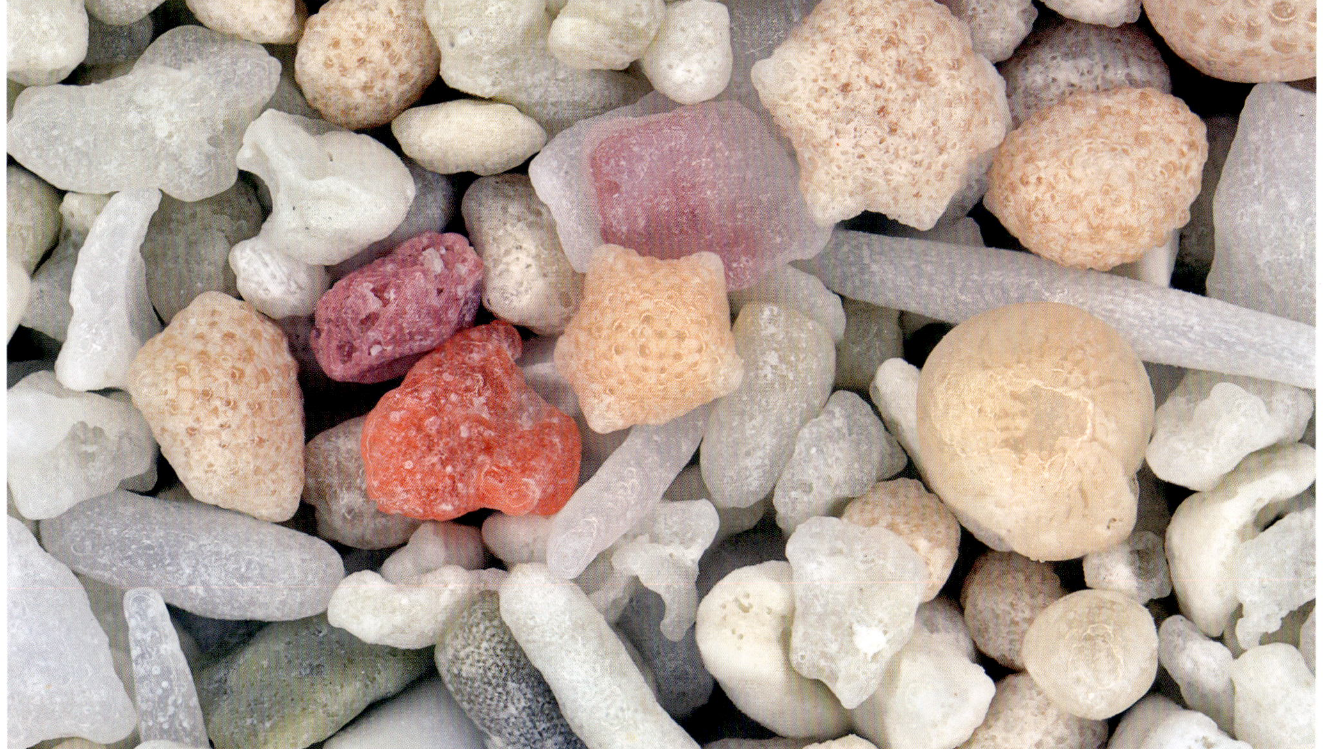

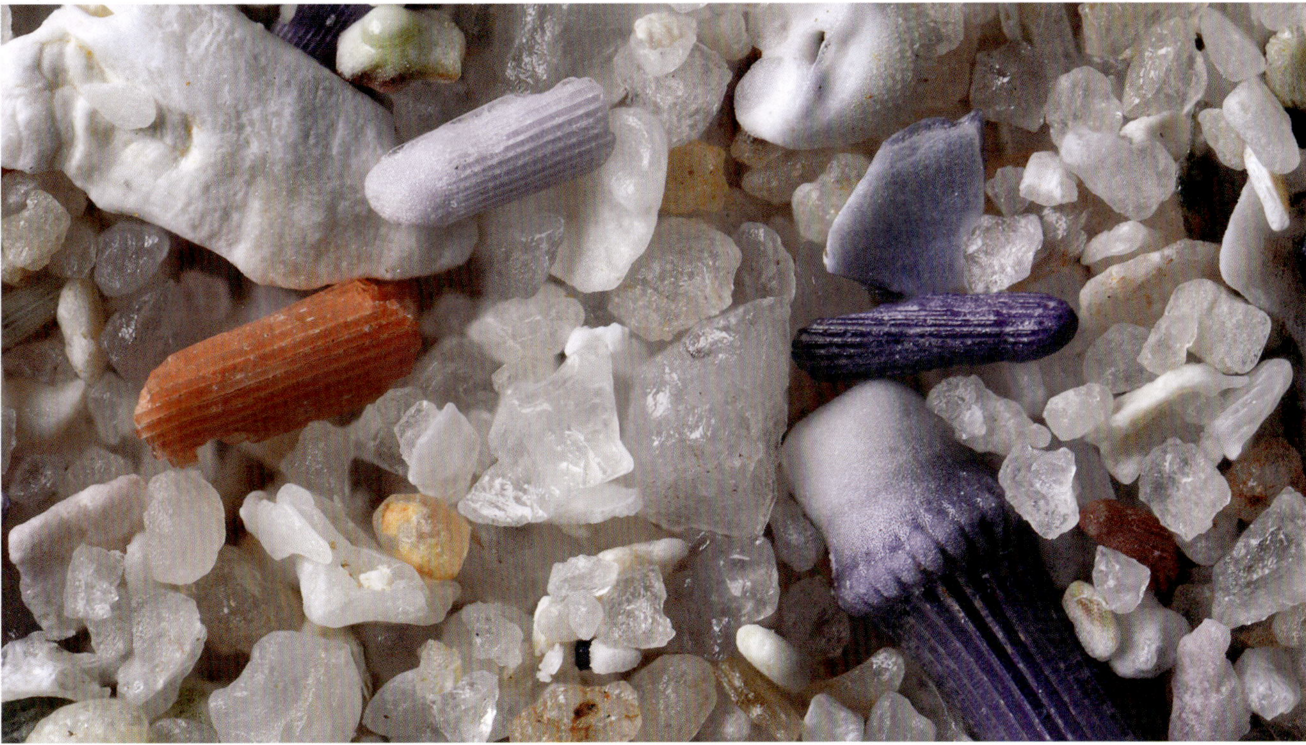

Figure 7.7 Typical sand of a tropical Pacific island, exclusively composed of biogenic sand grains. You can see coral and mussel remains along with elongated Gorgonian sclerites as well as red and cream-colored foraminifers. Michaelmas Cay, Australia. Image width 6.7 mm.

Figure 7.8 Continentally formed Atlantic sand from Boulders Beach on the Cape of Good Hope, South Africa. Boulders from the mountain base reach into the bay and leave behind the many clear, still-edged quartz grains. Biogenic grains include a barnacle fragment on the upper left and very prominent colored sea urchin spines. Image width 6.6 mm.

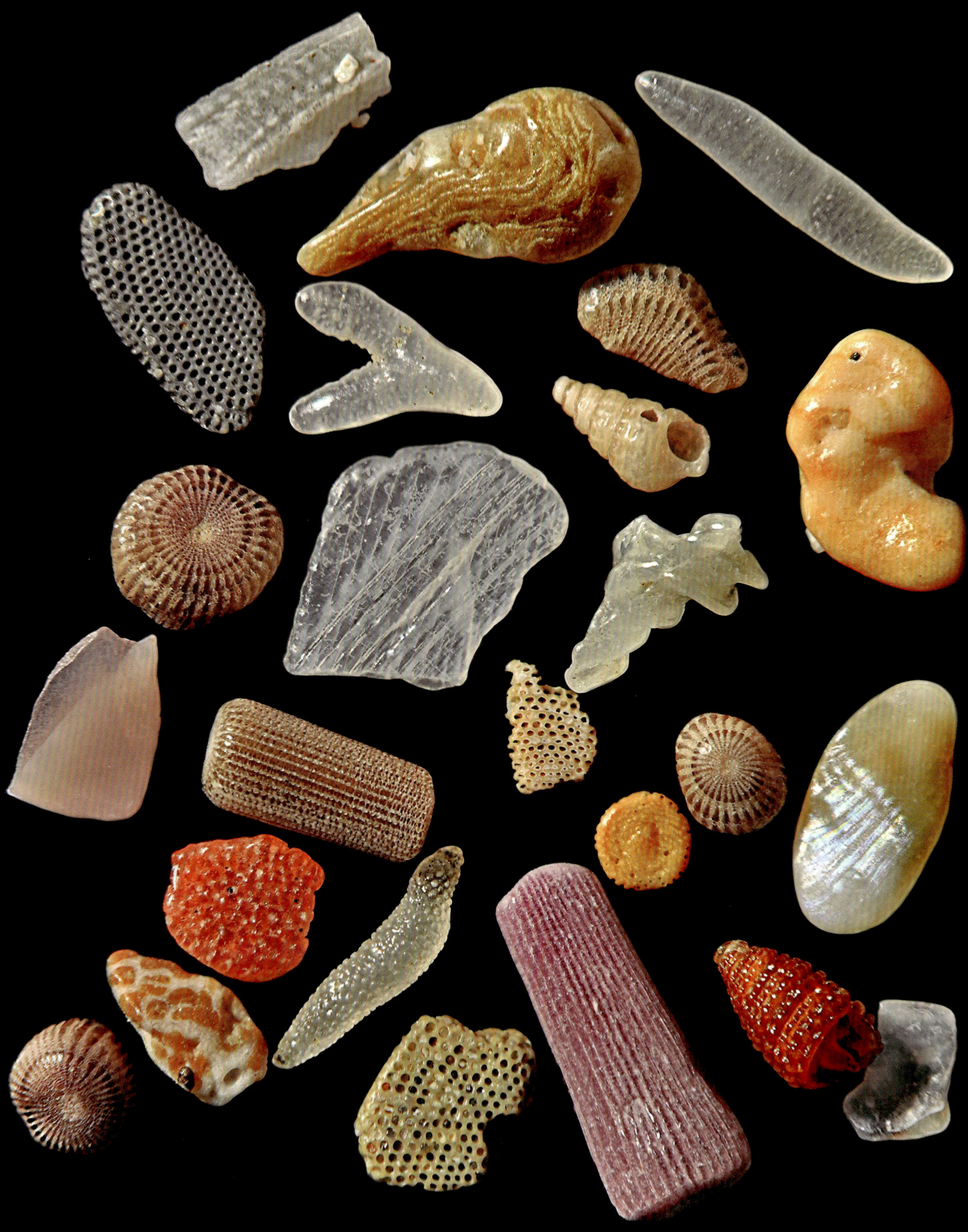

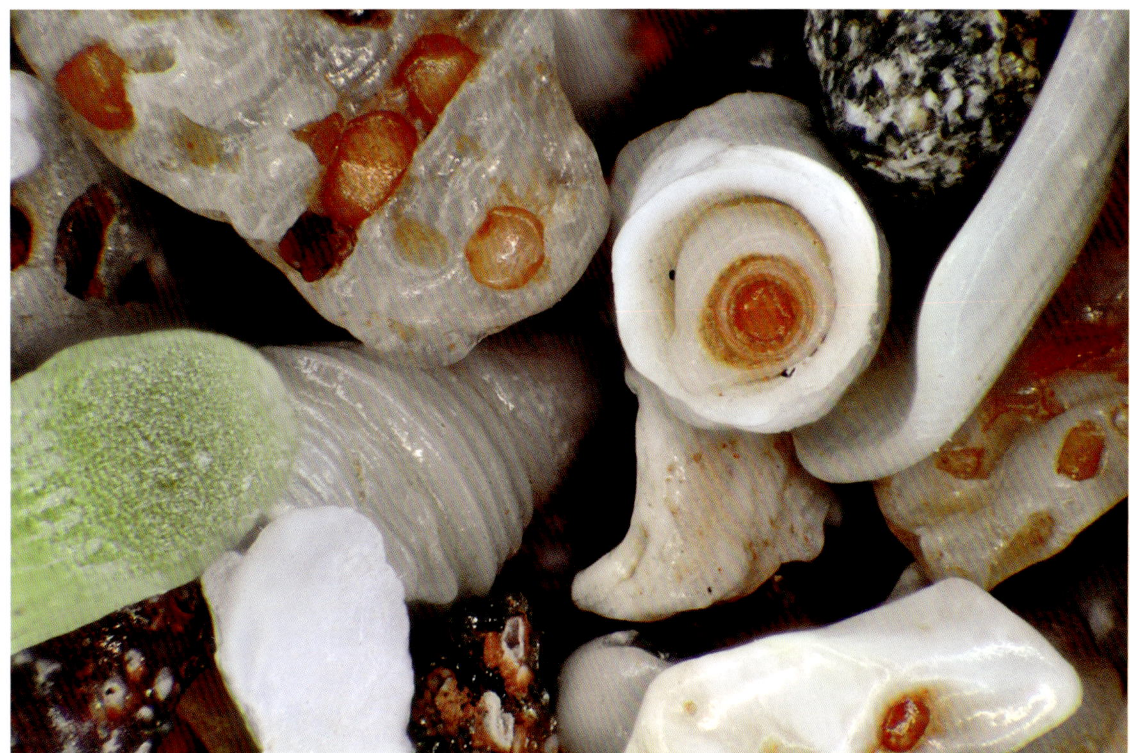

In addition to food in the form of plankton, detritus, and others, the waves also carry fragments of the coral reef communities that predominate in the deep zones. Both mix on the beach, whose sand thus documents and archives a broad and colorful spectrum of all life-forms around the tropical islands or coastal courses in question. The mechanisms of sand formation are quite different from those of the continental beaches. While there we stroll along the beach and see the remains of once-proud mountain massifs under our feet, here we see the remains of billions of tiny creatures. There is also a haptic difference: the thermal conductivity of lime is much lower than that of quartz and silicates. Lime-rich sands therefore feel warmer and more pleasant even when damp because they do not dissipate body heat as quickly as quartz sands do.

Figure 7.9 *Left side*: The small, tropical volcanic island Nosy Be lies on the northwest tip of Madagascar, a large island of crystalline bedrock in the Indian Ocean. The selection of biogenic sand grains shows fragments of mollusks, sea urchin spines, sclerite, foraminifers, and mussel shells. Image width around 9.5 mm.

Figure 7.10 The tiny island Isla Bartholome, with an area of only 0.46 square miles (1.2 km^2), lies east of the Galápagos Island of San Salvador, close to the equator. It is one of the youngest formations of the volcanic, well-known Galápagos Islands. The sand sample shows pieces of mollusks and gastropods next to green sea urchin spines, as well as glassy and bubby volcanic sand grains. Image width around 3 mm.

We can make another observation during our beach walk: in the dry area away from the waterline, our feet sink in up to our ankles. If we approach the wet, partially flooded strip at the waterline, it is as if we are walking on concrete; the sand seems hard and firm and contoured, and flat footprints remain. A little farther, already in the completely flooded zone, the sand seems to become liquid; it barely holds us, and we sink in. This remarkable behavior can be explained by the interplay of grain roundness, sphericity, and surface properties of the grains. The beach we are looking at, as we have seen, consists of calcareous fragments that are predominantly angular; that is, angular and angular in nature. Furthermore,most of the lime particles have a relatively rough surface. In the wet state, a thick film of water forms around the grains, and water bridges are formed between the grains. Surface tensions in the liquid, and interfacial tensions between water, air, and grains, make them stick together. These forces are greater than gravitational forces, as can be easily seen by carefully lifting two ping-pong balls only with the aid of a droplet of water between them. Even by adding 1 percent water, sand becomes a quite solid material and maintains these properties even up to a water content of about 10%.[139] However, if the water content is increased above this, the length of the boundary line between the grains decreases and the capillary strength of the sand also decreases, causing the sand to become fluid. Anyone who has ever tried to build a sandcastle underwater will quickly be convinced of the impossibility of this endeavor. Both completely dry and completely wet sand have no bond strength whatsoever.

Let's do another experiment while we are out on the beach. However, the experiment is much more effective on coasts with well-rounded quartz grains. If we place our foot on an exposed but water-soaked sandbank in front of the waterline and begin to vibrate slightly, water does not come to the surface, as one might think, but the sand suddenly appears dry.[140] How can this be explained? The constant wave motion had previously stacked the sand grains into each other in an optimal way. The densest possible packing of ideal spheres, which is never quite reached in nature, is 74% as a limit value, which means that a gap volume of at least 26% remains. The application of foot or hand has locally disturbed this order and thus increased the gap volume. As a result, the existing water is sucked back into the enlarged intergranular spaces, so the sand surface becomes drier for a short time, until finally capillary forces draw water from the immediate surroundings. That concludes the complex and interesting interaction of granular matter with water, and some suggestions for experiments on the beach.

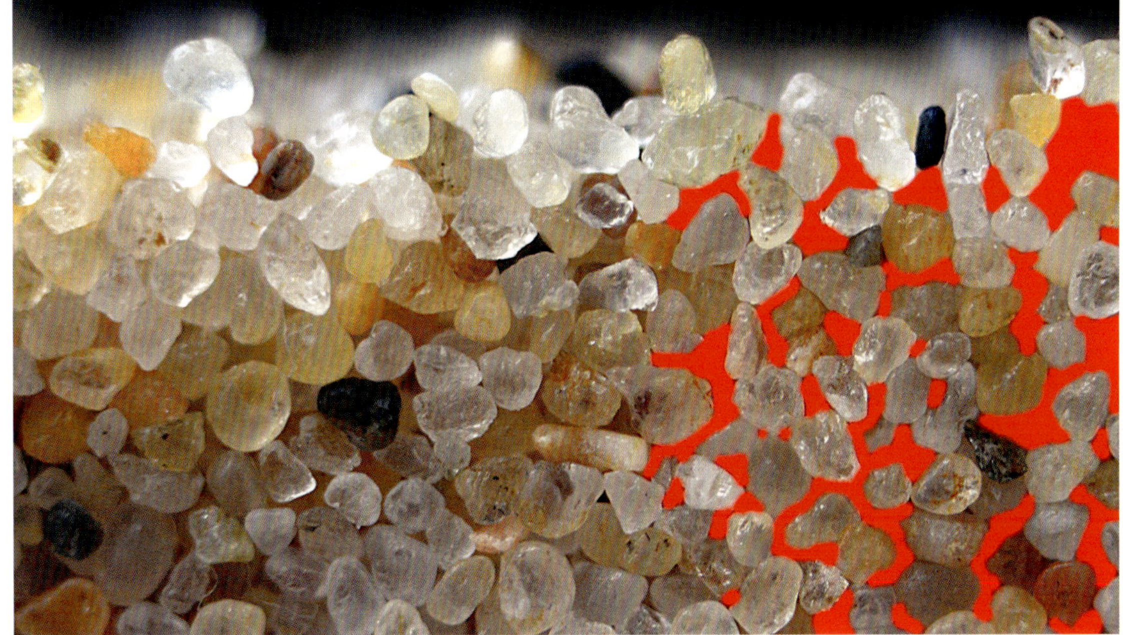

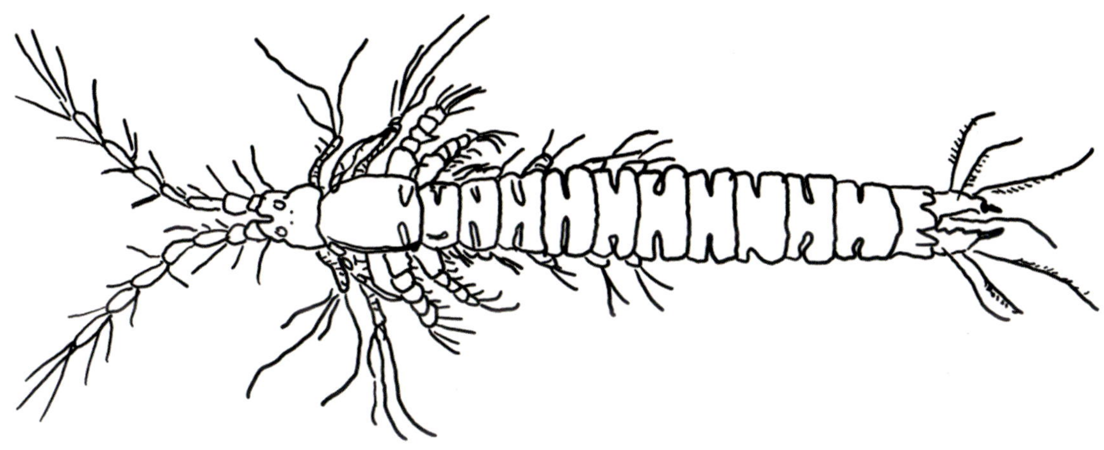

So, what does this have to do with our story? The explanation is astonishing. Due to the ever-present gaps, which, as we have seen, do not disappear even at maximum packing density, the sandy beach under our feet forms a gigantic labyrinthine gap system. Seawater on one side and groundwater seeping on the other side mix in coastal areas and form a subterranean brackish-water biotope that provides ideal conditions for a particular type of creature, the meiofauna.[141]

This refers to tiny animals between 63 *μm* and 1 mm in size that live between the sand grains and find shelter and food there. Meiofauna includes nematodes, copepods, ostracods, mites, tardigrades, abdominal hymenoptera (Gastrotricha), echinoderms (Kinorhyncha), and, in some parts of the world, a group of crustaceans called Mystacocarida, which has not been known for long.

Figure 7.11 Cross section through a sand sample, using an inserted glass pane. The view is directed to the vertical structure of the sand gap system near the bottom. On the right-hand side of the image, the three-dimensional gaps in the surface are colored red. In this branched system, brackish water would be found in nature, as well as inhabitants of the meiofauna, which move between the grains and search for their food. Sediment researchers press synthetic resin into isolated samples to obtain a real three-dimensional cast of the interstitial spaces. Sand from the Cote Sauvage, La Palmyre, France. Image width around 5 mm.

Figure 7.12 Adult, about sand-grain-sized Mystacocardia organism of the type *Derocheilocaris remanei*, with a view of the back. The animal has two pairs of lenticular eyes on its head and can sense light stimuli as well as smells with the help of the front antennae. Body length around 380 *μm* (depiction according to Gad 2010).

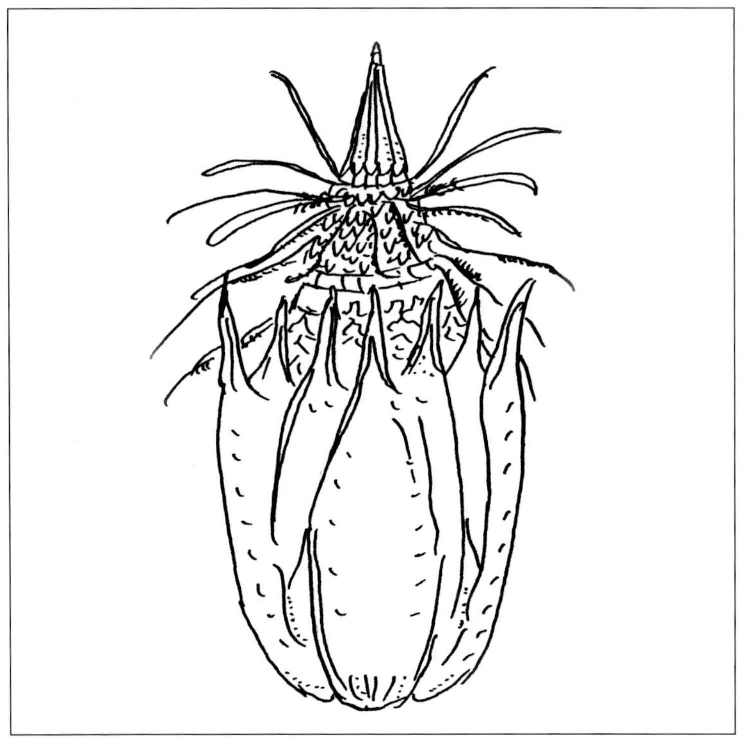

Mystacocarida are actually one of the most exciting zoological discoveries of the last century. The first organisms were discovered by the Americans Pennak and Zinn in 1943 in the wet bank sands on the eastern coast of the USA but were first described much later. There are twelve types of these organisms that are known worldwide.

Even more recent is the sensational discovery of a completely new group of animals by the Copenhagen researcher Mobjerg Kristensen. During a research stay on the Atlantic coast of Brittany in 1983, he discovered the Loricifera in the sand of the ocean floor. Loricefera are among the smallest multicellular organisms known, with an average length of only 250 *μm*. Even more than thirty years after their discovery, not much is known about them. Belonging to the Locifera, Nanaloricidae are inhabitants of the more oxygenated upper sedimentary layers. They live as true members of the meiofauna among the sand grains and are always found near the coast and at a maximum water depth of 164 feet (50 m). The animals consist of over 10,000 cells and even possess a fully developed organ system with nerve cells, a digestive tract, and musculature.[142]

Figure 7.13 Fully grown Loricifera organism from *Nanaloricus mysticus*, belly view. At the very top, the mouth cone can be seen, and below it spinoscalids (antennae-like), thorax, neck, lorica spine, and lorica with ventral plate. The large abdomen area to the right is covered with a strong cuticle that splits into plate-shaped areas. Diameter of the animal only about 100 *μm* (depiction according to Gad 2005).

Their cells, with dimensions under 2 *μm*, are probably the smallest in the known animal kingdom. Observing them is quite tricky because they are hardly separable from the small sand grains they live among, and upon which they can also attach themselves, thanks to special glandular secretions.

The sand gap system holds even more wonders. Probably the strangest microorganism of all, which can be found in our sand gaps but also in many other biotopes, are Tardigrades (from Latin *tardus* = slow and *gradere* = stride). They are always found in connection with water, but only in the sea—also and especially in damp mosses and lichen beds. They even like to inhabit the small moss cushions between the tiles of our roofs. Tardigrades have been known since the eighteenth century and are not at all rare. However, only a few nonspecialists will ever have seen them alive and in action. Very early it became obvious that these little animals, which got their name because of their cute appearance, have extremely special peculiarities. Many of them possess an almost unbelievable ability to fall into a state of cryptobiosis. If the living conditions deteriorate (for example, due to lack of water or low temperatures), the animal gradually shrinks into a regularly shaped dry "barrel." It thus transforms into a kind of sand grain. In this barrel stage, the tardigrade effectively ceases all active and observable life processes. If certain conditions are met, the animal can survive in dry rigidity for up to ten years. If it comes into contact with a drop of water during this time, it unfolds again within a few minutes into a lively and intact organism; the "grain of sand" awakens to new life. To top it all off, the described process of dehydration (i.e., formation of tardigrades and reawakening [rehydration]) can be repeated several times. Tardigrades, living in such cycles, can reach an age of over sixty years. One can well imagine the impression this apparent awakening from the dead left on the researchers of earlier times, and even today one cannot escape the special fascination that arises when one discovers a crawling tardigrade among lifeless grains of sand.

One could describe the sand gap system as a large, underground city, but this comparison is not precise enough. The branching and fragmented web of city streets is probably just as complexly structured on a corresponding scale, but it essentially remains constant over time and continues to develop in a more or less organized manner. Our subterranean city in the sand behaves quite differently. The gap system is in permanent change. The sand grains move and are swirled relative to each other almost without pause by tidal forces, capillary forces, and the flow effects of the upflowing and downflowing water.

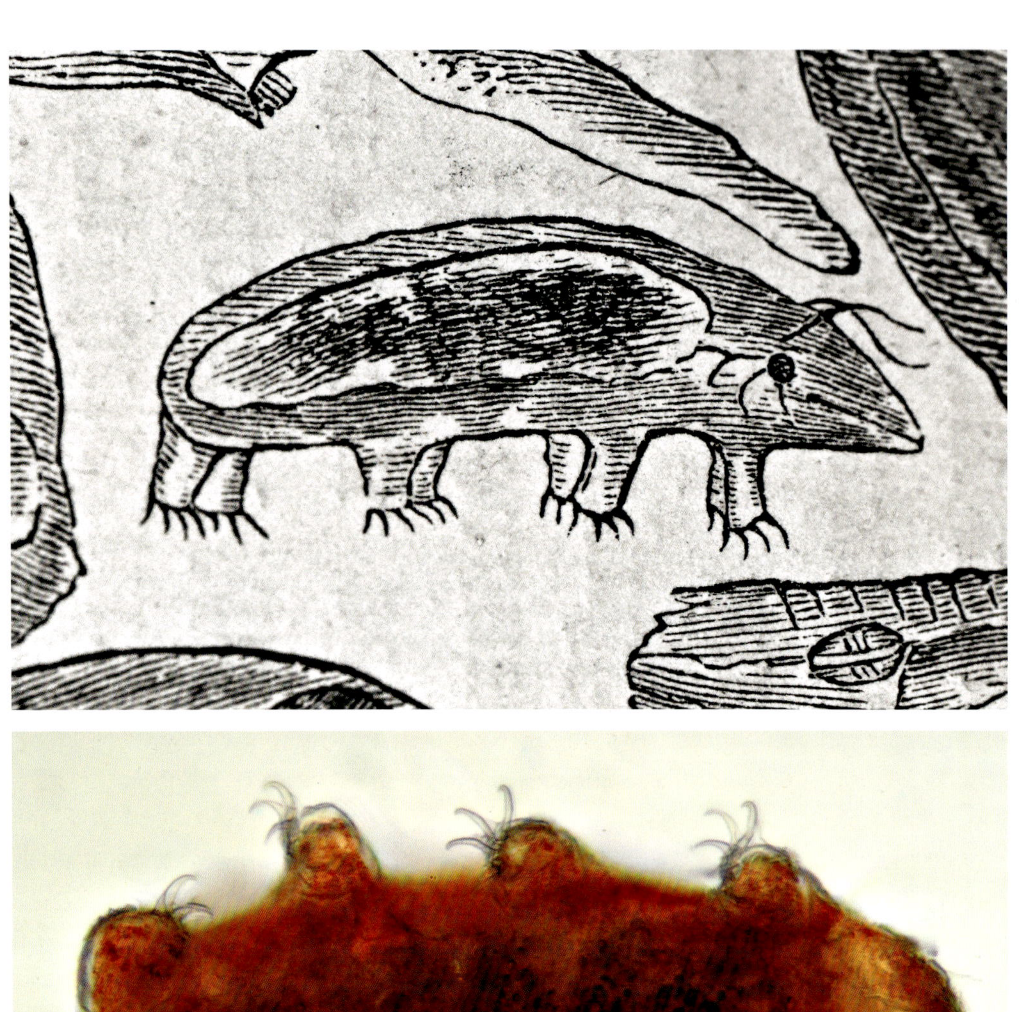

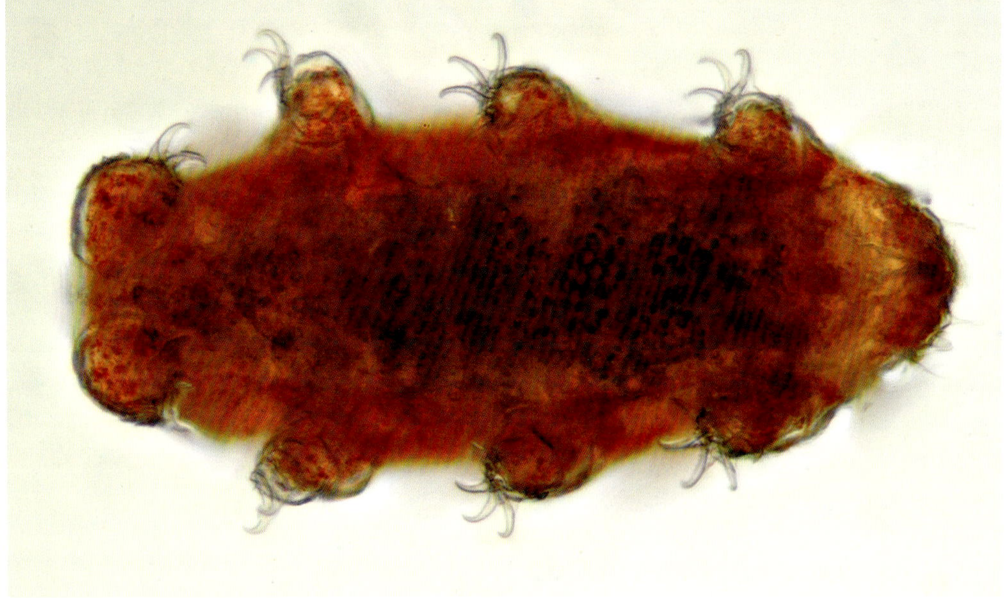

Figure 7.14 *Above*: Wood carving of a "little bear-animal" in a side view (probably genus *Echiniscus*) from Moritz Willkomms's *Wunder des Mikroscops* from 1878. The four clawed pairs of legs can be seen. *Below*: An armored *Echiniscus* tardigrade. View of the stomach, with eight clawed legs. Sachsenheim, Germany. Animal length 360 *μm*.

The subterranean inhabitants must put up with and endure all this, not to mention the fluctuations in salinity, and the already-described fluctuations in temperature and humidity. Without suitable adaptation to these highly dynamic and demanding conditions, the inhabitants could not survive. Many of them therefore have armor of cuticular plates, for example, which protect the sensitive interior of the animals. Other species are elongated in structure to move more easily in the gaps or have small feet or adhesive plates to cling to the grains. For example, the Mystacocarida, with the single species *Derocheilocaris remanei*, measuring just under 0.4 mm, prefers sand grain sizes between 0.1 and 1.0 mm. In contrast, the somewhat larger species *D. typica*, with 0.69 mm, requires grains between 0.25 and 1.5 mm.[143] It is fascinating how exactly the occurrence of the species is matched to the specific characteristics of the respective sand biotope. It is particularly impressive how finely optimized the locomotion system of the animals is to the three-dimensionality of their extraordinary habitat in the sand. If one were to examine the sand of Copacabana instead of that of Easter Island, Hawaii, or Bani, the results would be different. The sand here consists of well-rounded quartz grains with consequently higher packing densities. The animals find much less room in the interstitial spaces, and the water exchange with the sea or groundwater is lower. In addition, certain bacterial turfs, which are important for the nutrition of the meiofauna, adhere less well to the much-smoother quartz grains. The habitat is therefore completely different in nature and tends to be less suitable, so that the meiofauna there has a correspondingly lower species diversity than that in the coarser coral and calcareous sands of the coasts of tropical islands.

Overall, we can be very happy with our eager and abundant peoples in the gaps of sand. Without them, the beaches would not be so clean and inviting. Underground, the animals do their work by feeding on all the organic waste that accumulates and lands in this zone every day. Perhaps they can be compared to the life-forms in the leaves strewn in the forests of temperate climates. We own them a lot as well, at least when it comes to the beauty and freshness of the forest.

We have now looked, by way of example, at some of the animals that live their hidden but remarkable lives among grains of sand. There is another equally amazing way to use sand as a habitat, although it is hard to imagine that even a single grain can harbor life. Admittedly, we are dealing here with very primitive and, by necessity, very small forms of life, the algae and the fungi. In any case, life can be found in grains only if they are carbonate sands; that is, grains of sand made of lime. Here we do not speak of living beings that reside in lime shells, as we will get to know later, among other things, with the foraminifera, but we consider living beings that penetrate into rocks and also sand grains by drilling tiny holes into the lime. Such "endolithic microholes" lead to the phenomenon of bioerosion, one of the most important factors in the decomposition of carbonate grains in the marine environment.[144]

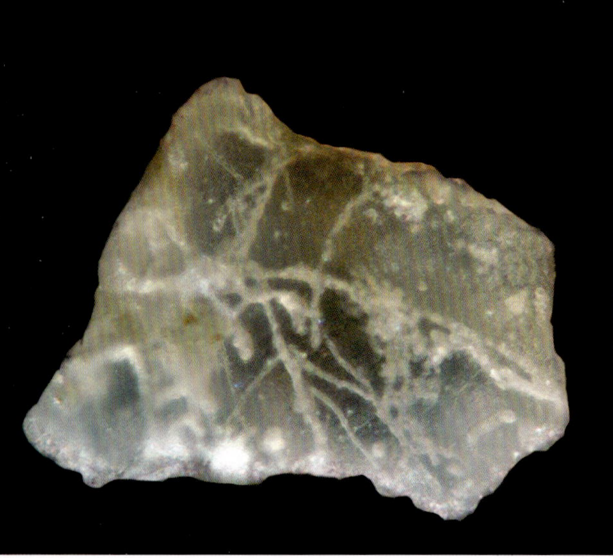

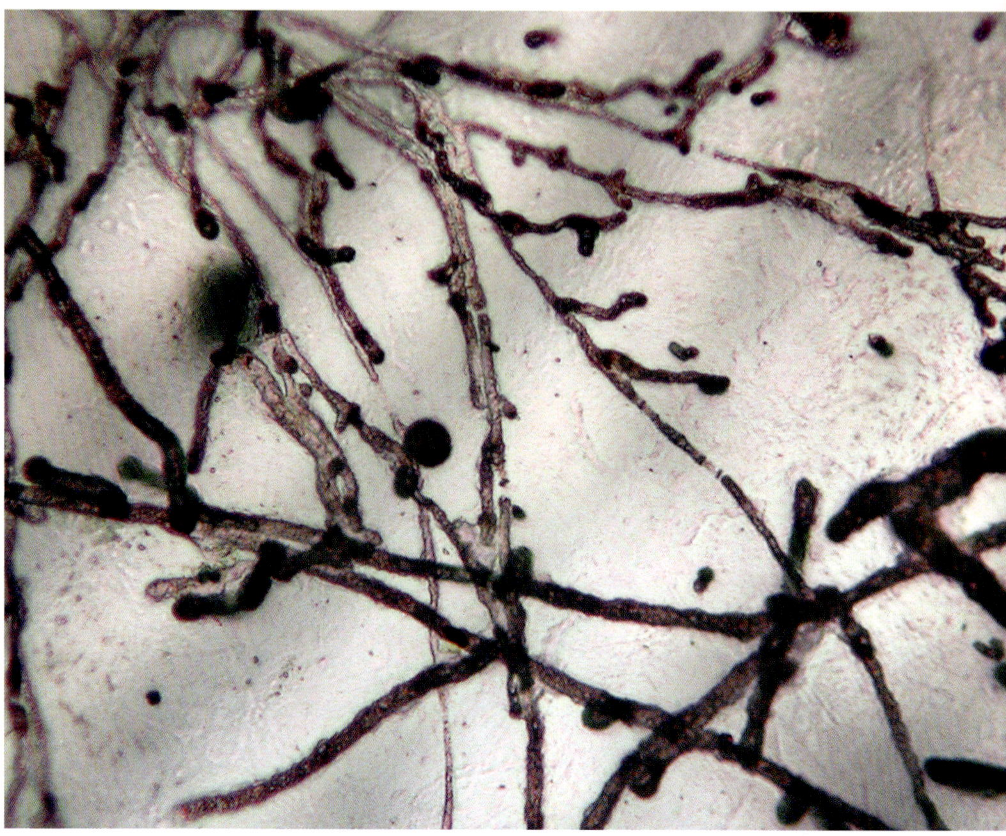

In contrast to algae, which photosynthesize, fungi do not need any light. Whereas endolithic algae (i.e., algae that bore into rocks) can be found in water depths from 328 to 1,214 feet (100–370 m), fungi, not dependent on light, can be found in far-deeper locations in the ocean. Algae and fungi bore their tiny holes into the sand grains following the principle of chemical corrosion: the lime is chemically decomposed with the help of corresponding acids—no abrasion occurs during this process. Blue, green, and red algae leave behind a network of extremely fine boreholes, the structure and shape of which are largely characteristic for the respective type.

Figure 7.15 Sand grain from Anse Vata, Noumea, New Caledonia. The translucent grain shows tiny microholes that were probably created from endolithic algae. Reflected light. Grain size 150 × 210 *μm*.

Figure 7.16 Translucent shell fragment of a mussel, Baie de Citrons, Noumea, New Calcedonia. System of tiny boreholes that were probably caused by endolithic algae. Diameter of the holes around 4 *μm*. Transmitted light. Image width around 350 *μm*.

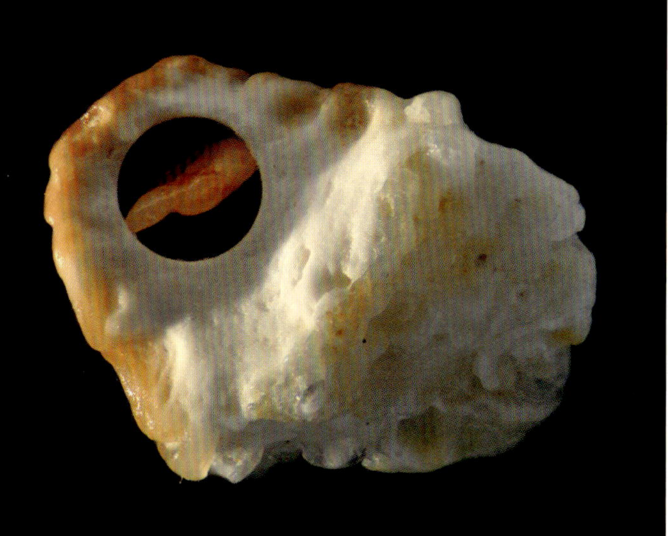

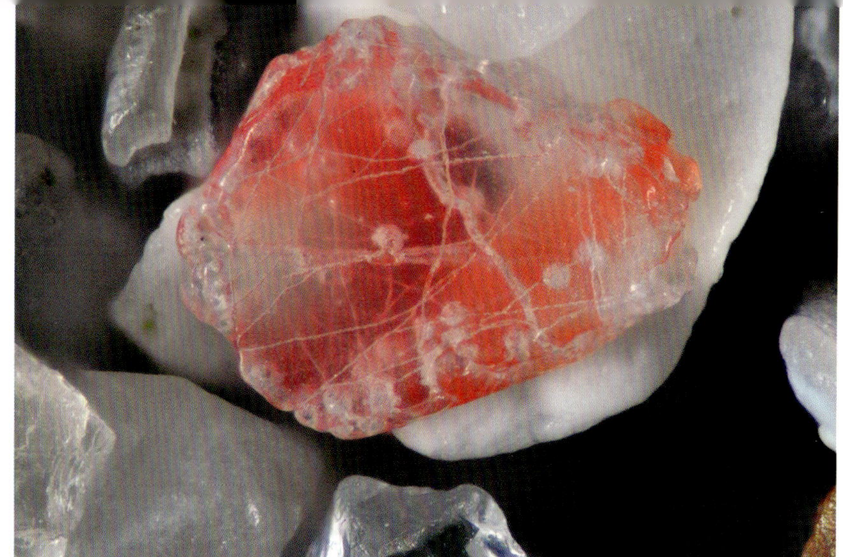

Professionals can identify the algae simply on the basis of their traces. Most of them have boreholes with a diameter of between 2 and 10 *μm*. Fungi are similar, but their holes can be even finer and otherwise occur in the same diameter range and cannot be distinguished from those of algae.

Endolithic algae can bore distances of about 0.3–36 *μm* per day. Only rarely do they go deeper than the substrate, since they are dependent on light. That is probably why we most often find their boreholes in sand grains made of translucent fragments of mussel shells that are permeable to light. Here, algae are still exposed to light even in larger depths. On rocks, the bioerosive ablation rate can be up to 1 mm each year. Endolithic boreholes as old as the Precambrian period can be seen on rocks.

Sponges, especially the tropical species *Cliona caribea*, are also able to break down carbonate sediments and leave quite typical traces in them, which have been the subject of a lot of research. In its drilling activities, the sponge proceeds quite differently from the algae and fungi described above. First, tiny gaps of only 0.2 *μm* are made with acids produced by certain cells. Subsequently, the areas loosened in this way can finally be separated from the substrate as particles. *Cliona* thus generates microparticles. Studies in the Cayman Islands of the West Indies yielded significant removal rates of 1.6 pounds per square foot (8 kg per m^2) per year. We mention *Cliona* sponges here only as a sidenote, since they do not live in sand grains but produce them.

Finally, we come back to an algae group that we had already introduced in connection with mineral quartz: the diatoms. They also do not colonize the interior of the sand grains but can be found adhering to the surface.

Figure 7.17 Biogenic sand grain from Isla Bartholome, Galápagos, Ecuador. An even-smaller grain of sand has gotten stuck in the perfectly round borehole, of unknown origin, with a diameter of 380 *μm*.

Figure 7.18 Microboreholes in a very small, red shell fragment of a mussel or snail. Vatersay, Outer Hebrides, Scotland. Reflected light. Image width 780 *μm*.

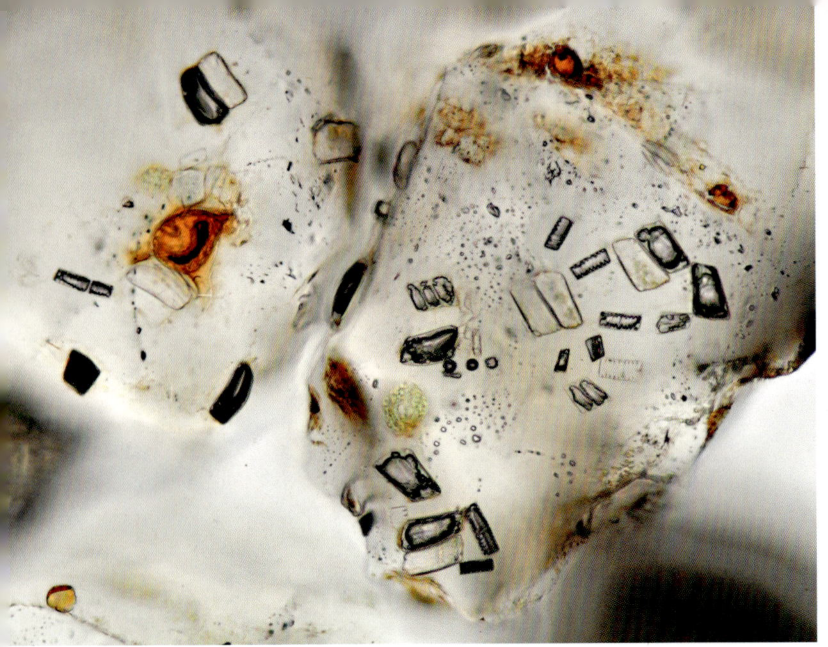

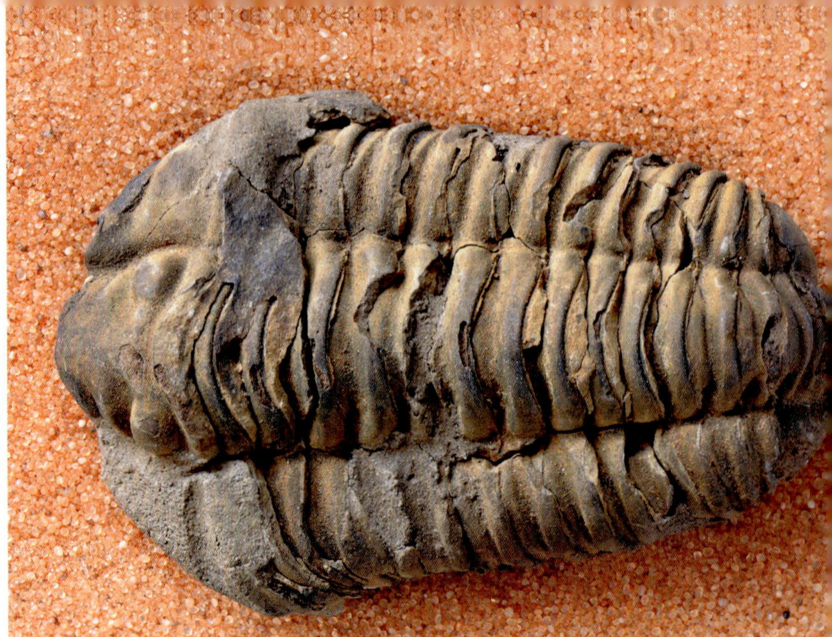

Characteristic is their mostly very nicely and symmetrically structured, box-shaped cell envelope (frustule). During microscopic observation in transmitted light, we must be careful not to confuse the algae with the mineral inclusions of the sand grains, which are partially also symmetrically structured.

Let's summarize in what way small life-forms stand in connection with sand. Four groups form as follows:

1. Life-forms that live in the interstitial spaces of sand, the sand gap communities, or meiofauna
2. Life forms that live inside or on the surface of sand grains
3. Life-forms whose mineral components turn to sand after their death, mostly
   - outer skeletons and inner skeletons as mechanical supports,
   - protective shells and armor such as mussel shells or snail shells, and
   - defensive instruments such as spines or even weapons, such as crab claws.
4. Life-forms that contribute to the formation of organogenic lime after death. Sand can then be formed from these limestones after uplift and erosive processes.

We want to limit ourselves to these four cases. Much could be added, such as sand fishes, ant bears, desert ants, and all the highly specialized and equally interesting animals and plants that have made sand, dunes, and the desert their home. They are recognizable even without a magnifying glass, so they fall into the macroworld and not into the microworld, to which we originally dedicated ourselves.

Figure 7.19 Sand grains from a Norwegian mountain lake near Valle. The stuck-on, rectangular algae shells next to the somewhat smaller and rounder inclusions of the grain can be observed, especially on the right-side sand grain of quartz. Light field, transmitted light. Image width 600 *μm*.

Figure 7.20. Trilobite *Flexicalymene ouzregui* from the Ordovician period, 450 million years old. The well-preserved fossilization comes from the Ktaoua Formation near Erfoud, Morocco. The three-part structure of the torso with its ribbed arches is clearly visible, as are, on the left side, the two eyeballs. Length 82 mm.

Therefore, we come to the next group on our list: animals that can produce inorganic crystals in a controlled manner (so-called biologically controlled mineralization). Little known in this context is that more than sixty different biominerals have been found on Earth that are specifically produced by living beings and are useful and advantageous to them in very different ways. By the Archean period, some species of bacteria possessed the ability to biomineralize. It was not until much later, in the Cambrian period, that the first animals with mineral compound eyes lived: three-lobed crustaceans or trilobites.[145] Fossil eyes have been found in two of these 534-million-year-old Cambrian species in China. Their construction is remarkable, since they consist of rows of biconvex lenses of the biomineral calcite (i.e., transparent double spar). The microlenses focused the parallel incident light onto an organic shell behind them, which was simultaneously capable of correcting optical aberrations. The lateral arrangement of both eyes on the cheeks of the animals gave them a horizontal field of view of about 70°. In addition, this provided the possibility for spatial detection of the surroundings.

What strange cyclical connections can be deduced from this. The calcareous remains of former life sinking down in the dim light of the primeval ocean enabled the trilobite, as a Cambrian animal half a billion years ago, to develop a lens for the first time through biomineral processes from the substance with which he will be able to see those animal remains that suppled the components of this substance: calcite. For the first time in the primeval ocean, the world was not only sensed but perceived. Admittedly, the rather coarse act of perception of the early times will have differed from the modern optical sensory impression of the human eye. What and how the animals may have felt remains hidden to us.

But calcite already helped prehistoric animals in the form of another equally amazing construction: not only seeing, but hearing. Hearing generally means to perceive a pressure fluctuation. Fish and amphibians possess (even today) special sensors for this, a kind of vibration organ. This consists of a bladder filled with a gelatinous fluid and several sand grains of calcite or tiny stones, the ear rocks. Incident sound waves cause a relative displacement of the sand grains and the bladder wall, which is then lined with sensitive hair cells. When the hair cells are bent, they send out an electrical impulse that is transmitted to the brain via an auditory nerve. This produces the sensory impression of hearing. The organ is also active in case of displacements, as it reacts to changed position in space with the described mechanism. So again, calcite is involved, and hypothetically even the tiny hearing sand grains (otoliths) could be discovered on a beach. So sand grains, which once conveyed sensory impressions: a beautiful but also disconcerting thought.

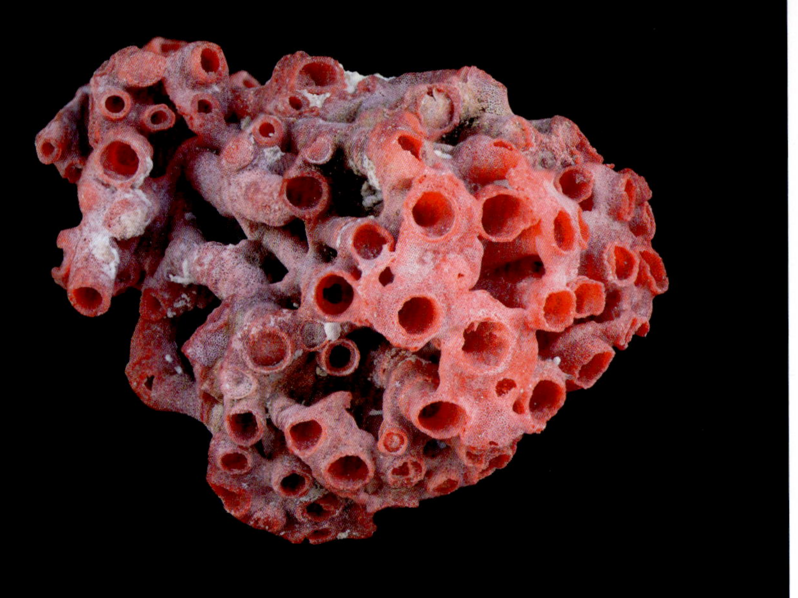

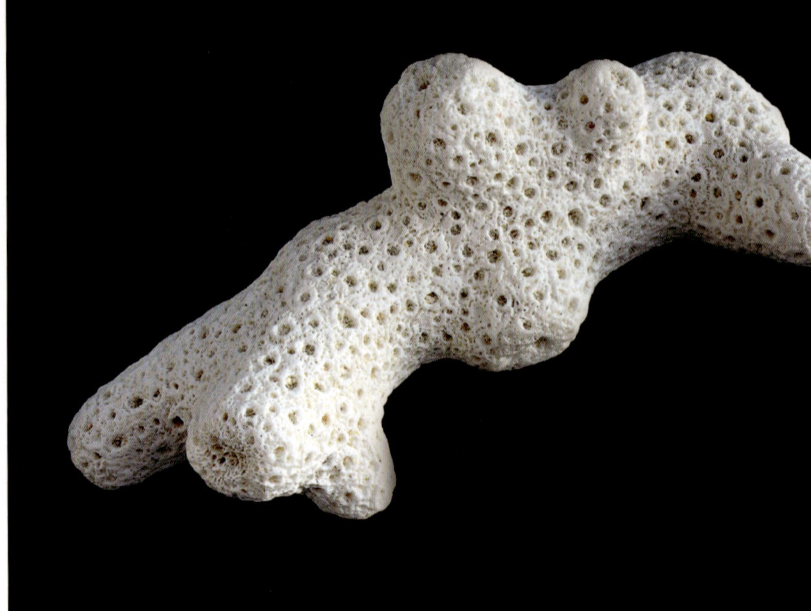

7.2 Corals

Let us continue with the sand variety of the basic building material calcium carbonate, the building material of corals, calcareous algae, sea urchins, sponges, foraminifera, and many others. Some of them (group 3 according to the classification above) will now be discussed in more detail.

If one looks at the treelike branched forms of the sedentary corals, one could get the impression that corals belong to the plants; after all, they do not move from their spot. Until the eighteenth century, researchers held this view, but it is wrong. Corals are animals. Like jellyfish and sea anemones, they belong to the phylum of cnidarians and further to the floral animals (Arthozoa). Corals can be roughly distinguished between octocorals and hexacorals on the basis of the number of tentacles. The latter divide themselves into rugose corals (Rugosa), tabulate corals (Tabulata), and stony corals (Scleractinia). Over the course of the earth's history, rugose and tabulate corals appeared during the Ordovician period and disappeared together in the Permian period. Stony corals are still around today, which developed later, in the Triassic period.

Corals exclusively inhabit the sea. Thereby they can live as single individuals or form coral stocks and finally huge colonies through budding. Only the stony corals were also able to build up reefs with their calcareous basic framework, which we have evidence for from the Triassic period through today. Corals essentially consist of a calcareous shell, which the young animal (polyp) builds up as a tube- or cup-shaped structure by secreting lime on the outer cell walls. During its growth, the animal deposits horizontal plates (tabulae) to separate the animal from where it was below.

In this way, the polyp always resides in the uppermost chamber of its shell, from which it holds its tentacles into the water flowing past, to catch food.

Figure 7.21 Red coral (organ pipe coral) from the beach near Jambiani, Ungula (Zanzibar), Tanzania. Image width around 40 mm.

Figure 7.22 White coral from Belmare Beach on the eastern coast of Mauritius. Image width around 60 mm.

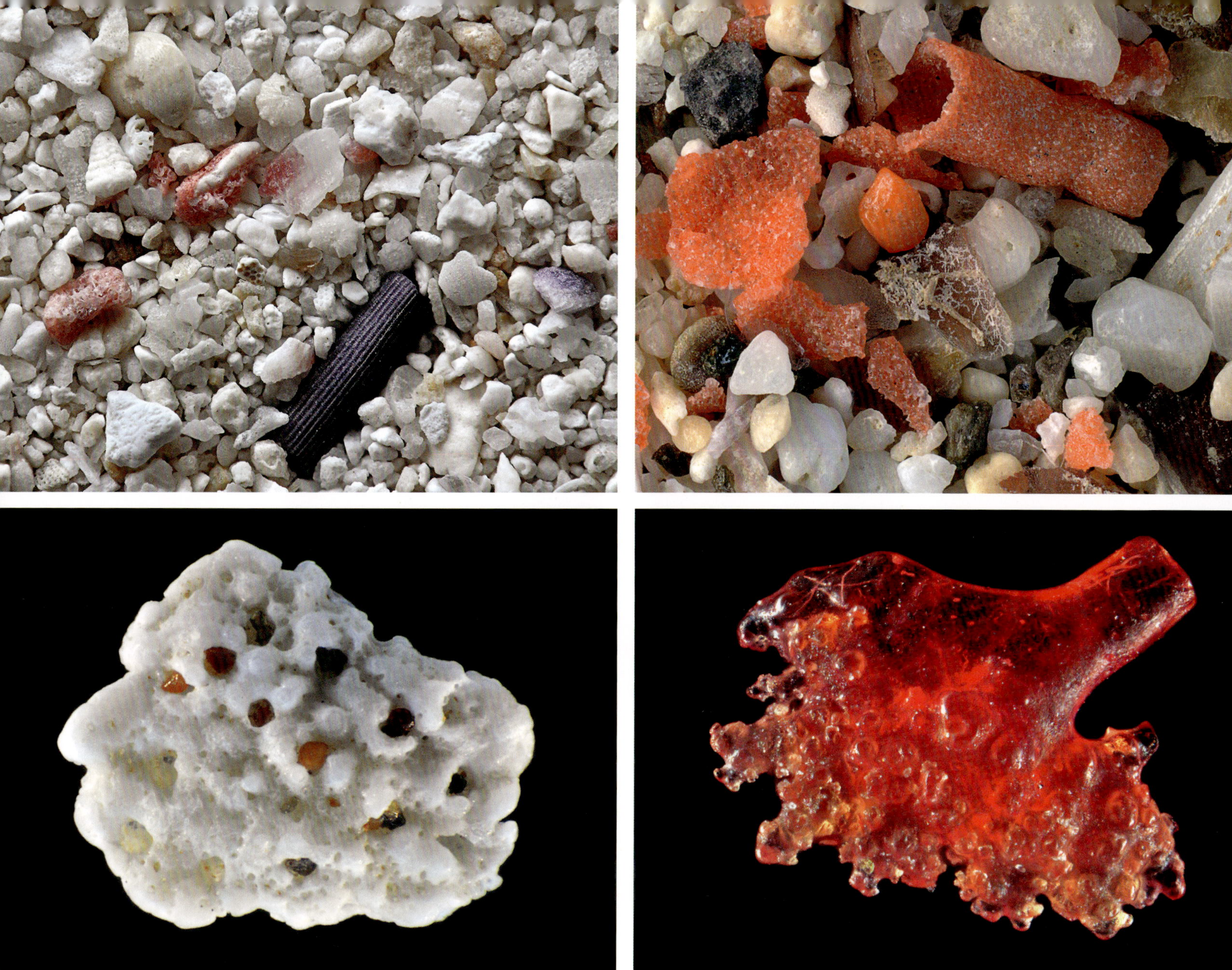

Figure 7.23 Coral sand from the eastern coast of Mauritius, Belmare Beach. Along with coral debris, white- and pink-colored foraminifera as well as blue and green sea urchin spines and many angular mussel remnants can be seen. Image width around 10 mm.

Figure 7.24 Coral sands from Luzon Island, Philippines. Corals, broken mussel pieces, and a sclerite (*to the right, below the hollow coral branch*) can be identified. The translucent shell fragment below the center of the picture already shows significant and branched endolithic microboreholes. Image width 6.8 mm.

Figure 7.25 Sand grain from a coral fragment. Isla Bartholome, Galápagos, Ecuador. Small sand grains have been deposited in the hollow spaces. Image width 2.8 mm.

Figure 7.26 Sand grain formed from a coral branch. Anghthong Marine National Park, Thailand. In the upper area, microholes can be seen. Diameter stem base 85 *μm*. Reflected light.

Tropical corals emit an organic mucus that helps them catch plant particles and sediment pieces. Young polyps build their calcareous shell on the remains of their predecessors, so that generation upon generation builds layer by layer, causing the reef to grow. Next to the well-known, spectacularly colorful reef-forming water corals of the tropical oceans, there are also cold-water types that inhabit the Atlantic up to Norway.

Coral breathes through its body surface, and tropical species also colonize endosymbiotic, microscopic, free-swimming algae within their outer walls. These algae carry out photosynthesis (i.e., they extract carbon dioxide from the seawater and then supply the coral polyps with glucose as a nutrient). In return, they receive protection from predators. Since these symbiotic communities depend on light, they prefer ocean depths of maximally 98–164 feet (30–50 m).

Their relatives, the cold-water corals, filter their own food from the seawater.[147] Coral reefs not only harbor corals but also form exceptionally rich biotopes that provide habitats for a wide variety of species. Many of these species, such as bivalves, gastropods, and red algae, precipitate significant amounts of calcium carbonate, which contributes to the height growth of the reef. Entire mountains can form underwater in this way. Probably the most prominent recent example is the Great Barrier Reef off the northeast coast of Australia, which is currently very endangered in terms of its ecological stability. Apart from their inestimable value as irreplaceable habitats, reefs form complex sedimentation areas that, depending on their location, accumulate continental mineral components and muds landward and then, seaward behind the actual reef crest, usually slope quite steeply toward the open sea.

The subclass of octocorals also includes another group: the horn corals, or gorgonians. They can be found almost everywhere in the world's oceans, from shallow water to the deep sea. They are of special interest to us because they have colorful and multiform tissue needles or sclerites. These sclerites can be found in many sands, especially in the tropical seas. Their variety of shapes is remarkable; they occur as spindles, needles, clubs; are cylindrical or dumbbell shaped; and are studded with small warty spines or are also smooth. Their dimensions are in the submillimeter range. In sand literature, they find almost no mention but contribute significantly to the richness of shapes and to the characterization of tropical sands.

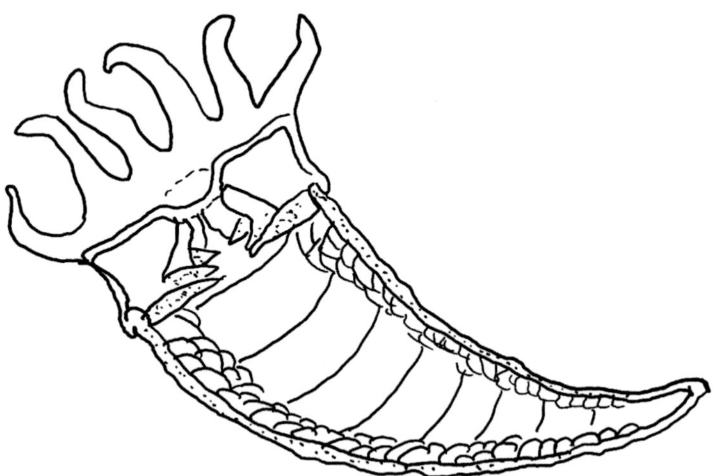

Figure 7.27 The tentacles of the polyp (coral animal) living in the upper section can be seen reaching out of the shell on the upper left. In the middle between the tentacles is the dashed line indicating the pharyngeal opening. In further sequence downward, intermediate legs are drawn in, which the animal retracts as it grows. The shell itself is made of lime (Rhode 2008).

Figure 7.28 Sclerite of a gorgonian, Darwin, Australia. Darwin is in northern Australia, on the tropical Timor Sea. Reflected light. Diameter of sclerite 240 *μm*. Image width 1.4 mm.

Figure 7.29 Sclerite of a gorgonian, Baie des Citrons, Noumea, New Calcedonia. Transmitted light, crossed polarizers with lambda compensator. Diameter in the center of the image is 140 *μm*.

Figure 7.30. Selected biogenic sand grains from Darwin, Australia. The six sclerites in red, orange, and light yellow; diverse foraminifera; and snail shells, as well as two quartz grains, can be seen to compare. Reflected light. Image width around 6.5 mm.

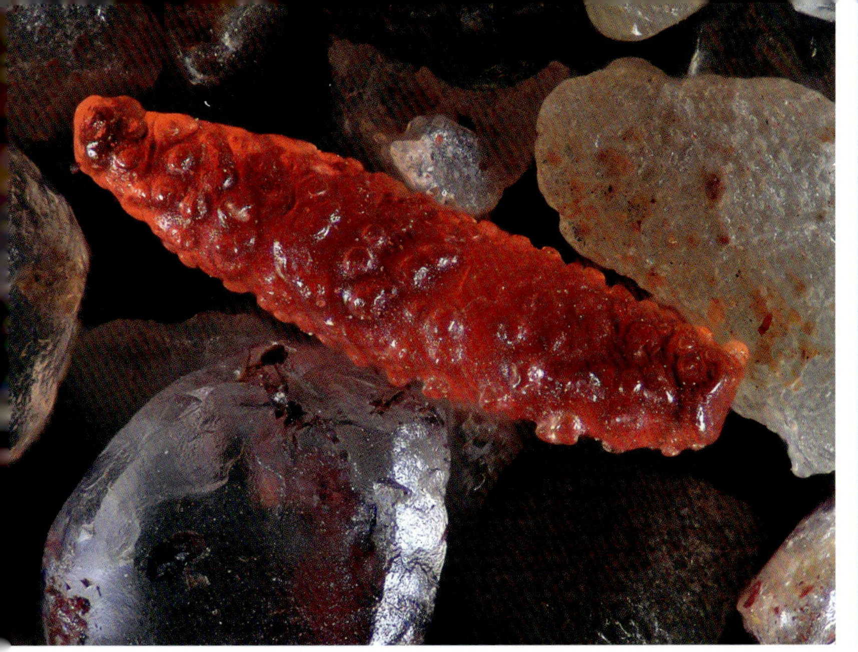

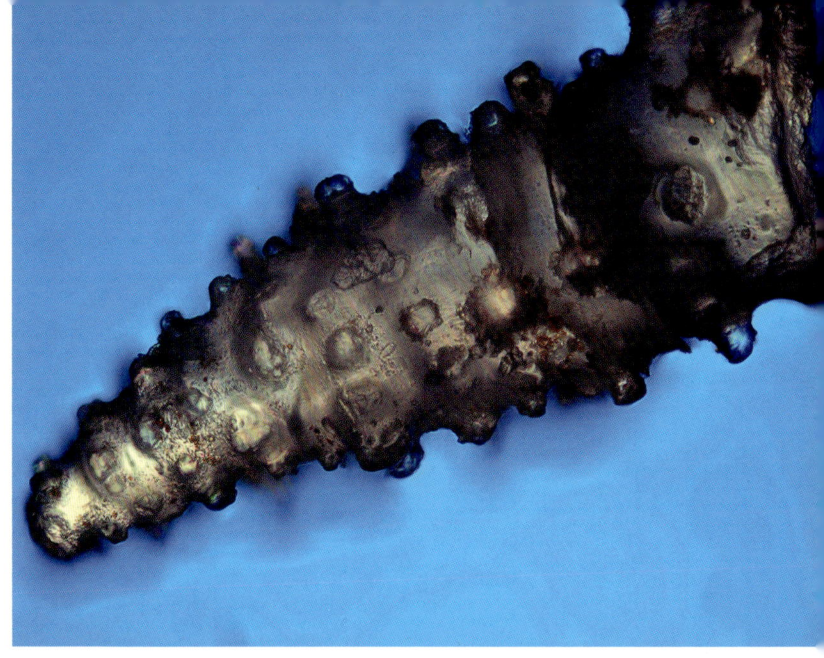

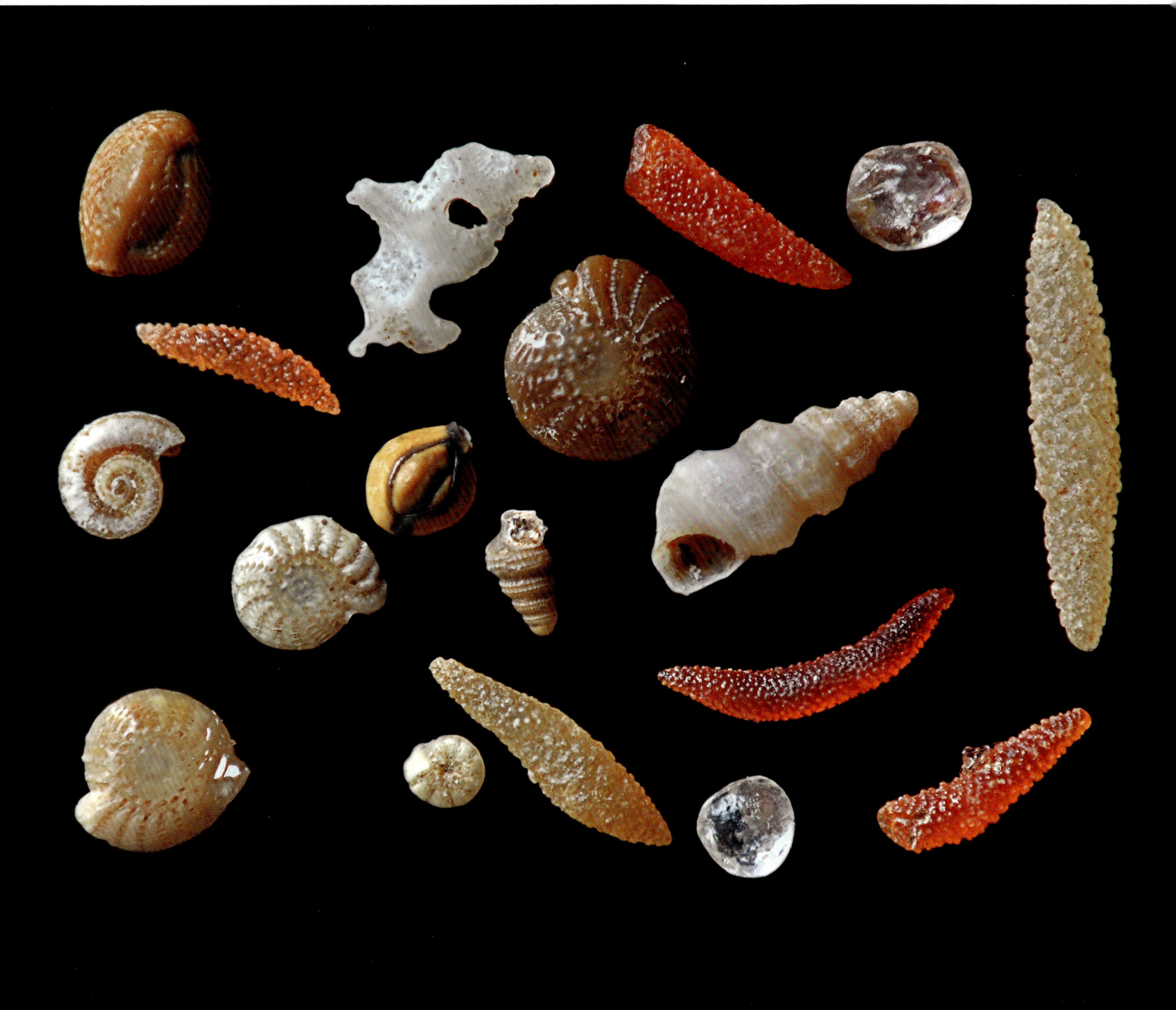

7.3 Maerl

A type of sediment that at first glance appears very similar to the coral sands but has remained largely unknown bears the name “rhodolite,” or, somewhat more commonly, “maerl.” Maerl is a sand that consists of more than half of the living or dead branches of the red algae (Corallinacea), which belong to the coralline algae. Two species dominate, *Lithothamnion corralioides* and *Phymatolithon calcareum*.

Maerl does not denote a grouping of certain species or genus but is to be understood as a collective term for the loose association of subtidal, nodular, and coralline branches of calcified algae. Other remnants of marine fauna and flora can also be found in the dense yet loosely footed and intertwined network of often-bizarre-looking algal branches, which retain particles like a sieve and form a very special, unique habitat through this specific mixture. Maerl habitats are found scattered along the European Atlantic coasts all the way up to northern Norway. The main bioindicator is temperature, with the optimum for growth conditions at 59°F (15°C) and the minimum at about 36°F (2°C). The growth rate is very low, averaging 1 mm per year. Since recent maerl beds have a high percentage of dead organisms, it can be assumed that more-favorable conditions apparently existed in earlier times. Maerl beds can reach up to 33 feet (10 m) in thickness and penetrate to water depths up to about 98 feet (30 m); greater depths have also been observed in isolated cases. Generally, the algae are bound to photosynthesis and therefore to light incidence. It is a great pleasure to trace the adventurous shapes of the grains and branches on the beach, imagining mythical creatures and other things in them.

Phenotypically, the individual, very diverse maerl “grains” can be roughly described as spheroidal, ellipsoidal, and discoidal in shape, with all transitions. Corresponding classifications had already been made by Mikael Foslie in Norway at the beginning of the twentieth century. Strictly speaking, maerl is not classified as sand, because the algal branches exceed the definitional limit dimension of 2.0 mm. Typical “grains” have dimensions between 10 and 25 mm.

Until modern times, the existing stocks were mined and used for various purposes such as an additive for soil improvement. This practice should be rejected outright, since the stocks, which are restricted to the Atlantic and the Mediterranean, grown back very slowly and form a rich ecosystem worthy of protection, which is irretrievably endangered by mining quantities of up to 5,000 tons per year.

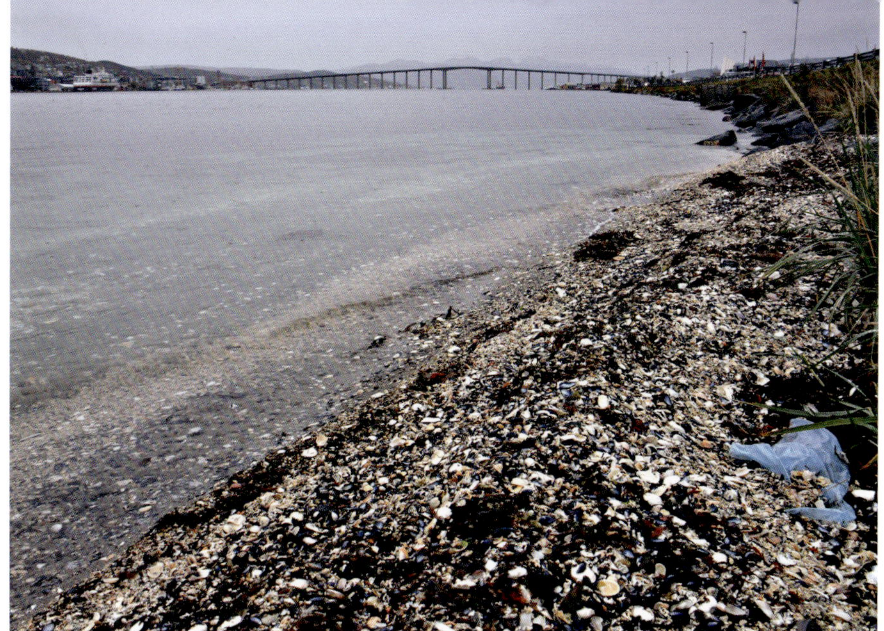

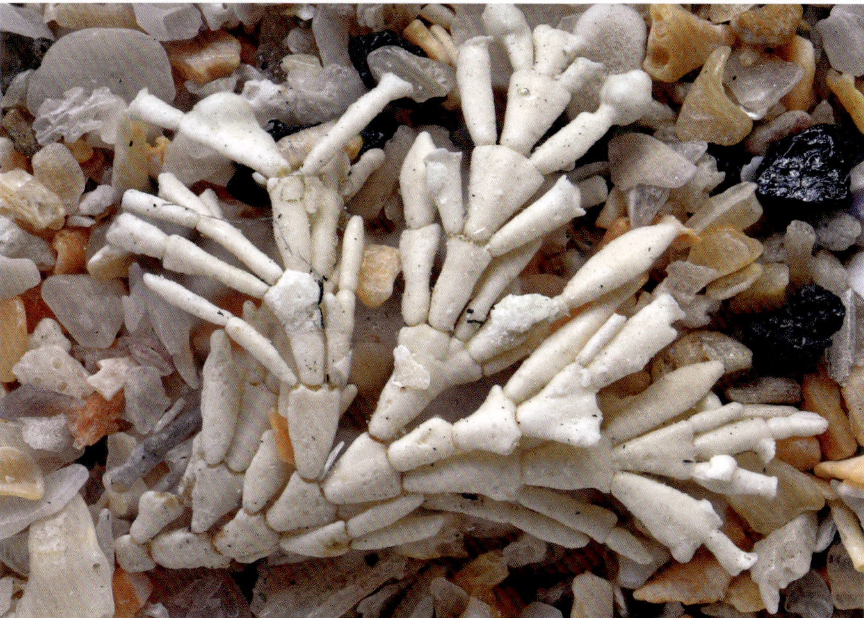

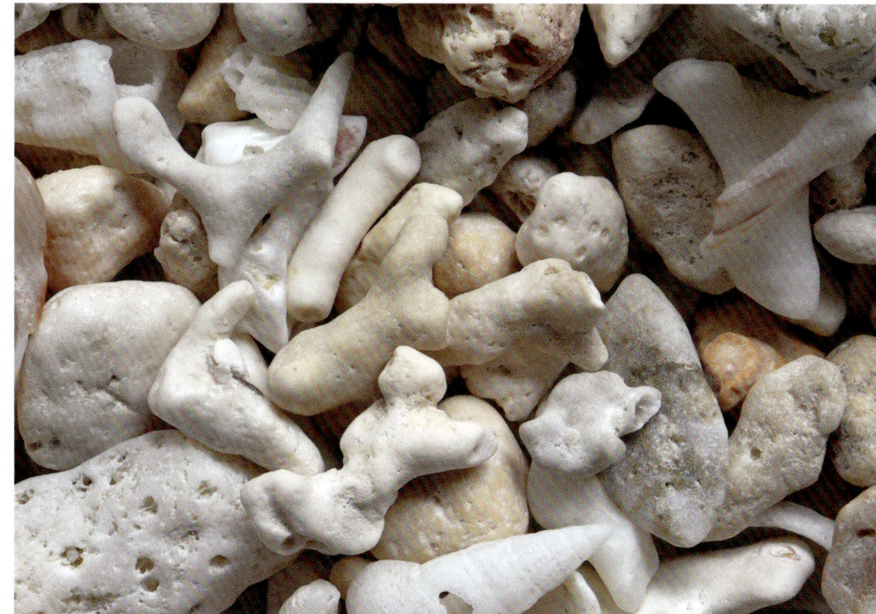

Figure 7.31 Rhodolite beach (maerl) near Tromsø, Norway. Tromsø is north of the Arctic Circle. *On the left*, the town with harbor facilities and ship dock for the cruises; *on the far right*, the famous 1965-built Ice Sea Cathedral. Unfortunately, this remote beach in the Far North is also a victim of the widespread pollution of the seas with plastic waste.

Figure 7.32 Sand from the Carbis Bay in Cornwall, southern England. The red algal branch is still whole and will soon disintegrate into individual sand grains. Image width 9.5 mm.

Figure 7.33 Sand with portions of calcified red algae as well as coral parts (*bottom left in image*) and snail shells. Flic en Flac, Mauritius, in the tropical area of the Indian Ocean. Image width 25 mm.

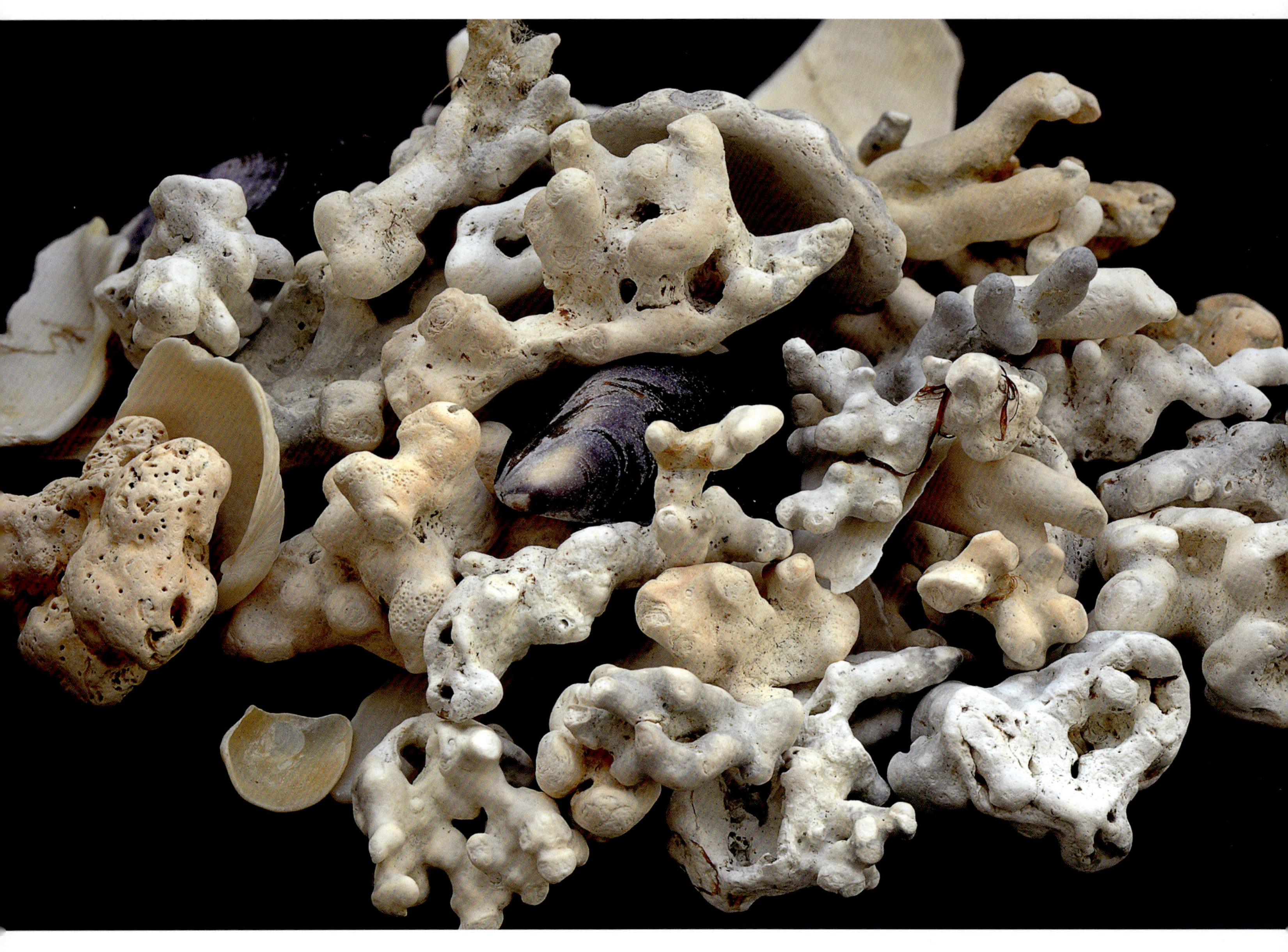

Figure 7.34 Maerl "sand" from Tromsø, Norway. Maerl or rhodolite consists of calcified fragments of red algae. In the grain on the left edge of the image, many microbore-holes can be seen—bioerosion sets in. Image width 80 mm.

7.4 Sponges

If we go back a bit on the time scale, we encounter one of the most primitive and oldest multicellular life forms, the sponge (Porifera). Their origin lies in the Cryogenian period, about seven hundred million years ago. Sponges have been around since then up to today, and they are found in the ocean except for a few freshwater species. Their size varies from a few millimeters to specimens 2 meters high. Over seven thousand species are known. Sponges are sessile aquatic animals; they live firmly attached to the substrate, never occur on land, and in extreme cases can live very long—up to several thousand years. Their basic shape is cup shaped, pear shaped, spherical, or cylindrical, and they are made of very simple tissue. Three basic blueprints exist, differing only in the design of the gastral cavity, a type of stomach. Characteristic of sponges is their porous exterior, which is explained by their life processes. Water enters a central cavity (paragaster) through the many small external pores, flows through it, and then exits through a centrally located opening (osculum) on the top. That's it. The steady flow of water is brought about by many collar flagellar cells (choanocytes), which are located on the inner wall of the paragaster. In the process, the flagellar cells filter out food particles such as plankton, detritus, and the like from the water flowing past. Waste material also leaves the sponge in this way through the osculum. Of particular interest in the sponge tissue are the freely moving cells, the archeocytes, which distribute nutrients in the tissue and can in turn transform into other cell types. The very loose tissue is supported by a number of sponge needles (spiculae, sclerae), which are species specific and are found preserved in the sand after the animal dies.

Depending on the material of the needles, lime sponges with calcite needles (Calcarea) and siliceous sponges (glass sponges, Hexactinellida) with needles made of silica can be distinguished. The latter are technically extremely interesting, since the high-strength yet flexible needles are synthesized from a vitreous material without the latter passing through the state of melting, as is otherwise required for glass. A third species, the horny sponges (Demospongia), has needles made of spongin, which decays rapidly in the sediment and is also not fossilized. With the naked eye, the partly very tiny and transparent needles are not visible. If you look at them under a magnifying glass or a microscope, you are immediately taken in by their beauty. Not only are sclerites formed by sponges, but they also support the tissues of other animal groups, such as soft corals (Alcyonacea), tunicates, and sea cucumbers (*Holothuria*). For the layman, they are difficult to distinguish but can be found in a very colorful mixture in the sands of tropical coasts. The "rattle stones," which are very popular with collectors, are fossil siliceous sponges of the species *Plinthosella squamosa*, around which siliceous concretions (flint balls) have formed. These special spherical flint nodules can be recognized by their openings.

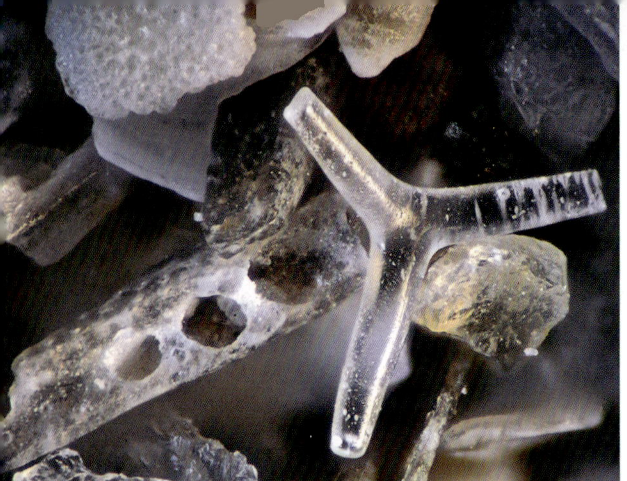

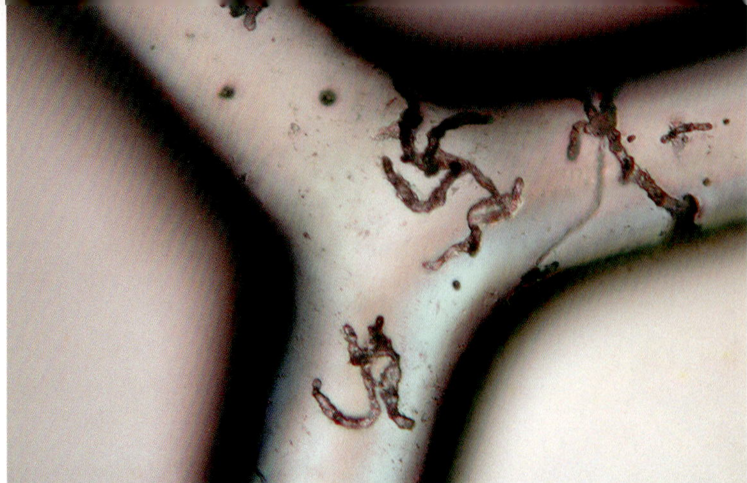

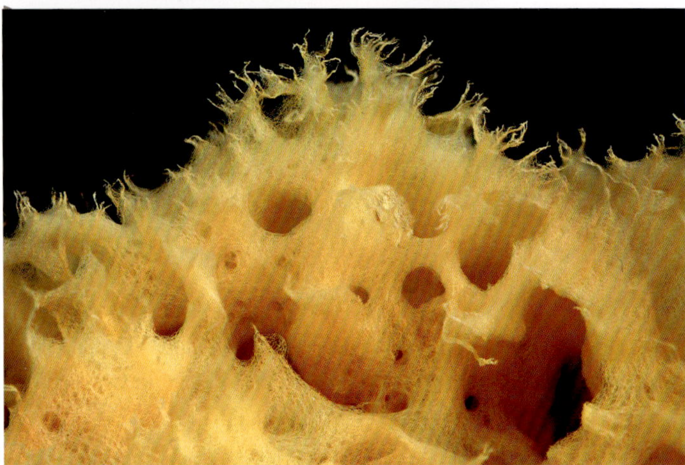

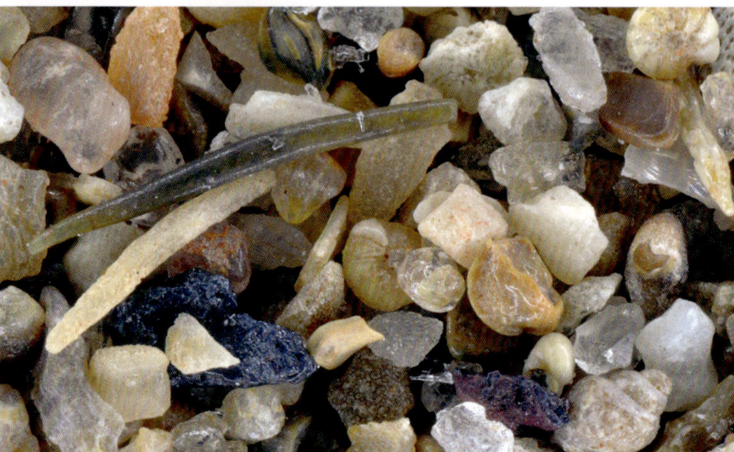

Through these openings, which were originally penetrations of rootlike outgrowths of the sponge, seawater entered after the release of the balls from the Cretaceous-period sediments. Over time, the seawater loosened the junctions between the inner sponge skeleton and the outer flint sphere. If one shakes the sphere, one notices a slight rattling, since the inner sponge can move in relation to the outer shell after the dissolution for the joints. The rattling stones originated from the formation of the Upper Cretaceous period. In the rock, of course, both shells are still connected, so that no rattling is possible. Only as a boulder among debris from the Quaternary period did the separation of both finally take place. However, you need to be very lucky to discover such a fascinating fossil on the beach.

The evolutionarily oldest representatives of the sponges are the glass sponges. At the time of their formation, the earth was between two major glaciation periods. The cold tolerance required at that time has preserved the glass sponge as a "living fossil" through today.

Figure 7.35 *Top left*: Plage Saint Efflam, Côtes-d'Armor, France. Below the image center is a Bryozoa. To the top right in the image is a three-armed sponge needle. Diameter of the arms is 54 μm each. *Right*: Center of a three-armed sponge needle form Mauritius, with bore traces of endolithic algae. Transmitted-light inference contrast. Diameter of the arms is 140 μm.

Figure 7.36 *Bottom left*: Sponge, "bath sponge." This is the sponge skeleton of the horn silica sponge. Note the filigree structures, which look almost like a web. The horny filaments are surrounded by tissue on the living animal. Tissue and sclerites are removed after "harvesting." This specimen comes from the Mediterranean Sea. Image detail 25 mm.

Figure 7.37 *Bottom right*: Sand from the mainland coast near Darwin, Australia. Continental tropical marine sand with siliceous components and numerous biogenic fragments. *Above center*, a narrow, green sponge needle; *below*, a yellow version. Image width 4.0 mm.

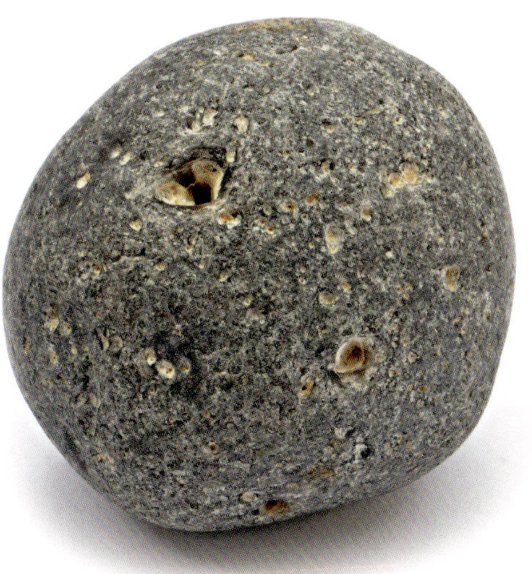

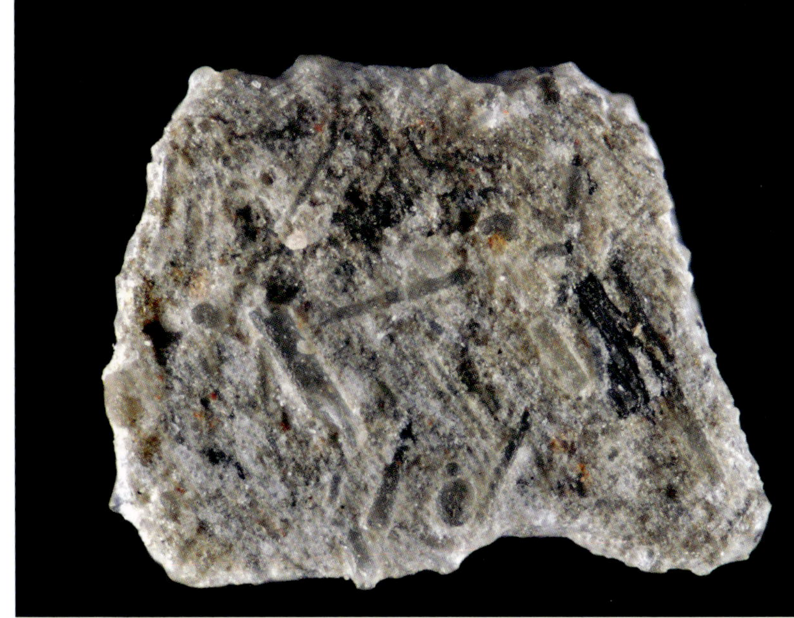

But not only that: the cells of the glass sponges possess an initial differentiation but can still transform into any other cell form. They are totipotent. This means that individual parts of the animal are viable on their own. If a sponge is divided, the parts not only can continue to live but also reproduce.

The possession of a rudimentary skeleton identifies the sponges as animals even in their early days. The formation of such a skeleton from silica and later calcite was made possible by the flushing of mineral sediments into the seas through glacial streams.[148] The first biomineral processes began. With a length of 8.2 feet (2.5 m), the sponge *Monorhaphis chuni* today forms the world's longest biomineral silicate structure. In addition to having the function of supporting the soft tissue, the transparent skeletal crystals are also capable of conducting light and feeding it into the algae living symbiotically with them, which represented a considerable selection advantage under the conditions from seven hundred million years ago.

Sponges do not show bilateral symmetry—they do not differ between right and left but, for the first time in a living being, between top and bottom, between head and foot or root and crown. The sponges developed this blueprint. We humans are descended largely from these ancestors, except that in humans (and vertebrates), what was the anchorage in the sponge now becomes the head during embryonic development. Humans and sponges are thus structurally mutually twisted by 180 degrees, so to speak, but go back to the same blueprint. The most-important neurotransmitters of conduction, acetylcholine and serotonin, also play a crucial role in (recent) sponges by inhibiting and stimulating the flagellar beat. Glutamate signals the presence of food. We humans today are still highly controlled by these same substances, which serve as neurotransmitters in our brains. In this respect, too, we carry the heritage of the sponges within us—a strange notion.

Figure 7.38 Flint sphere with an internal fossilized silicate sponge *Plinthosella squamosa* ("rattle stone"). The openings described in the text can be easily recognized. Marienleuchte, Fehmarn, Germany. Diameter 40 mm.

Figure 7.39 Very rare example of a sand grain with a microfossil within the grain. The long rods are sponge needles with a diameter of around 60 µm. Lake Como, Villa Geno, Italy. Dimensions of the grain are 970 × 1,100 µm.

7.5 Sea Urchins

Next, we turn to the animals whose relics are sometimes quite prominently represented in the sand and have a highly interesting structure—the sea urchins. We find them from the tropics to the polar seas and register them as quite striking and painful because of their pointed spines, which get stuck in skin after beach walks. Sea urchins belong to the echinoderms and have been on Earth since the Ordovician period. We think of them as preferring to live in the sand of shallow marine coastal areas, avoiding mud, but they can also be found in the deep sea. However, they will never be found in fresh water.

Sea urchin bodies are composed mainly of small calcareous platelets and show fivefold symmetry. Sea urchins can be divided into regular and irregular forms—only the regular ones are strictly symmetrical, while the irregular forms show a break of symmetry (e.g., by the asymmetrical position of the mouth), but apart from that they also show fivefold symmetry. The calcareous platelets of the shell show a remarkable peculiarity; they are monocrystalline. So, they consist of a single crystal. In addition, the position and orientation of the optical axis of the respective monocrystals are species specific.[149] The spines of the animal also consist of monocrystalline calcite, whose optical axis coincides with the longitudinal axis of the spine. They are connected by a kind of ball-and-socket joint to the exterior of the calcite platelets, which in turn are covered with small, spherical articular tubercles (spiny warts).

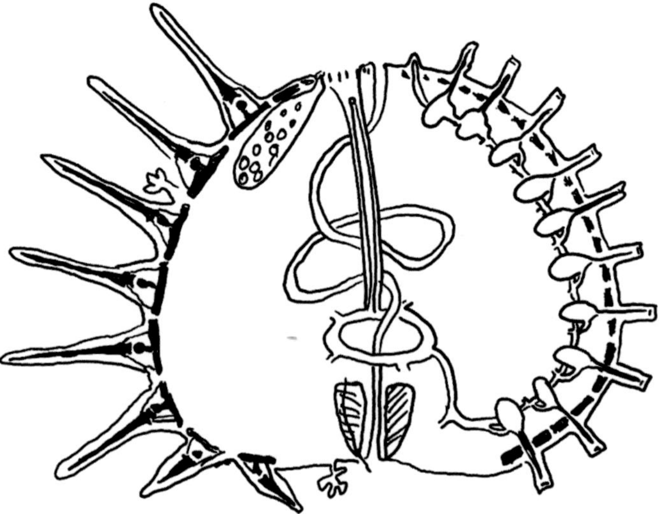

Figure 7.40. Schematic cross section of a sea urchin. The left half shows, exemplarily, the spines mounted on ball-and-socket joints and their muscular articulation. Between them is a three-armed pedicellaria. The right side shows the extended ambulacral vessels, with internal-pressure-equalization ampoules and the associated connecting lines. On the lower side is the mouth opening, with the jaw apparatus indicated on both sides, the so-called Lantern of Aristotle (depiction according to Rhode 2008).

Figure 7.41 Detail of a limestone slab from a sea urchin carving. Sea urchins live in the sea from the surf zone to the deep sea. On the spherical joints of the spiny warts, which can be seen well here, sit the spines of the living sea urchin, which can be moved voluntarily by the animal with its muscles and are used to defend against enemies. Northern Sardinia, Italy.

Figure 7.42 Fossilized sea urchin spine in club shape. Easy to see at the lower left end is the ball socket as part of the jointed arrangement of the spines on the shell plates. White cliff near Dover, England.

Bottom right side:

Figure 7.43 Body plate of a sea urchin from a sandy beach near Tromsø, Norway. In addition to the round spiny warts, ambulacral openings (*see text*) can be seen on the right half, from which small feet protrude in the living animal. Image width 7.3 mm.

Figure 7.44 Sand grain from Cape Trafalgar, Spain. This is a body plate of a most likely still-young sea urchin, which can be determined by the bubble-like structures (diameter around 140 μm). Reflected light. Image width 1.15 mm.

The spines carry concave, spherical half shells at their basal thickening, which, held by muscles, are articulated on the spiny warts. By directing the muscles, the animal can move the spines at will for various purposes.150 Thus, in addition to locomotion, the spines also serve as a defense against enemies. Through appropriate positioning, the sea urchin can even jam itself into rock crevices in order not to be carried away by water currents.

Some species also use their spines as drilling tools. The quite-hard spines can apparently even penetrate rock. What is sure is the ability of sea urchins to perform drilling activities with their complex jaw apparatus. It consists of five very hard teeth and forty flexibly arranged skeletal plates and was named after the Greek naturalist and polymath Aristotle, who first described it. More-recent investigations revealed that the tooth tips have an increased magnesium content, which explains the greater hardness.

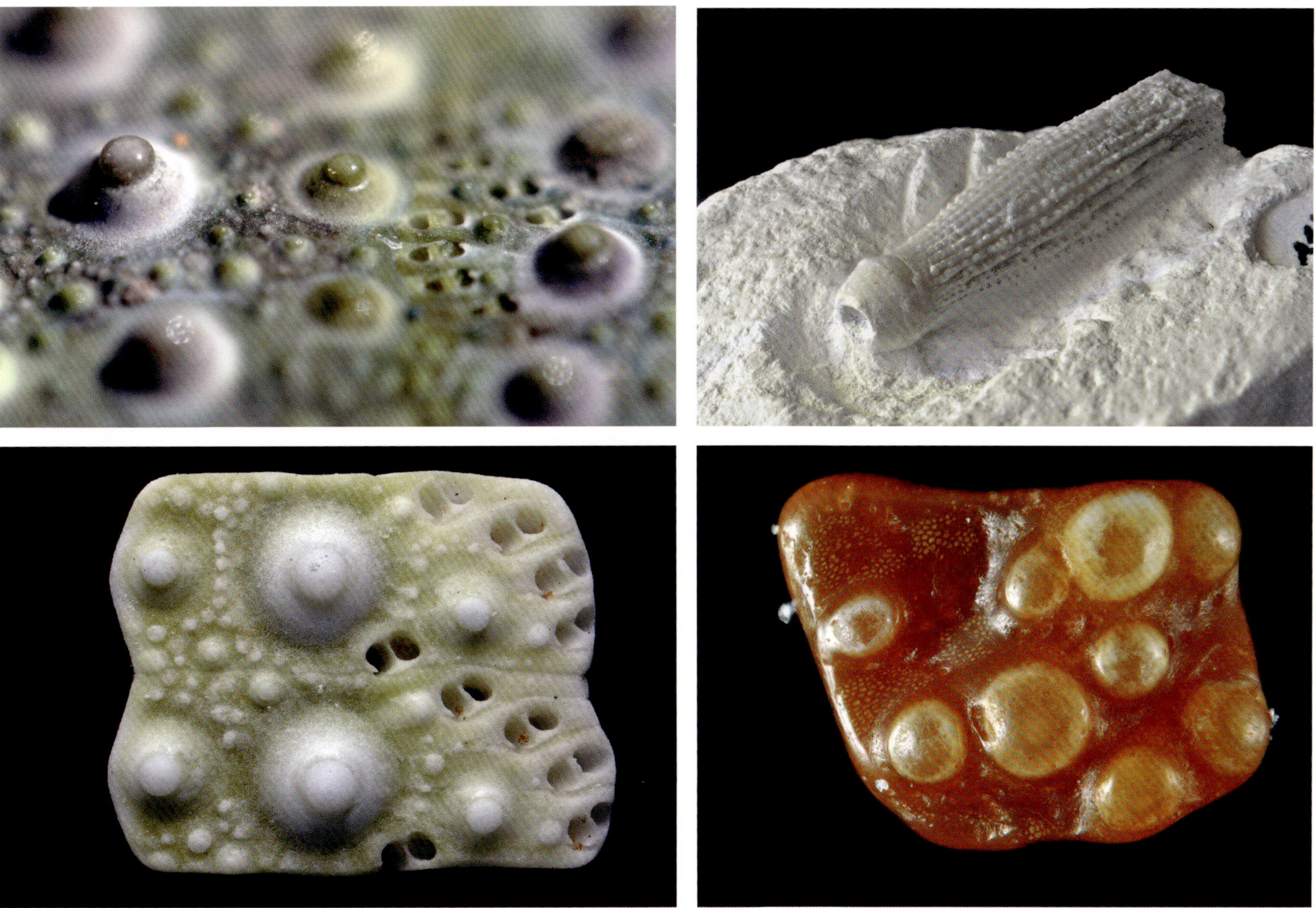

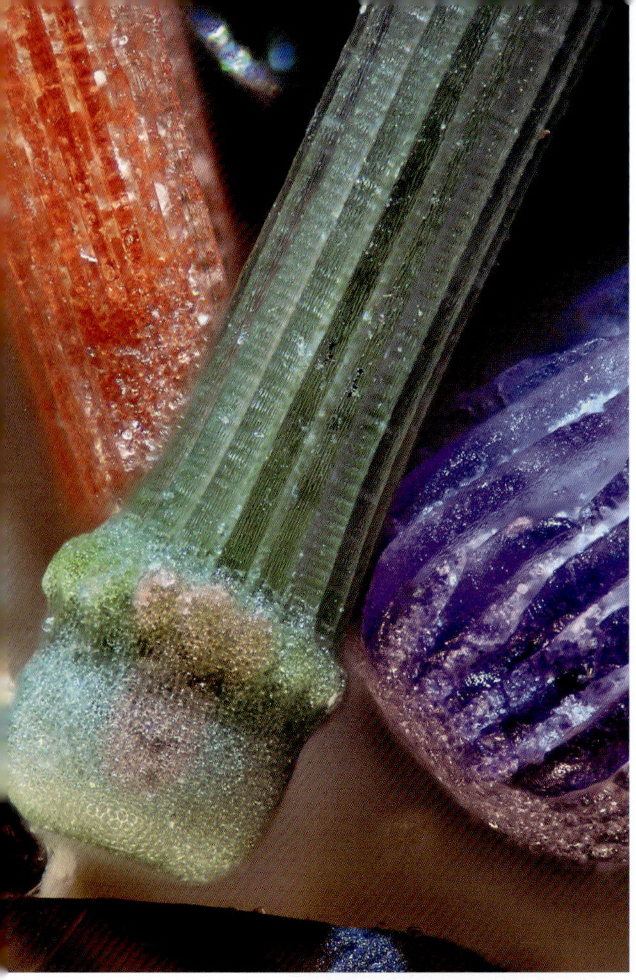

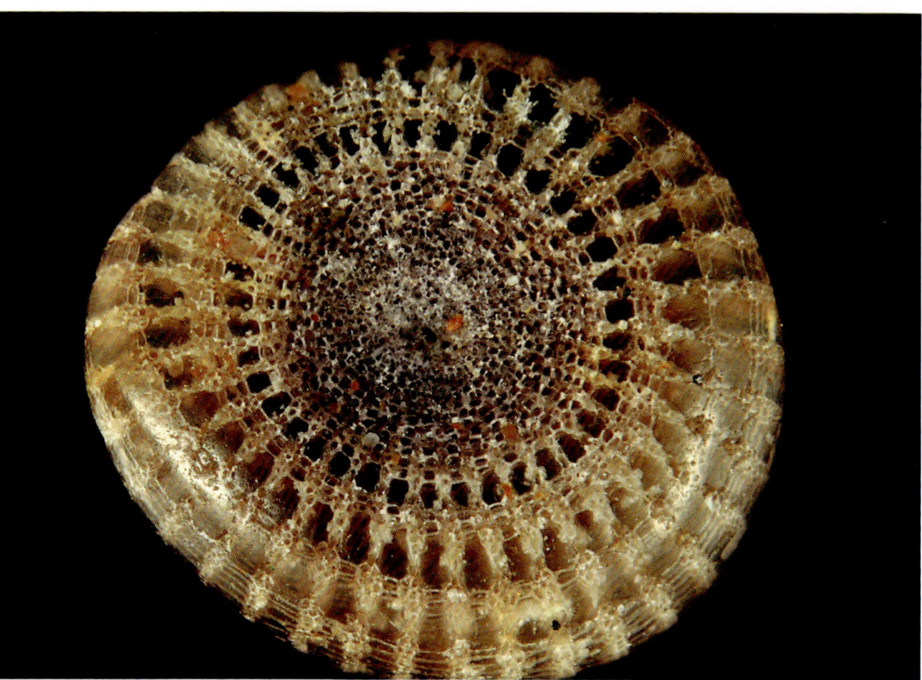

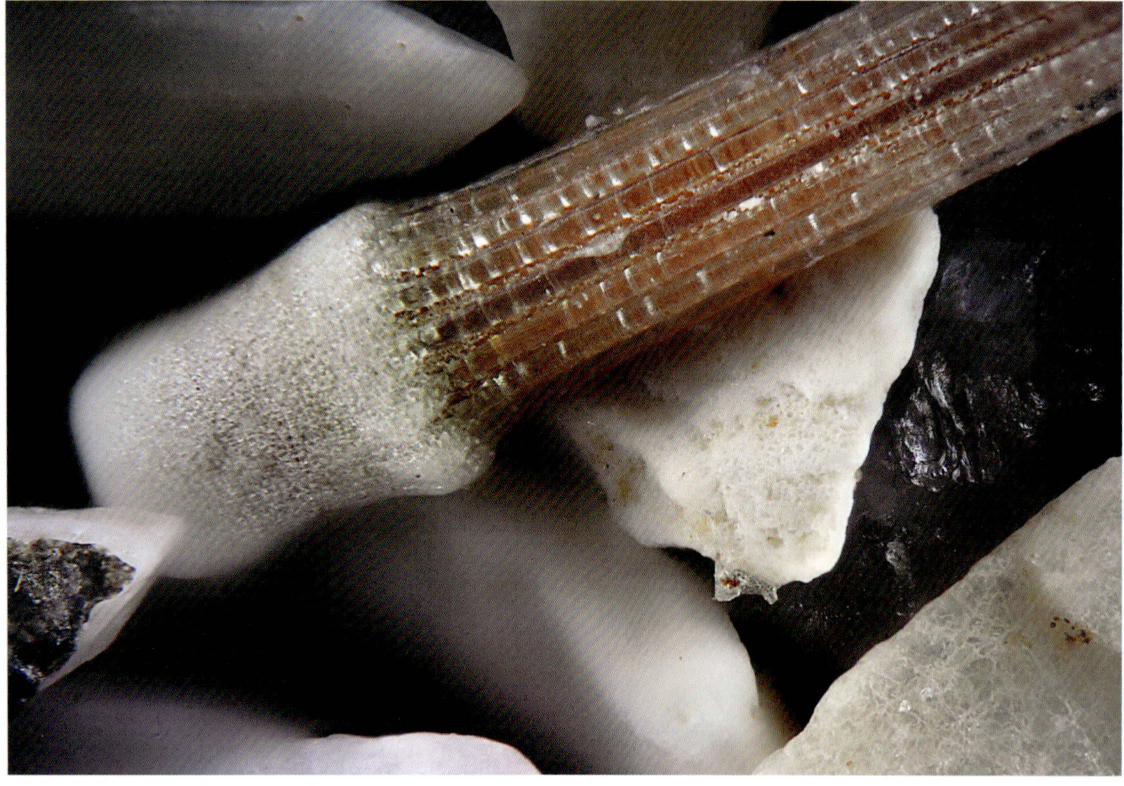

The increase in hardness results from blocked slip planes in the crystal lattice of the calcite, a mechanism that engineers use analogously to increase the hardness of steel materials.[151] The sea urchin spines reveal fascinating fine structures, shapes, and colors, whose beauty is especially apparent under a microscope or a strong magnifying glass. After the animal has passed away, the spines and the small lime particles of the shell are found in the sand.

In addition to the spines described above, sea urchins have another variety of spine-like protuberances, the ambulacral feet. These are part of a tiny system of water vessels interconnected by equally tiny channels. Mini ampullae serve as pressure equalization vessels. The ambulacral feet are also of versatile use to the sea urchin. With them, the animal can catch prey and transport food to the mouth opening, move around, and even climb. For this purpose, the feet can be moved and controlled by pressure changes. The water entry required for this hydraulic system is through a sieve-like plate (madreporic plate) opposite the mouth opening. The ambulacral feet themselves emerge from the body in five semi-meridian-shaped zones, the ambulacral fields. The plates therefore carry small groups of openings in these zones, which can be seen with the naked eye.

Between the spines of the sea urchin are other very special and pincer-shaped organs, the pedicellariae. These are small grasping pincers, bony elements with two or three arms, which can be controlled by the animal and are suitable for cleaning the outer skin, transporting food, and even defense. Some even carry venom glands. The pedicellaria also fall off after the death of the animal, and the calcareous components become part of the surrounding sediment. So, the sea urchin gifts us with a myriad of highly interesting relics to discover. Sea urchins like to burrow and therefore often occur as well-preserved fossils.

Figure 7.45 Sand from Boulders Beach, South Africa. Sea urchin spines of different colors and dimensions. The fine interior structure of the monocrystalline spines is easily recognizable. Reflected light. Image width 900 µm.

Figure 7.46 Sand grain from Nosy Be, Madagascar. Here we can see a cross section of a sea urchin. The remarkably complex inner structure of this small marvel can be particularly well seen, which is distantly reminiscent of a mandala.

Figure 7.47 Sand with sea urchin spines from Vardø, Norway. Shaft diameter of the spine 190 µm. Image width 1.0 mm.

7.6 Foraminifera

Let us now move from the quite large and conspicuous sea urchins to a group of animals or, better, to a group of tiny living beings with the characteristics of animals, which are probably known to very few people through their own observation: the foraminifera, or "chamberlings," as they used to be called.

Foraminifera are quite primitive organisms, which, in contrast to the living organisms discussed so far, consist of only one cell. They are protists, which belong to the rhizaria. Like the unicellular diatoms (diatoms) they form a shell, which of course interests us in connection with sand. Foraminifera have existed since the Cambrian period in smaller numbers, and since the Upper Carboniferous they have appeared in huge quantities and are even rock forming. With few exceptions, they live in the ocean, where they provide up to 90% of deep-sea biomass.[152] The vast majority of the 10,000 current species and over 40,000 fossil species have shells measuring less than 2 mm, but current species several centimeters in size also exist.

Fundamentally, the foraminifera can be divided into two groups depending on their preferred way of life:

1. the planktonic (i.e., suspended in water) species, occurring since the Triassic period
2. the benthonic, bottom-dwelling species, found mainly in warm, shallow water but also extending into the deep sea; this group represents the most species

While planktonic foraminifera usually remain well below 1 mm in diameter and form rather round shapes, we find a wider range of shapes and dimensions among benthonic species.

Planktonic foraminifera live partly in symbiosis with zooanthellae, which explains why they are found predominantly in shallow water flooded with light, and, as a result, the shells of the dead animals are easily scattered to the sandy beaches, where they can be found. In surface water, population densities of up to two hundred individuals per cubic meter can be reached.[153]

In contrast to the intensive preoccupation of researchers with their diverse shell forms and their geological significance as rock formers, very little is known about the living animal. Yet, foraminifera reach a relatively remarkable age for unicellular organisms, from several months to several years. The protoplasm of their single cell is located both inside the chambered shell and outside, enclosing the shell. Cushman therefore recommends considering the animal as living both inside and outside its shell. The protoplasm forms tentacle-like outgrowths, with whose help the animal can move and feed itself; for example, with algae.

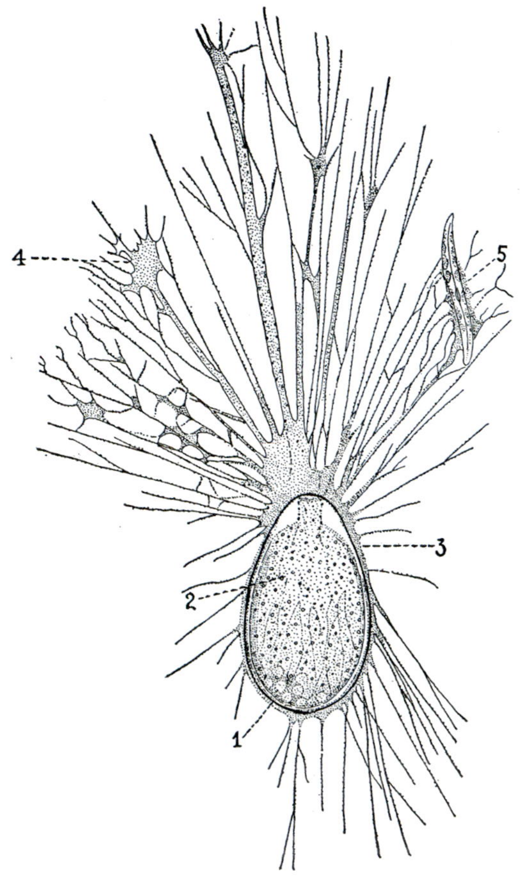

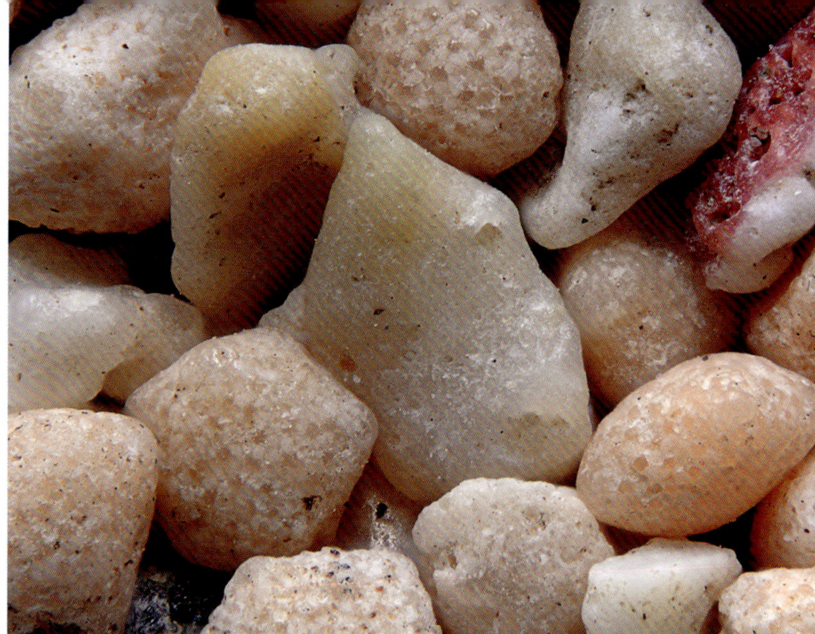

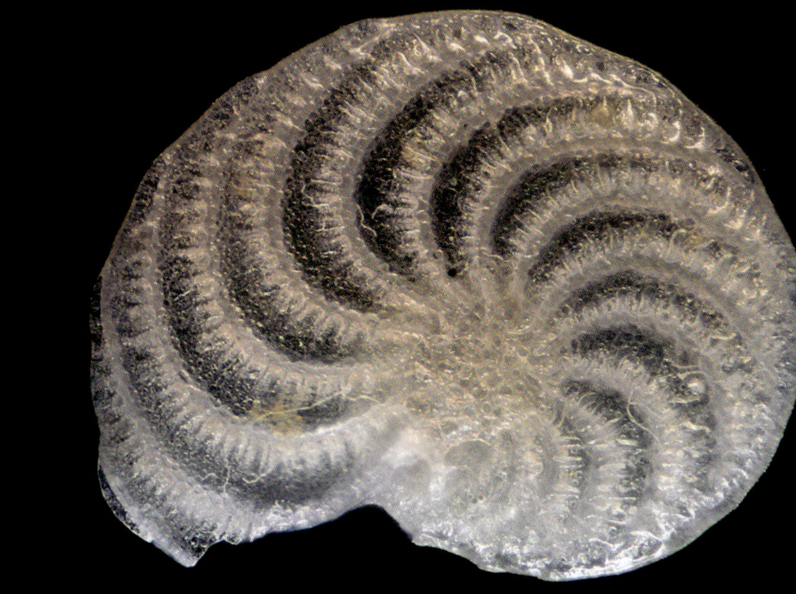

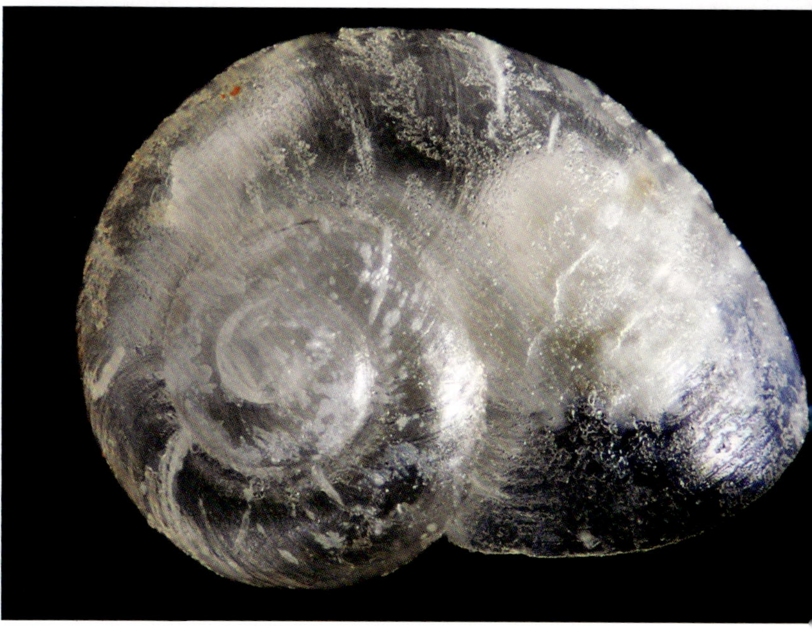

Figure 7.48 Foraminifer, schematic depiction of a living animal (according to Cushman 1950). 1: Shell, 2: protoplasm inside the shell, 3: protoplasm covering and enclosing the shell, 4: anastomosing pseudopodia, and 5: a diatom that has become entangled in the pseudopodia and will later be incorporated as food for the foraminifer.

Figure 7.49 Sand from Mamanuca, Navini Island, Fidschi. The sand is almost exclusively composed of bodies of the slightly salmon-colored foraminifera Baculogypsina sphaerulata that have been grounded into a round form. To the top right is a piece of the red-colored foraminifera *Homotrema rubrum*. Reflected light. Image width 3.35 mm.

Figure 7.50. Foraminifer from Praio do Vau, Portugal. The graceful shell is divided into individual curved and interconnected chambers, which clearly distinguishes it from a snail shell with its single continuous chamber (for comparison, see the following picture).

Figure 7.51 For comparison: snail shell from Boulders Beach, South Africa. In contrast to the foraminifer in the previous picture, the snail shell has a single continuous interior without chambers. Reflected light. Image width 1.5 mm.

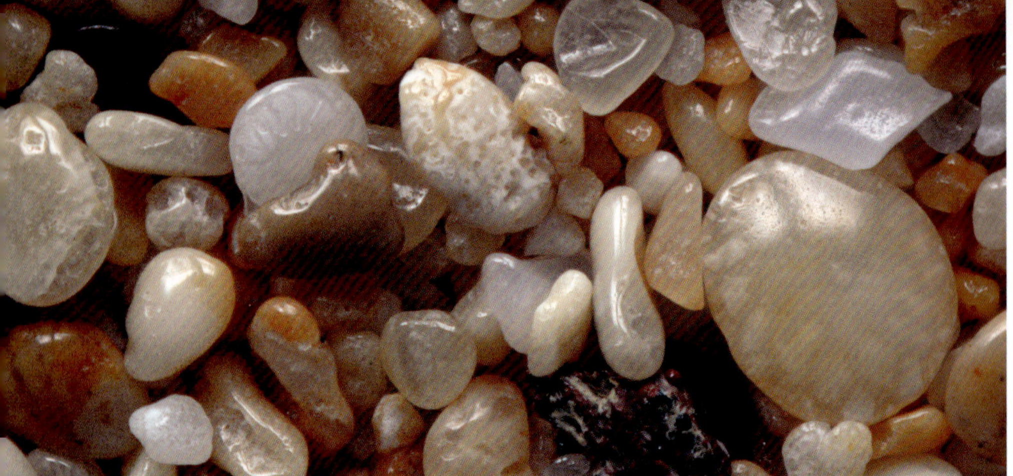

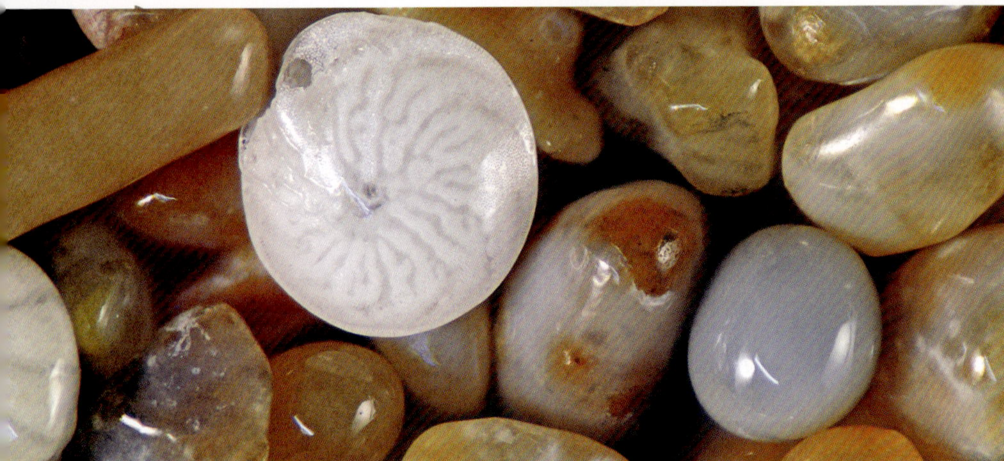

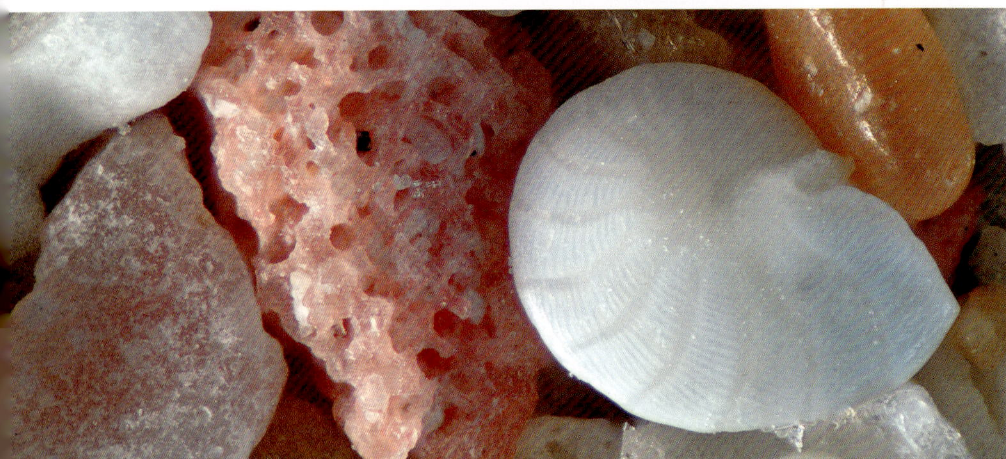

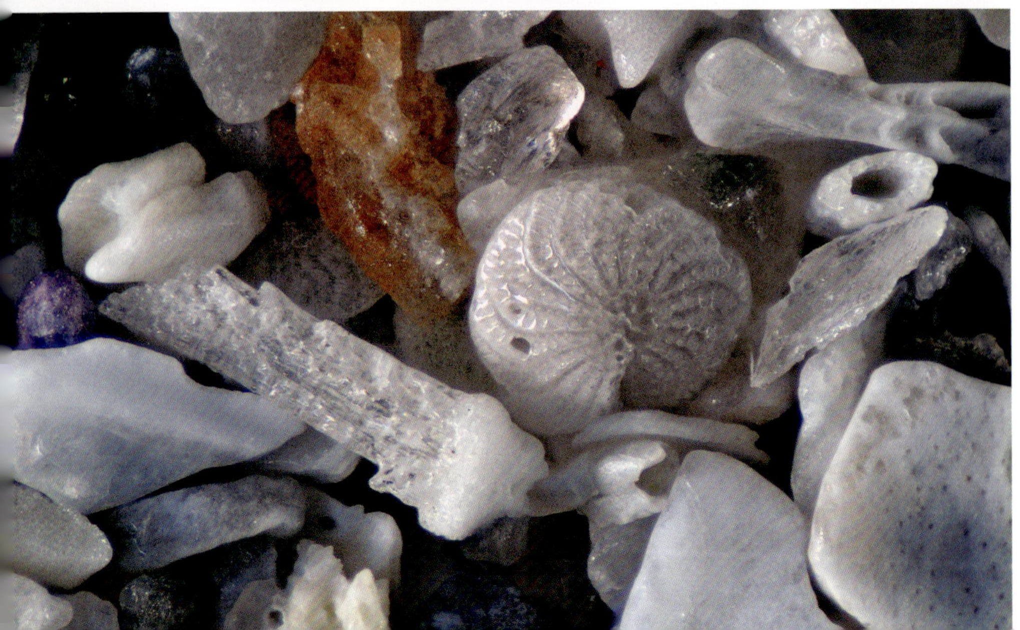

Figure 7.52 Sand rich in foraminifera from Wailea, Big Beach on Maui, Hawaii, USA. A large discus-shaped shell is recognizable at first glance to the right of center, as well as a white, smaller shell to the left above the middle. Upon further observation, there are eight more foraminifera to be found. The purple sand grain points to the volcanic origin of the island beach. Reflected light. Image width 3.6 mm.

Figure 7.53 Sand from Golden Beach, Cyprus. The bright-white foraminifera *Amphistegina lessoni* has a diameter of 670 *μm* and has been just as finely polished from the waves and surf as the ooid sand grains around it. Reflected light. Image width 2.2 mm.

Figure 7.54 Sand from Coral Bay, Ningaloo National Park, Australia. The slightly pink-colored *Homtrema rubra* is one of the few colored foraminifera and is additionally irregularly and porously formed. Next to it to the right is a pure-white and flat-spiraled foraminifera, probably from the genus *Elphidum*. Reflected light. Image width 2.2 mm.

Figure 7.55 Very fine sand from Vatersay in the Outer Hebrides, Scotland. Foraminifera can also be found in bodies of water farther north, the one here probably from the genus *Elphidium*. Reflected light. Image width 2.2 mm.

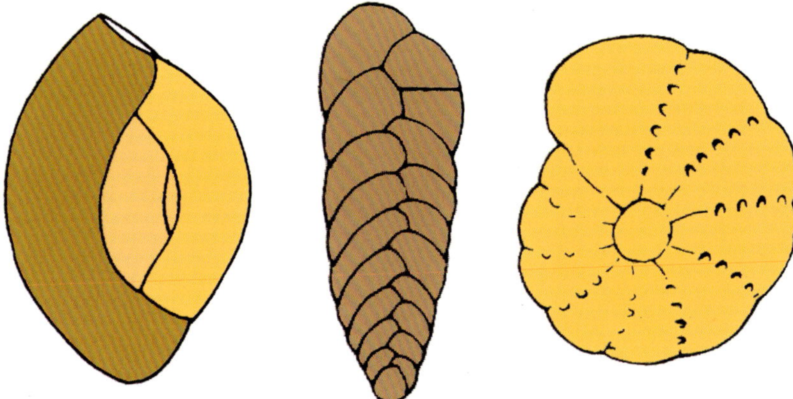

In the process, suitable particles become entangled in the amoeboidal arms of the protoplasm, which is always changing, and are eventually ingested by the animal. The small arms can change continually and fuse with each other and be retracted again (i.e., they can anastomose). The shells usually consist of several chambers (septae), which communicate with each other through small openings in the partitions (foramen), allowing the plasma to access all chambers. The last chamber (aperture) gives the animal access to the environment. Some shells are additionally equipped with tiny openings on the sides; these serve for gas exchange and the exchange of nutrients.

Foraminifera move only very slowly along the ocean floor; on average, speeds around 10 mm per hour can be reached.[154] Like the unicellular diatoms, they are able to build their own shells, an amazing feat for a microscopic unicellular organism. Some species, however, manage without a shell at all and therefore consist only of the naked cell. We are interested primarily in the shells themselves, because they can be found in large quantities and with a remarkable richness of shapes in the sand of many beaches. Yes, even more: with the naked eye they are—at least with most species—indistinguishable from other sand grains; the shells are sand and—as we will see—in some species they even consist of it. What a beautiful and circular thought. Grains of sand that agglomerate into new grains of sand with the help of living beings and decay back into sand as soon as the life in them ends.

For us multicellular creatures, it is difficult, actually impossible, to put ourselves in the place of a being that consists of only one cell.

Figure 7.56 Some schematic basic shell shapes of foraminifera of shelf seas. *Left*: Miliolina in quinquelocular form, *middle*: Textulariina with biserial mound, and *right*: Rotaliina with plano-spiral shell. The chambers are separated by sutures. Many other groups of forms exist; only examples are shown here.

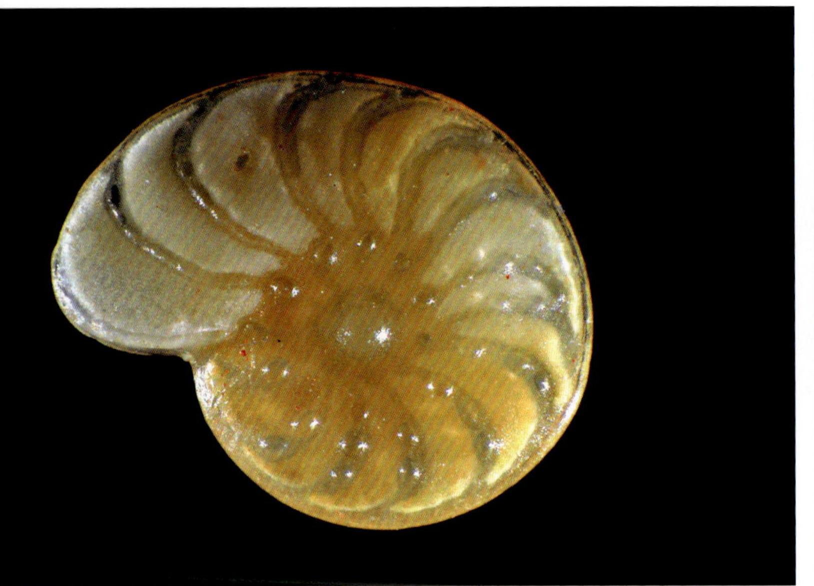

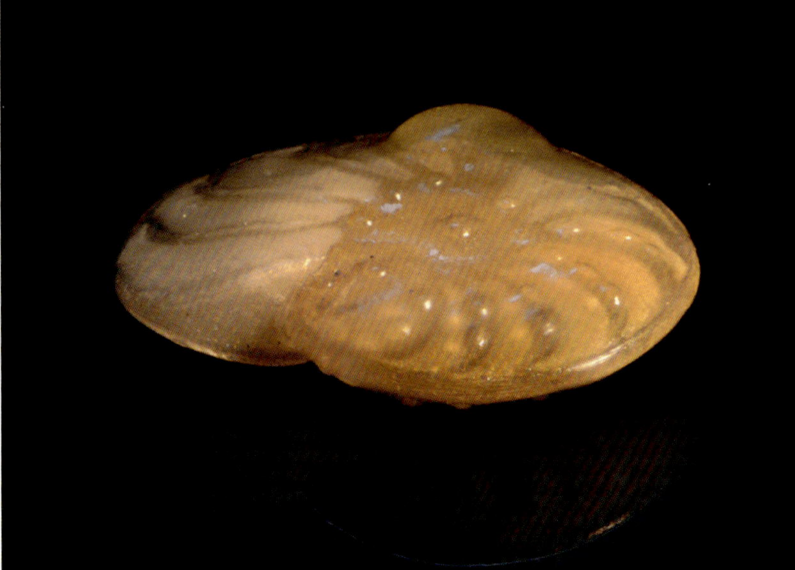

And it may seem strange to refer to such a creature by using the term "animal," which we are much more used to using when talking about higher organized life-forms. How might it feel to be a foraminifer—does it feel like anything? The question itself is perhaps already wrong, because "to feel" already presupposes that corresponding sensory perceptions are present and could of course be felt by "someone." But unicellular organisms have neither specific sensory organs nor even a consciousness, which could take up sense stimuli, process them, and bring them into relation with themselves. Does this mean at the same time that such a being lives without relationship in its surrounding world? Certainly not. There are living conditions that are more favorable to the being than other, possibly hostile or threatening, ones. Will it then not strive toward the favorable conditions, ultimately like we humans do? Is it wise to dismiss such an interaction between animal and environment as blindly controlled? We cannot say with certainty, but it is worth thinking about.

The transition from abiotic to living systems, the process of the origin of life, is still largely uncertain; research in various disciplines approaches the phenomenon asymptotically. This may be because definitions are also at play here; no strict definition for "life," for example, has yet been generally accepted (and neither has one for "consciousness").

Figure 7.57 *Left*: Foraminifer, Rotaliina type. Darwin, Australia. Reflected light, dimensions of the flat spiral shell 940 × 800 *μm*. *Right*: The same foraminifer lying on a black glass plate and taken from the side. From this perspective, the difference from snail shells becomes clear. The last chamber is connected to the outside world only through a small (however, not visible in the picture) opening (aperture). On the other hand, the open cross section of a snail shell increases toward the exit and therefore has the largest opening at the end of the coils.

Five factors seem to stand up to in-depth discussion:[155]

1. An inside and outside must be established; for example, by a membrane.
2. A directed, controlled transport of substances must exist (metabolism by absorption and excretion of substances).
3. Living systems have lower entropy (thermodynamic disorder) than their environment and have to maintain it by using energy.
4. Living systems can grow and reproduce in certain phases.
5. There is an exchange of information with the environment, which can be used to the advantage of the system (stimulus, reaction, selection).

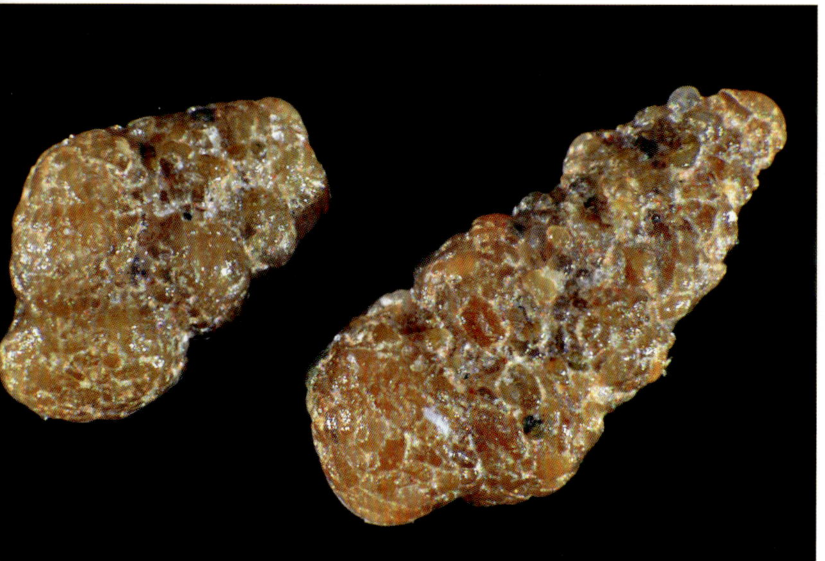

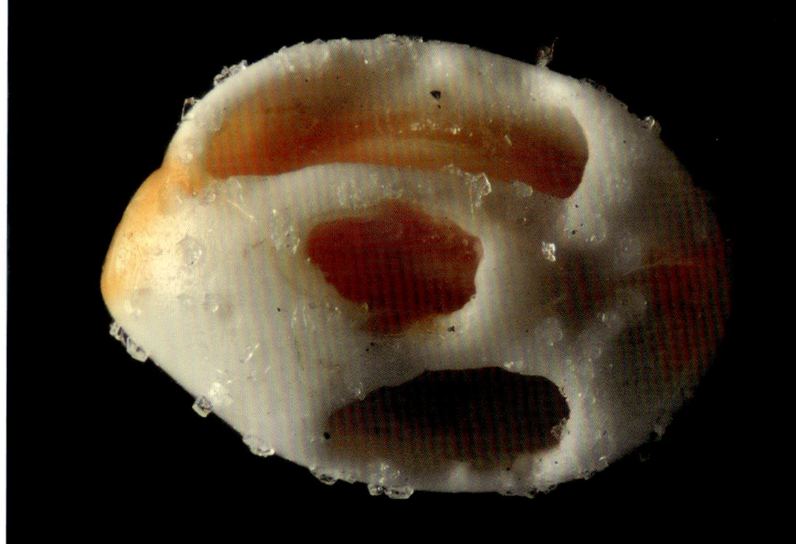

Figure 7.58 Darwin, Australia. Two foraminifera of the type Textulariina. Especially on the specimen on the left, the biserial, as if braided, structure of the chambering can be seen. It is an agglutinating species, which interestingly incorporates sand grains into its shell structure. Image width 2 mm.

Figure 7.59 Three-chambered foraminifer from the family Milionidae. The shell is eroded, leading to the chambers being opened and visible. Sand grain from Carbis Bay, Cornwall, England. Image width 700 *μm*.

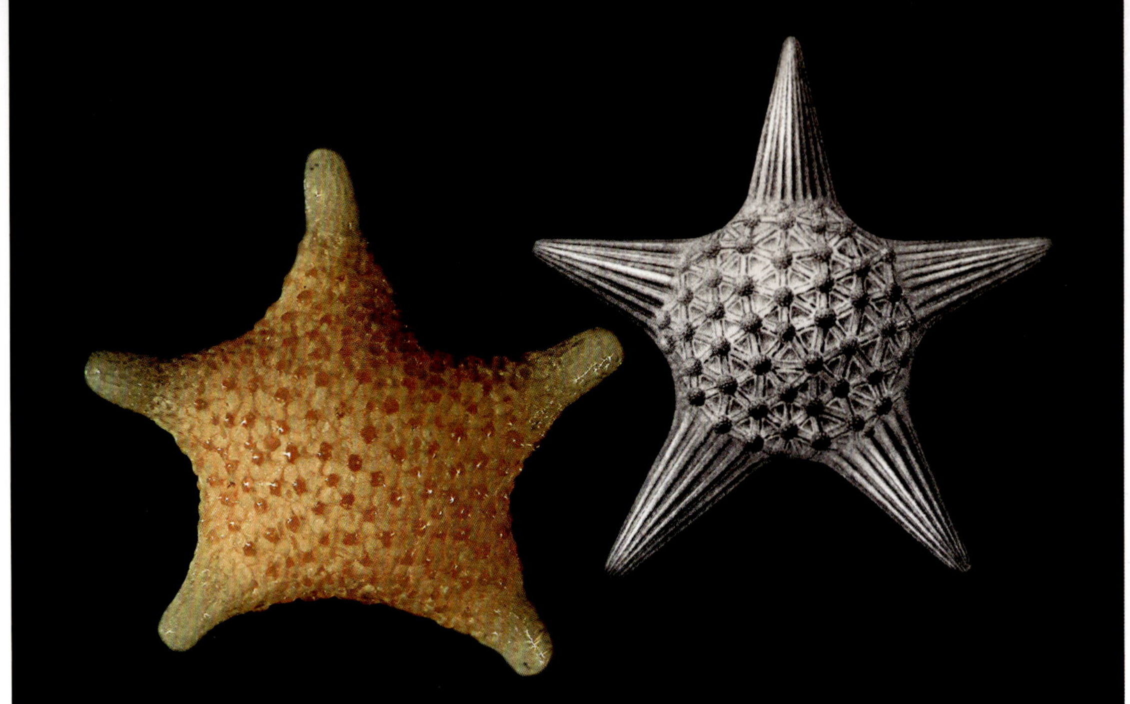

To what extent these aspects are absolute, complete, or even correct remains open and tends to be a question of the setting, the pure definition, because there is not an external right/wrong in this case. In addition, not all criteria have to be filled, but to the same extent, one factor is not enough. For example, crystals show a high level of structure and order and can even, depending on how you see it, reproduce. But one would still not count it as a living thing.

Living systems must in any case escape the state of thermodynamic equilibrium. If such an equilibrium with the environment is established, the system is dead, because no more activities can take place. But does life emerge by increasing the complexity of a system more and more as a side effect and gradually, so to speak, or is it suddenly there? Or is life an emergent property of a system, so it must come to it? We call "emergent" a phenomenon that cannot be explained directly from the original elements of a system and, as such, is a new and unidentifiable addition. This requires an explanation. If, for example, we were asked to analyze the second part of Goethe's *Faust*, what would we find? About a third of a pound (150 grams) of cellulose and fillers, as well as a few milligrams of printing ink. Nothing more. Everything else, all the meanings and interpretations of the text, would be—in terms of the purely material—emergent properties. Nothing in the chemistry of the paper points to the agonies of the soul of the man Faust, which are recorded on this very paper by means of printer's ink. Emergent properties, as the example shows, cannot be deduced or extrapolated from the properties of the individual components of the system under consideration; they are additional. Should "life" be perceived as emergent?

Figure 7.60. Michaelmas Cay, Australia. *Baculogypsina sphaerulatus*. To the right is a depiction of Ernst Haeckel's *Art Forms in Nature* (*Kunstformen der Natur*) for comparison. One marvels at the amazing detail of Haeckel's picture, which points to the use of very good optical instruments. The massive presence of these beautiful foraminifera on a few Japanese beaches led to the formation of the coveted "star sand." Diameter about 2.2 mm.

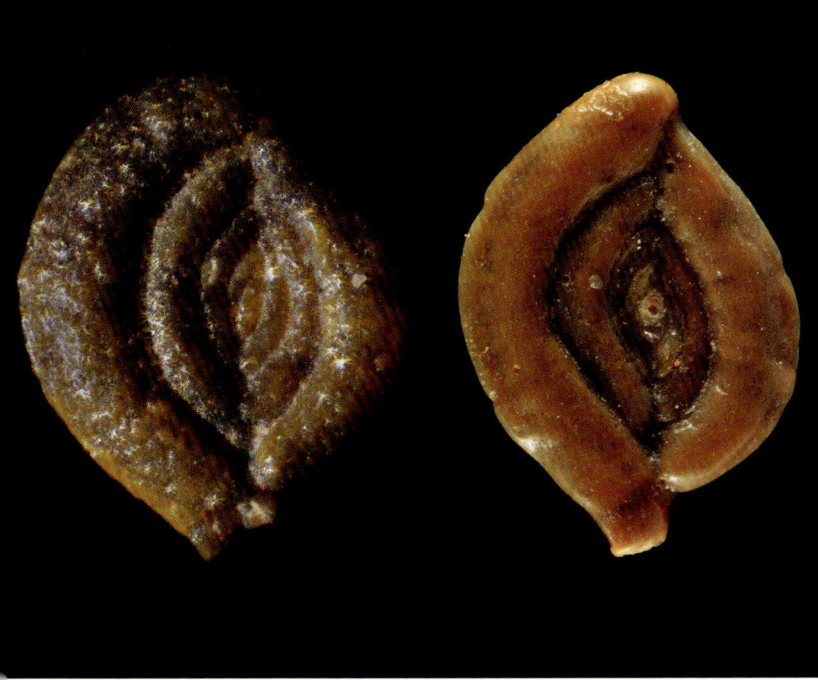

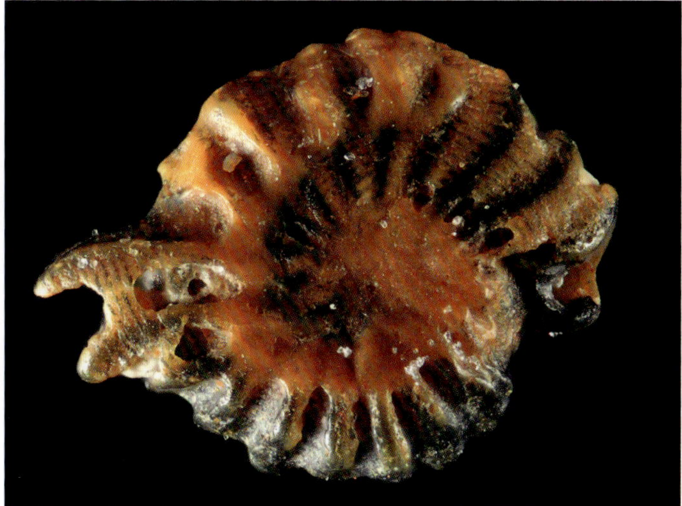

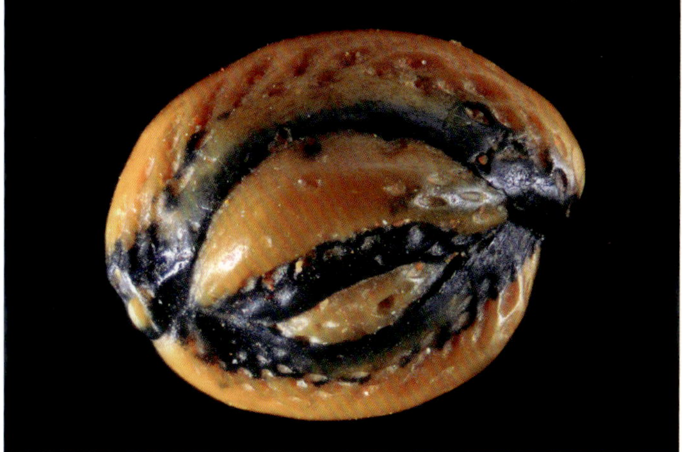

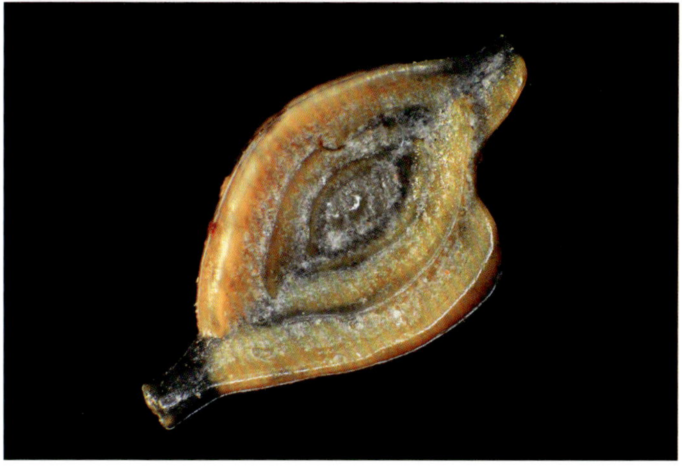

Figure 7.61 Two foraminifera from the sandy beach of Darwin, Australia. The examples to the left come from the Milionidae family, ev. *Massilina secans*. Image width 1.5 mm.

Figure 7.62 Sand grain and foraminiferer from Darwin, Australia. Some of the chambers arranged in spiral form are open. Reflected light. Image width 1.4 mm.

Figure 7.63 Sand grain and intact foraminifer from Darwin, Australia. The aperture (opening) is on the right side, and the individual chambers are in an elongated, curved arrangement.

Figure 7.64 Foraminifer from the family Milionidae, genus *Spiroloculina* or *Massilina*, both of which are difficult to distinguish from each other. Darwin, Australia. Shell width 570 *μm*, image width 1.35 mm.

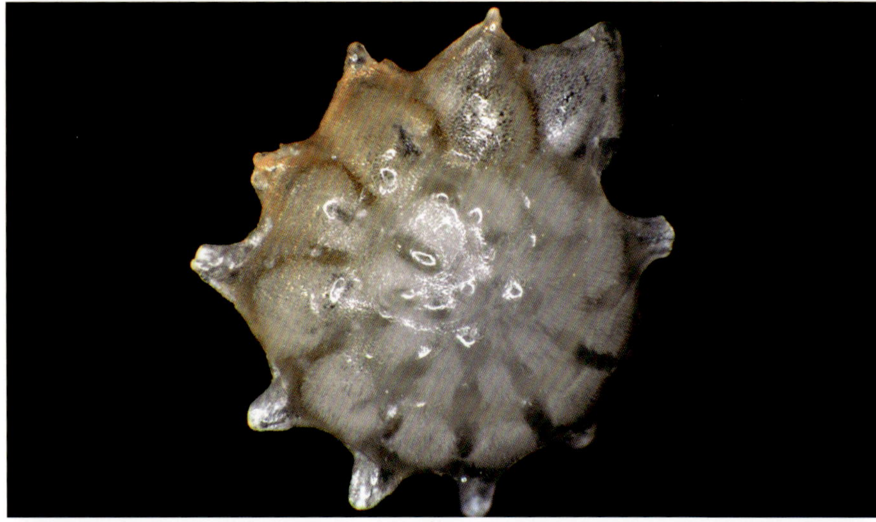

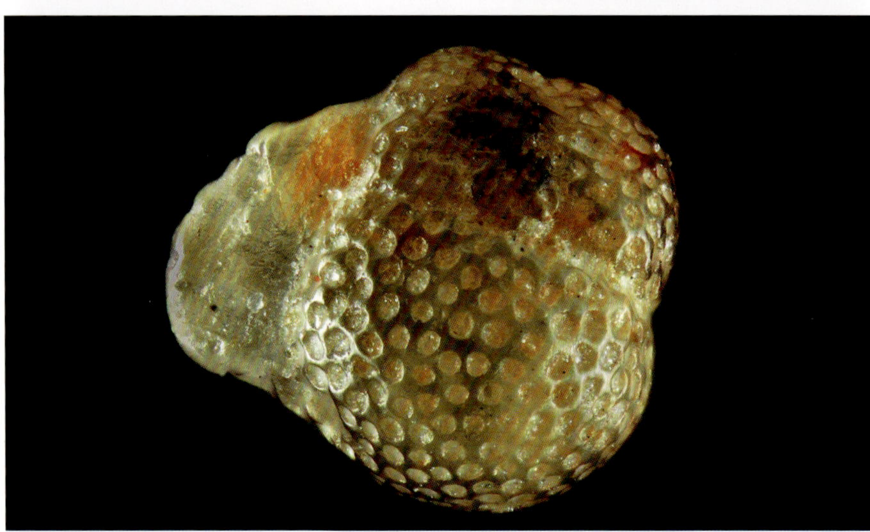

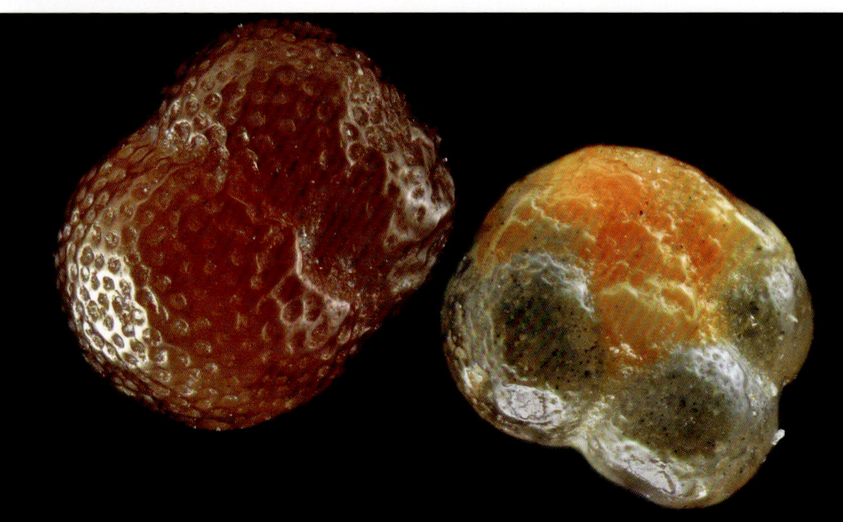

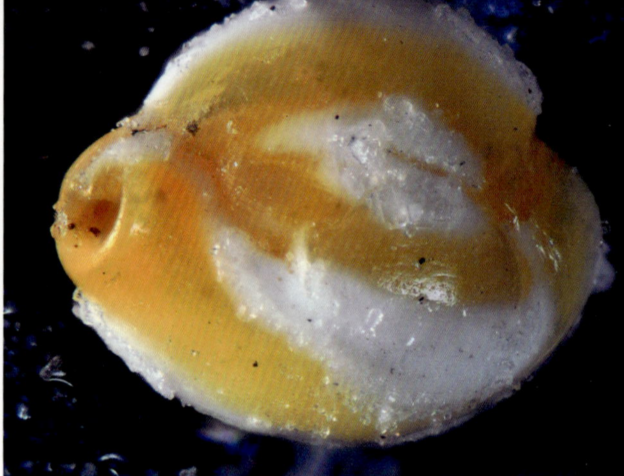

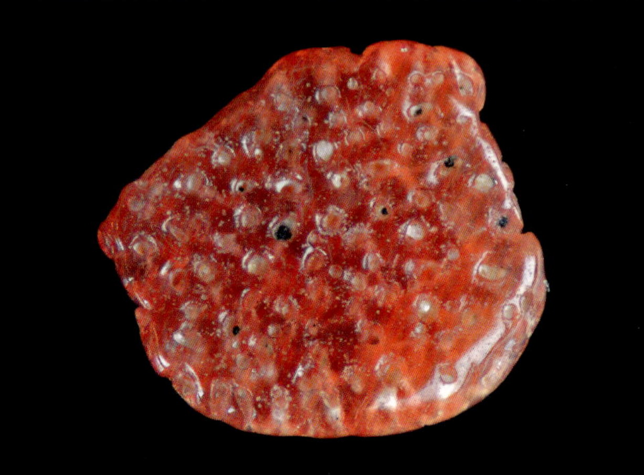

Figure 7.65 Sand grain and foraminifer from Belmare Beach on Mauritius. A rare example with spiny appendixes. Reflected light, image width 1.24 mm.

Figure 7.66 Sand grain from Corfu, Greece. This is a foraminifer from the Globigerinidae family. These foraminifera occur pelagically in the sea (i.e., far from the shore zones) and are therefore rarely found in the sand without some luck. In addition, the shells are very small. Reflected light. Image width 550 *μm*.

Figure 7.67 Two representatives of the Globigerinidae family of foraminifera from Corfu, Greece. *To the left*, probably *Globigerina rubra*. Members of this family have appeared as fossils since the Cretaceous period. Reflected light. Image width 550 μm.

Figure 7.68 Foraminifer from the sand of Carbis Bay, Cornwall, England. The animal belongs to the Miliolidae family and has dimensions of 600 × 490 *μm*. On the left side, the characteristic opening (aperture) can be seen. Reflected light.

Figure 7.69 Fragment of the *Homotrema rubrum* foraminifera from Nosy Be, Madagascar. *H. rubrum* attaches to the shells of other ocean creatures such as mussels or corals. *H. rubrum* is an important element of the so-called coral sands and is responsible for their pink color. Reflected light. Image width 1.9 mm.

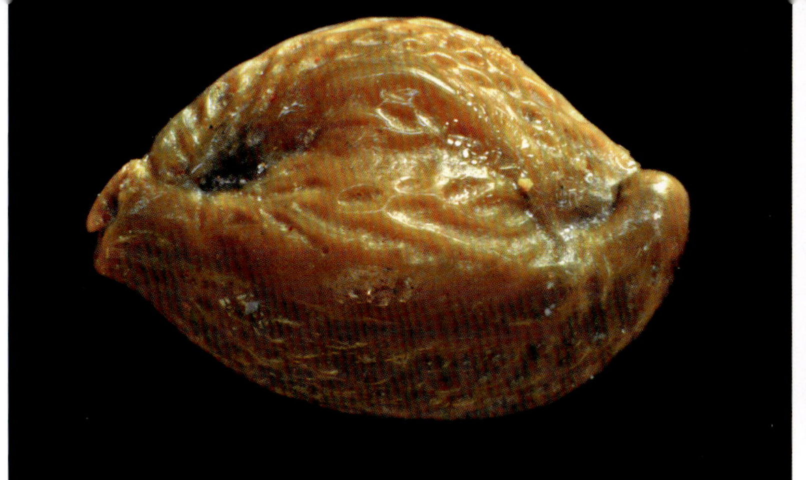

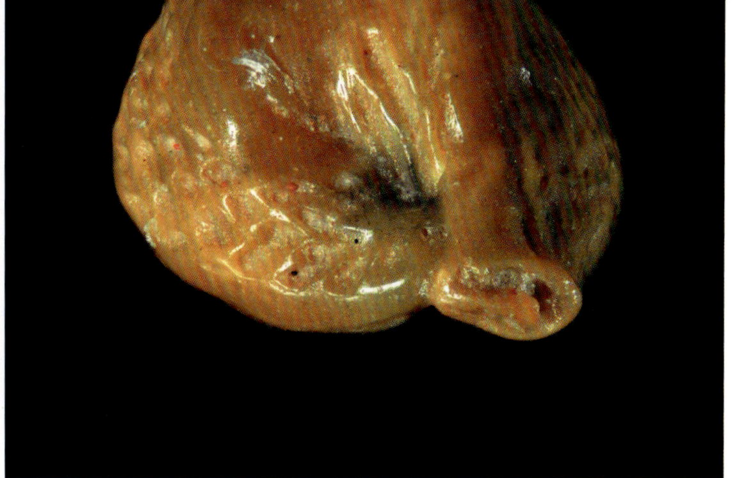

This question can most likely be answered only from an ideological point of view and on an individual basis.

Let's get back to the foraminifera and their shells. Generally, six different shell structures can be distinguished.[156]

1. Siliceous shells, which affect only a few species
2. Tektin shells, which are poorly preserved as fossils and have been around since the Cambrian period
3. So-called sand shells, foraminifera whose shells consist of granular material. This is taken from the surrounding sediment and strictly selected according to size and structure.157 There are species that use sand exclusively, while others have specialized in sponge needles, mica, or smaller foraminifera. The suborder Textulariina shows this behavior among "agglutinating" species especially, which have been around since the Cambrian period. Sand shells are most often found in cooler and deeper water.
4. Porcelain-like shells. Under reflected light the shells look like porcelain. Foraminifera of this type have no perforation of the shell and have been preferentially represented by the suborder Miliolina since the Carboniferous period.
5. Microgranular shells, which are composed of tiny calcite grains and are represented by the suborder Fusulinia, which are rock forming in the Carboniferous and Permian periods and used for standard zoning.
6. Glassy, hyaline shells, represented by the suborder Rotaliina, which are pale in transmitted light and dark and glassy in reflected light. The shells are perforated with tiny holes (diameter between 0.5 and 15 μm). Most planktonic species belong to the hyaline forms.

Figure 7.70. *Left*: Darwin, Australia. Portrait of a foraminifer from the family Milionidae with a porcelain-like, shimmering shell. The opening (aperture) can be found on the left side. Reflected light. Image width 1.25 mm. *Right*: The same animal from another perspective. We are now looking directly at the aperture, which has a so-called tooth for Milionidael, which in this case lies across the aperture. The tooth shape depends on the type.

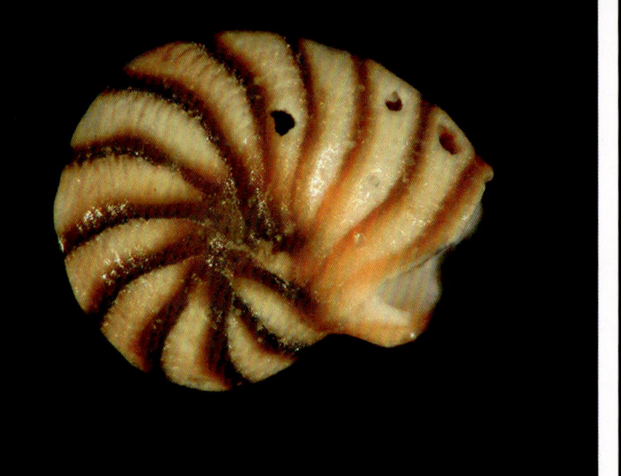

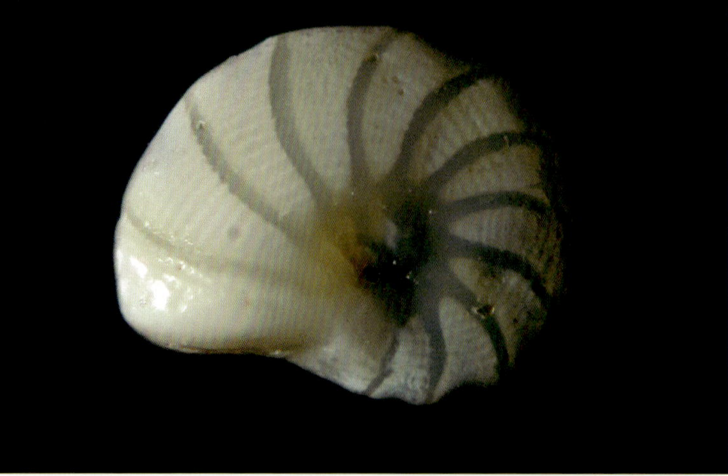

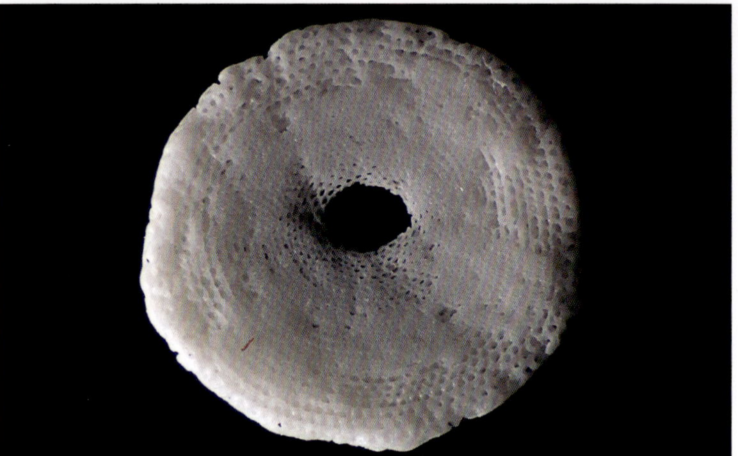

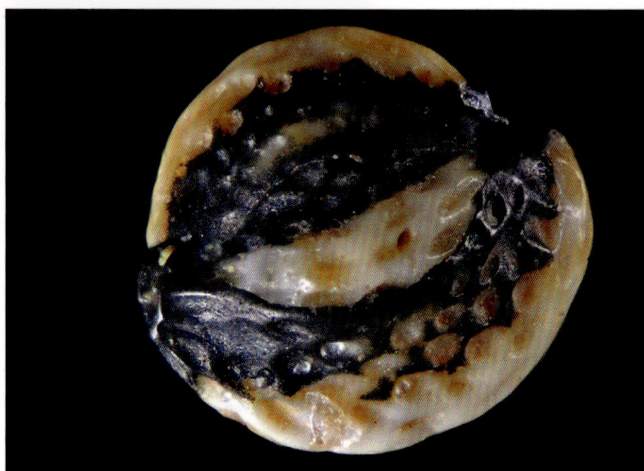

We should always keep in mind what complex and admirable life-forms already exist at the level of single-celled organisms. There, where the boundaries among the great kingdoms of fungi, plants, and animals become blurred. We can see this clearly in the example of the unicellular foraminifera.

Figure 7.71 *Top left*: Foraminifer from the sand of the eastern coast of Crete, Greece. Image width 1.35 mm.

Figure 7.72 *Top right*: Foraminifer from Noumea, New Caledonia. Some foraminifera have shell structures suitable for conducting light inside for endosymbiotic algae. These can be fine ribs, as in the specimen shown, or light-conducting tiny columns.[158] Reflected light. Image width 710 *μm*.

Figure 7.73 *Bottom left*: Flat, plate-shaped foraminifer from Michelmas Cay, Australia. The shell is eroded—when the animal is living, the chambers are closed. Probably from the family Penerioplidae. Image width 4.4 mm.

Figure 7.74 *Bottom right*: Sand grain and foraminifer from Darwin, Australia. Image width 880 *μm*.

7.7 Moss Animals or Bryozoans

It doesn't happen very often, but on some beaches we can observe sand grains that look like small branches and are chambered. In this case, we are most likely dealing with moss animal representatives. As is so often the case, the name is misleading, because the animals only sometimes look similar to moss. Bryozoans belong to an undefinable subgroup of prehistoric mosses, and there are about 5,600 current and more than three times as many fossilized species on Earth, which they have inhabited since Ordovician times. Most species live in the sea, only a few in freshwater biotopes, so in river sand we are not likely to encounter bryozoans. Bryozoans are very popular with geologists because they serve as index fossils. This means that the occurrence of defined species is characteristic for certain geological times and vice versa. If fossils are found in an initially unknown sedimentary layer, whose temporal and spatial occurrence is known (i.e., they are index fossils), the age of the layer can be consequently classified. Of course, other than bryozoans, many more index fossils of quite different species are used.

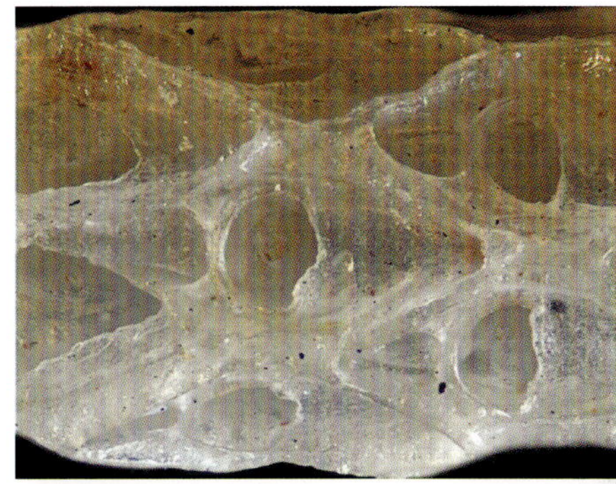

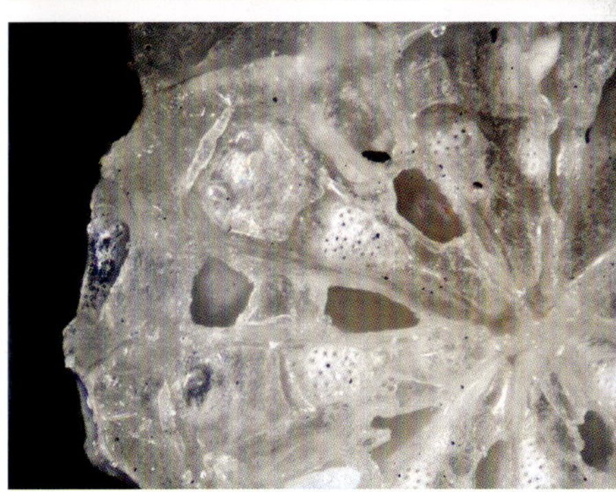

Bryozoans form colonies that are composed of many individual animals, zooids. Each of the individual animals consists of a front body with a ring of ciliated tentacles, which swirl water and food to a mouth opening, as well as a box- or cylinder-shaped hind body, which forms a calcareous skeleton and into which the animal can withdraw with the help of a central muscle. The entirety of the calcareous shells of the individual animals is connected to form a so-called zoarium, which can assume very different shapes depending on the species. This can be imagined as a kind of residential community. There are a variety of branch-like, lobed, treelike, fan-shaped, and crusty forms, on which the individual animals sit in rows. The individual shells are interconnected by wall pores. Tentacle crowns used to protrude from the laterally recognizable openings of the bryozoan we found in the sand.

Remarkable is the fact that polymorphism occurs among bryozoans; certain individual animals thus differ from others and take over very special tasks in the colony. This is very reminiscent of some ant genera (including *Pheidole*, *Messor*), which also show polymorphism; for example, some produce "soldiers," which are able to provide for the defense of the colony due to their significantly strengthened muscles of the mouthparts (mandible muscles).

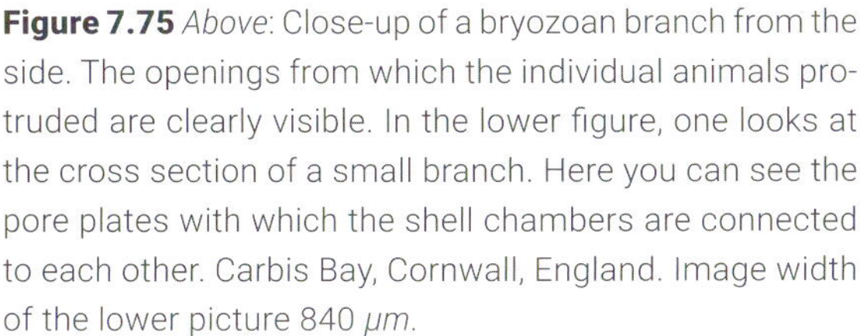

Figure 7.75 *Above*: Close-up of a bryozoan branch from the side. The openings from which the individual animals protruded are clearly visible. In the lower figure, one looks at the cross section of a small branch. Here you can see the pore plates with which the shell chambers are connected to each other. Carbis Bay, Cornwall, England. Image width of the lower picture 840 *μm*.

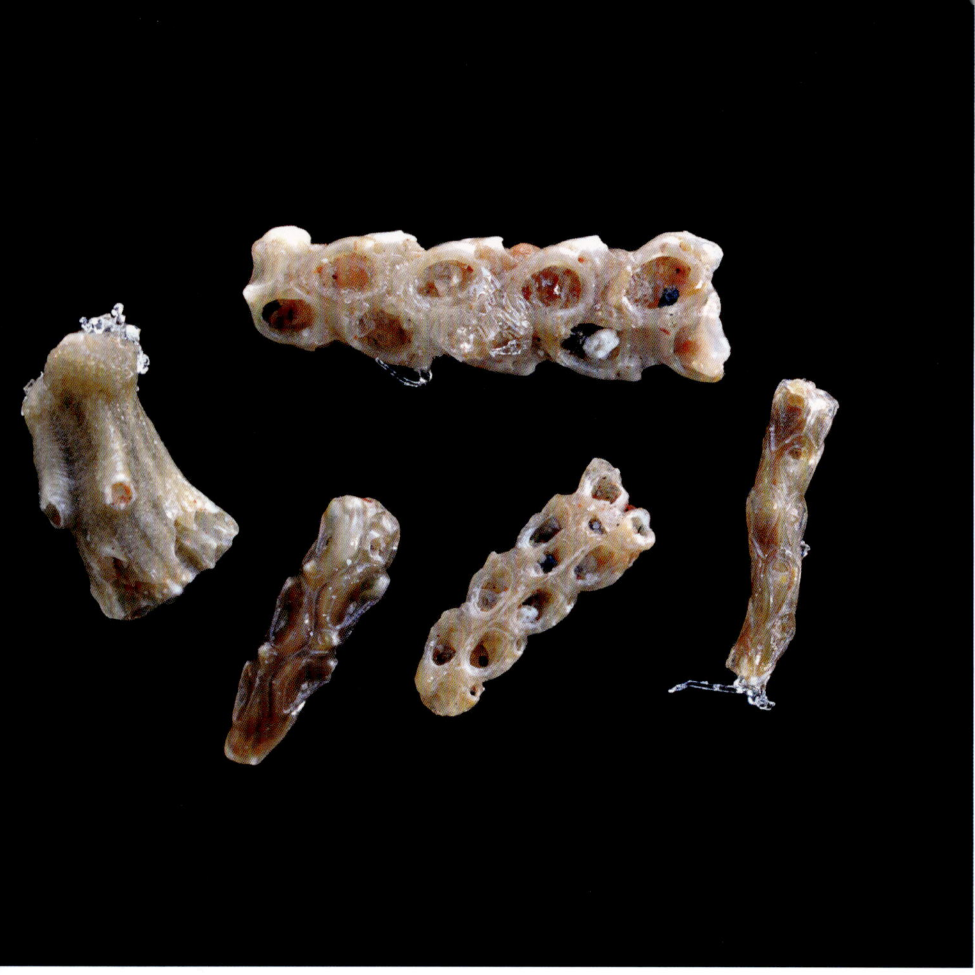

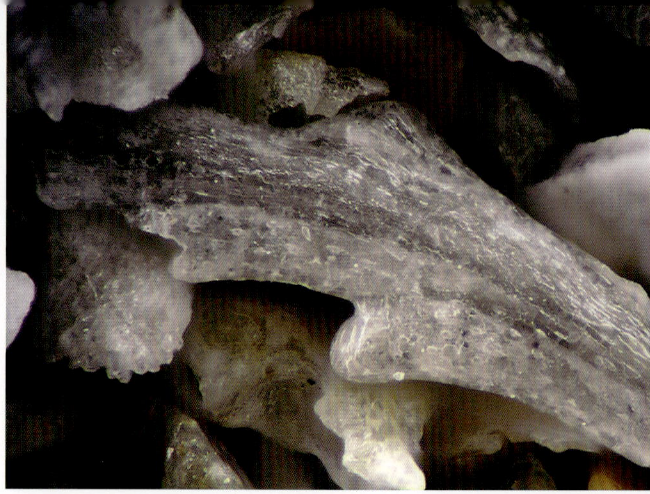

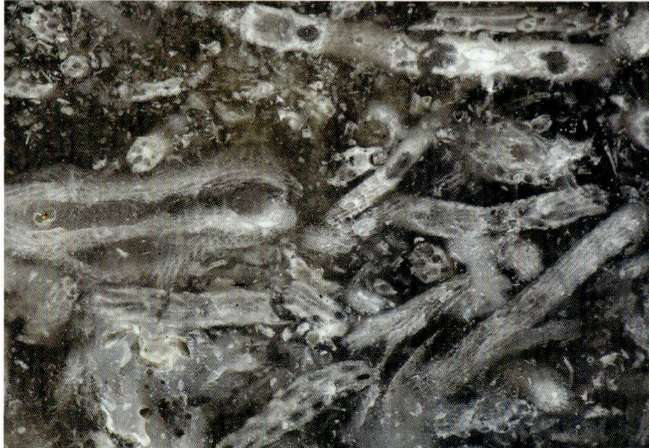

There is a similar situation with the bryozoans. Specialized individual animals take over the cleaning of the surface or also the anchoring of the colony to the substrate.[159] There are even individual animals, the gonozooids, which are responsible for larval rearing; bryozoans thus engage in brood care. Bryozoans, incidentally, can reproduce both sexually and asexually by budding. Bryozoans feed on plankton. The digestive system of the animals is quite simple; namely, U shaped. Food is fed in at the tentacled side, processed in a saclike stomach, and excreted at the other side of the U. It is quite amazing what small and remarkable, miraculous creatures we find in the sand, creatures whose remains have been in the sands of the world for many millions of years, often unchanged in structure and function over time.

Figure 7.76 Bryozoan from the sands of Darwin, Australia. A bryozoan of the colony lived in each of the individual openings. Image width 3.3 mm.

Figure 7.77 Sand with bryozoan branch, Plage Saint Efflam, Côtes-d'Armor, France. Width of the branch 190 *μm*.

Figure 7.78 Fossilized bryozoan in a ca. eighty-million-year-old flint from the beach debris. Hohwacht, Germany. Reflected light. Image width 13 mm.

Figure 7.79 *Right side*: Tower snail, mussel shells, and an eroded foraminifer (*above center right*) from the sands of Bartholome Island, Galápagos, Ecuador. Reflected light. Image width 4.75 mm.

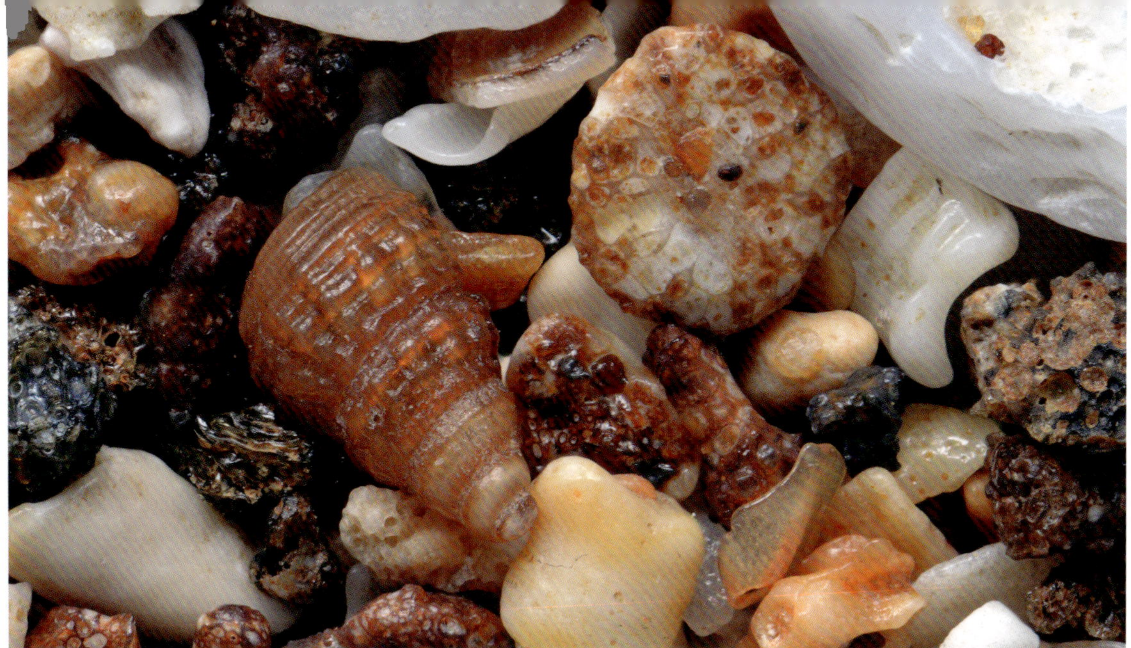

7.8 Snails and Mussels

Now that we have dealt with some lesser-known species, let's turn our attention to an animal species that probably everyone has had contact with in one way or another, either out of pure interest or in a culinary context: we're going to talk about mollusks.

Snails and mussels belong to the most-common mollusks, and their remains also represent the largest fraction of organic material in marine sands by volume. From the roughly 130,000 kinds of mollusks, snails make up the majority, with about 110,000 of them. Snails, or gastropods (which means nothing more than "abdominal feet"), are the only one of the mollusk classes to have also produced land-dwelling species. They are estimated at about 32,000 different species.[160] Snails cover a wide range of sizes. The smallest of them are under 0.5 mm, so in the realm of foraminifera, they should not be confused with others. The complete animal, including its shell, is about the size of a midsized sand grain with these miniature snails. On the other end of the spectrum, some species of snails can reach dimensions of several inches.

We are interested as always—unfortunately—only in the fragments, the remains of these highly interesting animals. In contrast to the two-part shells of the bivalves (Bivalvia), the snails have a spirally rolled-up, nonsymmetrical shell.

If the spirals lie in one plane, we speak of planispiral shells; if the spirals are wound over each other in a towerlike manner in the direction of elevation, we speak of trochispiral shells. The latter type of shell is known to us from the native Burgundy snails (*Helix pomatia*). Furthermore, all transitions between these basic forms exist, as well as cup shaped, elliptical, and others. Not all snails have a shell, but those do not leave any particles in the sand and shall not concern us further.

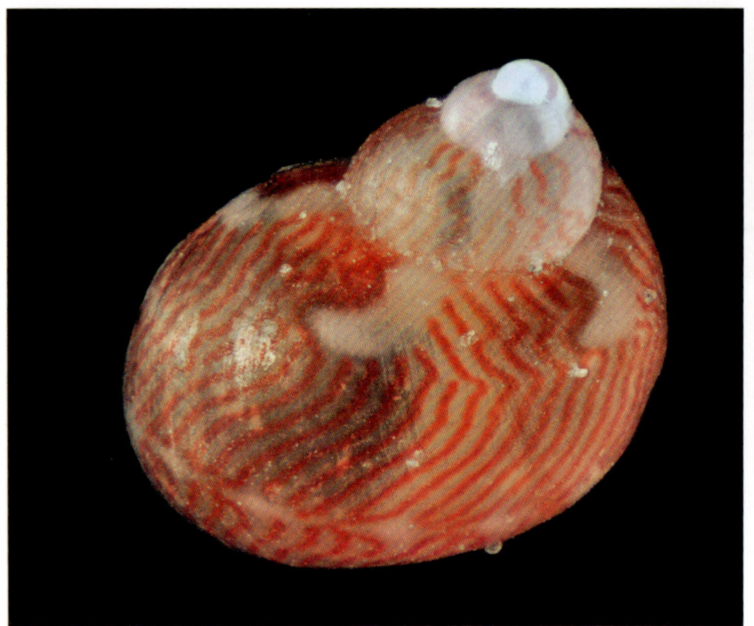

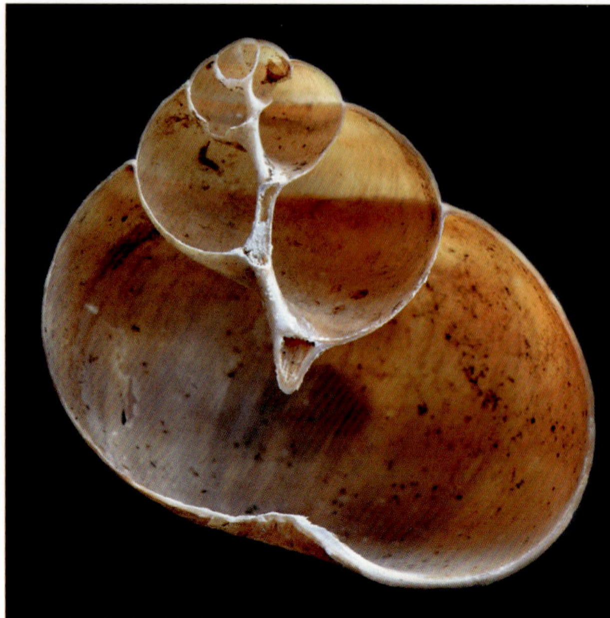

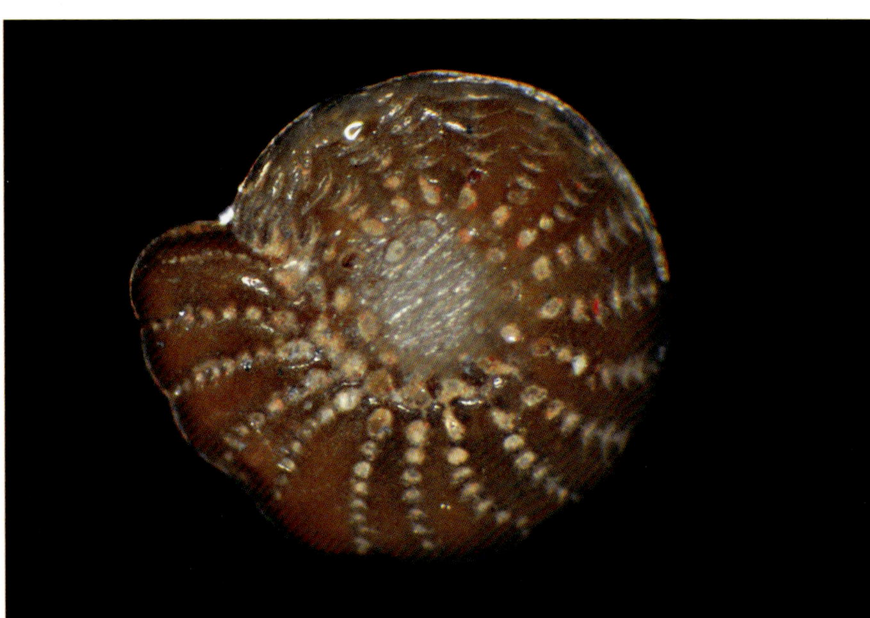

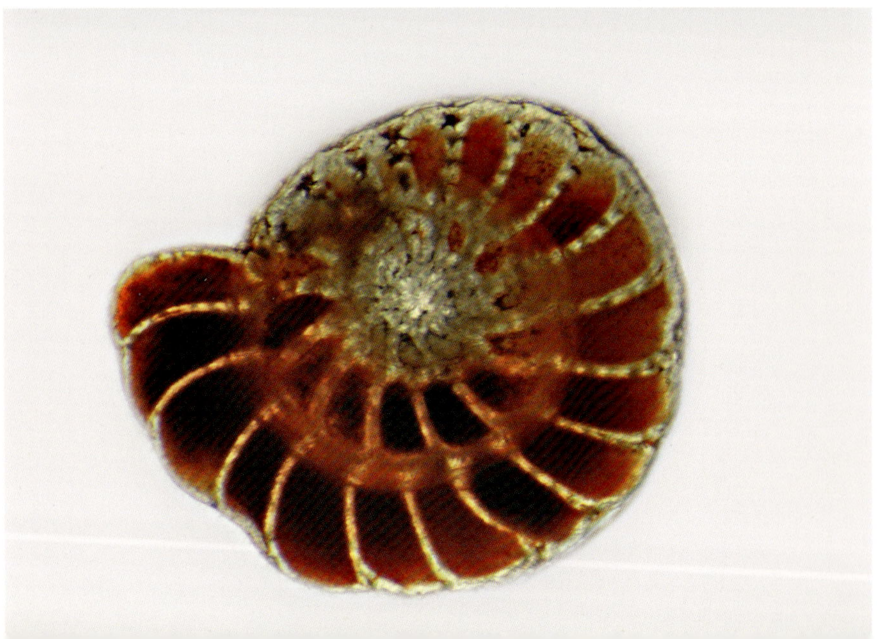

Figure 7.80. *Above*: Sand grain made from a snail shell, Big Island, Hawaii, USA. The very small shell, with a diameter of only 1.5 mm, has a trochispiral shell that can be easily identified from the outside. To the right is a cross section through another trochispiral shell of the common Burgundy snail (*Helix pomatia*). The shell, which winds all the way to the top, consists of a single continuous volume. Diameter at the very bottom is 38 mm.

Figure 7.81 *Middle and bottom*: A foraminifer from Darwin, Australia, for comparison. The shell on the picture to the left could be confused with a snail shell. It was ground down about halfway in the direction of thickness after the photographic image was taken. The lower image shows the result in transmitted light. The chambering of the coiled shell can be seen very clearly. In contrast to a snail shell, which consists of a single continuous and outwardly open space, the shell of the foraminifera is not continuous but, depending on the species, is divided into a number of small chambers ("chamberlings"), which, as can be clearly seen, are also connected to each other with openings. Diameter around 1 mm.

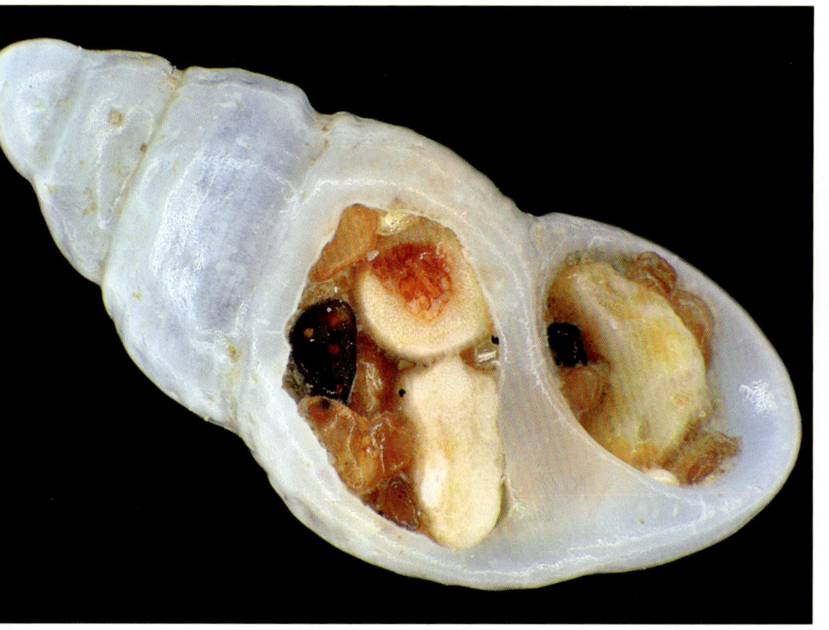

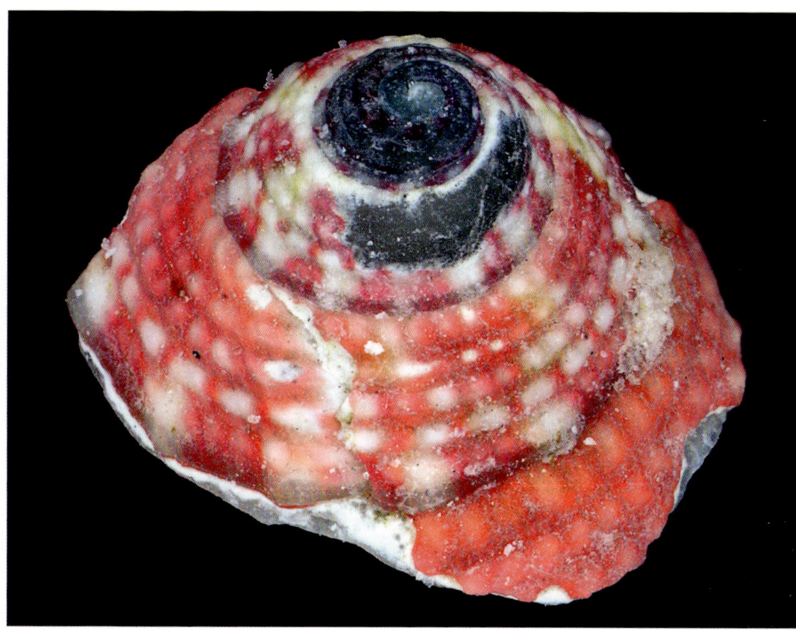

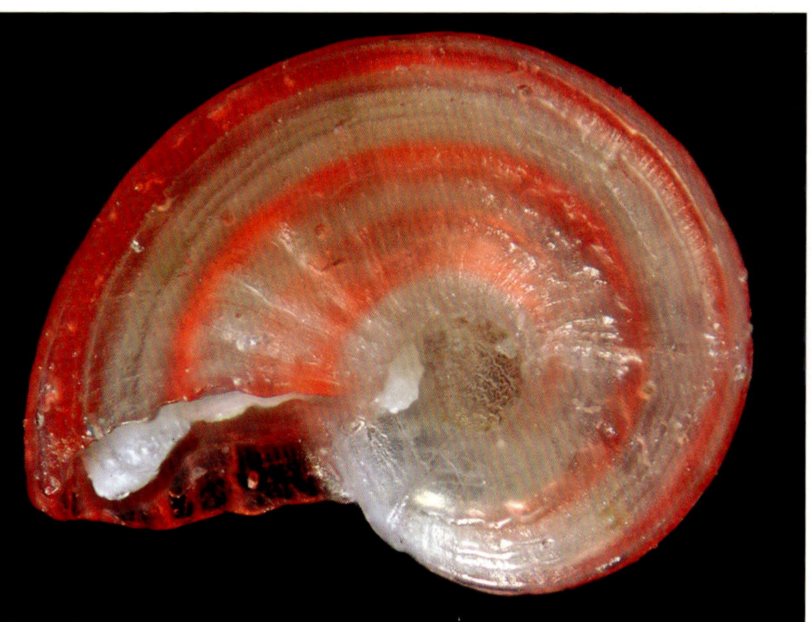

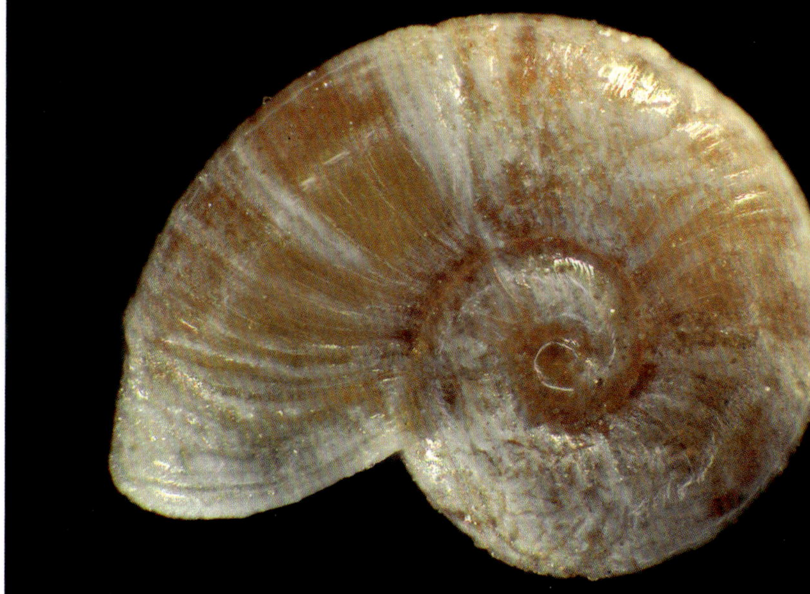

Figure 7.82 Isla Bartholome Galápagos, Ecuador. Tower-shaped snail shell filled with sand grains, "sand in sand." Diameter about 1,000 *μm*.

Figure 7.83 Puuhouna Beach, Hawaii, USA. Bowl-shaped snail shell. Diameter 2.8 mm.

Figure 7.84 Snail shell from Belmare Beach, Mauritius. Reflected light. Image width 770 *μm*.

Figure 7.85 Snail shell from the sand of Boulders Beach, South Africa. Reflected light. Image width 810 *μm*.

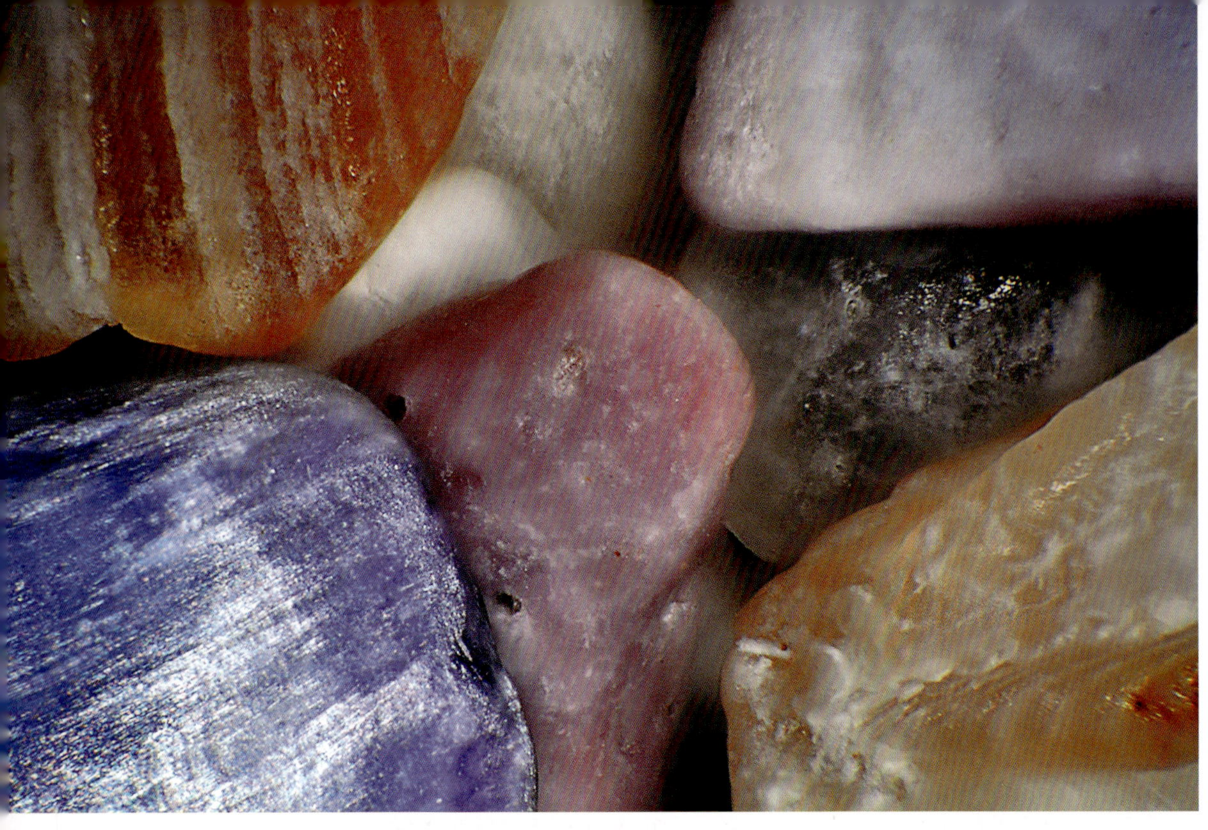

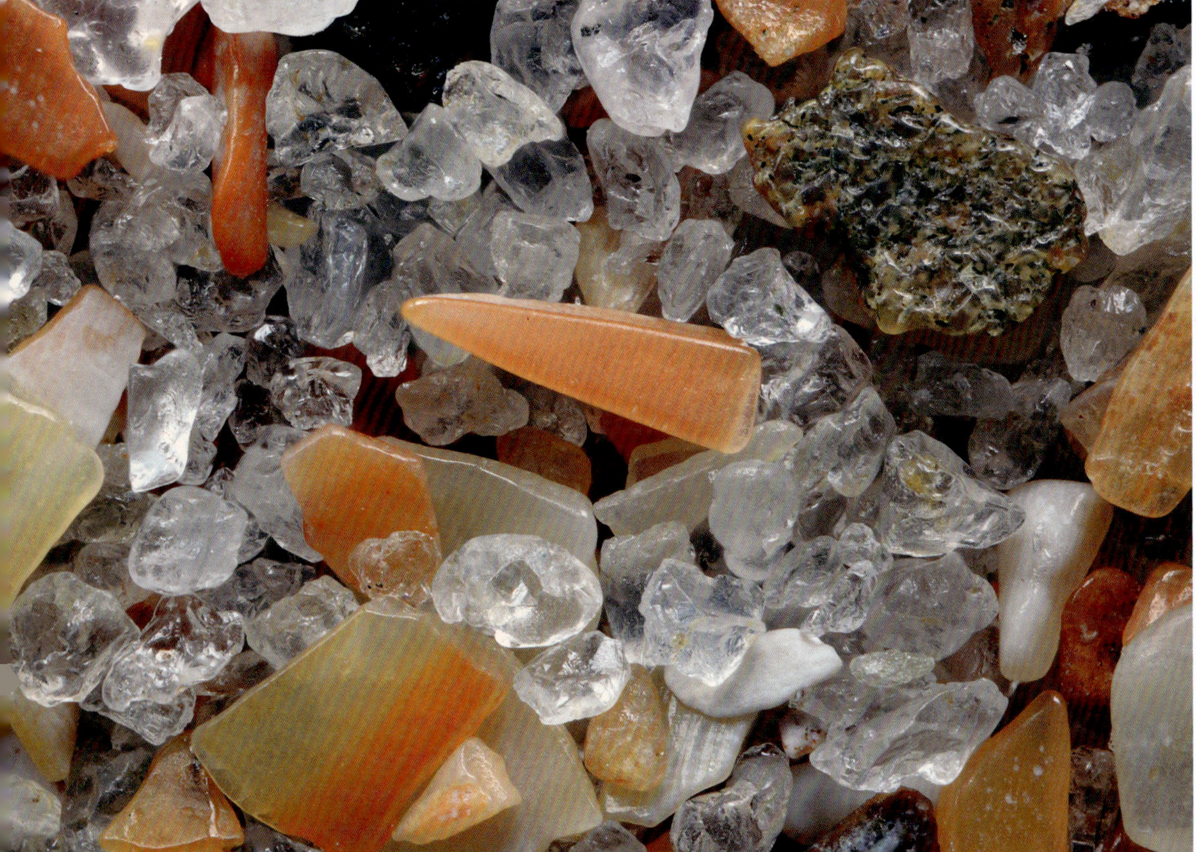

Figure 7.86 Colorful snail and mussel shell remains in the sand from Vardø, Norway. Image width 900 *μm*.

Figure 7.87 Sand from Flagler Beach on the eastern coast of Florida, USA. Orange-colored, graded mussel shells and quartz come together to form a very decorative and unique total impression. Reflected light. Image width 5.35 mm.

The snail shell, like the mussel shells, is composed of three layers. The outer layer of the protein conchiolin is followed by a middle layer of aragonite prisms and finally the inner layer of aragonite and conchiolin. Depending on the conchiolin content, this inner layer shimmers beautifully like porcelain or mother-of-pearl. This is particularly impressive to observe in porcelain snails or cowries, whose beauty has highlighted them as desirable objects for humans since time immemorial. Whether as jewelry, for cultic purposes, or even as currency, cowries would be worth a chapter of their own. Since neither conchiolin nor aragonite is very durable, the degree of preservation of fossil snail shells is poor; usually only a stone core of the shell remains.

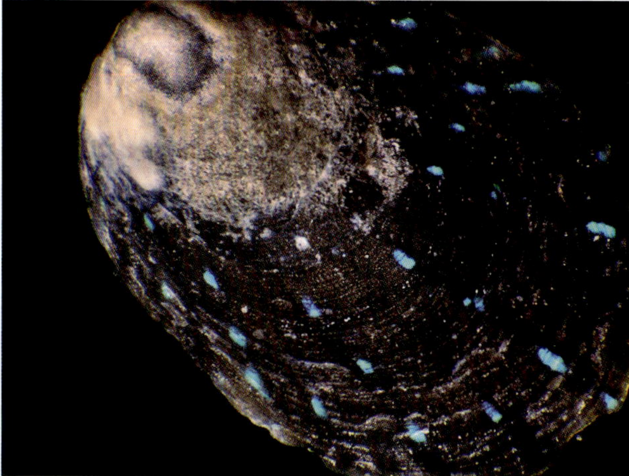

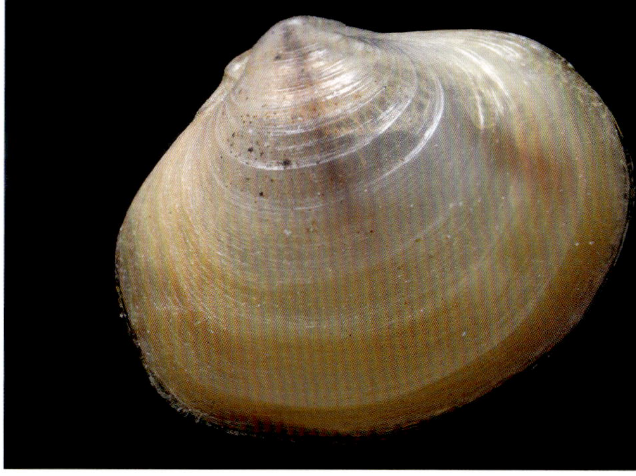

In contrast to the shells of the foraminifera, which are sometimes very similar in appearance, the shell of the snails consists of a continuously steadily increasing volume toward the outlet. That of the foraminifera is always chambered, which is sometimes difficult to recognize from the outside. Snails and foraminifera overlap in their dimensions, so that size alone does not provide a sufficient distinguishing characteristic, although the vast majority of snail shells are comparatively larger.

Very fittingly, we supplement our lineup here with a sand that once achieved world fame, the Steinheimer snail sand from the Pharion pit in the Nördlinger Ries. In the Tertiary Upper Miocene, about fourteen million years ago, a large, ring-shaped crater lake was formed in the impact basin of the already mentioned Ries meteorite. Only the central mountain, formed by fall-back masses, remained uncovered by water. In this relatively small and enclosed habitat, freshwater snails, the planorbids, developed in generational succession. Their fossil remains are exceptionally well preserved and can be found in large quantities in the area of the accessible Ries outcrops.

Figure 7.88 This only 1.6 mm wide mussel shell was found in the beach sand of Boulders Beach, South Africa. The turquoise-colored, almost neon-like, bright marks are only 38 *µm* wide. Reflected light.

Figure 7.89 This pretty river mussel comes from the sandbanks of the Elbe near Lenzen, Germany, and became a sand grain itself (in the size class of fine gravel). Reflected light. Image width 4.75 mm.

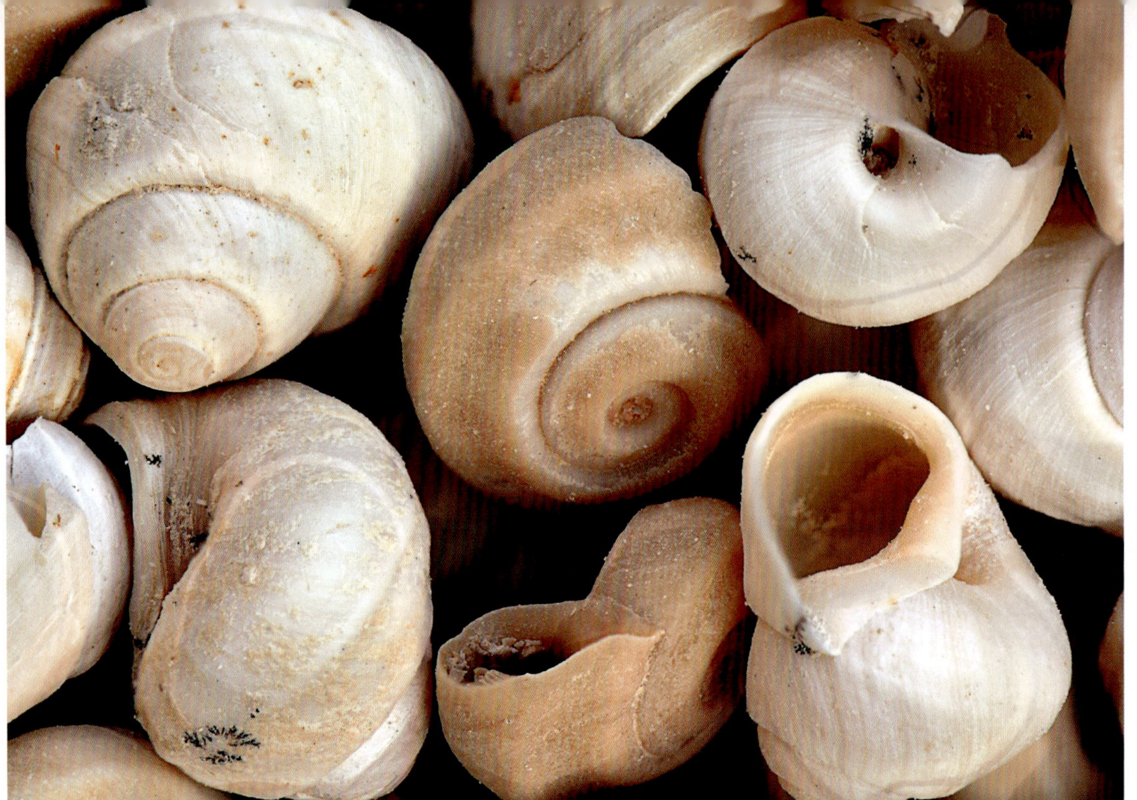

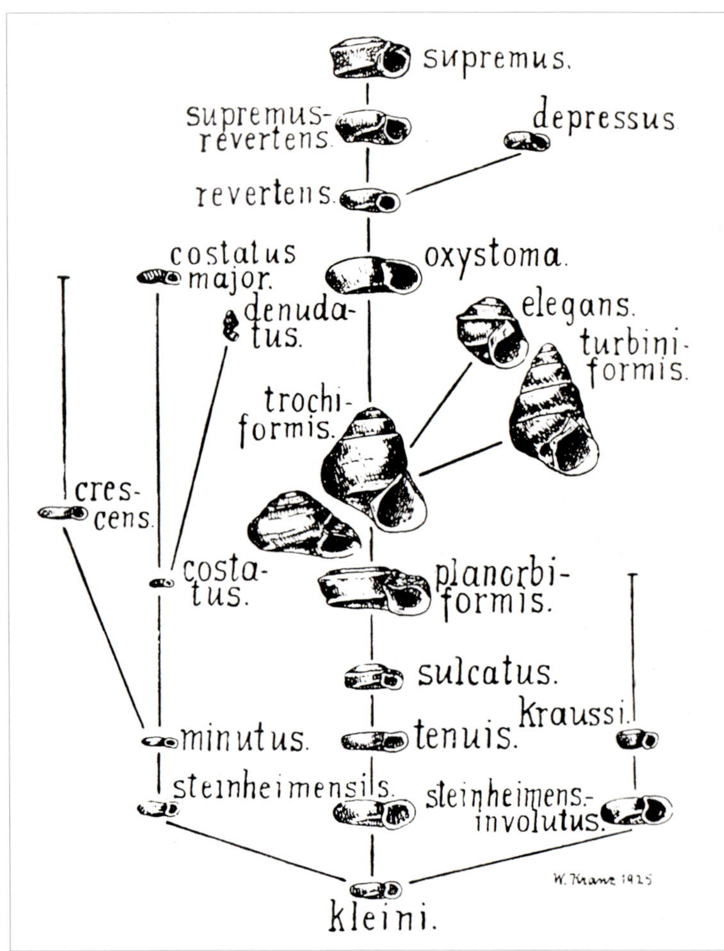

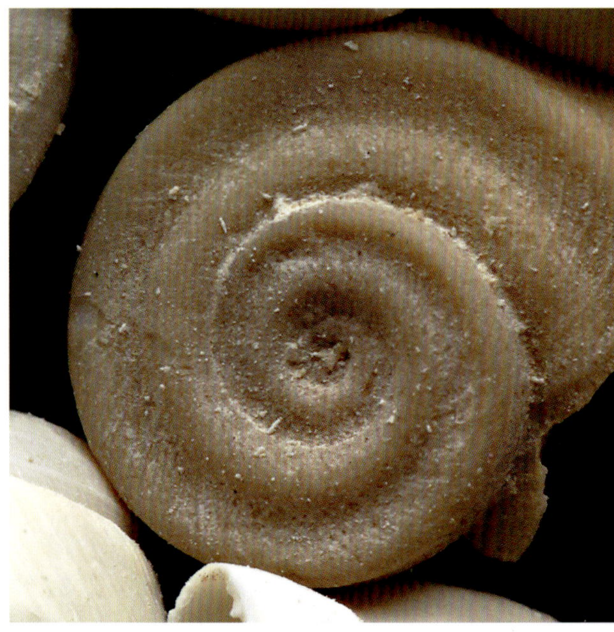

Figure 7.90. "Snail sand" from the Pharion pit, Steinheim crater, Germany. You can see many different coil types. The wonderful, cream-colored snail shells were sorted from other sediment parts and attachments before the picture was taken. Image width 15.6 mm.

Figure 7.91 *Left*: Planorbid progression according to research from Hilgendorf and Kranz 1915 (graphic from K. D. Adam, *Das Steinheimer Becken* ["The Steinheim crater"]). The transitions from planispiral to trochispiral shell shapes in all their variations can be seen.

Figure 7.92 *Right*: Planispiral snail shell from the "snail sand" of the Pharion pit, Germany. Diameter 5.2 mm.

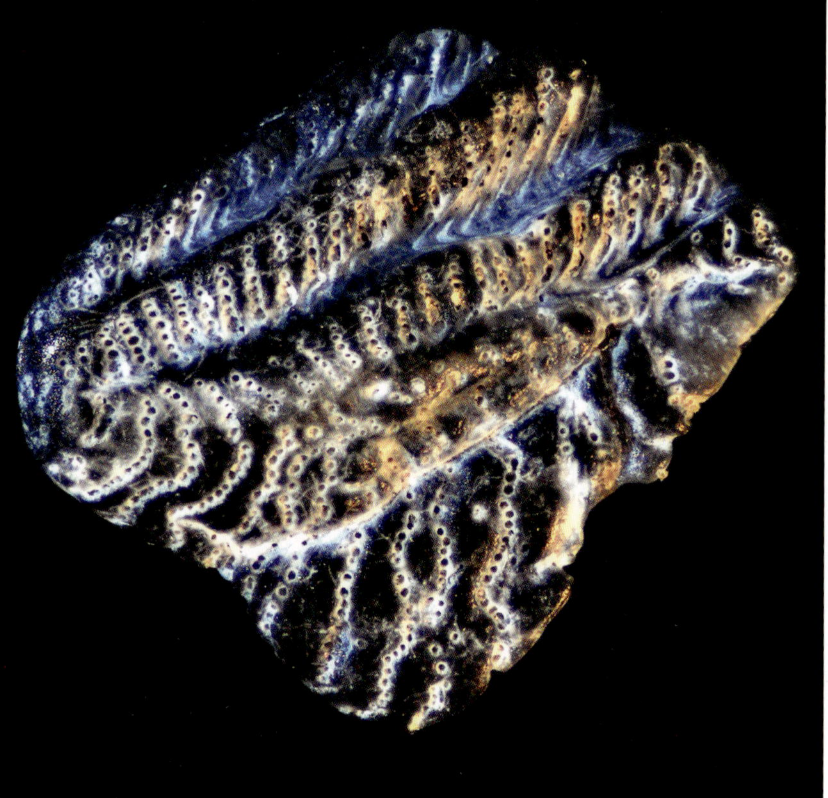

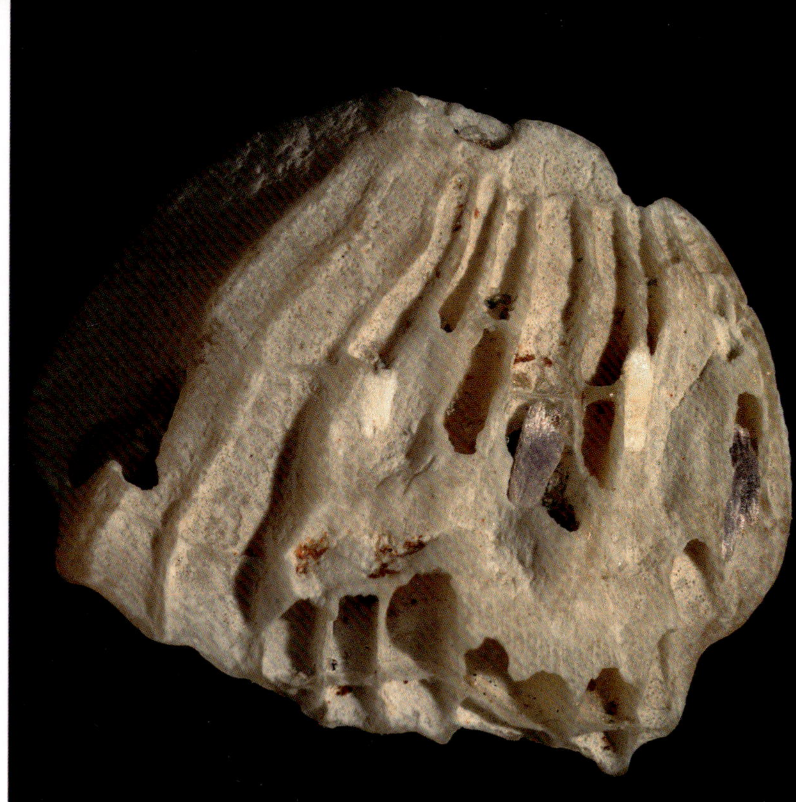

Fossilized snails and their variety of forms have attracted the interest of researchers for a long time. Franz Hilgendorf (1839–1904), student of the famous Professor Quenstedt from Tübingen, devoted his entire scientific life to the successive and overlapping forms of the snail *Planorbis multiformis*. He summarized his results in a dissertation on the subject. Hilgendorf's investigations represent the first direct evidence for Darwin's theory of evolution. Later, in 1925, Walter Kranz drew up the phylogenetic tree and the evolutionary series of all forms that had emerged from the snail *Gyraulus trochiformis kleinii*. He based this on the work of Hilgendorf and Gottschick.[161]

If one carefully frees the sediment from the Pharion pit from fine particles, one obtains a beautiful assemblage of sand grains consisting exclusively of fourteen-million-year-old snail shells of various shapes and coil types.

It is almost regrettable to stop at this point because of space—there are many more examples that could be shown. We have only thematically touched upon shells and snails: animal traces in the form of teeth could also be separately considered. However, their frequency is not so great: you have to be lucky and patient when searching in the sand, as well as when looking for the remains of barnacles marking the shoreline; the strange, ribbed tubes of calcareous tubeworms reminiscent of plastic tubes; or even the jaw plates of rays and stingrays. We therefore leave it at a few exemplary and characteristic illustrations.

Figure 7.93 Sand grain from Snaefellsnes, Iceland. Part of the jaw apparatus of a stingray. Reflected light. Image width 2.5 mm.

Figure 7.94 Shell fragment of a barnacle from Tromsø, Norway. Barnacles belong to the rudder crabs and are sessile. Their shells, with their characteristic length structure, can be found occasionally in coastal sand. Reflected light. Image width 4.9 mm.

Thus, we close group 3, listed at the beginning of chapter 4, which deals with the remnants of shells, spines, and skeletal parts in sand, and now turn to the second group, sands of organogenic limestones.

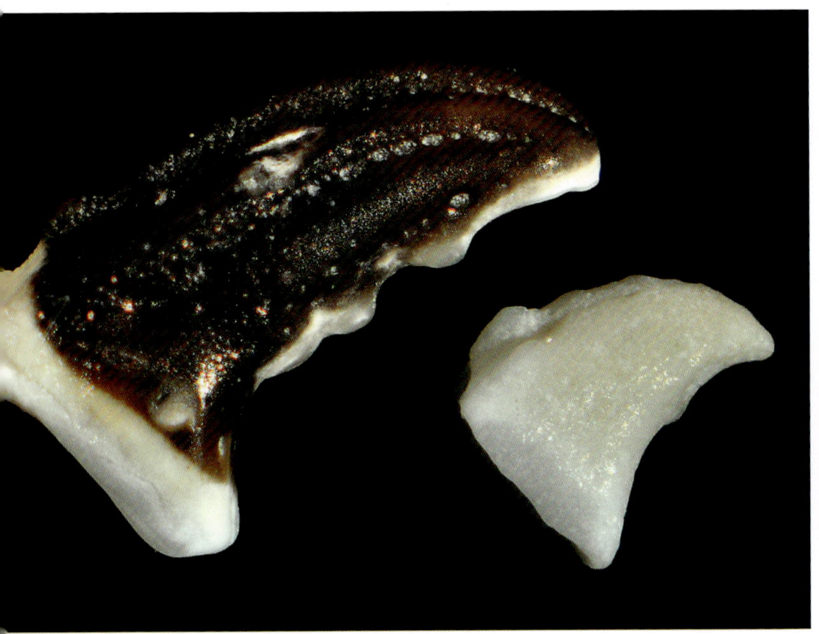

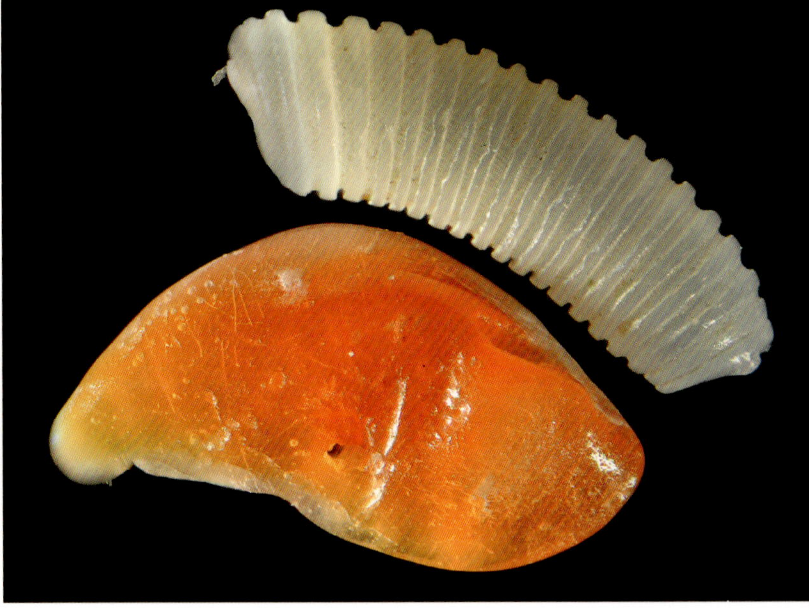

Figure 7.95 Flic en Flac, Mauritius. Teeth in the sand. Reflected light. Image width 6.2 mm.

Figure 7.96 Tubelike shell of a calcareous tube worm over the polished fragment of a coral out of the sand from Playa Espadilla Sur, Costa Rica. Reflected light. Image width 2.7 mm.

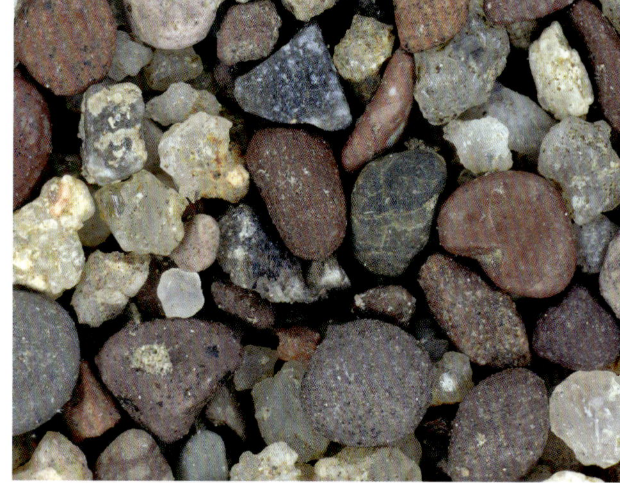

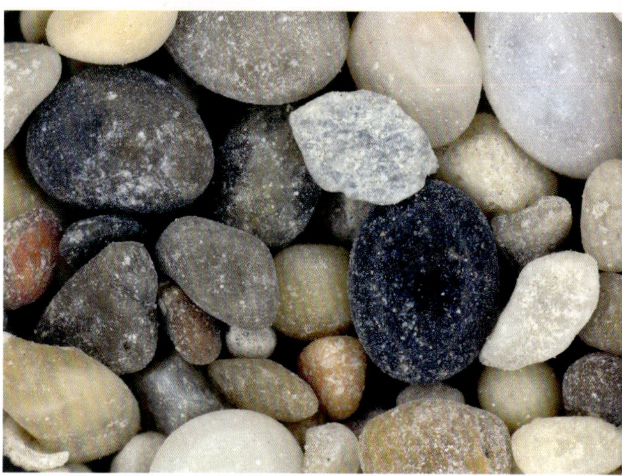

7.9 Lime

Carbonate rocks (i.e., limestones) are among the oldest sediments on Earth: they were already formed in the Archean period. In the Precambrian period, large reefs grew up, whose builders were identified as prokaryotic cyanobacteria. They did not provide the framework of the reefs, as the corals did later, but created reefs through biochemical carbonate precipitation. Cyanobacteria carry out photosynthesis and, as "extremophiles," are extraordinarily adaptable, so that they can colonize even seemingly hostile places. They had and have no predators. It is believed that cyanobacteria were among the first life forms on Earth. Today, as true all-rounders, they can be found in virtually any habitat endowed with light, from Antarctic ice to hypersaline waters and hot springs.

So-called stromatolites are visible signs of the first forms of colonization of the earth today. They are the product of the excretory processes of prokaryotic microbial communities. These consisted first of evolutionarily very early sulfur bacteria, which did not need light, but later mainly of cyanobacteria, which covered the shallow marine soil with gelatinous biofilms, in which the smallest mineral particles or also microparticles from the plankton got stuck. Along with precipitated calcareous mud, the gelatinous film became increasingly opaque, so that the light-sensitive bacteria again colonized it. In this way, layer after layer settled on top of each other at an average rate of 1 mm per year and grew into wall-like reef structures called stromatolites. The oldest fossil stromatolites are found in northwestern Australia near Marble Bar and have an age of about 3.4 billion years. They formed in a shallow water lagoon in the caldera of a prehistoric volcano. Their state of preservation is exceptionally good because they are located on the even-older Pilbara craton, which resisted all tectonic influences and folding as an artifact of the ancient earth's crust.[162] Even today, there are places on Earth, such as in Australia's Shark Bay, where recent stromatolite formation can be observed and studied. Fossil remains of stromatolites can be found in many places on Earth, including Europe, sometimes even in the form of river and sea debris. They catch the eye because of their wavy and fine lamellation, which is nothing more than a kind of ring depicting years gone by and is particularly beautiful when ground and polished.

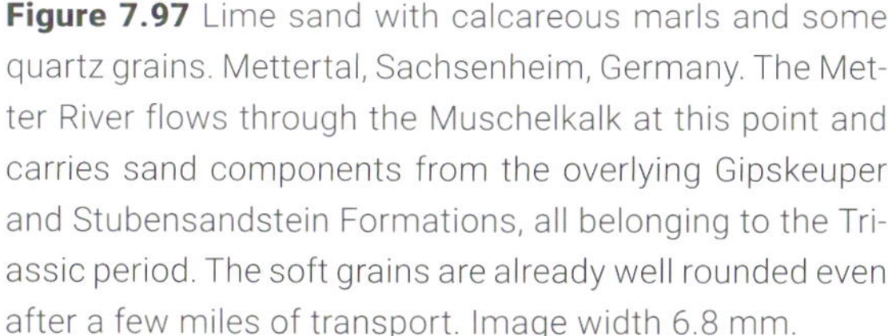

Figure 7.97 Lime sand with calcareous marls and some quartz grains. Mettertal, Sachsenheim, Germany. The Metter River flows through the Muschelkalk at this point and carries sand components from the overlying Gipskeuper and Stubensandstein Formations, all belonging to the Triassic period. The soft grains are already well rounded even after a few miles of transport. Image width 6.8 mm.

Figure 7.98 Lime sand from the coast of Pargas, Greece. Image width 6.8 mm.

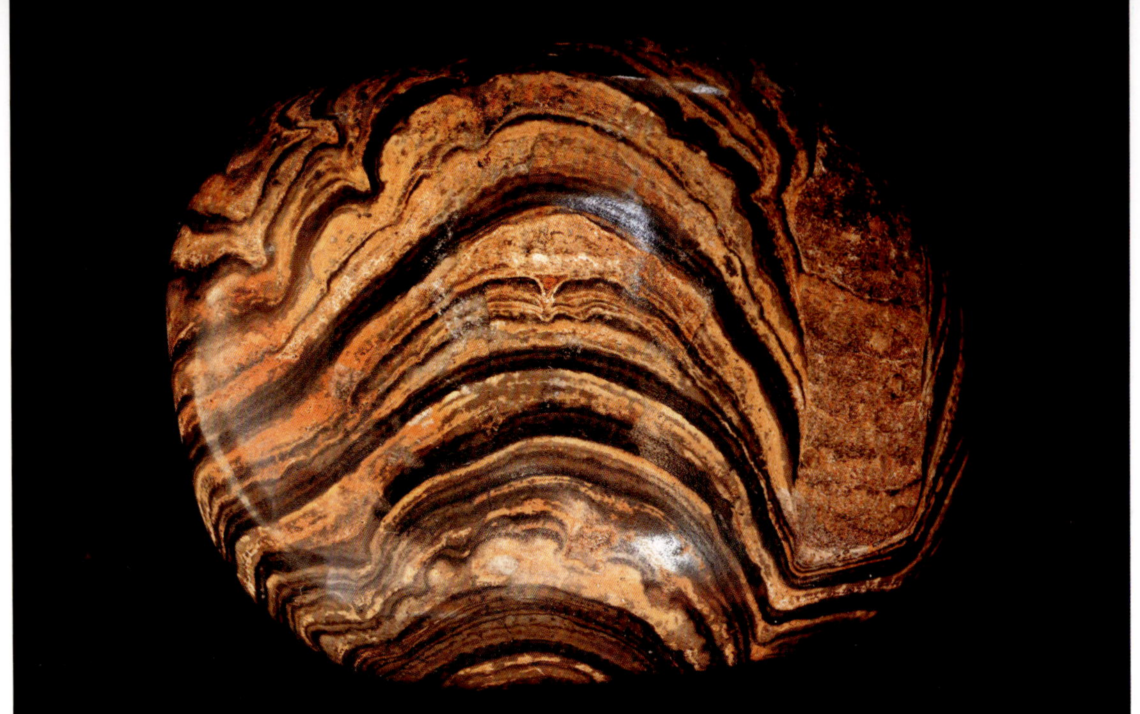

Stromatolites are therefore often made into jewelry or hand charms.

So, what exactly are we talking about when we say "lime"? Basically, we understand lime to be sediments from the group of carbonates (i.e., the salts of carbonic acid). Limestones consist of about 80% of these carbonates. A large part of lime is of biogenic origin, "omnis calx ex vivi," as Carl von Linne so aptly put it. This is also the reason why we are dealing with lime in this chapter, which is about formerly living things, and not in the section about rocks, where it should have been classified.

Calcite, aragonite, and dolomite belong to the most-important carbonates. We want to introduce them briefly and in the right order.

Calcite ($CaCO_3$) is by far the most important nonsilicate mineral on Earth. It belongs to the trigonal crystal system and has characteristically good cleavage in three oblique planes; otherwise it breaks brittlely and conchoidally. Calcite is the mineral with the most forms: hundreds of crystal forms and over a thousand surface combinations are known.[163] With a Mohs hardness of 3, calcite is very soft; lime sands fall quickly to erosion and are already rounded after just short transport distances. By adding cold and diluted hydrochloric acid, calcite can be easily identified by violent foaming.

Figure 7.99 Stromatolite. The stacked light and dark layers are very clear, depicting the progress of settlement in earlier times. Polished drum product without information on its place of discovery. Width 38 mm.

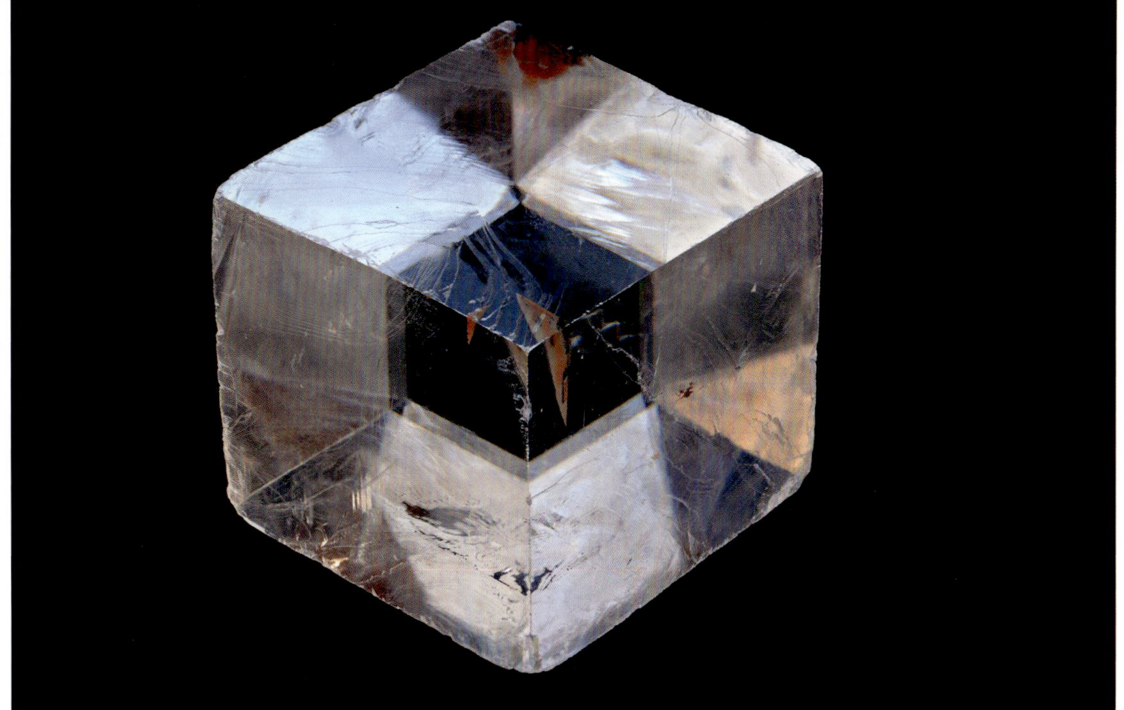

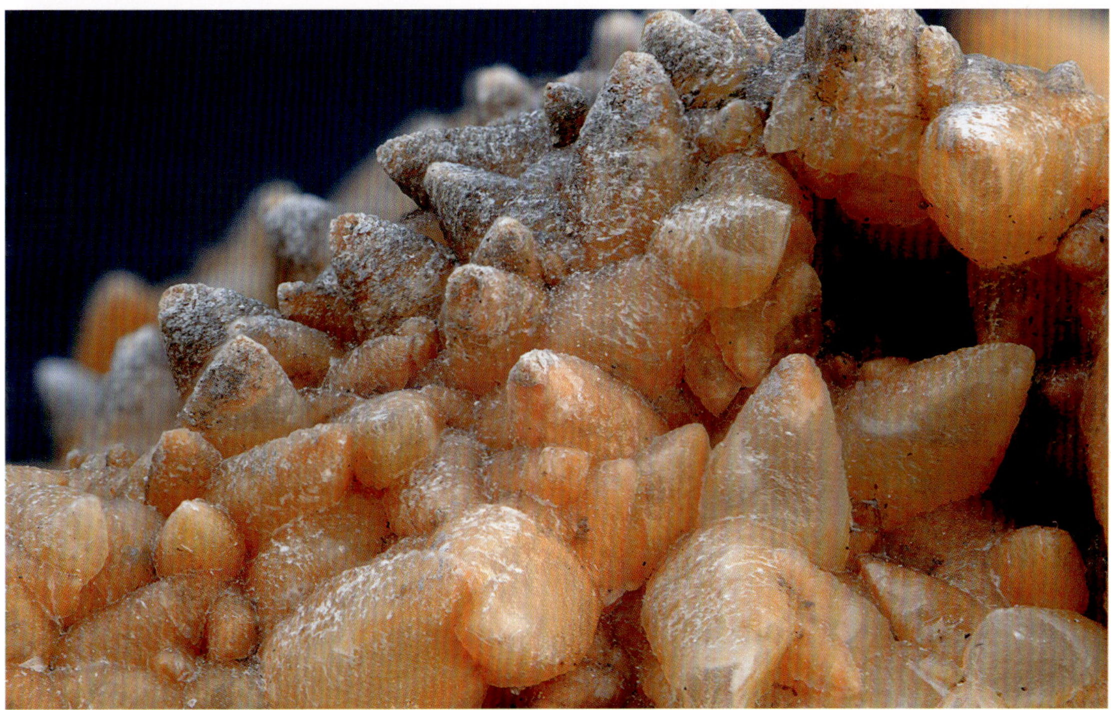

Figure 7.100. Rhombohedral cleavage piece of the sought-after Icelandic double spar, a pure and fully clear variety of the mineral calcite. This was used for optical purposes until modern times. Of special interest is its high birefringence. Reflected light on a black background. Side length 35 mm.

Figure 7.101 Scalenohedral calcite crystals from the Erfurt Formation, a formation of the Triassic period. Sachsenheim, Germany. Reflected light. Image width 20 mm.

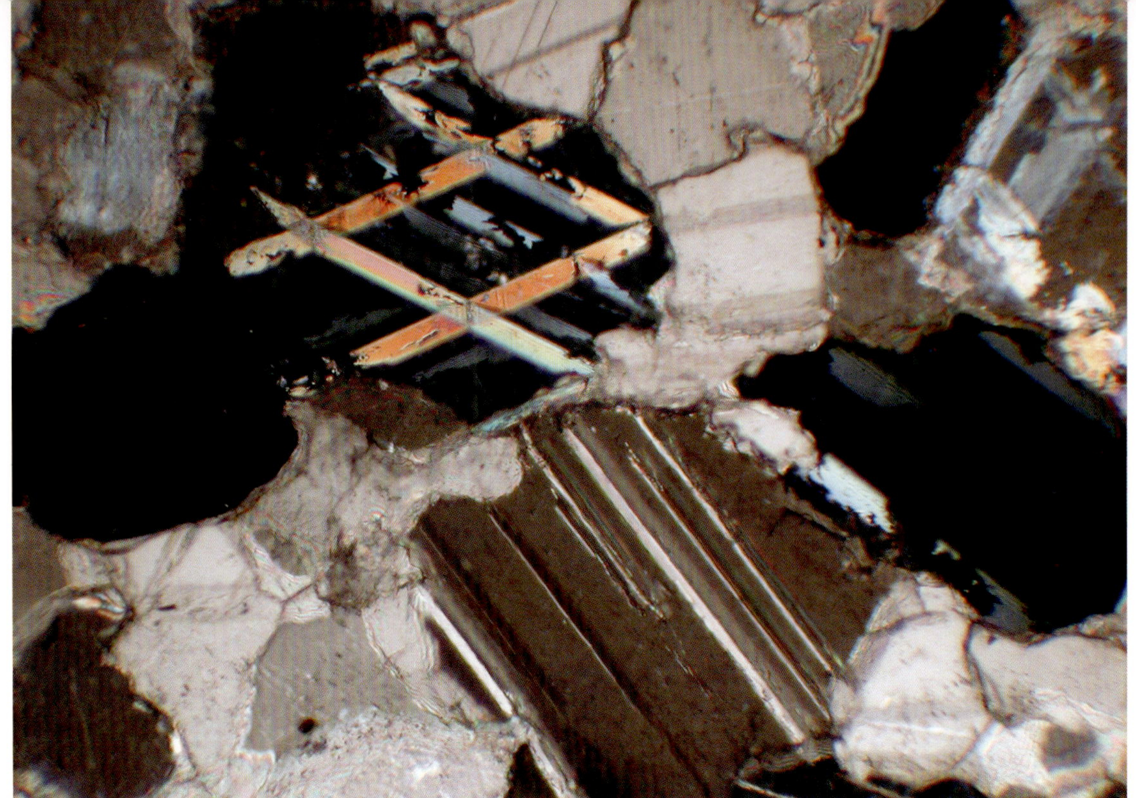

Unlike calcite, aragonite has orthorhombic symmetry, with the same chemical composition. Aragonite is stable only under higher pressures, but not on the earth surface. Therefore, it is not formed diagenetically but mainly hydrothermally in fissures or in cavities of volcanic rocks. A well-known example is the Karlsbader Sprudelstein. Aragonite is formed stably under high pressures, such as in subduction areas of plate margins. In contact with fresh water, aragonite is particularly unstable and decomposes easily. It is important in our context because the calcareous elements of many animals are composed of aragonite, such as in green algae, calcareous sponges, bryozoans, corals, and gastropods, to name only the most important. In contrast, ostracods, brachiopods, and barnacles have skeletons or shell elements of calcite. The calcareous, tubular structures of marine worms are also composed of calcite. Mussel shells have portions of aragonite and calcite. Often the aragonite skeletal elements are mixed with portions of magnesium calcite. This is a likewise metastable variety of calcite, in whose lattice magnesium ions are statistically distributed in large quantities.

The mineral dolomite contains half calcium and half magnesium ions. It has a crystalline structure similar to calcite, in which magnesium-bearing and calcium-bearing crystal layers alternate. Dolomite can be formed in sediments from calcite as well as from aragonite under supply of magnesium. Recent dolomite formations under the conditions of the earth's surface are currently a subject of research, the mechanisms of which are not fully explained, whereas it is easily formed at temperatures above 212°*F (*100°C).

Figure 7.102 Double section of a piece of Carrara marble (i.e., metamorphic calcite), under crossed polarizers in transmitted light. The rhombohedral cleavage faces can be seen in individual crystallites as well as the twinning parallel to the longitudinal edges. This distinguishes calcite from dolomite in the thin section. The latter shows twinning in the direction of the small diagonal in the rhombohedron; calcite, however, shows it only in the direction of the longitudinal edges or in the direction of the large diagonals.[164] Image width 800 *µm*.

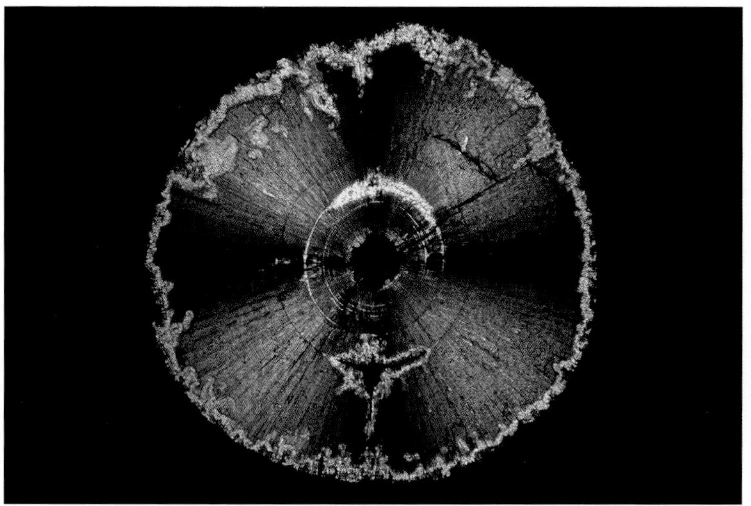

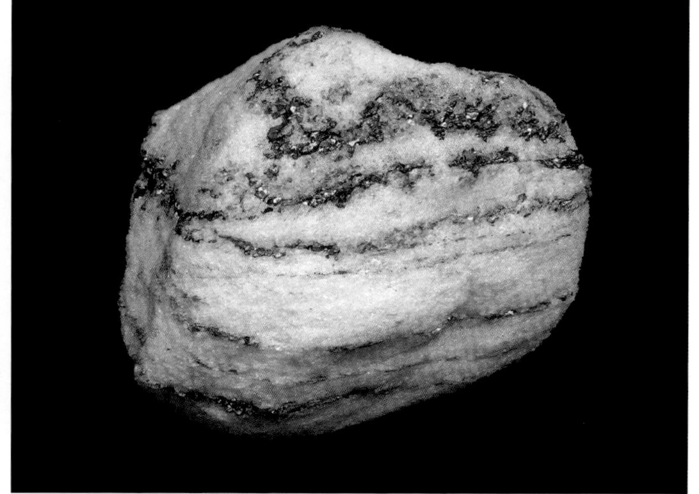

Metamorphic dolomites (i.e., dolomites formed under pressure) sometimes show a very beautiful white, sugar-grained appearance, which is, however, feared in civil engineering, since these dolomites are likely to crumble under slight pressure. Mineral and rock bear the same name in honor of the French geologist Dolomieu. Dolomite foams with warm hydrochloric acid, but not with cold, which can be used as a fairly good distinguishing feature from calcite. In direct comparison to limestone, dolomite rock is harder, heavier, and more brittle and looks yellowish gray to light gray, with a mostly sugar-grainy sheen. The main components of carbonate rocks are lime, clay, and (quartz) sands. Mixtures of lime and clay are summarized under the term "marl." If the lime portion predominates, we speak of lime marls, and in the opposite case, of clay marls. Marls were deposited in calm shelf areas of the seas, into which calcareous, fine-grained organism remains sank as lime mud and into which clay sediment loads were brought from surrounding rivers. Their very soft consistency as rocks results in ephemeral and rapidly eroding sands of only local importance. Due to their important function as raw material for the cement industry and soil improver in earlier times, quite a lot of good and geologically interesting outcrops can be found.

Figure 7.103 Thin section through the rostrum ("thunderbolt") of a belemnite fossil from the beach debris of Fehmarn, Germany. Rostrums consist of durable (here, radially crystallized) calcite. All other skeletal parts of the animals belonging to the cephalopods consisted of aragonite and have therefore long since decayed. Diameter 9.5 mm.

Figure 7.104 Sugar-grained dolomite marble with pyrite crystal layers from the world-famous Lengenbach mine in Binntal, Switzerland. From sediments two hundred million years old of the Tethys Sea, these beautiful rocks were formed by metamorphosis and recrystallization during the formation of the Alps. Stone width 35 mm.

Limestones are mostly light-colored rocks that are made primarily of carbonates that were deposited in the oceans in prehistoric times. The buildup of limestone sedimentation in today's oceans takes place at an average rate of 2–10 mm per 1,000 years.[165] On the earth's surface, the only limestone that appears is that which was raised tectonically. Limestone mountains are therefore always former oceanic sedimentary environments.

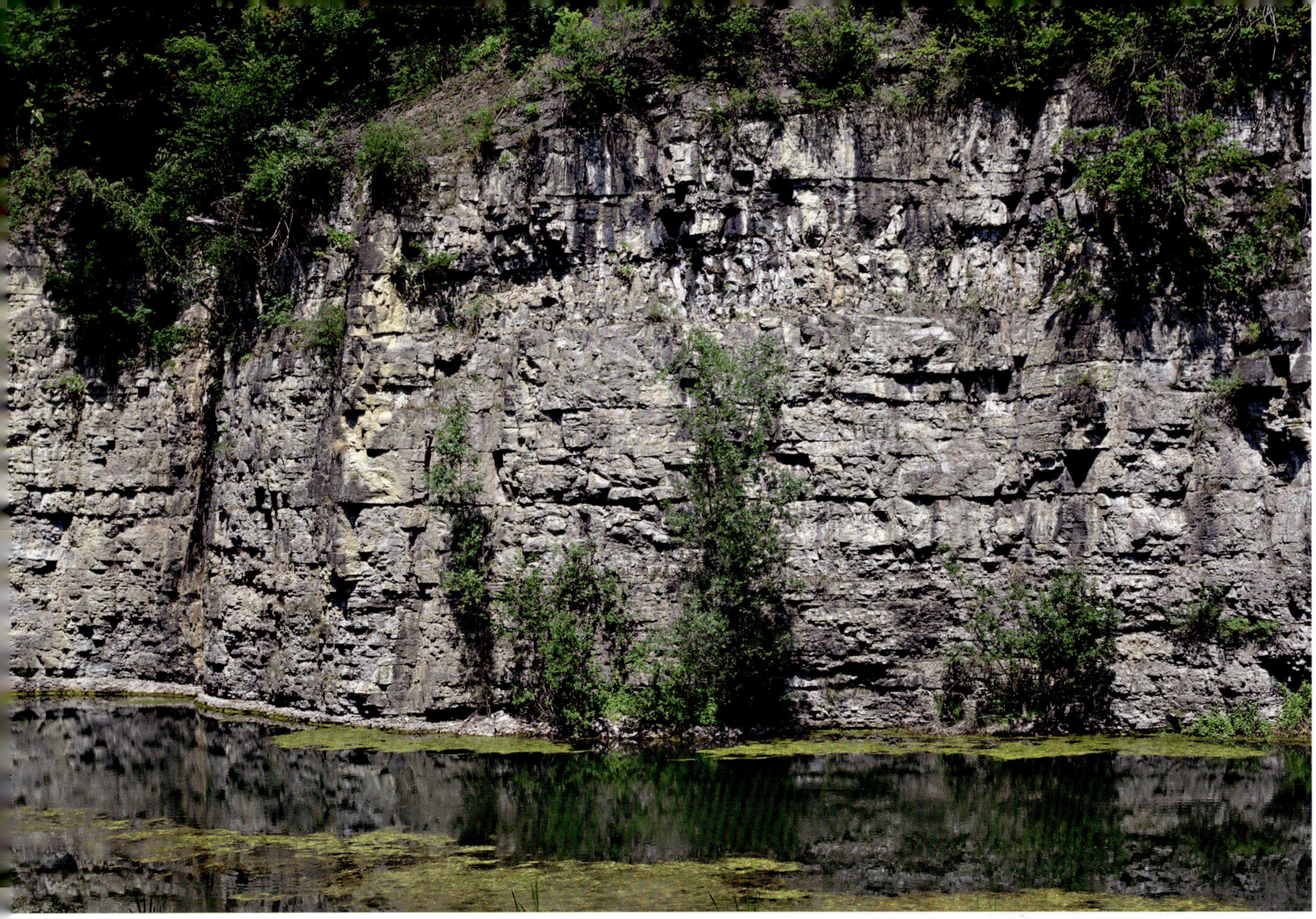

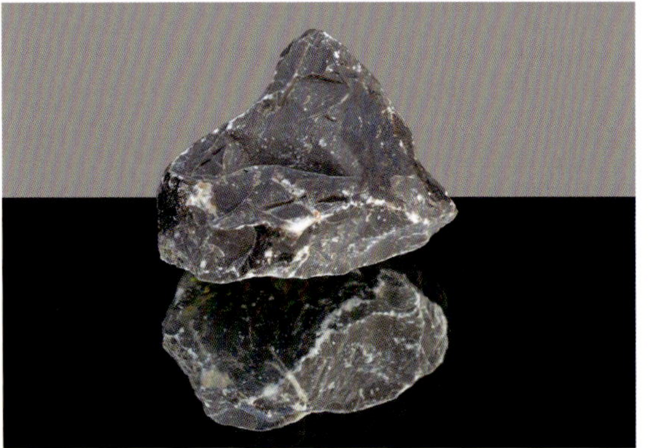

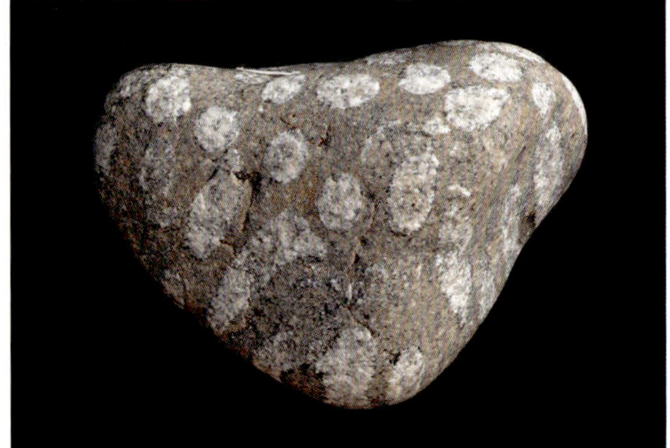

Figure 7.105 Bedded limestones of the uppermost Muschelkalk, with horizontal layering. The overlying layers of the green bank represent the boundary to the Lower Keuper (both Triassic). Sachsenheim, Germany.

Figure 7.106 A piece of shell limestone crushed as road gravel. Typical gray color, conchoidal, brittle fracture, and fine crystalline (micritic) structure are clearly recognizable. Sersheim, Germany. Image width 88 mm.

Figure 7.107 Coral limestone. The organisms, here fossilized corals, were sediment catchers in the ocean in prehistoric times. Lime mud was deposited between the branches of the coral and was then sedimented with them. Bedrock from the Welsh coast near Trevor, Great Britain. Image width 57 mm.

In addition to their enormous economic importance for the building-material industry, limestones are the main reservoir for the greenhouse gas carbon dioxide. According to estimates, the earth's limestones store a total of forty thousand times as much carbon dioxide as all the world's oceans, the atmosphere, and the biosphere combined.[166] They therefore play a crucial role in the global carbonate cycle. Living organisms that form calcareous shells or skeletons, from which rocks are formed after they die, have a high impact on the climate; they form a long-term carbon storage program. Annually, these organisms fix an estimated 0.1 gigatons of carbon worldwide.[167] During the weathering of limestones after mountain-building processes, carbon dioxide is fixed from the atmosphere and fed in the form of calcium carbonate by river transport to the oceans, where the organisms mentioned above in turn build their shells from these chemical weathering products of the mountains, which are subsequently sedimented. The circle is complete.

There are multiple approaches to the classification of limestones. We will simply designate two main types from Füchtbauer (1996).

1. Particle limestones. In these limestones there are large, biogenic particles formed by mussels, gastropods, echinoderms, cephalopods, and so on. The cavities between these clastic components can now be filled, on the one hand, by even-finer sediment, the matrix, or, on the other hand, by diagenetic precipitates, the so-called cement. Depending on the relationship between each other (i.e., between large particles and matrix or cement), different types of limestone are distinguished (mudstone, wackestone, grainstone, and boundstone). From the type and structure of the cement, the expert can even infer the associated formation conditions.

2. Structureless, fine- or coarse-crystalline limes without visible particles, also called calcilutite. Ultrafine crystalline limestones with particle size below 4 μm are called micrite. These limestones formed from lime muds and precipitates in marine basins above the calcite compensation depth. These limestones are very dense and virtually pore-free. Conversion of organogenic aragonite components to calcite is thought to have played an important role in the process of compaction.

Micritization is a special form of cementation. As we had already seen, there are fungi and algae that drill tiny holes in lime shells and lime particles in general. Usually this occurs to depths of 10–50 μm. After the organisms have died, these boreholes fill with cryptocrystalline Mg-calcite, the micrite. Over time, regular coatings are formed on the grains in this way. If the grains dissolve, only these micrite coatings remain, which collapse when the sediment is compacted. Here, too, one can infer the underlying formation conditions from the corresponding structures.

As we come to end of this chapter, let's take another look back. In tropical regions (e.g., in the beaches of many vacation regions we enjoy so much), sand consists of completely different material than what we are used to in central Europe. Quartz takes a back seat, since tropical islands have lime that dominates the scene and do not have granite, with a few exceptions. Shell remains, foraminifera shells, sea urchin spines, snail shells, sponge needles, and parts of corals form the basic material for tropical coastal sands, whose age thus clearly lags behind that of continental areas and whose origin is always regionally limited. Under the magnifying glass or even the microscope, an almost inexhaustible splendor and variety of forms and colors can be seen. The highly ordered structure of crystal lattices of mineral sand grains cannot be observed directly even with the light microscope. In contrast, the patterns of biogenic grains can be admired simply with good magnifying glasses. With a handful of sand from a tropical coast, one can occupy oneself for days. Beyond all scientific observation, the richness of shapes, colors, and structures is overwhelming. One finds spirals, circles, and patterns of all kinds, and even within tiny fragments of lime, one can make out road systems of feeding traces of still-tiny organisms whose habitat is limited to a single grain of sand.

Through the process of biomineralization, organisms have succeeded at what researchers in the field of nanotechnology have just begun to undertake: the construction of defined, complex, and reproducible structures out of simple building blocks. For example, the already mentioned magnetosomes in magnetotactic bacteria, with dimensions of only 0.045 μm, are completely free of lattice defects in the crystal.

Two small animals have the last word here as representatives of an insect order (Trichoptera) that I already marveled at as a child in mountain streams: caddisfly larvae. Their access to sand is totally practical. They build their living tube from selected sand grains of their stream (and from other components) with the help of a spiderlike secretion. Let's see what the two animals I asked to the photo session in the Alps selected for sand grains. Of course, a short time later—and probably a bit surprised—they were sitting again with their splendid shells in their familiar surroundings.

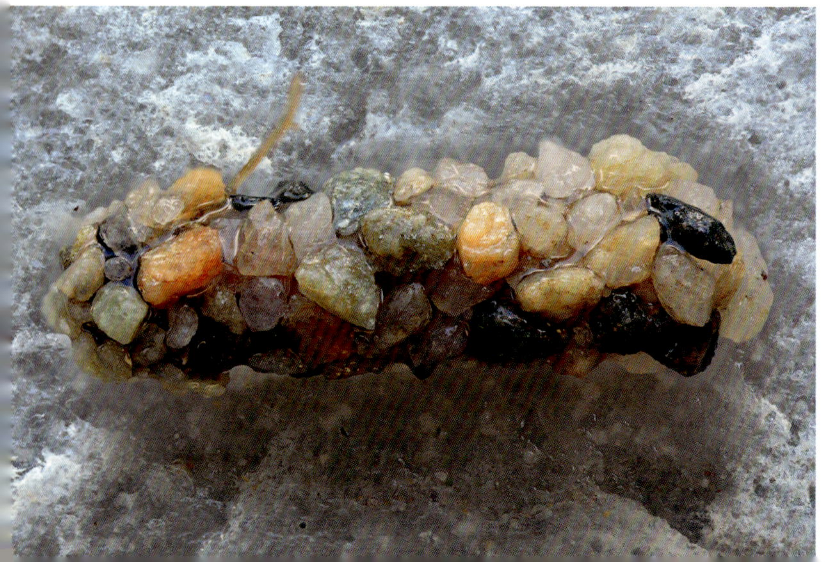

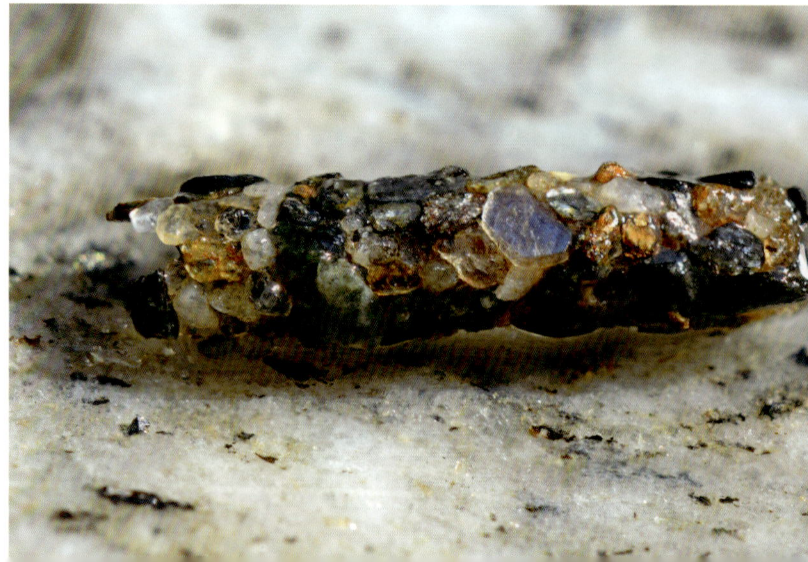

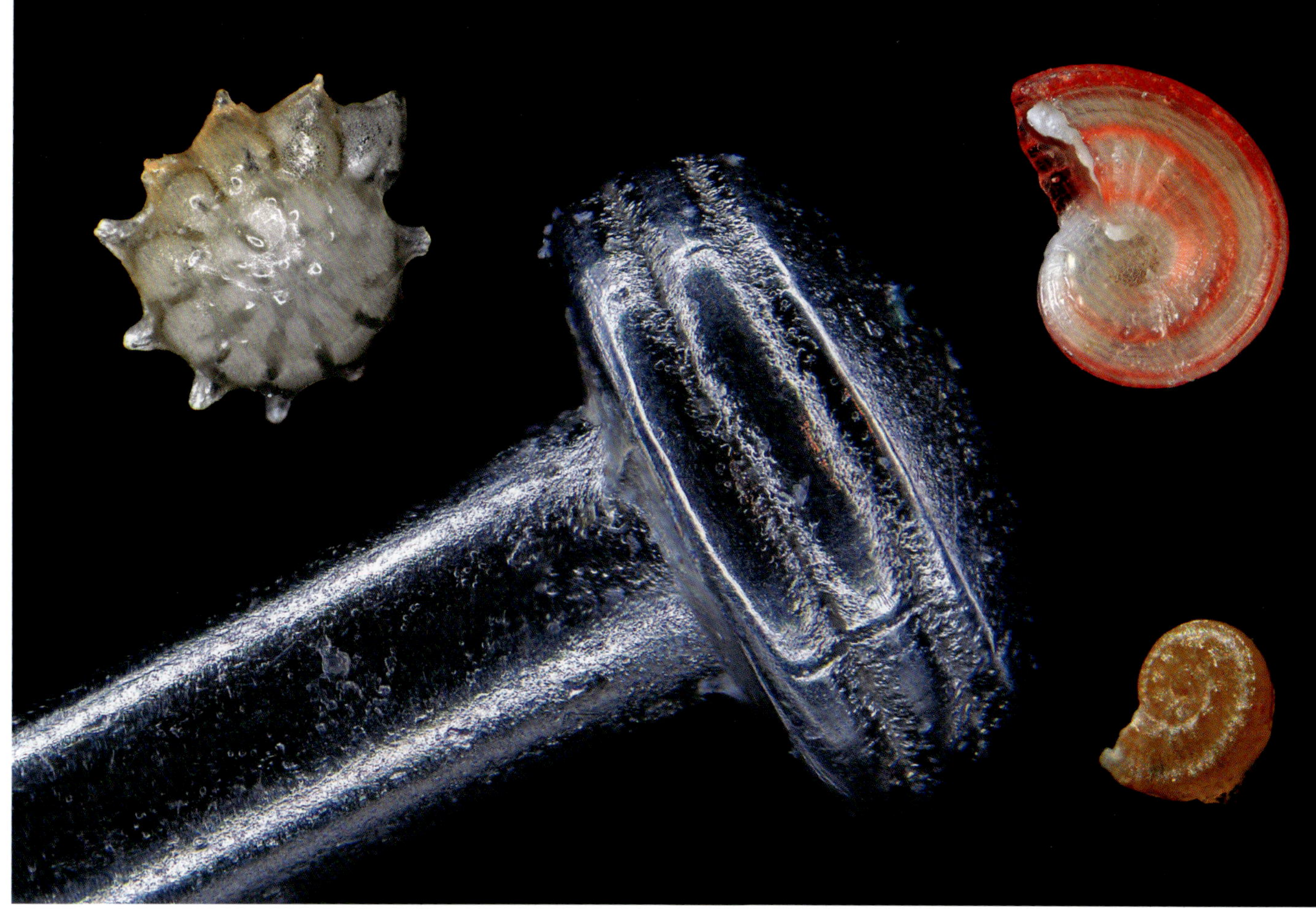

Before we now move on to the next chapter and leave the level of analysis of the individual components of the sand, perhaps a note for better context. In several places I had openly expressed my enthusiasm for the complexity and beauty of the smallest structures, which reveal themselves during sand observation especially under higher magnification. And the fact that we are dealing here with (formerly) living organisms, which influence their environment and vice versa are influenced by it—in short, interact with the world. In this context, "small" is, of course, the measurement of a property that can be understood only relatively: small in relation to us, the observers. This is self-evident, but I would like to illustrate to the reader in a last picture quite directly what might be lost in a photograph without a frame of reference, what the real dimensions are that we are talking about, although every illustration is always accompanied by a scale reference. In a direct, illustrative comparison, three of the photographs in the chapter are shown here in the same scale with an ordinary household pin. I think that the picture speaks for itself.

Figure 7.108 *Left side to the left*: Caddisfly larvae from a channel (water pipe) above Chandolin on the Illhorn, Switzerland. Length 17 mm. The animal is lying on a piece of Rhone debris out of light, metamorphic lime (marble).

Figure 7.109 *Left side to the right*: Caddisfly larvae from the Binna near Giessen in the Binn Valley, Switzerland. Pay attention to the careful selection of the sand grains used. The pretty, hexagonal mica crystal (muscovite) catches the eye as a jewelry piece for the animal. Length around 15 mm.

Figure 7.110 Depiction of the size relation between biogenic sand grains compared to a household pin (this one also reveals very interesting structures in the enlargement). *To the left*, a foraminifer; *to the right*, two snail shells).

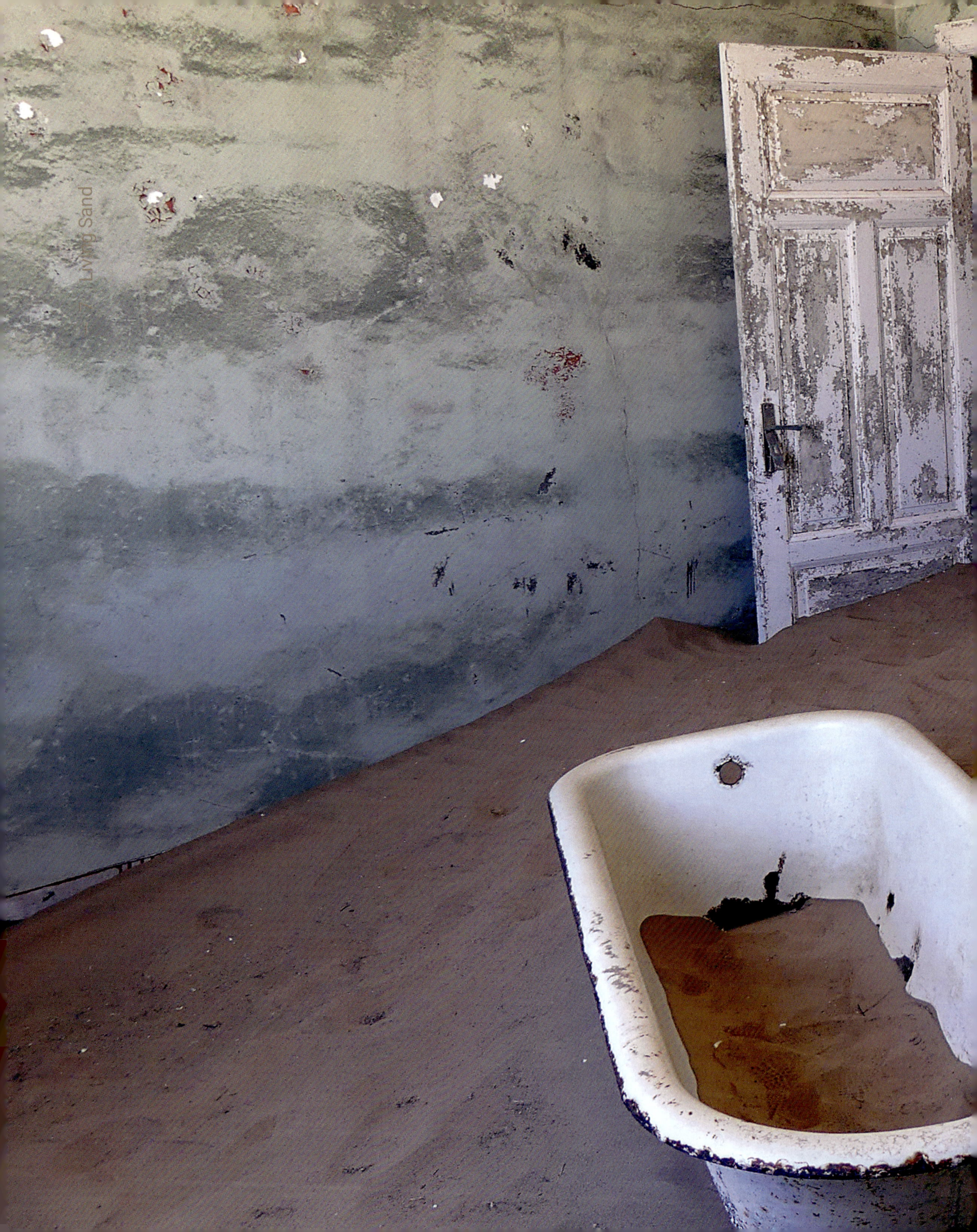

— 8 Thoughts on Sand, Time, and Numbers—an Interlude

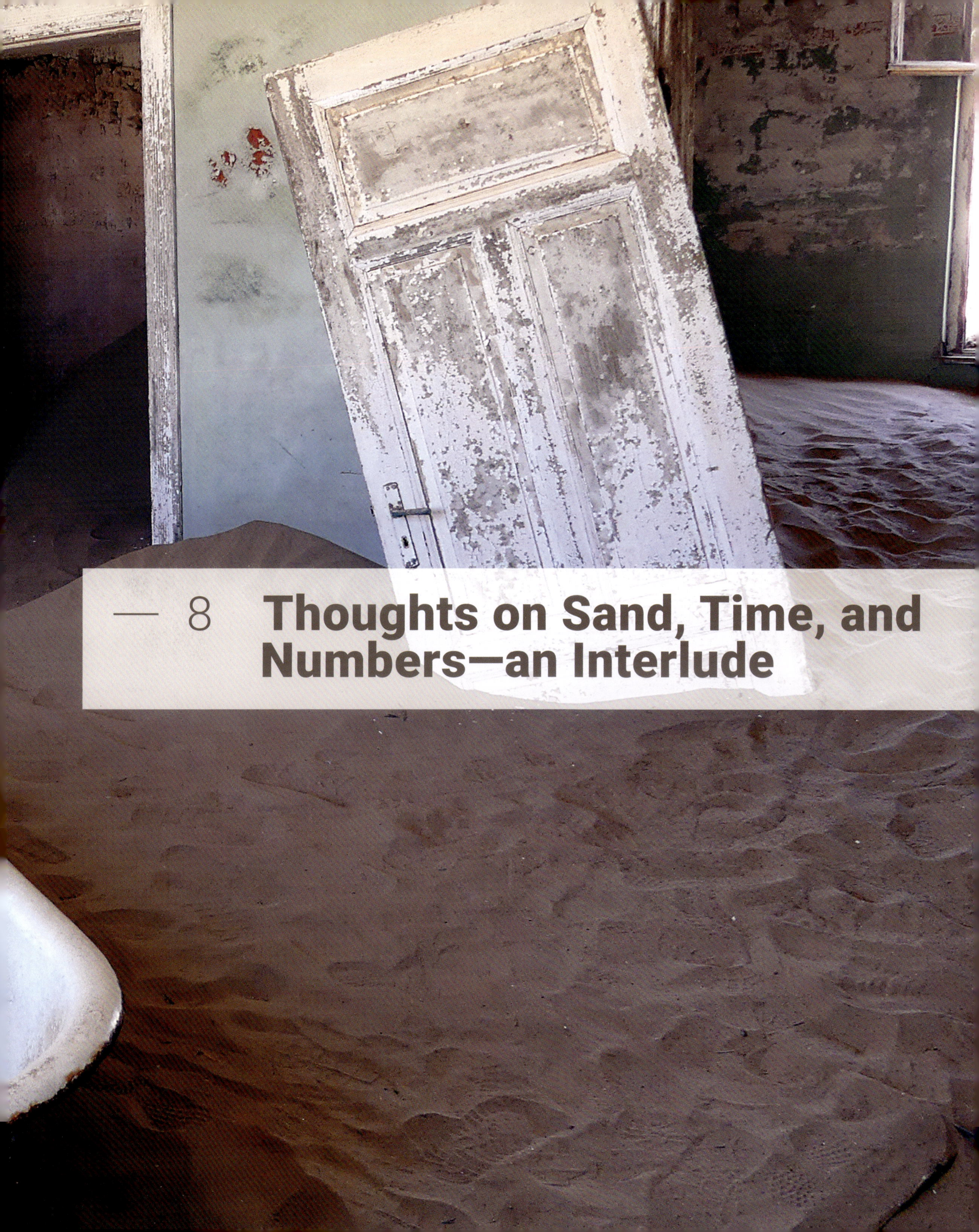

If thou wilt step into the infinite,
Just explore every path in the finite.

—J. W. v. Goethe, ***Collection of Sayings*** *(1815)*

An interlude, an epilogue (almost) to wrap it up? But have we not seen—at least for sand—that even the end of a cycle is only the beginning of a new one, so also an interlude? This chapter has to do with large numbers and long periods of time, a pretty extensive theme, but does it have something to do with sand? Quite a bit. Sand encompasses these two terms almost perfectly. It builds the bridge between the microcosm and the macrocosm in many ways. We will dive a little deeper.

"For all eternity! Not for a year or an age but forever. Try to imagine the awful meaning of this. You have often seen the sand on the seashore. How fine are its tiny grains! And how many of those tiny grains go to make up the small handful which a child grasps in its play. Now imagine a mountain of that sand, a million miles high, reaching from the earth to the farthest heavens, and a million miles broad, extending to remotest space, and a million miles in thickness, and imagine such an enormous mass of countless particles of sand multiplied as often as there are leaves in the forest, drops of water in the mighty ocean, feathers on birds, scales on fish, hairs on animals, atoms in the vast expanse of air. And imagine that at the end of every million years a little bird came to that mountain and carried away in its beak a tiny grain of that sand. How many millions upon millions of centuries would pass before that bird had carried away even a square foot of that mountain, how many eons upon eons of ages before it had carried away all. Yet at the end of that immense stretch time not even one instant of eternity could be said to have ended. At the end of all those billions and trillions of years eternity would have scarcely begun."[168]

—James Joyce

Previous double page:
Kolmanskop, Namibia

We humans use the term "infinite" without much thought in everyday life. One is "infinitely happy" or "infinitely sad," and poets especially as guardians of words and concepts obviously know and celebrate the "infinite pain" or the "infinite joys." As is so often the case, meaning is sacrificed for effect, and emphasis finally becomes a habit: "infinity" degenerates into "very much."

But let's take a closer look. From the point of view of physics, infinity is a rather problematic term. We don't even want to talk about mathematics, which even knows different gradations of infinity. It is certain that to each arbitrarily large number another number can be added, so that a still-larger one results. There is no reasonable reason to think of an end to such a sequence: there is obviously no final number. Consequently, infinitely many numbers are to be assumed. As plausible as this may sound to us, the Greek mathematician Archimedes of Syracuse pursued this question in regard to sand grains around 220 BCE in his treatise *The Sand Reckoner*:

> "Again there are some who, without regarding it as infinite, yet think that no number has been named which is great enough to exceed its magnitude."[169]

Basically, the number of numbers was apparently considered to be limited, and it was therefore important for Archimedes to argue against this. Of course, the connection between sand and numbers is particularly interesting here, which we will discuss later.

Does this also work with the concept of time? Is it possible to add another time interval to an arbitrarily imagined point of time or to determine the moment before? The answer is probably no. According to all of what is known today, the world accessible by us by observation—and only about this we can make statements, of course—has not existed forever but had a kind of starting point. Mathematically and physically, this starting point and everything that happened afterward can be described quite precisely, without, however, a vivid event that we could imagine perceiving. As much as the term "Big Bang" has penetrated the linguistic and thinking area of everyday life, so little is principally imaginable in this event. To say nothing of a moment before it. The statement that time begins with the Big Bang is plausibly derivable in the context of physical theory formation, but for the intellect striving to have a look, not much is helpful, and the question about a time before the Big Bang is as futile after today's state of the knowledge as finding the beginning of a circle. The conclusion that a lack of view is to be equated with a lack of truth is, however, inadmissible and just as unhelpful. Goethe still demanded to find a visual perception within the sciences:

> "The desire of knowledge is first stimulated in us when remarkable phenomena attract our attention."[170]

Almost two hundred years later, it is becoming increasingly difficult to fulfill this requirement, due to exponentially advancing specialization. Particularly in the natural sciences, research has advanced into areas that not only elude but actually oppose visualization.

Holism

The best proven of all physical theories we know, quantum theory, is profoundly counterintuitive, since it has, for example, disproved the basic human experience that things are at certain places at certain times for very small scales (the so-called nonlocality), and replaces all strict chains of cause and effect through statistical observation and probability of presence. The experimenter as the observer plays a principal role for the first time. In the Copenhagen interpretation of quantum mechanics,171 which today is still treated as a hypothesis because its conclusions are not shared by all physicists, measurement as an act of observation leads to the "collapse of the wave function" and makes a possible event a factual one. Again, it was Goethe who formulated a premonition of this in his *Maxims and Reflections*:

> "The manifestation of a phenomenon is not detached from the observer—it is caught up and entangled in his individuality."[172]

We read this idea, which was very central for Goethe, even more clearly in the "Historical Part" of his *Theory of Colours*. Everything is connected—phenomena cannot be completely isolated from their context without also changing the phenomena themselves in relation to their existence embedded in reality. This attitude let Goethe reject the experiments of Newton, who, of course, pursued a completely different path and, just as we are used to it today, prepared things to be proven strictly from the potentially influencing environment and made it accessible to the observation or measurement. Objectivity versus subjectivity. Goethe's approach, which we do not want to overinterpret at this point, approaches—of course not foreseeable for him at that time—quite rather the newer understanding of physics in the area of quantum phenomena: a phenomenon that is known as "entanglement" of states. It describes a nonlocal connection of several particles, which can be described together mathematically and whose properties correlate without delay with the properties of the other particles in the considered system, which remains completely inexplicable through classical theories, of course including Newton's theory.

"Everything in nature is intimately linked and connected, and even what is separated in nature, man likes to bring together and hold together. Therefore, certain individual natural phenomena are difficult to separate from the rest and are not easily solved didactically through intent."[173]

And elsewhere even more clearly, we quote this passage in full since it very compactly summarizes what we might actually take to be a call for a holistic view of science:

> "All the actions, of whatever kind, which we notice through experience, are connected in the most constant way; they merge into each other; they undulate from the first to the last. That they are separated from each other, that they are opposed to each other, that they are mixed with each other is unavoidable; but therefore, in the sciences a boundless conflict had to arise. Rigid, divorced pedantry and deluded mysticism both bring the same disaster. But those activities, from the basest to the highest, from the brick that falls from the roof to the luminous spiritual vision that comes to you and that you share, they move against each other. We try to express it: random, mechanical, physical, chemical, organic, psychic, ethical, religious, genius."[174]

But how would it be if we took this idea even further in a much more consequentual way? A hint in this direction is given by C. F. v. Weizsäcker:

> "The world is not composed of objects, only the finite mind of man dissects the whole to which he himself belongs into objects in order to find his way. To understand this is the beginning of the ascent."[175]

However that may be meant, it holds a thought that has the power to fundamentally change our empirical approach to the world. It is as if a curtain were briefly lifted and allowed an equally brief glimpse into what reality might be.

Usually this thought cannot be carried very far, but let's try it anyway. In short, the completely unquestioned basis of thinking par excellence—namely, that what we think about are objects that are sensually directly perceptible and interpersonally consistent to have an existence—is doubted here or at least brought up for discussion. One may argue about quality, quantity, and other categories in individual cases, but we always assume that there are objects that exist outside our own subject. Yes, it is we ourselves, after all, who conceive ourselves as an "I," an "I" that is precisely not a "you" or at least different from something that is understood by this "I" from the outside.

Here we see very clearly the basic principle of the cell, which is first constituted and separated from the outside world by the presence of its cell wall. In humans and animals, the skin assumes a comparable function. System boundaries can be of a representational or nonrepresentational nature.

But what if all this were only fiction, a setting? The cell wall would still exist, as would the skin in the material sense—but the context would change. There would be the physical border, but it is not explicitly thought of as such. The cell merges into the cell wall, which merges into the next cell, and into the fluids surrounding it, and so on. There would be no more separation of objects; dividing lines would be contingent, unessential, and arbitrary—purely linguistic settings.

What would happen then? It must be emphasized quite explicitly that the described approach is not to be confused with the one that doubts from the (always circular) constructivist point of view if there is a real world around us at all.

So, what would change? Physically, nothing would change at all; grain of sand would lie against grain of sand, as before. Individuals walk across a square covered with paving stones, shade, and light. The effects in our approach to the world, on the other hand, can hardly be overestimated. We would see ourselves in what Plato calls "the One." There is no twoness, since we, as cognizing subjects, also belong to this One. This One is an infinite quantity of spirit and matter, unseparated, permanently rearranging itself, without limits, without elements of the individual. Rather to be compared to a fluid or to the light. The thought is not to be grasped in an analytical way, since the language itself already presupposes analysis and delimitation—it could otherwise not exist. If I say, "I see an apple," already the linguistic structure presupposes that there is both me and the apple, which exists separately from me, as well as the mediating process of seeing. There is no linguistic construct that would correspond to an approach to the world that is holistic and that implies that objects are arbitrary settings.

Consequently, besides the demarcation of analysis and synthesis, in this case also the conceptual pair of cause and effect would have to be interpreted differently.

Figure 8.1 Cells are constituted by their cellular walls. There is an outside and an inside. Cells of the onion scale epidermis with cellular walls and nucleus. Differential inference contrast, transmitted light. Image width around 160 *μm*.

Figure 8.2 Discrete objects in space as well as a presumed history, a causal and temporal sequence

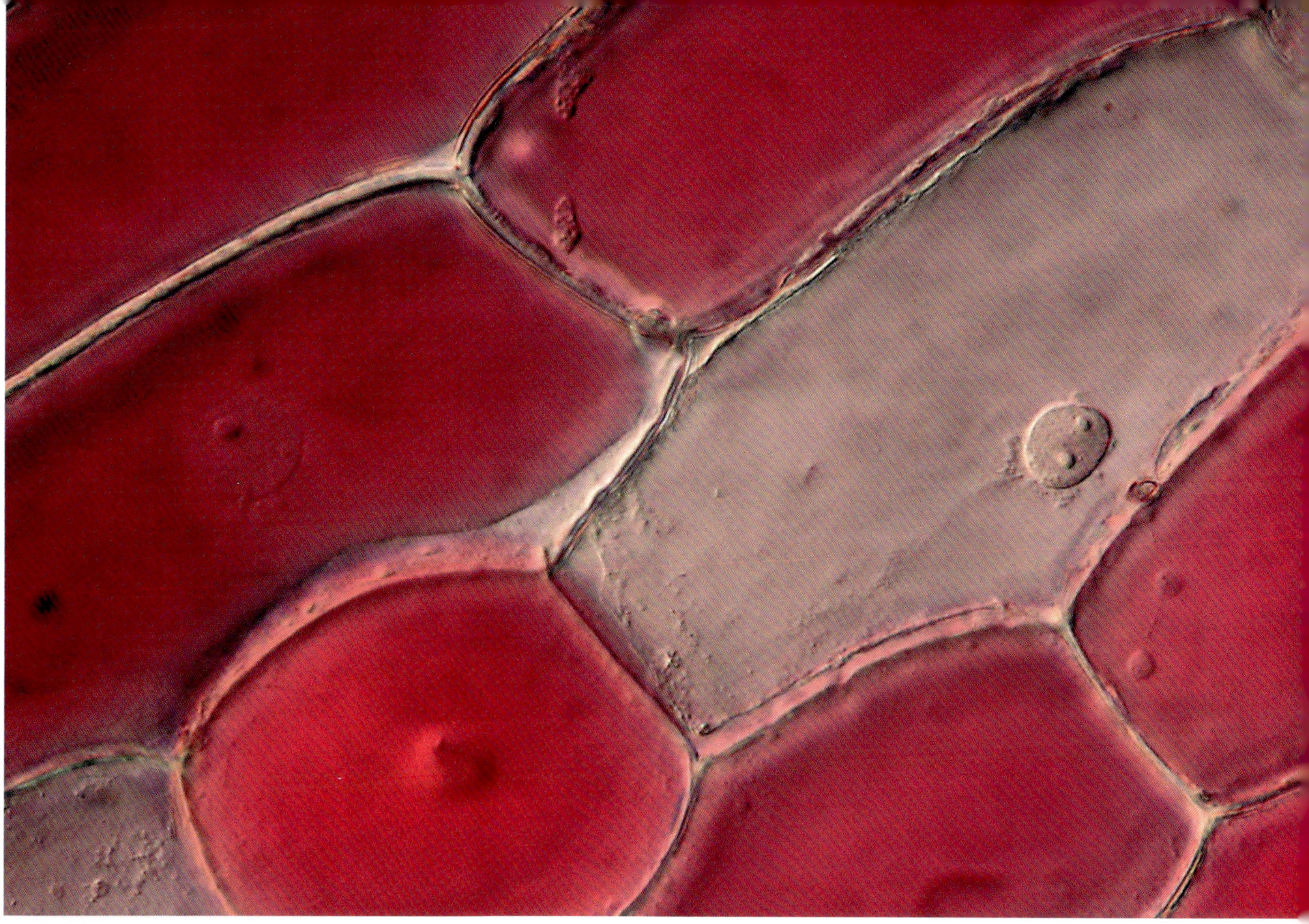

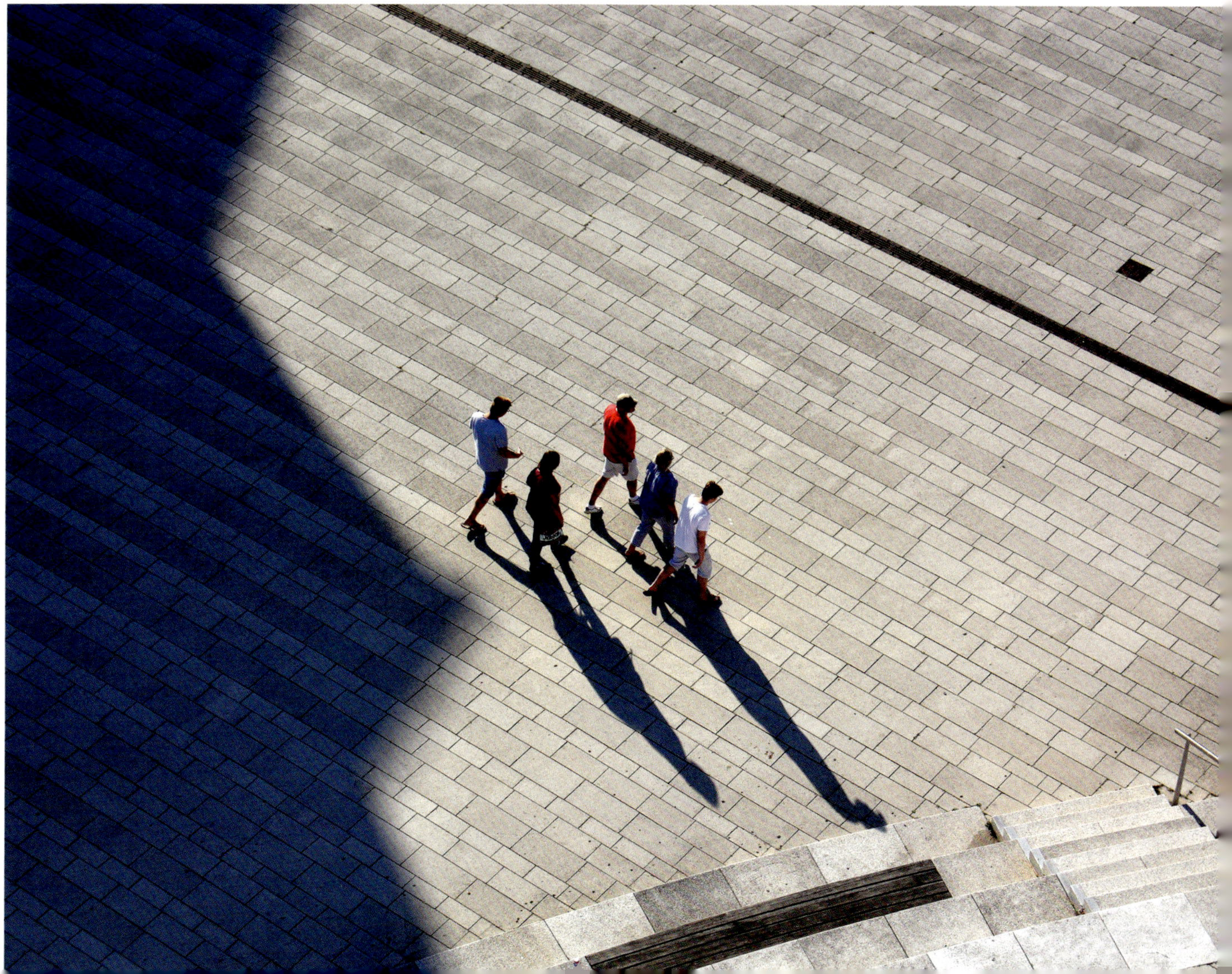

The cause cannot precede its effect if both are connected as one and the definition-measured system borders are not considered in each case. This can be explained briefly by the example of light separation through a prism. If one sends white light through a prism, it is divided into its spectral colors. Goethe has denied this very simple and everywhere easily verifiable fact of Newtonian analytics all his life. It remains to assume that he by no means denied the spectrum as such, which is visible for everybody, but the simplistic interpretation of the experiment.

> "The thinker makes a great mistake when he asks after cause and effect. They both together make up the indivisible phenomenon."[176]

Perhaps we are reaching too far, but Goethe seems to suggest a holistic view of the experiment here. The whole phenomenon is described, also without using causality, provided that the world is no longer thought of as objects. Everything is inseparably entangled and one. Thus, consequently, the categorial separation of cause and effect is also dropped, also without negating the entropy-conditioned arrow of time. Light falls on a prism; behind the prism the light shows spectral colors. Even simpler: light, prism, colors—a linguistic description without object categories will not succeed. But the attempt would be equal to a dumb observation of the phenomenon "in itself," which does not distinguish but understands everything as a synopsis. It is very difficult to mentally—even briefly—understand this. The reader may try to do so. Should it succeed, the danger of a radically changed worldview threatens. Let us let Weizsäcker have his say again at the end of our thought experiment:

> "Indeed, the world of the figures is immeasurable, but it is connected everywhere. The connecting of what was previously separated only traces the lines of the real connection. To separate is an operation necessary for the human spirit, but all mere separation is artificial. The discrete, countable is only thought; continuity is a characteristic of reality."[177]

In the submicroscopic area, the area of validity of quantum theory, it has been known for a long time that the concept of object loses its meaning or does not stand up to the assignment of meaning in everyday life. The impact of the results of quantum theory on our thinking and on modern philosophy is still underestimated. The process of using quantum-mechanical findings has just begun.

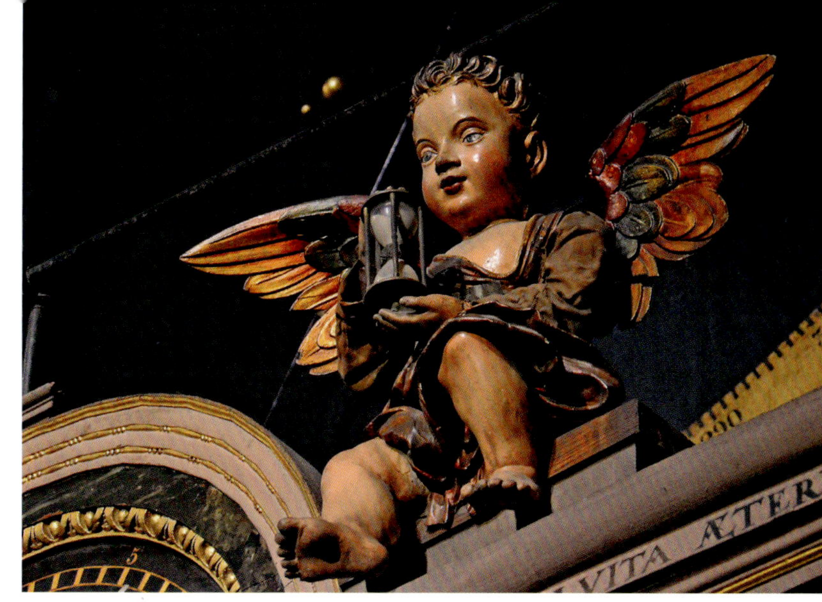

Figure 8.3 Angel with hourglass. A relic of the great astronomical clock in Strasbourg Cathedral, completed around 1574 by Konrad Dasypodius and the Habrecht brothers, the angel turns the hourglass each full hour.

Time, Movement, and Their Denial

For all our considerations, time plays a decisive role. Even though we can give clear information about what is understood under the terms "space" and "things in space" due to the sufficient sensory organs that nature has bestowed us with, the situation is different in regard to time. "But what is our organ for sensing time?" asks Hans Castorp in Thomas Mann's *The Magic Mountain*. Apart from the brain, we have none, which at the same time also processes spatial information. No state of mind is so close to us and yet so incomprehensible in its essence as the perception of time.

In all attempts to define time or at least to describe it, the concept of movement is present. Making time manageable in the form of its measurement relates to the description or representation of movements. If we remain in absolute silence and stillness, it is very difficult, almost impossible, to feel time in an *appropriate* form, precisely because we lack a direct sense organ for this. For this purpose, human culture has invented time-measuring devices, clocks. A clock is nothing else than a device that executes a uniform—this term is quite circular—and a periodical movement that makes time visible. For this purpose, very different operating principles and operating media are available. The oldest clocks, known since antiquity, consisted of a vertical rod, the gnomon, which implements the basic principle of a sundial. In ancient Greece, *klepsydren*, water clocks, were in use, which had to be periodically refilled and did not read very accurately. The first "glass clocks," hourglasses, finally appeared in the fourteenth century. Today, we are largely influenced by the circular dials of the wheel clocks, which were also invented in the fourteenth century and, in Europe, were used to provide the official time, accessible to all, via the church clocks. After gaining only temporary popularity, the representation of time using only numerals in digital watches is now mostly used only in scientific display instruments.

The hourglasses interest us particularly with their extensive connotations to sand. Is it not remarkable that just sand—and this still in its well-rounded form—was used.

Quartz sand, which probably reflects the millions of years out of itself like no other natural material, surviving from one geological cycle to the next, perfecting its roundness, as we explained in chapter 5. The grain as such thus represents an albeit indecipherable clock. Jorge Luis Borges put it in a concise form in a verse of his poem "Happiness":

> "Whoever looks at an hourglass sees the dissolution of an empire."

Decaying mountains, eroded in quartz grains of increasing sphericity, display the passing of time in glass casings. We don't want to be too subtle and must also mention the quartz of the rock within the glass. In modern hourglasses, of course, artificially produced and perfectly rounded grains are used, which allow for a smoother flow and higher precision. The hourglass contains this uniformity in a special way as an operating principle. Time, which we perceive as flowing, finds direct expression in the hourglass. The periodicity, however, can be produced only by additional external intervention: the turning of the glass. Columbus noted in his ship's log on September 13, 1492, that the hourglass had to be turned twenty times from sunset to sunrise.[178] He thus measured a ten-hour night; at the time, half-hour glasses were common on board to divide the watches. Hourglasses were popular on board for such measurement because they proved to be largely robust against the fluctuations and vibrations of ships at sea.

Without movement, time cannot be made visible. Here we refrain from the electronic time measurement, although even here a movement—namely, the oscillations of the quartz rod or the cesium atom—is used for the measurement and for the standardization of the time unit of a second.

The school around the philosopher Zeno of Elea denied exactly this possibility of movement and, with it, time altogether. For Parmenides there was only being; he excluded nonbeing. Since movement can be understood as a transition from being to nonbeing, the consideration of Parmenides finally led to a complete denial of the possibility of change and thus also of movement. Before the falling grain of sand reaches the ground, it must first have covered half the distance; before it measures this half distance, it must have first covered its half, a quarter of the way, and before that an eighth, a sixteenth, and so on.

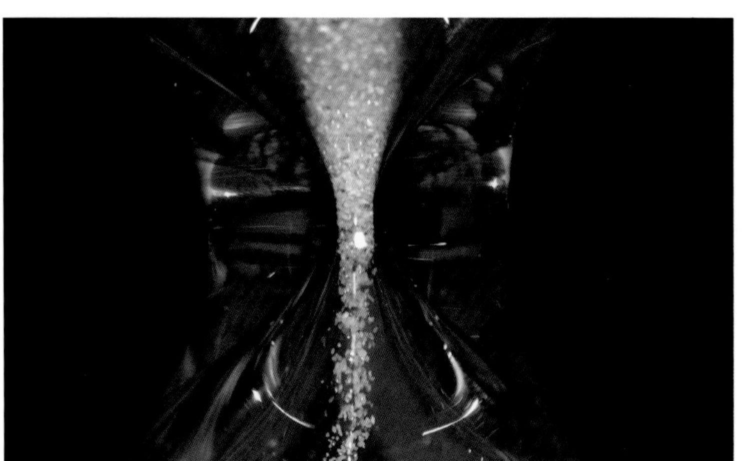

Figure 8.4 The mouthpiece of an hourglass. Sand follows gravitation as in nature and appears in the hourglass only subdued. The mouthpiece gives space to movement. Movement of the sand, which gives expression to the time.

The sand grain remains in the observed position for eternity, as described in the beginning, so the thought of Parmenides and the paradox made by his student Zeno (Achilles and the turtle). The mathematical ideas needed for the resolution of the paradoxes did not exist in ancient Greece but are school material today, but that does not mean that the concept of time has become much more transparent or accessible as a result.[179]

Time, Duration, and Points in Time

Independent of what is discussed above, time has another aspect, that of a relevant resource. The available, but not the experienced, time is homogeneous and independent of the perspective of the person: twenty-four hours in a day.

We name the immediately felt time of our existence with the term "now." Everyone lives exclusively in this "now," but without being constantly aware of it. Schopenhauer goes so far as to attribute reality only to the experienced now, the present; if the future and the past are relativized or even negated, time is thus denied the recognizability that we subjectively feel is a given.

> "Above all, we must clearly recognize that the form of the phenomenon of the will, and hence the form of life or of reality, is really only the ***present***, not the future or the past. Future and past are only in the concept, exist only in the connexion and continuity of knowledge in so far as this follows the principle of sufficient reason. No man has lived in the past, and none will ever live in the future; the ***present*** alone is the form of all life, but it is also life's sure possession which can never be torn from it. The present always exists together with its content; both stand firm without wavering, like the rainbow over the waterfall."[180]

The Roman emperor and philosopher Marcus Aurelius, an embodiment of the good statesman whom Plato longed for in his *Politeia*, becomes even more forceful in the self-reflections, a small work close to the stoa and of timeless insightfulness, which he wrote during his numerous stays in field camps:

> "Were you to live three thousand years, or even thirty thousand, remember that the sole life which a man can lose is that which he is living at the moment; and

> furthermore, that he can have no other life except the one he loses. This means that the longest life and the shortest amount to the same thing. For the passing minute is every man's equal possession, but what has once gone by is not ours. Our loss, therefore, is limited to that one fleeting instant, since no one can lose what is already past, nor yet what is still to come. . . . For the sole thing of which any man can be deprived is the present; since this is all he owns, and nobody can lose what is not his.[181]

It was also Marcus Aurelius who remarked elsewhere that one should be aware that in one hundred years, none of the people in the world now, our random selection of contemporaries, will be alive anymore. The thought remains the same—even if one increased the life expectancy to 150 years. The statement is not complicated, but of captivating clarity with regard to the relativization and classification of importance to which we assign events in our everyday lives, mostly without reflection.

Wittgenstein formulated it somewhat more concisely about one hundred years after Schopenhauer in his *Tractatus logico-philosophicus* by giving the concept of eternity a shift in meaning. He distinguishes an infinite duration of time from infinity itself, which is thereby without the thought of a temporal course or context and becomes a sort of timelessness.

> "If we take eternity to mean not infinite temporal duration but timelessness, then eternal life belongs to those who live in the present."[182]

If one thinks this through to the end, one comes to the question of whether individuals, be it animals or also, under certain conditions, humans, who lack the ability to perceive their existence in a temporal course, who are thus completely fixed in the now, are able to stay in the state of an eternity perceived as such. In connection with Helena, who embodies the ideal of happiness and life in the second part of Faust's tragedy, and only with her, Faust comes to the insight that only in unquestioningly letting oneself fall into the completely present can happiness be felt and even be perceived as timeless. This, of course, is in direct contrast to his wager with Mephistopheles, in which he renounces the pleasurable perception of the moment. Only in connection with Helena is Faust able to overcome this destructive negation—at least temporarily. This decisive sentence completion, of course, has to be spoken by Helena and not by Faust:

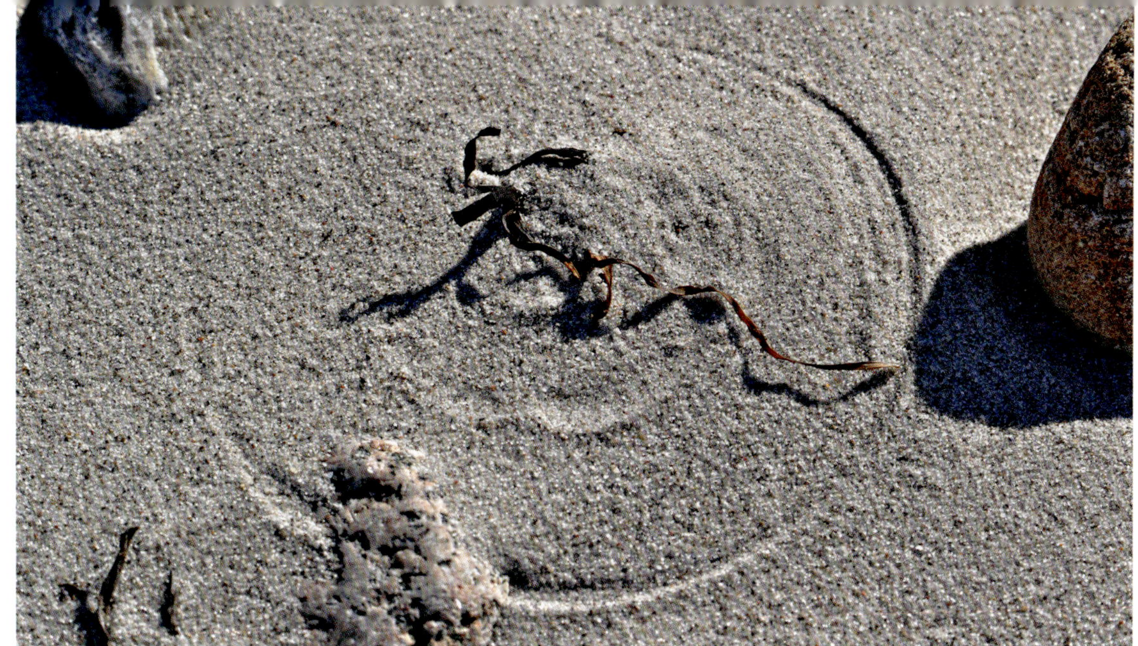

> FAUST:
> Not backward, forward is the spirit's sight,
> This moment now, alone,—
> HELEN: —is our delight.

In contrast, Schopenhauer's previously quoted consideration reminds us of Plato, who in his dialogue *Parmenides* is probably the first to name the bizarre point between the times as such and to incorporate the concept of movement—as mentioned before in connection with Zeno of Elea—into consideration.

> "This incomprehensible being, the moment, lies between the movement and the stillness as not being in any time, and into it and out of it the movement goes over to the stillness and the stillness to the movement."[183]

The present becomes a still-tangent point and thus subjectively time-free. The concept of change is meaningless with respect to a single point in time and thus comes quite close to Wittgenstein's "infinity" without denying the existence of Zeno's movement.

It should be mentioned in passing that the relativization up to the negation of time has always been one of the central elements in Zen Buddhism. Through meditation, the subject immerses himself in the purely time-free present, excluding polar, analyzing, judgmental, or wandering thoughts. The Chinese Zen master Huang-Po expressed it quite aptly in the ninth century:

Figure 8.5 Sand, time, space, movement, and their traces. The tip of the dried beach grass draws a circle in the fine sand. Baltic Sea, Hohwacht, Germany.

> "As soon as thought or sensation arises, you fall into dualism. Beginning-less time and the present moment are the same. There is no this and no that. To understand this truth is called complete and unexcelled

Along with the absence of time sequences, asceticism plays a special role in any kind of evaluation of what is sensually experienced. The Chinese Zen master Han-shan formulated the act of unintentional awareness in the seventeenth century with the beautiful image of a pumpkin floating by on the river:

> "The ever-flowing thoughts are without solid ground and unreal. . . . Let them happen and look at them like a pumpkin floating down a river."

Time and Subject

Beyond the pure now feeling, the past is present to us in the form of found facts. The future is present to us in the form of possibility. Time is thus the constant transition from possibility into factuality. Factuality becomes the presence of the past. Facts and possibilities meet in the now.

"Time passes—hark—time passes."[185] We subjectively feel the passing of time—a metaphor derived from the use of the hourglass. But, Heidegger asked, "Why do we say: time passes and not just as emphatically: it comes into existence? With regard to the pure sequence of now, both can be said with the same right."[186] Nothing speaks against speaking of a constant new emergence of time instead of a passing of time, which turns the eschatological "passing" into the much more creative "come into being out of nothing." Also, the sand in the hourglass remains there and does not pass away. As we have seen, this view corresponds much more to reality: The past does not pass, and neither does time. Every single moment as a space-time coordinate remains factual forever and can never pass away or change; it only remains inaccessible to the immediate experience, also forever. An example: if you look at the North Cape on September 9, 1994, at about 10:00 CEST, you will have done this for all time; yes, you still do this, because the event corresponds to an unchangeable location in the space-time continuum. This fact is very comforting. The representation of the moment persisting for all times is physically superior to the analogy of a flowing time. A certain moment of time can never flow; otherwise it would be another one.[187]

The perception of time is cultural and shaped by convention. We speak of "looking into the future," thinking that the future lies "ahead" of us. The people of early Mesopotamia, on the other hand, perceived time completely differently. They meant to look into the past "in front" and thus to turn their back to the future. This seems strange but is very consistent, because the view is turned to the known and not to what is hidden in the future, as it is so natural to us today.[188] However, no one has ever observed—and this is independent of all convention—that the sand in an hourglass rises from the lower storage container to the top, and nothing in the world can bring a broken glass back to the edge of a table. Time obviously has a fixed direction and probably also a beginning.

Neither of these points is trivial. There is only one elementary equation in physics that implies an arrow of time (i.e., that gives time a direction). It says that heat always flows only from the warm body to the cold body, and never vice versa. With this, and only with this, a difference between past and future is clearly given. Modern physics has made further discoveries about time that are very difficult to imagine but are well confirmed. Accordingly, time passes more slowly in the proximity of masses; for example, the earth. A clock that stands on the cupboard goes faster than one that stands on the ground, admittedly only by very little, but measurable for atomic clocks. Likewise, the faster one moves relative to a frame of reference, the more slowly time passes. If one could fly at light speed, time would stand still. Both are insights of Einstein, who could furthermore show that the concept of simultaneity, which is so familiar to us, can be applied meaningfully only in our immediate environment but is simply meaningless outside it. The physical reality of the world can be described sufficiently only by abstract fields. The space-time continuum, as we have come to know it, is, for example, equivalent to the gravitational field and would exist even without masses. Only changes of these fields describe the world, not, for example, things around us, which are basically nothing more than long-lasting events.[189] Before our head starts to spin, and we also learn that all quantities known to us, such as space and time, are quantized in the tiny scale, and thus granular, we now quickly enlarge the scale and take a look at the universe.

On the Universe

It is presumed that the world and, with it, everything that exists around us has originated in a singular event. But will the world become infinitely old—will there be an infinity of temporal nature? Is it possible to add another moment to every moment, as is possible with numbers? Probably not. The distant fate of the universe with its matter, to which we also belong, is not yet clarified—too little is yet known about the strange "dark energy." The most probable scenario other than the Big Rip approach, a kind of cosmic apocalypse, is the bleakest at the same time. The universe will continue to expand and expand, and the matter and energy in it will become more and more rare. Stars and galaxies will burn out, and collections of matter will dissolve and decay. Elements cease to exist and change bit by bit into the stable element iron, until this also decays. It becomes dark, and the coldness of space prevails shortly before absolute thermal zero. Everything comes to a standstill: even the atoms will decay, and finally, after huge but finite periods of time, the protons and their components will decay as well.

There is neither light nor motion. It becomes senseless to speak of time, because of course nobody can perceive it anymore, but also because no events take place anymore to be seen in a temporal order.

And even this description is circular, because the word "slow" is defined only within the context of time. The state of pure being in the Parmenidean sense is reached. Time stands still.

But until that happens, the little bird from James Joyce's story will have to have finished its laborious work.

Amount of Sand

The concept of infinity has always been linked to that of sand. Sand as a substance is the numerically largest accumulation of natural individual parts from everyday experiences that is accessible to the imagination and verifiable. It is obvious that every single grain of sand consists of many atoms, but atoms are not comprehensible in the literal sense and contribute only little to the perception. Sand grains are comprehensible—one can see them and touch them, and they crunch between the teeth. You can count them, even if there are as many as you can fit in one hand. As always with large numbers, however, the human imagination is put to the test when it comes to sand. For example, how many grains of sand fit into a thimble?

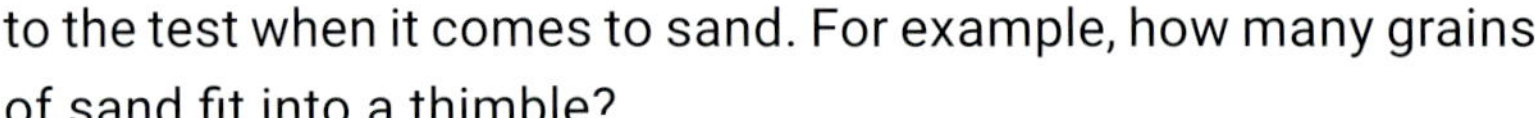

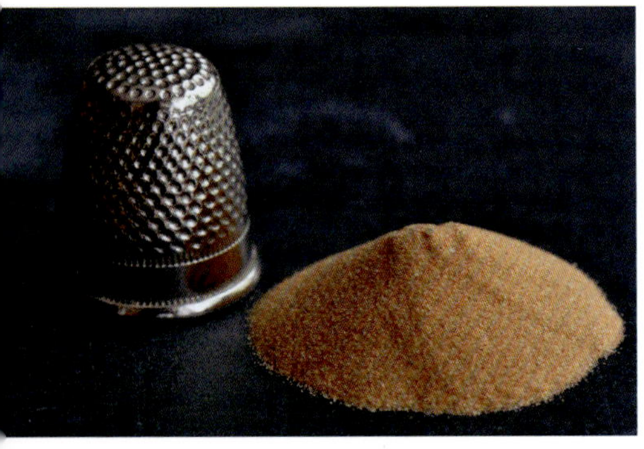

If you take the typical (fine) sand from the dunes of the Sahara, there are more than a million pieces. For a billion sand grains, you need about a liter jar. Nowhere else can the difference between these two large numbers, one million (10^6) and one billion (10^9), which is considered so small in everyday language, be illustrated more aptly. The idea of how many grains of sand fit into a whole dune already becomes frightening, and the question suggests itself: How many grains of sand might there be on the whole earth? This can indeed be estimated mathematically, although, as expected, different researchers have come to different conclusions, which nevertheless largely agree on the order of magnitude. For the calculation, the first volume of the total sediments on Earth is estimated. Here, the thickness of the sediment layers on the continents and the continental shelves as well as the sediment volumes in the suboceanic areas is determined. The estimated values are between 3 and 13×10^8 km^3; that is, on average, about eight hundred million cubes with an edge length of 1 kilometer. From observations and empirical values, it can be estimated that between a quarter and a third of this sediment volume consists of sand.[190] Once the volume is known, the number of sand grains can be calculated by an assumed mean grain size and mean specific gravity.

Figure 8.6 About a million sand grains (fine sand) fit into a thimble.

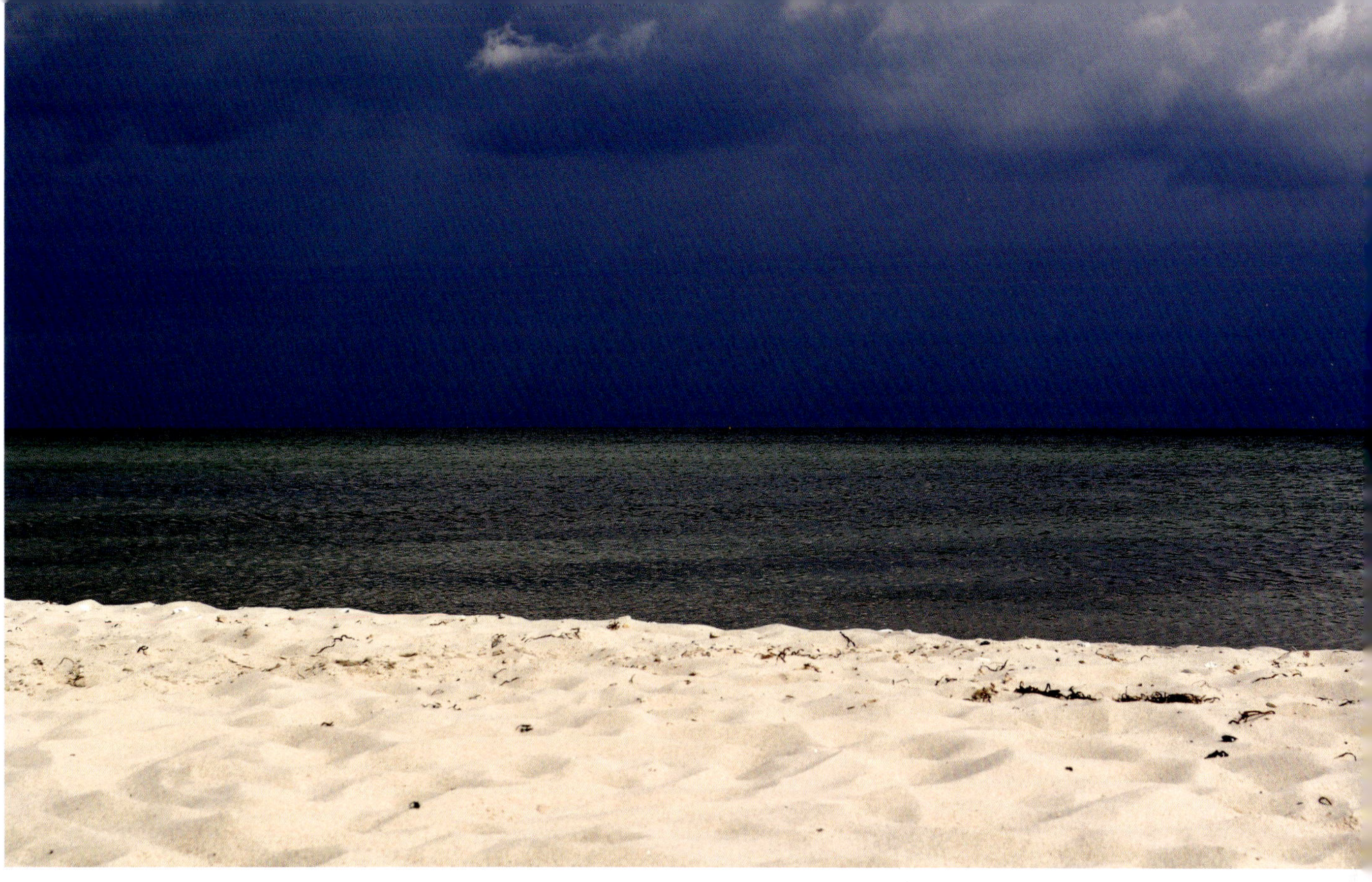

In addition, there are estimates of the mean shape of the sand grains. According to Kuenen (1959), as we have seen, deserts are primarily involved in causing the rounding of quartz grains.[191] Through a combination of experimental investigations and extrapolations, he concludes that 2 million square kilometers of desert would be necessary to keep the average degree of rounding of quartz grains existing worldwide constant on average. This corresponds to one-third of the current stock, so the roundness of quartz grains should increase worldwide in geologically current times, which would theoretically have an impact on packing density and number of grains.

Each of these assumptions and estimates naturally leads to an accumulation of mistakes and uncertainties. Poldervaart[192] comes to a number of 85.7 or 85.6 × 10^{24} pieces, but even this seemingly inconsequential difference comes to the gigantic sum of a hundred trillion grains. It is completely futile to try to imagine numbers of this magnitude. If we were able to count the impossibly large number of one million grains every second, we would still need almost three trillion years to count all the grains.

It has often been attempted to answer the question of if there are more grains of sand on the earth than stars in the universe. According to newer estimates, there are about seventy trillion stars accessible to the modern telescope, whereas the total number of stars in the universe is higher many times over.

Figure 8.7 Seashore with approaching thunderstorm, with the elements earth/sand, water, air, clouds, and electric fields, representing a cycle

Thus, one can assume that there are more stars in the universe than grains of sand on Earth, which neither is surprising nor significantly extends the understanding.

Much more interesting is the question of if the amount of sediment or, in our case, the amount of sand grains tends to increase in sum or decrease. However, it needs to be taken into account that current processes make this consideration very difficult to formulate. Since the discovery of plate tectonics, it has been known that the continents are consistently moving closer/farther apart from each other. The speed of this movement, the so-called drift speed of the continental plates, between America and Europe, for example, is about 1.2 inches (3 cm) per year and on the Eastern Pacific Ridge about 6.3 inches (16 cm) per year and is therefore quite perceptible. Europe is therefore drifting at the speed of fingernail growth from America. As the plates are moving apart from each other, new magma is brought to the surface, and at the same time the edges of the continental plates are pushing down into the subduction zone (e.g., San Andreas fault).

In this process, which takes millions of years, the entire plate edge of the continent, including the mountains and surface rocks, is slowly pushed toward the interior of the earth and melted under high pressures and temperatures. The continental margin disappears, and with it also the existing sediments, including sand. Besides this, there are other geological effects to which our sand grains can fall victim. On the other side, the erosion taking place everywhere provides for a durable replenishment of sand grains. It can be estimated[193] that an annual volume of 0.01 cubic miles (0.05 km^3) more quartz grains is released through erosion than is destroyed through other processes. Therefore, the number of sand grains is actually growing today and has been for a long time, with each second producing about one billion new grains worldwide.

The first scholar to deal intensively with sand and numbers, using sand as a metaphor for "much" but without wanting to renounce exact quantification, was Archimedes of Syracuse, as we mentioned in the beginning. And he brought further terms into play in his treatise. One was the poppy seed as the usual comparison unit for sizes as well as the concept of the cosmos, which he took over from the Greek astronomer and mathematician Aristarchus of Samos (310–230 BCE). He addressed his writing *The Sand Reckoner* to King Gelon of Syracuse and began as follows:

> "There are some, King Gelon, who think that the number of the sand is infinite in multitude; and I mean by the sand not only that which exists about Syracuse and the rest of Sicily but also that which i s found in every region whether inhabited or uninhabited."

And he continues after further explanation:

> "But I will try to show you by means of geometrical proofs, which you will be able to follow, that, of the numbers named by me and given in the work which I sent to Zeuxippus, some exceed not only the number of the mass of sand equal in magnitude to the earth filled up in the way described, but also that of a mass equal in magnitude to the universe."[194]

This is where everything comes together: sand, number, cosmos, and infinity.

At that time the idea was that the universe was approximately the size of the assumed solar system (i.e., the diameter of the earth's orbit). He approaches this size step by step and starts with the size of a poppy seed, which is set to one twenty-fifth of the width of a finger. The circumference of the earth is assumed to be 3 million stadiums, and the diameter of the orbit around the sun to be 10,000 times the diameter of the earth. Also, the size of the sun itself was estimated and determined to be 10 billion stadiums. A stadium thereby corresponds approximately to the length of 185 meters. The reader may calculate with pleasure himself to what extent the data of Archimedes were correct.

Power calculation was not known at that time—this explains the very cumbersome representation and development of large numbers. At the time of Archimedes, the number 10,000 (a "myriad") was already considered unimaginably large; everything above it was already on the edge of infinity. Archimedes developed a number system based on 10^8 when calculating his "sand number." Step by step he approached the final result, the hypothetical number of grains of sand that would fit into the whole universe. If you recalculate his cascade, you end up with about 10^{64}. In the commentary to "Ostwald's Classics," it is said illustratively, "This number is a one with so many zeros that a scribe who wrote ten zeros every second and worked without any interruption would have to work for more than 85 million years."

Here, the circular dance closes to Joyce's small bird, which must carry the sand off the mountain that reaches up to the highest heavens. And when he finished with it, eternity would not even have scarcely begun.

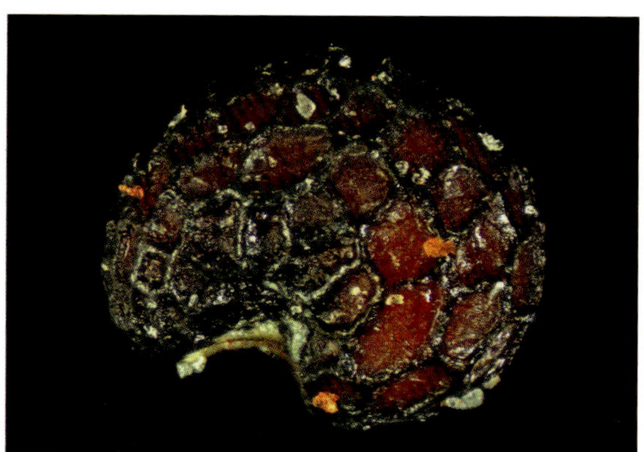

Figure 8.8 A poppyseed, which Archimedes probably used as a comparison. Reflected light. Dimensions 1.15 × 1.34 mm. About 10,000 grains of the finest sand could fit in a sphere with the dimensions of a poppyseed.

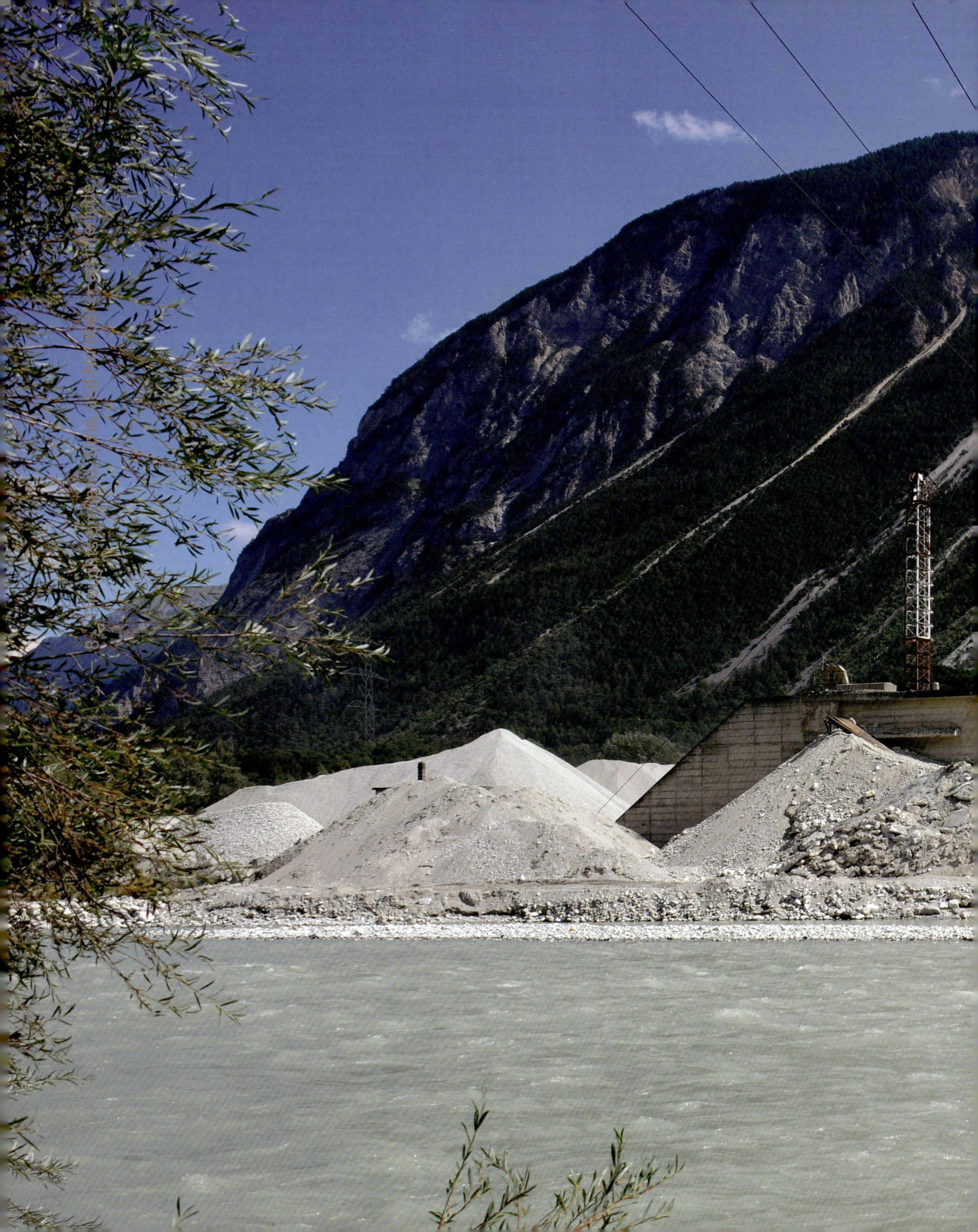

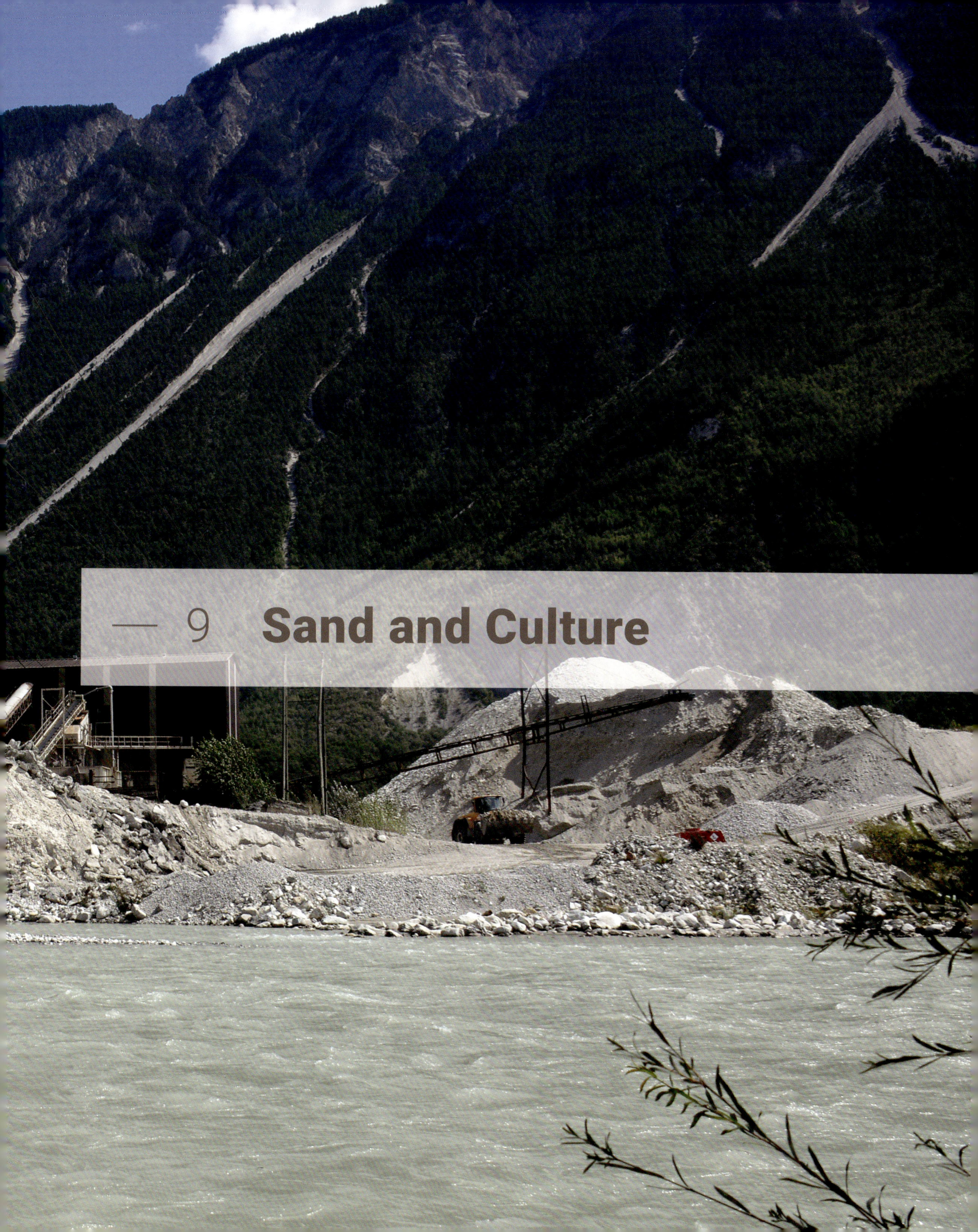

— 9 Sand and Culture

What used to please my eyes, below,
Has vanished with the centuries.

—J. W. v. Goethe, ***Faust****, II 11336*

Before there were humans, sand was a product of movement, of transport. But how have humans changed the appearance and composition of sand with their invasive mechanization? Are we sure that the sand grain that we are looking at also comes from the place where we picked it up? Culture and technology influence nature and influence sand and sedimentation. Sand also leaves traces in culture and technology. We want to take a look at a few of them.

"On the shore here I found, for the first time in my life, a starfish, and an echinus thrown up by the sea; a beautiful green leaf, smooth as the finest bath paper, and other remarkable rubble-stones, the most common being limestone, but occasionally also serpentine, jasper quartz, granite, breccian pebbles, porphyry, marble of different kinds, and glass of a blue and green color. The two last-mentioned specimens are scarcely productions of the neighborhood. They are probably the debris of ancient buildings; and thus, we have seen the waves before our eyes playing with the splendors of the ancient world."[195]

Anyone walking along the beach of the small Danish town of Hirtshals on a quiet day and looking down at the ground will probably enjoy the colorful juxtaposition of sand and ice-age debris. In the north of Denmark, as well as on the German Baltic coast, there are, as we know, fragments of Scandinavian rocks everywhere, mostly deep rocks, which have been deposited there in random assemblages from the most-diverse areas of the Far North after glacial transport. Granites, gneisses, porphyries, ignimbrites, sandstones, limestones, and many other types of rocks lie close to each other in the moraine debris. The sands, formed directly on-site by weathering and abrasion, consequently contain an equally abundant mixture of all the numerous minerals of which the rocks were composed. At this place in Denmark, therefore, a part of the Scandinavian mainland plate, ground to sand and slowly rising from the depths since the ice ages, lies at our feet. But there is something more.

Previous double page:
Rhone near Salgesch, Switzerland

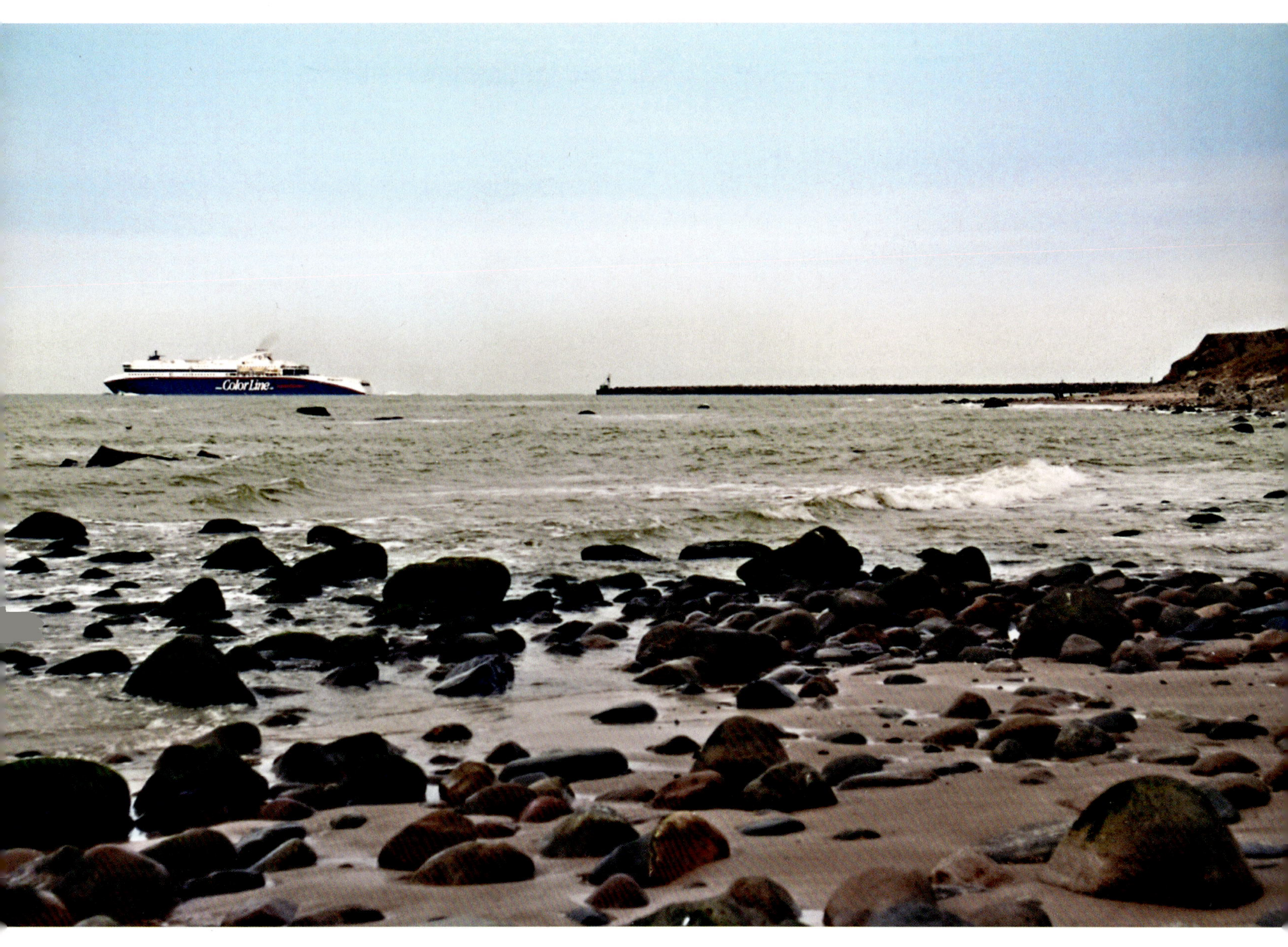

Figure 9.1 Beach view of Hirtshals in Denmark with sand and ice-age debris. *To the right*, in the background, are the historic bunker facilities, and, *to the left*, a ferry to Norway.

If we walk along the beach a bit farther toward the ferry port, fragments of a completely different kind catch our eye. Ugly, massive wall remnants rise from the sand and reference the recent history of the area. Protective bunkers from World War II, now exposed to the rough weather and the sea, are slowly but inexorably disintegrating into their component parts. In addition to reinforcing steels, which are almost completely oxidizing into small crumbs of rust, it is the very resistant concrete that, as a building material of our time, is weathering before our feet and being absorbed directly into the mineral stock of the beach. But what is concrete made of? Well, there are probably more books about concrete, its composition, and its specific properties than about sand. In summary, concrete consists of cement, water, and aggregates, the latter being composed of gravel and sand, each in a different mixture. Sand, too! During the process of weathering, the concrete gradually disintegrates into small aggregates of cemented sand grains, which eventually dissociate completely and finally become indistinguishable from the naturally existing sand surrounding them.

If we take a sample of sand from the beach to investigate, we find a curious object that does not come from the location we collected it from, but, for example, from a sand pit that a construction company chose to build near the beach. You cannot tell by a single grain if it was transported as bulk material by railroad cargo or by icebergs as moraine cargo.

Wind, waves, and rocks persistently crush everything that is brought to a beach area, in terms of cultural assets. And this has been the case ever since culture existed. The information about origin is completely and irretrievably lost when it is weathered out of an artificially produced compound. Culturally influenced beach regions can therefore almost always contain a changing but unknown proportion of sand grains, which are of unknown origin, the result of a migration of whatever kind. In this context, one also thinks of completely artificially filled beaches or beach fillings for the purpose of shoreline stabilization, land reclamation, or increasing tourism attractiveness. In the last forty years alone, Singapore has increased its area by 50 square miles (130 km^3) through sand fillings, the sand being imported for this purpose. Another example, Virginia Beach, has been replenished more than fifty times with foreign sand in the fight against beach displacement by ocean currents. The inland sand pits in the US, which are now being developed everywhere, are able, with the help of modern plant technology, to mix the appropriate desired sand according to shape, grain size, and color, depending on the customer's needs, and supply it for coastal replenishment. All these sands, which may then delight us at first glance on the beach, come from completely different regions and are basically signs of considerable environmental destruction. In the worst case, the sand was extracted from the seabed with huge floating suction dredgers.

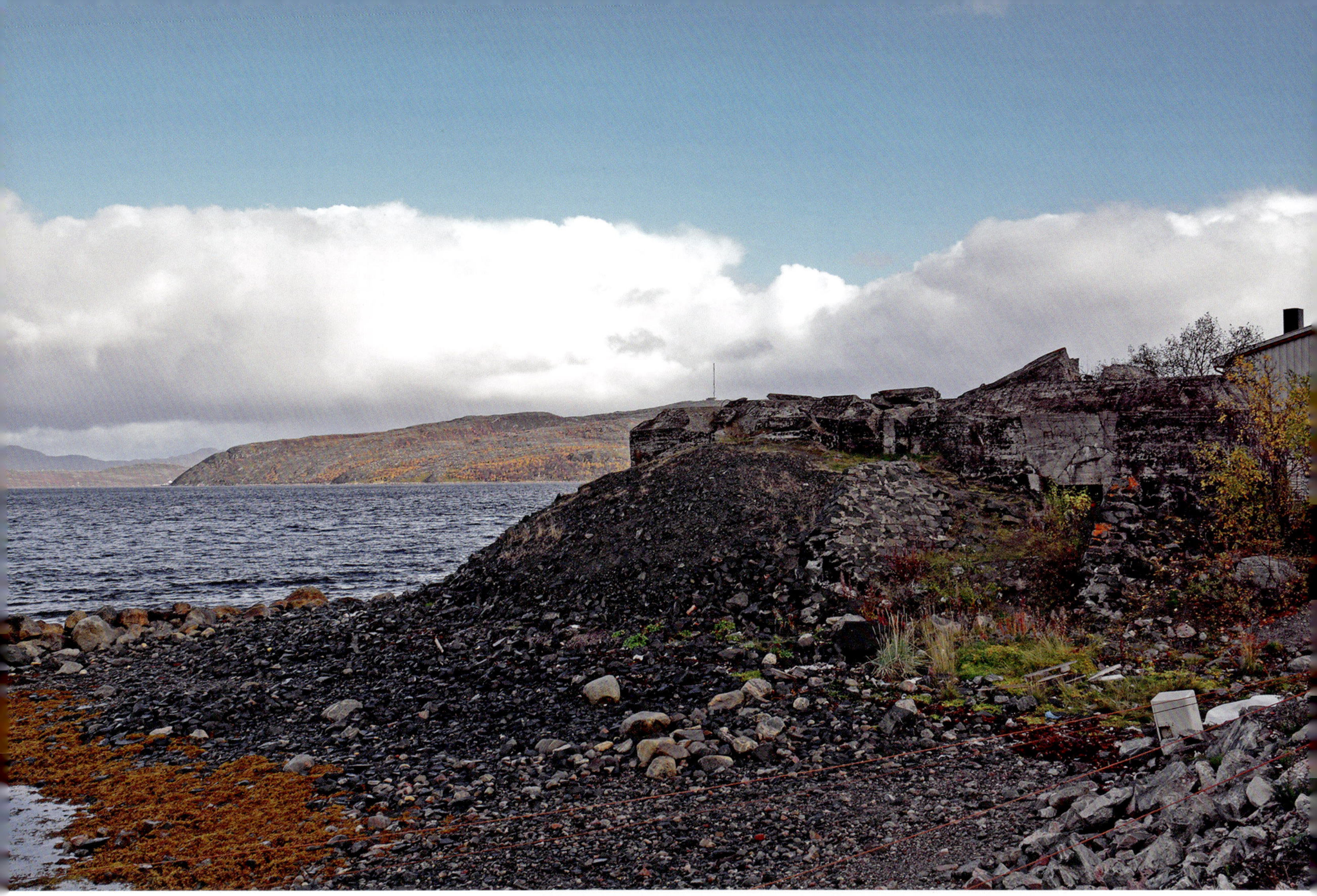

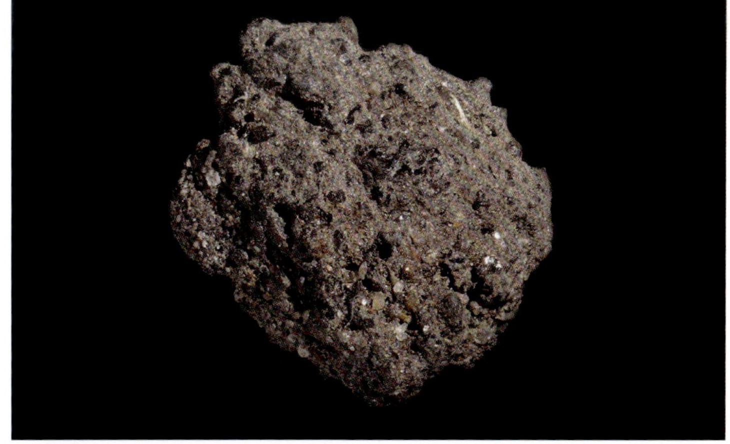

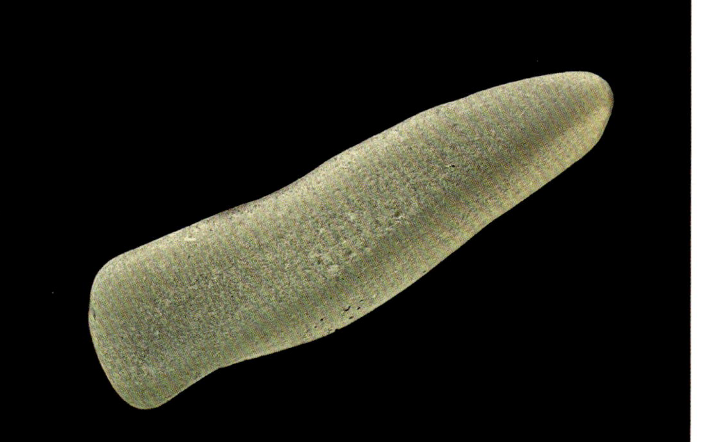

Figure 9.2 Bank area near the port of Kirkenes, Norway. Rubble, building debris, and rock brought in to secure the bank mix along the waterline. Concrete pieces slowly decay back to sand.

Figure 9.3 Sand grain (coarse sand) from the banks of Lake Ontario, Toronto, Canada. Eroded concrete particle, composed of cemented fine-sand grains. Reflected light. Image width about 10 mm.

Figure 9.4 Glass shard formed by the sea and matted from the sandy beach of Cala Bitta in North Sardinia, Italy. Reflected light. Image width 43 mm.

Water turbidity of up to 10,000 tons per hour is only one of the harmful factors causing great ecological damage.[196] Not only as sand lovers do we find the large-scale industrial transport of sand masses disturbing, which is increasing worldwide without regard to sustainability concerns. In Goethe's time, of course, there was no talk of this. Goethe included the enrichment of the natural sediments with fragments of cultural decay benevolently in his reflections, the "glories of the previous world." And indeed, if you look closely, you can find quite pretty fragments of the cultural and engineered world. Tiny, round cut-glass fragments that shine out in wondrous colors of green, blue, red, or white between neighboring grains. Most of them have an interesting matte surface and are quite easy to identify. Moreover, glass has no birefringence, so glass particles the size of sand grains remain invisible (i.e., black) under the microscope under crossed polarizing filters (such as garnet grains). Likewise, with luck, reddish-earth brick grains and oddly iridescent colored pieces of former ceramic pottery can be found. On beaches, black slag grains, coal particles, or even remnants of thickened heavy oil can be distinguished with the naked eye from the otherwise mostly light beach sand. Silent witnesses from all over the world can also be found along such routes, be it packaging with exotic inscriptions, shoes, or clothing remnants, whose stories remain hidden from us. When entire sea containers go overboard, which can happen during heavy storms, their contents spread over very wide stretches of coastline and, depending on their composition, are crushed over the years to centuries and finally added to the sediments. Environmental pollution on a new and large scale.

Along with the rather local effects of human activities on the distribution and occurrence of sands, as they have taken place in the past centuries up to modern times, the overexploitation of sands taking place today seems almost monstrous. With the global introduction of concrete as the most important building and construction material of our century, the original picture has changed drastically. Sand, as a significant fossil resource, has become an internationally sought-after commodity, second only to water, needed and consumed in enormous quantities. Sand is used in very many industries and for a very wide range of products. It can be found finely ground in paper, toothpaste, cosmetics, smartphone displays, instant teas, and powdered sugar, as well as glass, bricks, water filters, and computer chips, to name just a few examples.

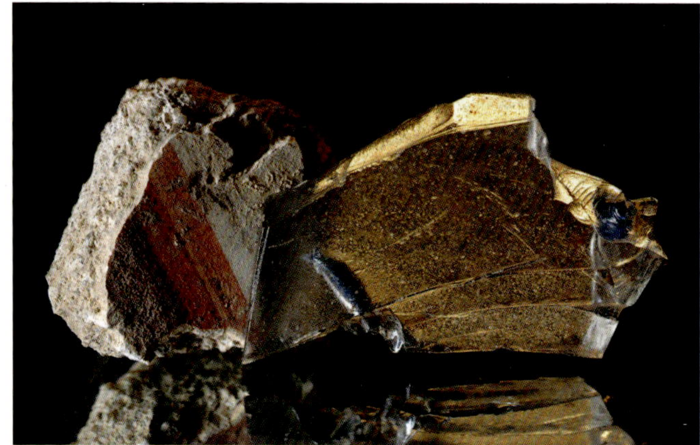

Figure 9.5 Painted piece of plaster with sandy components (*left*) and glass shard with enamel inclusions from a Roman estate near Saschsenheim at Holderbueschle (ca. 220 CE), Germany. These remains of human activity will also be included in the sedimentary record. Image width around 50 mm.

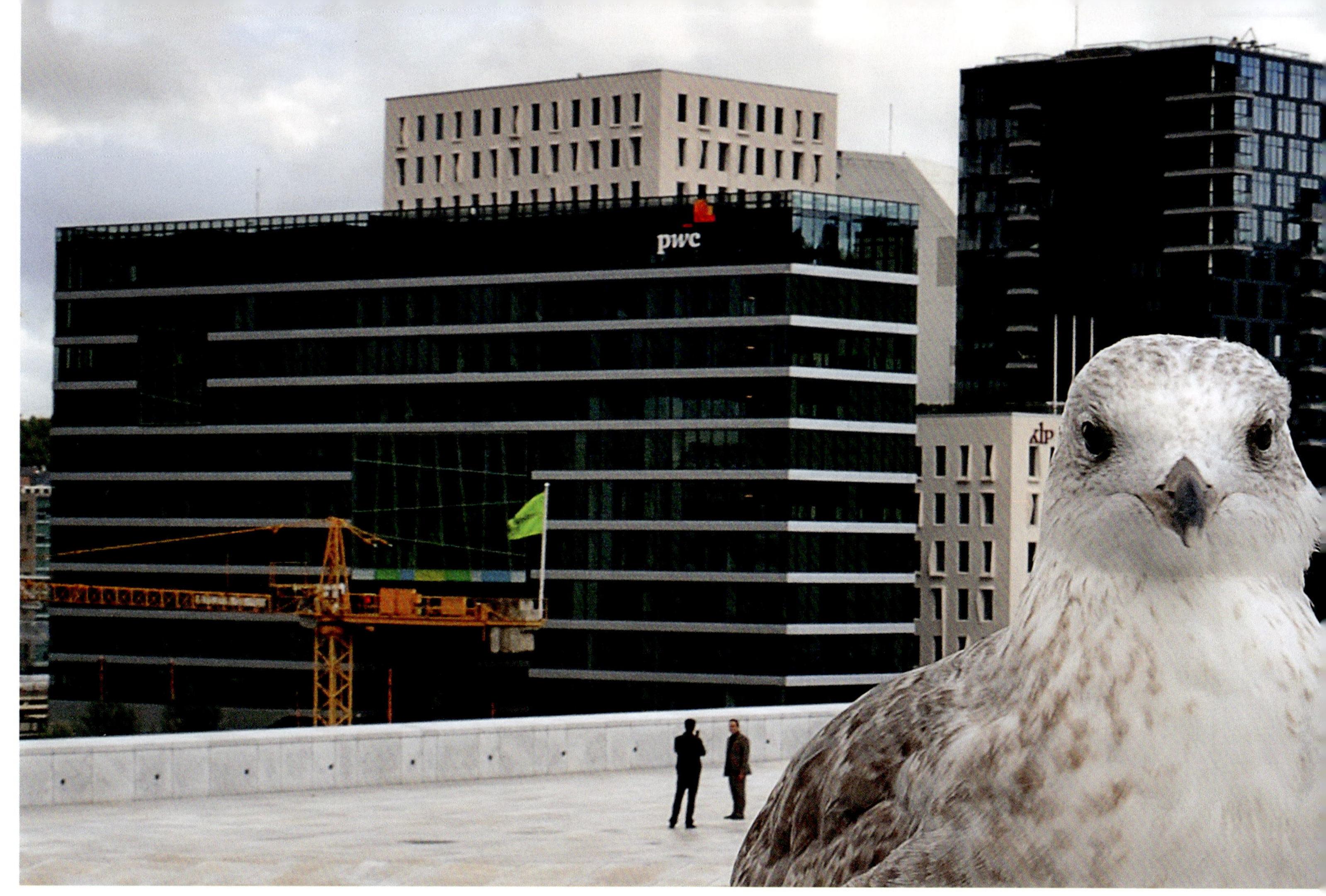

Three-quarters of all mined raw minerals worldwide are sand and gravel.[197] According to recent estimates, 50 billion tons of sand are mined worldwide each year. Globally, humans are now moving more sediment and rock than are all the world's rivers combined, through direct activities such as mining and urbanization, as well as indirect effects such as erosion acceleration through agriculture.[198]

In Germany, each citizen consumes around 9 tons of sand per capita and year, and total extraction amounts to around 240 million tons per year, with an upward trend.[199] Sand is already becoming scarce in Germany too; around two thousand sand and gravel pits are currently in operation, and many of the other sand deposits are located out of reach in protected areas. For example, 1 cubic meter of concrete of strength class C25/30 contains 661 lbs. (300 kg) of cement, 48 gallons (180 liters) of water, and 4,167 lbs. (1,890 kg) of aggregates, such as gravel and sand.

Figure 9.6 Newly built high-rise district in Oslo, near the opera house (2013). The main construction materials are concrete and glass, from which thousands of tons were used, and contain quartz sand as the main component. Each cubic meter of concrete needs over 2,200 lbs. (1,000 kg).

One single-family home needs around 200 tons of sand, and 0.6 miles (1 km) of highway needs 30,000 tons. A nonnegligible proportion of the earth's sand resources is already embedded in concrete today. The construction boom in China and the Asian countries as well as the Arabian desert states is devouring the majority of this. China has become the largest consumer of sand, consuming around 7.8 billion tons of construction sand in 2016 alone.[200] In some places, overexploitation is becoming a problem. So far, this has not been penalized because sand as such is not protected as a raw material. Concrete recycling is currently still in its infancy but could perhaps make a small contribution to the solution in the future.

Some countries are now beginning to ban the collection of sand (and other natural objects) on tourist beaches, since taking even what are usually small drinking bottles full of sand as souvenirs can have problematic effects with hundreds of thousands of visitors over time. This is especially true for beaches with rare and decorative sands. Sand is a popular feature and for many people is an ideal souvenir for the time spent on vacation that refers to the land and is a better time anchor than the imported wares you can buy. In relation to the industrial overexploitation by sand-pumping ships, however, the tourist extraction is only a small and local problem.

The question now arises as to why the vast desert areas of Africa and the Arabian Peninsula are not able to cover a large part of the world's demand for construction sand without any significant impact on the respective ecosystem. In the case of the booming desert state of Dubai, the sand source would be right on its doorstep. In chapter 6, "Looking for Traces," we had dealt with the subject necessary to answer the question—it is unfortunately not possible.

Compared to marine sand, desert sand is not suitable as an aggregate for higher-strength concrete types due to its shape and surface structure. In arid deserts, the aeolian transport of the sand predominates, which, as shown, causes a more energetic and almost uncushioned collision of the sand grains with each other and thus leads to a strong abrasive wear. The sand grains round off faster than in transport paths in water, which has a cushioning effect. Compared to desert sands, sea sands or river sands have a lower sphericity, are more angular, and produce higher compressive strengths as aggregates in concrete than rounded grains, which are unsuitable for this reason. Recent developments with polymer concrete are likely to allow the use of desert sand in the future. Research in this regard looks promising. Another use is becoming increasingly important and of concern, especially in the USA: the fracking process for oil extraction. This involves fracturing oil-bearing but oil-impermeable shale rock via boreholes, using high-pressure water containing chemicals. To prevent the fractures and tiny cracks created in the shale from closing again, quartz sand with specific properties is added to the water.

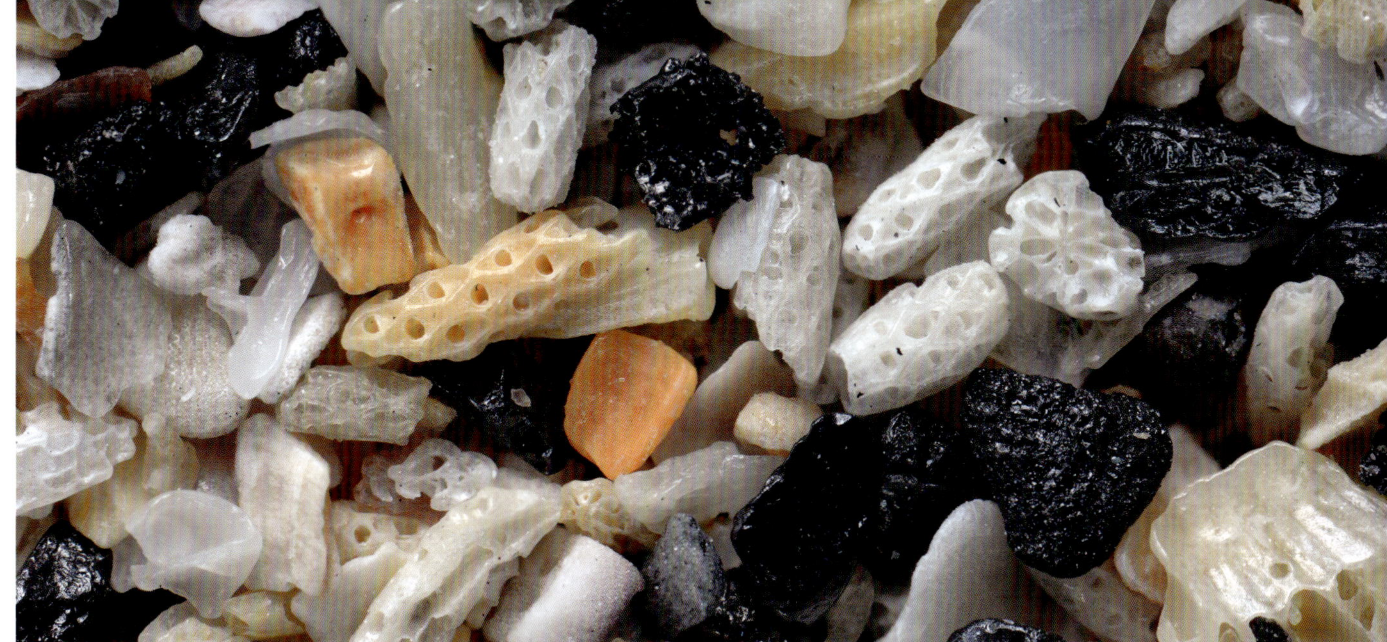

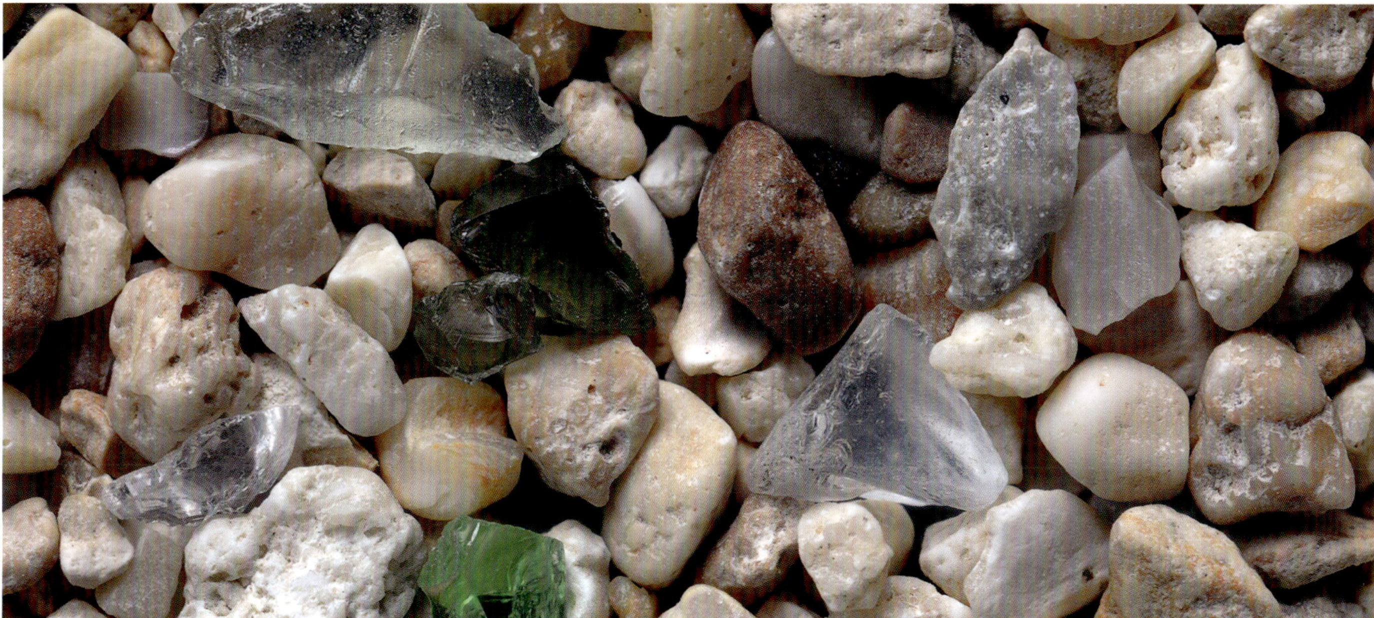

The sand must be sharp edged and fine grained, a rare combination of characteristics, and its extraction is therefore also very profitable. Each borehole can consume up to 25,000 tons of sand. Almost 70 million tons of fracking sand were mined in the USA alone in 2016.[201] Angular and therefore adhesive quartz sands are therefore also required and in demand for this application.

Figure 9.7 Carbis Bay, Cornwall, England. The sand consists mostly of very diverse biogenic material. Bryozoan branches, barnacle remains, and many more can be seen. The pitch-black particles, which are of artificial origin and originate from overboard coal and cinders of passing ships, are particularly noticeable. Image width 8.5 mm.

Figure 9.8 St. Paul's Bay, Burgibba, Malta. The sand is "contaminated" with numerous grains of glass. Image width 22.5 mm.

An extreme example of adhesion, by the way, is lunar sediments. There is no atmosphere on the moon, and neither wind nor water can transport the extremely sharp-edged particles, which are also electrostatically charged by the solar wind.

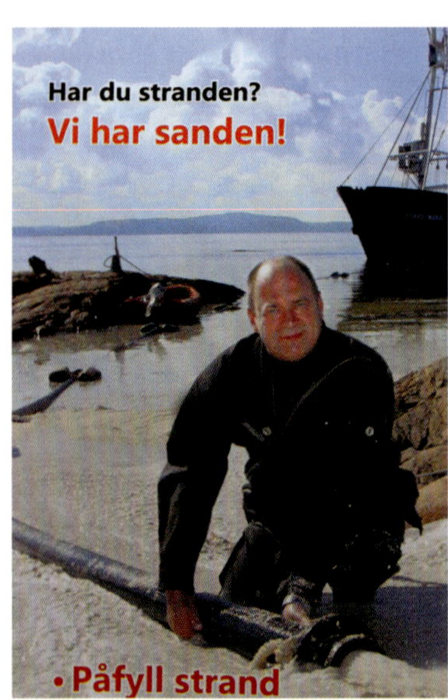

They adhere very strongly to each other as well as to tissues and objects, as the Apollo missions showed.[202] A footprint compacts the sediment, which has an extremely fine average grain size of 70 *μm* and makes it adhere to each other in this form (almost) for all eternity. The relevant pictures for this are known from the media.

The demand for quartz sands with specific suitability profiles has fatal consequences. Dubai, for example, had 4.2 billion cubic feet (120 million cubic meters) of sand used for the artificial filling of islands in the middle of the sea as part of the "Palm Jumeirah" project.[203] Illegal mining of sandy beaches is increasing worldwide. Dredgers pump marine sands to the surface along with all the creatures they contain: ecosystems are damaged or destroyed on a large scale, and beaches lose their natural protective functions. Sand is also a very important filter for groundwater resources. Habitats and ecosystems of these sensitive zones are endangered, especially when you consider that 66% of the world's population lives in strips only 31 miles (50 km) wide along the coasts. According to the Indonesian government, eighty of their islands have now disappeared due to sand theft. On the Cape Verde Islands, sand must be imported to meet construction needs. Sand theft has led to coastal erosion: salt water is penetrating inland and making soil infertile.[204]

Apart from the abovementioned effects on nature, the finding and analysis of grains in the affected areas is no longer comprehensible. Ocean currents and winds transport sand that has been artificially piled up or released by the erosion of building materials on a large scale. Thousands of tons of sand end up in systems outside their origin. So, there may be a large proportion of sand from Australia in a sample collected in Dubai.

In general, we can divide the influence of anthropogenic activities on the occurrence and composition of sands into three groups. These cannot be distinctly separated, because mixed forms occur, but they summarize what has been previously presented somewhat more systematically.

The first group includes sands that were formed naturally but were subsequently transported to a new location through technical processes. For example, during the Elbe flood in the summer of 2013, large quantities of sandbags were transported to the banks of endangered rivers and dikes. When this sand ends up in the affected river, it will be deposited in places that it would never have reached naturally and will probably no longer be readily distinguishable from the sands of its new landfall area.

Figure 9.9 Extreme example of the transportation of sand to coastal areas where there was previously no sand. The advertisement comes from a Norwegian newspaper and promotes the project with the words "Do you have a beach? We have the sand!" A water-sand mixture is spouted into suitable bays by means of pumps. In addition, according to the advertisement, already-existing sand beaches can be refilled. Nothing is said about the origin of the sand. Since such activity does not require labeling, sediment studies at these locations, or at locations in the direction of the ocean current, can be completely misleading. Source: hytteliv 7/18.

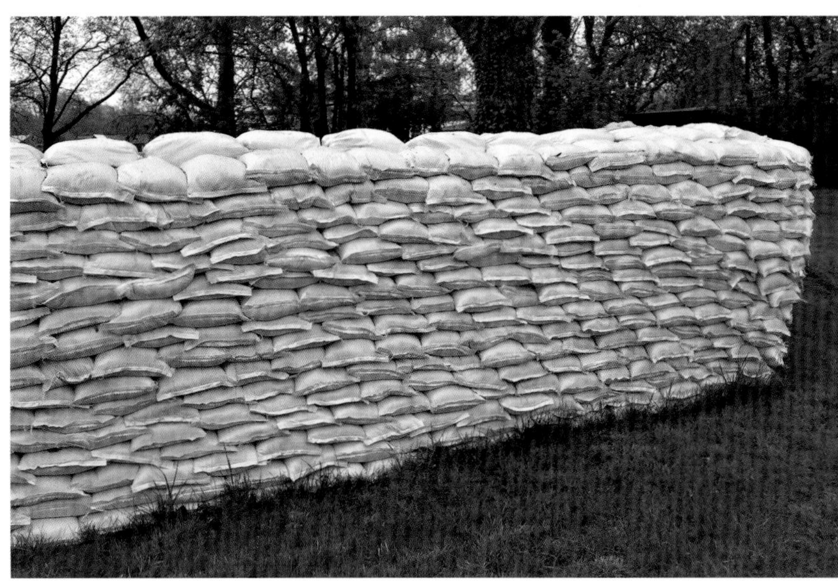

The beach fillings already described can also be classified in this context. The ocean currents along the coastlines of the North and East Frisian Islands, for example, permanently carry away large masses of sand on the windward side and deposit them again on the leeward side. This reshaping of the island contours is counteracted by a wide variety of measures, including the sand piles. Although, in contrast to the practices described in the example in the USA, mostly locally occurring sands are used here, which are harmless from an ecological point of view, but the natural deposition process is altered. Beach fillings are now common practice along vulnerable coasts around the world.

The second group we can define as sands that were created on-site but are a weathering product of materials that were artificially produced. Our example mentioned at the beginning stands for this group; weathering ruins, wall remains, buildings in general, rubble left to the action of erosion, and the like. All this is eventually crushed on-site in one way or another and returned to the natural cycle as the resulting particles of unknown origin gradually mix with the local sediments and themselves become part of the sedimentation.

The third group includes sands or sand components that consist of materials that originated from a purely artificial or technical production and have reached the place of deposition in whole or in part as a result of human activity. We include industrial waste in the appropriate size fraction of slags, granules, and ground products that did not reach their intended location as transport losses, mineral wool particles, firework residue, particles from roofing tiles, ash, and much more.

This also includes the alarming increase in plastic waste worldwide. Microplastics have a very large number of sources and, according to a study by the Fraunhofer Institute for the Environment, Safety and Energy Technology (2018), reach an order of magnitude of 330,000 tons per year in Germany alone, corresponding to as much as 9 pounds (4 kg) per capita.

Figure 9.10 Sandbags for the reinforcement of dikes or to generally prevent floods bring natural sands to areas where they did not originate.

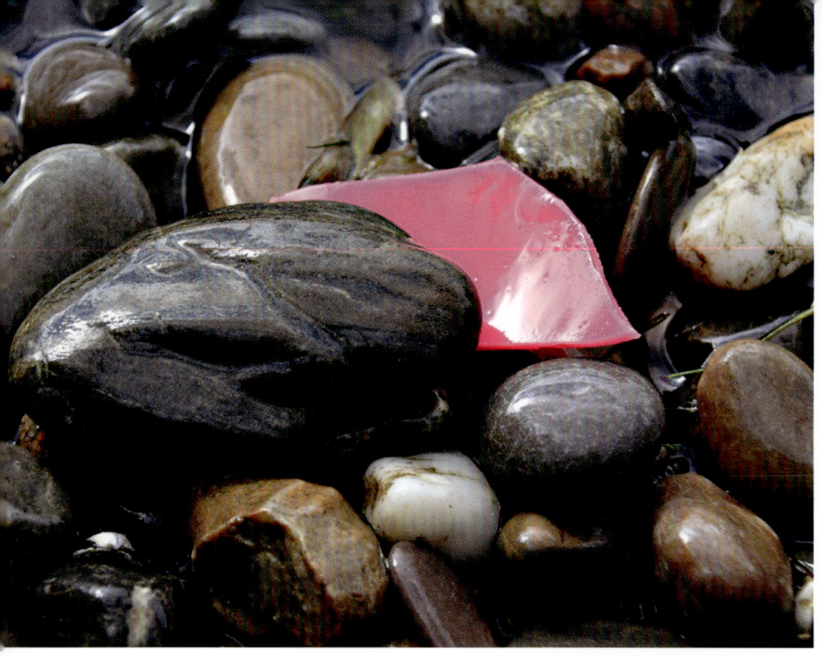

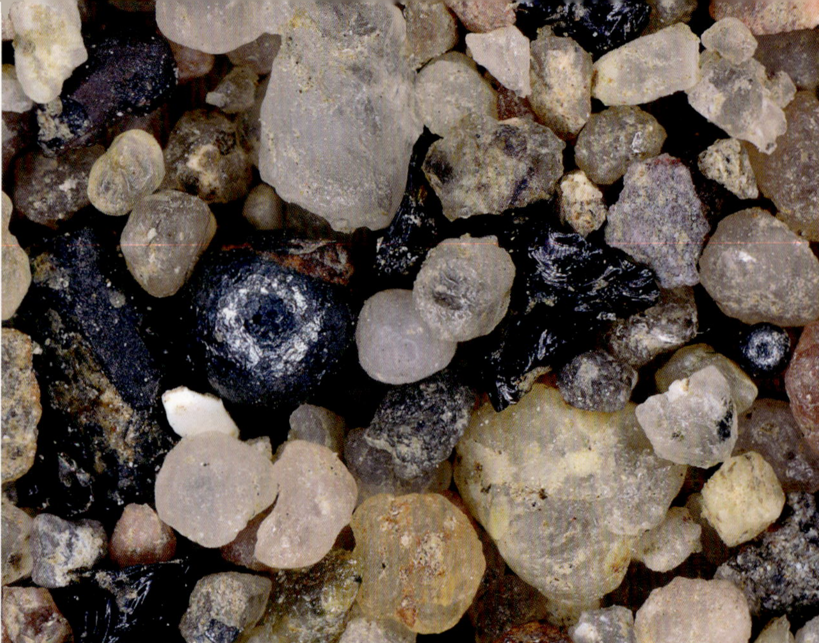

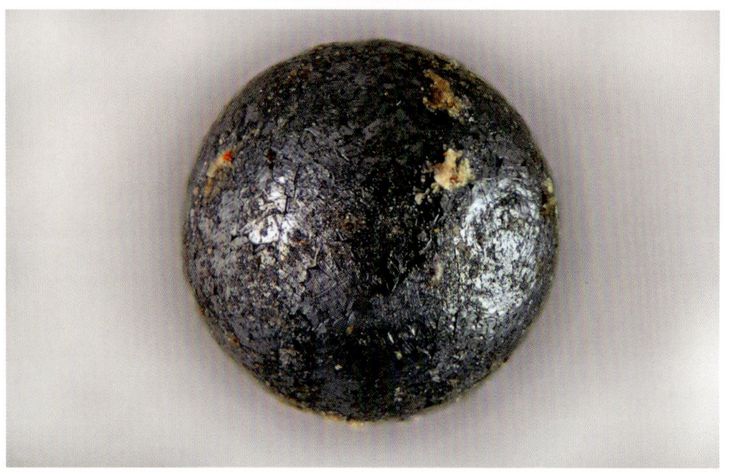

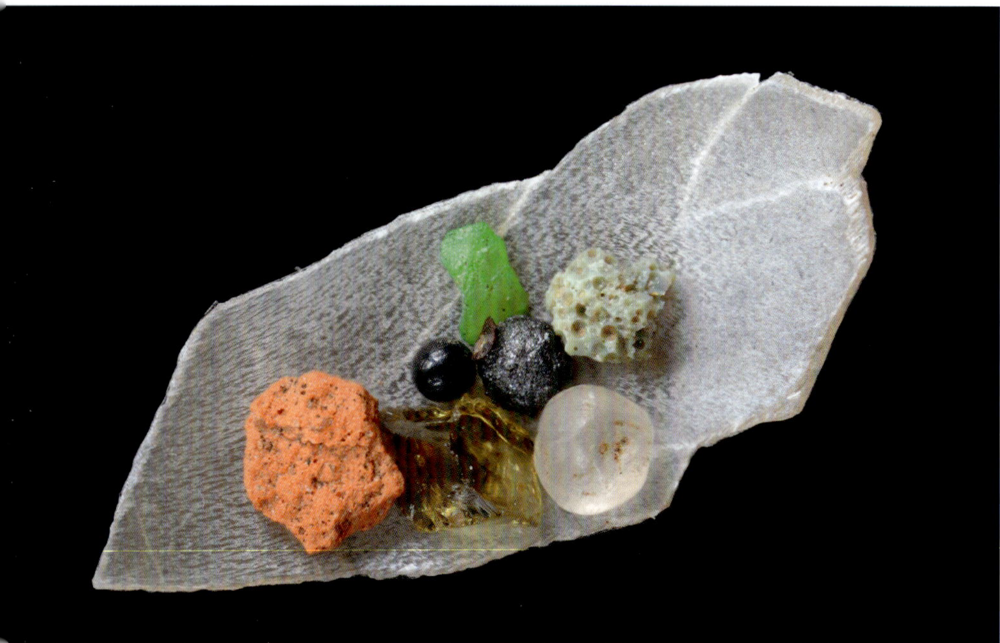

Figure 9.11 Rhine bank near Kehl, Germany. It will probably take more than a century for this piece of plastic to fully decompose and be absorbed into the cycle of materials.

Figure 9.12 Rhine bank near Kehl, Germany, directly on the Europa Bridge. Among unsorted, clayey sand, metal balls can be seen, which probably originated from the welding of the bridge. Image width 7.5 mm.

Figure 9.13 Portrait of a grain from the Rhine bank near Kehl by the Europa Bridge (see fig. 9.12). The metal sand grain is of anthropogenic origin and probably from welding work. Micrometeorites have a similar appearance. Reflected light. Diameter 485 *µm*.

Figure 9.14 Sand grains from the bank near Bad Kreuznach, Germany. Only the very round quartz grain (*bottom right*) is of natural origin—all other grains are the result of technical processes (shingles, glass, slags, glaze, plastic). Image width 4.8 mm.

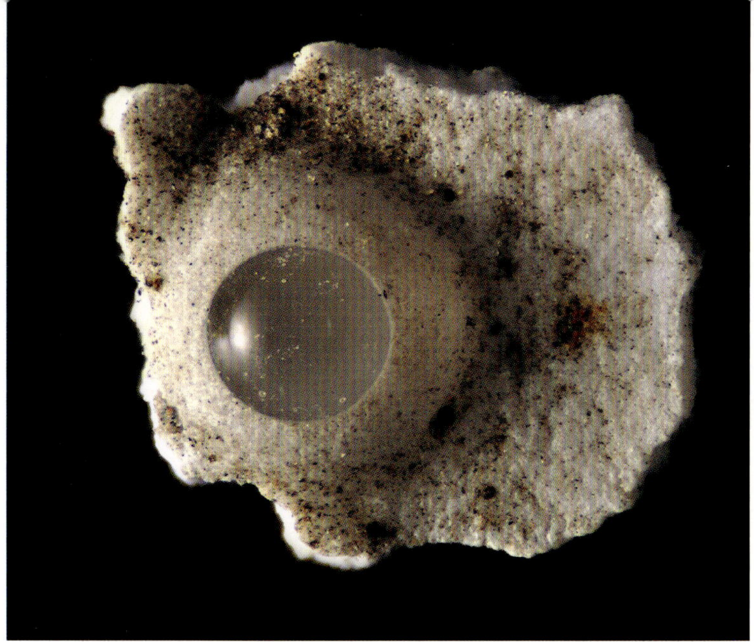

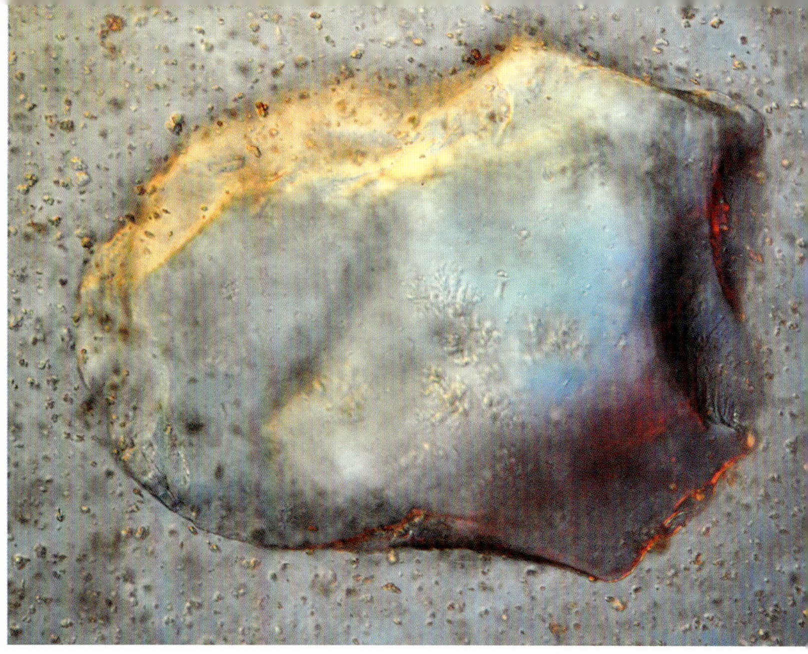

This is a large part of the total estimated annual emission of 460,000 tons of plastics in Germany. A distinction is made between two types of microplastic emissions. The first includes all particles that have been added directly to products during their manufacture; for example, microplastic particles in cosmetics or cleaning agents. The second type of particle is created during use, through abrasion. In first place is tire abrasion, which accounts for almost 2.9 lbs. (1.3 kg) per capita and year. In second place, at 0.7 lbs. (300 g) per capita per year, is the number of plastic particles released into the environment during waste disposal. The abrasion of road markings (fig. 9.15) is still in ninth place, far ahead of additives in cosmetics, which appear in only seventeenth place but could be reduced most easily. Some of these emissions, however, are in a size class below that of sand, but this in no way diminishes their hazardousness. On the contrary, the small dimensions of the particles make their existence seem far less present than larger waste residues, which are thus subjectively perceived as more troublesome. Respirable fine dust poses a particular hazard.

In the manufacturing industry, there is a wide range of demand for sand, even beyond its use as a pure raw material. One example is the use of molding-sand mixtures in casting technology. Natural sand is mixed with a two-component binder based on phenolic resin in the so-called cold-box process and then is processed into sand molds or sand cores (fig. 9.16). With the addition of compressed air containing amine, the binder hardens and gives the sand molds the strength they need to withstand the stress caused by the inflow of molten metal during the casting process.

Figure 9.15 *Left*: Sand grain, collected on the side of the road near Kleinsachsenheim, Germany. It is a piece that comes from road markings. The small, embedded plastic ball can easily be seen, which reflects the light of automobile headlights at night. In the sand sample there was a multitude of other and single balls of different diameters. Reflected light. Grain dimensions 1,370 × 1,240 *μm*.

Right: Microplastic particle from a shower scrub. The particle is approximately the size of a grain of sand (dimensions 360 × 265 *μm*). The creamy matrix also has particles that are significantly smaller, at about 2–4 *μm*. Differential inference contrast with *λ* compensator, bright field.

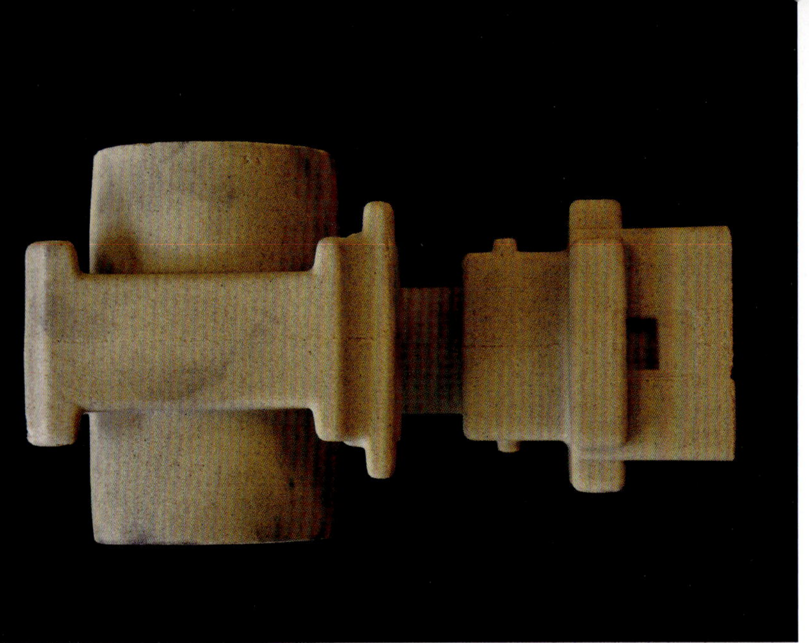

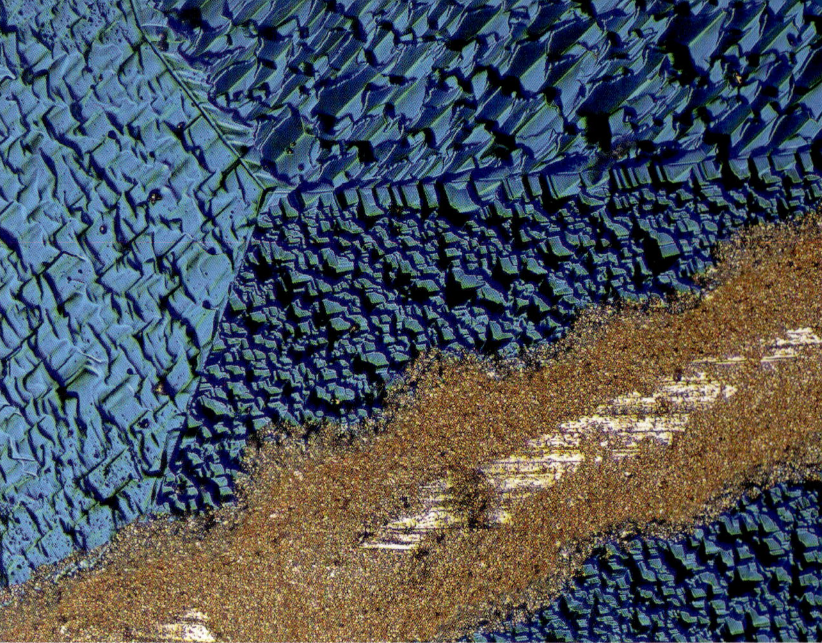

Many tens of thousands of tons of molding sand are produced in this way every year and have to be recycled at great expense by using suitable processes.

A nice connotation to the sand is the solar cell. Fig. 9.17 shows the microscopic image of the surface of such a solar cell used for photovoltaics. For polycrystalline cells, the silicon extracted from sand does not have to be quite as pure as is required for monocrystalline cells. Nevertheless, the required purity level is over 99.999999%. In the production process, a block of liquid silicon is first cast, which is then cut into slices after cooling and crystallization. The cells are then provided with gold contacts. Sand crystals are turned into silicon crystals by using energy, which in turn gain energy from sunlight, a nice circle. Even higher-purity grades of quartz sand are needed for the production of microchips. The high-purity quartz sand from Spruce Pine, USA, from which the 20-ton mirror of the Mt. Palomar Observatory was also made, is particularly sought after.[205]

Some of the grains of technical origin found in sands have a strikingly spherical shape and are almost ideally round. This is easy to explain in the case of molten products that fly through the air while still in a liquid state—such as sparks in grinding or welding processes. The melt assumes the state of the optimum ratio between volume and surface area (i.e., it becomes spherical shortly before solidification). The vast majority of these small metallic magnetic or even metallic nonmagnetic or glassy spheres are of anthropogenic origin. If you are very lucky and search long and systematically, you can also find extraterrestrial material. These are then micrometeorites, which continually hit the earth's atmosphere from space at high speed, fifty times faster than a rifle bullet. Only a few of them make it to the ground; we had discussed similar things with meteorites.

Figure 9.16 *Left*: Hardened sand core for the casting of an aluminum gear case. Image width about 30 cm.

Figure 9.17 *Right*: Surface of a polycrystalline solar cell with gold contact. The differently structured fields of silicon crystals can be seen. Reflected-light bright field. Image width 690 *μm*.

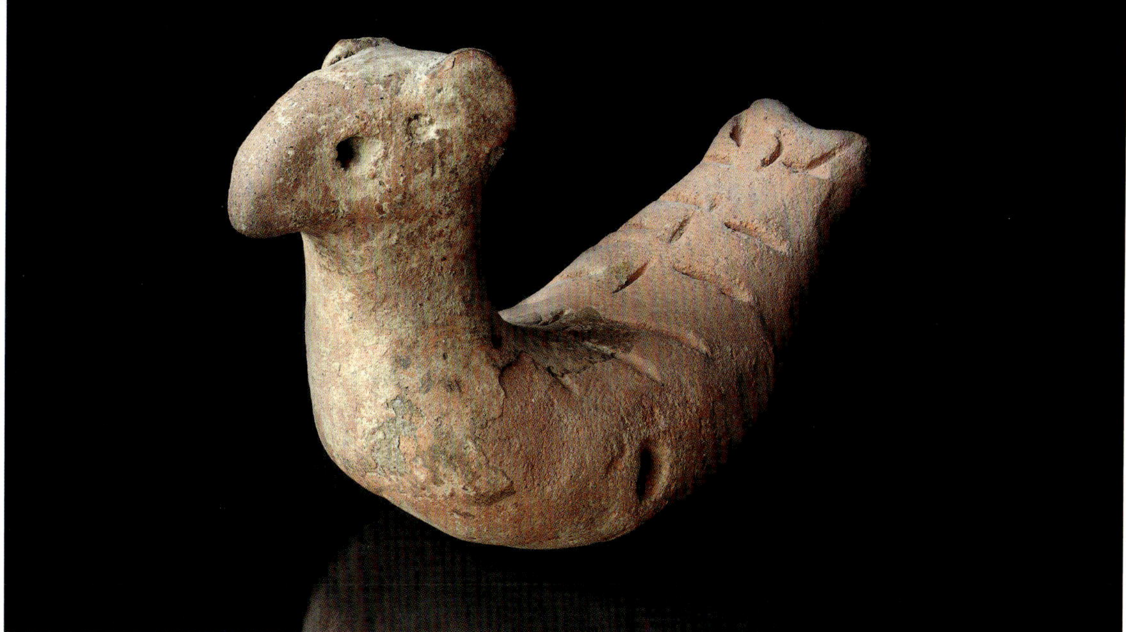

It is estimated that the average annual impact rate is one particle (with 100 μm diameter) per square meter. Since micrometeorites have an average size of 300 μm, this gives a realistic value of two spherules (as the small extraterrestrial spheres are called) per 50 m^2 per year.[206] This is not very much, especially because the small spherules oxidize easily. Should one go searching, the chances are greatest on large, flat roofs. However, the distinction to terrestrial matter can be done only with a lot of experience, or measurements under the electron microscope with additional equipment (EDX). Increased nickel content and a shell of magnetite with characteristic fir tree structures are characteristic, among other features.

Much more could be added. Sand is closely connected with the culture of the people. This applies not only to the technical applications described above and to sand as a building material, but also to sand as a raw material for art objects. Sand, clay, and loam were the basic mineral materials from which, as we saw in chapter 2, people in ancient Mesopotamia, the Mesopotamia between the Euphrates and the Tigris Rivers, created their houses, their everyday objects, and their works of art and cult objects. With an object from this time, we want to close our unfortunately extremely incomplete collection of topics.

The small figurine in fig. 9.18 was made over four thousand years ago from the sand and clay of the plain between these two great streams. Probably it served as a toy; we do not know. After all these years, however, it has lost none of its enchanting charm, although it was created with the simplest means imaginable, the hands, and from the simplest material imaginable, the sand and dust of the dry earth, with precious water, and was fired with the help of fire for posterity and thus permanently preserved over time.

Figure 9.18 Sumerian terra-cotta figurine, ca. 2000 BCE. The figurine probably depicts a small bird or a sheep. It seemed to have had an axis with two wheels and could have been pulled with a rope. The figure originally had a coating that can still be seen around the head and neck. With the microscope, partly rounded, partly edged sand grains can be seen in the fine matrix. Length 103 mm.

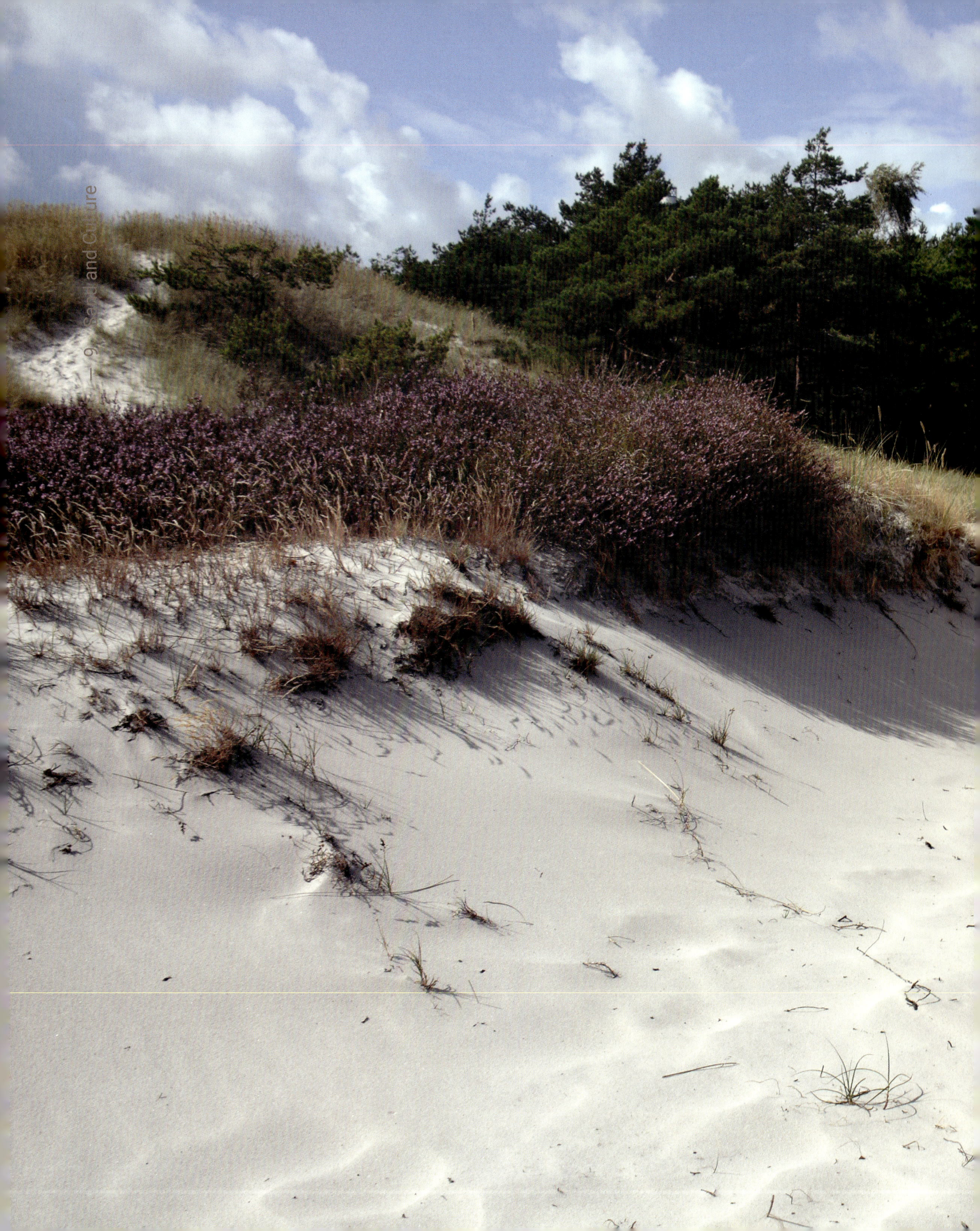

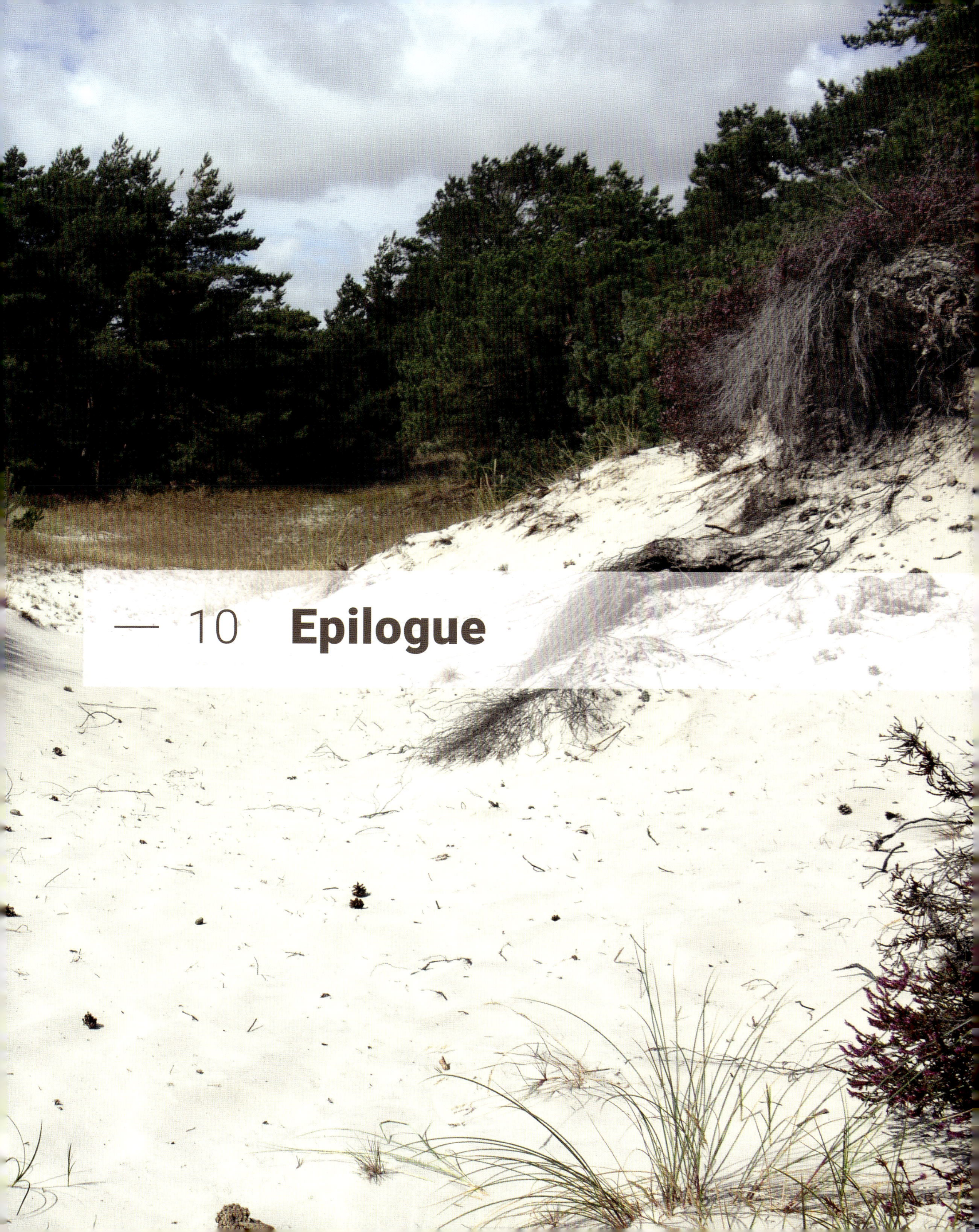

— 10 Epilogue

De hoc lapide multi multa, omnis aliqid, nemo satis.

—Inscription on the meteorite from Ensisheim from 1492 by Kaiser Maximilian I[207]

It had not rained for a while. I was in a hurry but wanted to quickly take a sand sample from a spot that had caught my eye from a distance due to its dark coloration. Maybe I was lucky, and it was a mineral soap of interesting composition. With my hiking boots, I made good progress in the dry and loose sand. I attributed a soft squeak that seemed to accompany me first to the gusty wind, then to distant seagull calls, before it penetrated my consciousness more and more clearly. When I finally paid attention, the strange squeak grew louder, coming directly from the sand below me, step by step. I had stumbled upon a "squeaky sand beach," the phenomenon of "squeaky sand," here, on a stretch of coast in western Bornholm.

Such squeaking sands, which must not be confused with the "singing sands" of deserts, are known from some places in the world. From Australia to America to the British Isles, they have been described in the literature. One of the world's former leading sand researchers, Ralph A. Bagnold (1896–1990), devoted a small chapter of his own to "whistling beach sand" in his investigations into the physics of sand.[208] The phenomenon was recently described, but not explained, in the journal *Spektrum der Wissenschaft*.[209] And this is what is actually fascinating about "squeaking sand." No one has yet been able to satisfactorily explain what causes the squeaking, by which parameters it is influenced, and how it might be reproduced in a deliberate manner.

In the previous chapters, we have described some properties of the sands that could be considered in principle. Grain size, grading, mineralogical composition, degree of rounding, and surface structure could play a role in squeaking. If one goes into detail, it becomes very complex, since moisture, bulk density, loading velocity, and resonance chambers of the direct environment are added as influencing variables; all this, however, only as a statistically valuable combination of properties, because, as we know, no two sand grains, let alone two sand samples, exist that are completely identical and could therefore behave identically in a reproducible manner.

Bagnold found in his investigations that apparently neither temperature nor depth of the dry sand layer plays a prominent role. He described that the sand samples of squeaking beaches consist of predominantly well-sorted and well-rounded quartz cores in the diameter range around 300 µm. He also said that minor impurities from other minerals were not significant for squeaking. Laboratory tests did not yield any meaningful answers. On the contrary, Bagnold observed that some sand samples squeaking on the beach, which were then taken to the laboratory and isolated, did not squeak anymore.

Previous double page:
Dueodde, Bornholm, Denmark

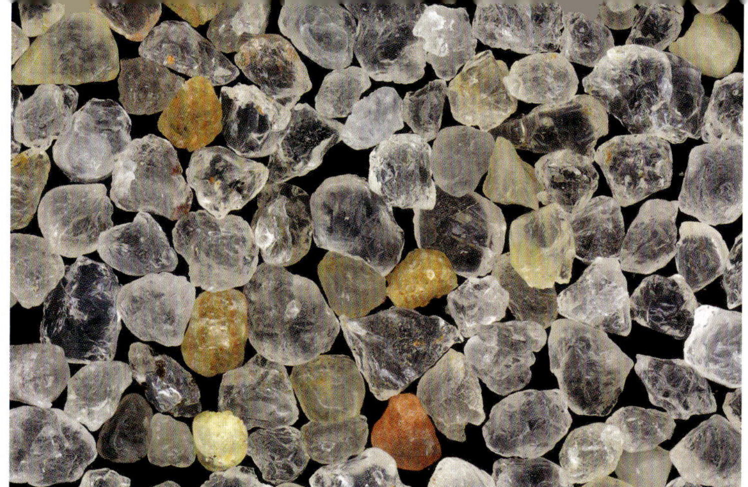

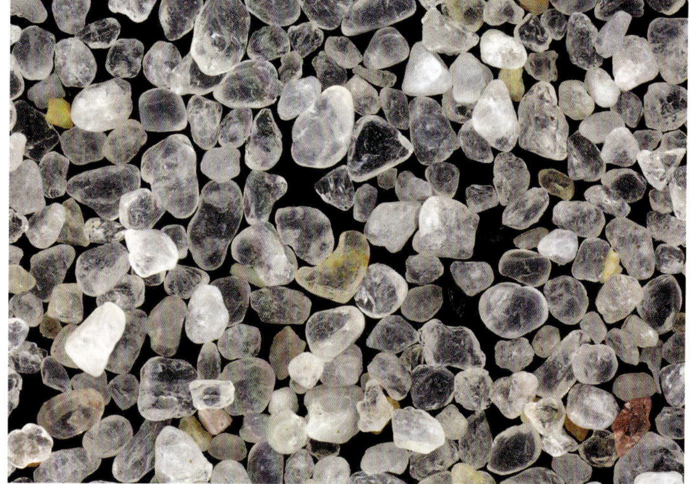

Even more than sixty years later, not much progress has been made. In short, there is still no useful explanation for the remarkable squeaking of some sand. And presumably no funding program or even a private sponsor will ever provide research funds for this. No further benefit could be expected from the knowledge to be gained, even upon longer reflection. Apparently, the sand seems to be able to keep this secret for quite a while longer yet.

This brings us back to the beginning. "What is useful is only a part of what is important," declared old Jarno in Wilhelm Meister's *Journeyman Years*, who later appears in the novel as Montan, a man of the mountains, of the earth sciences. Goethe probably gave this character some of his own traits, since the earth sciences were especially close to his heart, though admittedly in a different way than we are used to today. For him, a rock was not just a rock, a dead object, but rather a product and result of an intrinsically living natural process. A symbol of this process even.

"In living nature, nothing happens that is not in connection with a whole. When experiences appear to us in isolation or when we look at experiments as presenting only isolated facts, that is not to say that the facts are indeed isolated. The question is: how do we find the connections between phenomena or within a given situation?"[210]

Besides the reference to modern natural sciences made elsewhere to this thought, another aspect is to be emphasized. Instead of striding mentally from the individual to the whole, the whole, the universal, can appear in the individual, the particular. The individual already contains the universal, through whose individual expression it can be seen.

Figure 10.1 *Left*: Sand from Hasle, Bornholm. The sand comes directly from a squeaking-sand area. Reflected light. Image width 5.7 mm.

Figure 10.2 *Right*: Sand from the "Squeaky Beach," Wilson Promontory National Park, Australia. Neither sample, taken under identical conditions, shows any significant similarities at first glance or characteristics that would point to the squeaky phenomenon. Reflected light. Image width 5.7 mm.

The individual as a symbol of the whole, but inescapably connected—the connection is always there; you just have to look for it. Against this background we had understood sand and sand grains. As the result of a living dynamic natural process. Light, matter, and life shine out in the sand. Every single grain as part of the natural phenomenon of "granular matter" carries its own history in uncountable variety. Be it grains, whose emergence we can observe directly and immediately "live" when weathering out of a rock face, or be it grains, whose emergence on Earth dates back over a thousand million years in distant times. We have tried to shed light on them and some of the mysteries hidden within them. Transmitted light and reflected light.

Sand takes on a kind of intermediate role among the substances of nature. Molecules and atoms arrange themselves into maybugs, sunflowers, and giraffes all by themselves: we can only watch them. However, nobody can produce a giraffe, but they can produce a computer or a house. Even Aristotle sarcastically remarked that if one buries a bedstead from willow wood, a willow develops from it, but not a new bedstead. It is different with sand. Due to its definition by size and granularity, it can be produced simply by using a hammer or mortar. Sand is therefore somewhere between a natural and a manufacturable material, between singularity and plurality as well. We have tried to show this, in that the photographic portrait of a single grain, by lifting it out of the mass of grains, gives a face to precisely this one, whose traces suggest its history and whose trace is now manifested by the portrait in history, even if the grain of the almost infinite mass of sand of the earth is once again and finally covered, forever now untraceably hidden in the anonymity of the multitude.

And sand, as we understand it, is a mediator between the worlds of the macrocosm and the microcosm, thematically linking the formation and passing of enormous mountain masses with the microscopically small, iridescently colored interference figures of light as it passes through crystalline grains on our microscope slide. The randomly chaotic arrangement of sand grains in any given stretch of beach, permanently rearranged and rearranged by ocean surf and wind, contrasts with the strict, immutably fixed crystallographic lattice structure of each individual grain. The synopsis of macrocosm on a large scale and microcosm on a small scale becomes particularly prominent if we realize again that we had already met the oldest and originally most-distant grains of sand of the solar system (chapter 2).

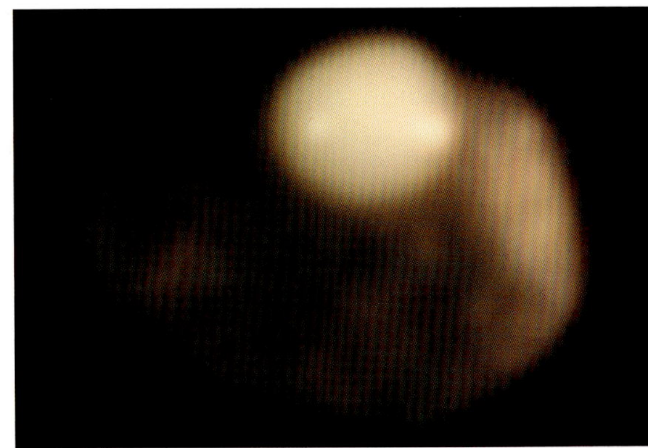

Figure 10.3 Sand grain photographed through a pinhole microscope without an optical lens, where the size of the pinhole is similar in size to the observed grain of sand. Imaging was accomplished without any mediation solely by light, an aperture, and the light-sensitive surface of the silicon image sensor. The strangely dreamlike shape of the grain may stand for the metamorphic level of the sand as a mediator between different levels of meaning, as an intermediate between macrocosm and microcosm. Snaefjellsnes, Iceland.

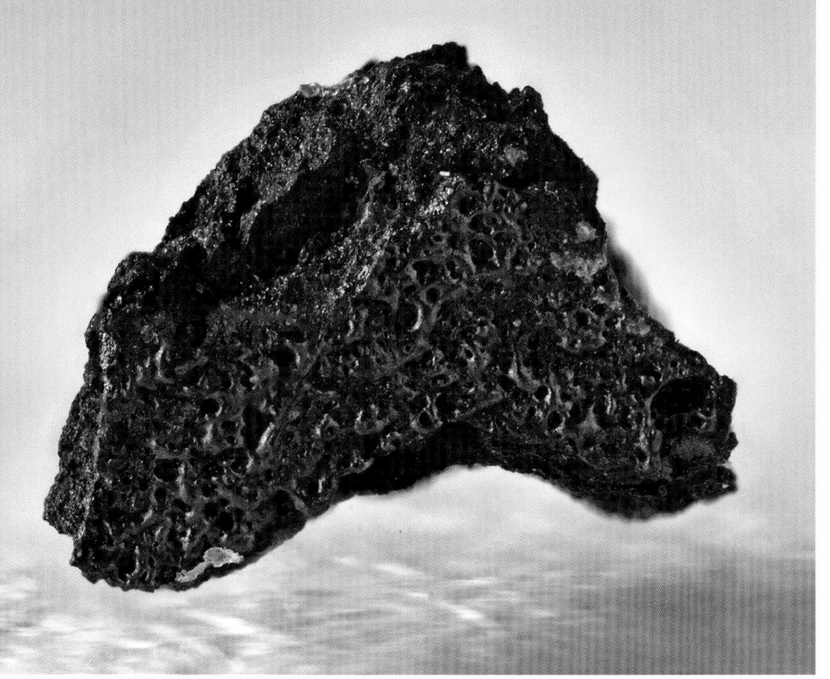

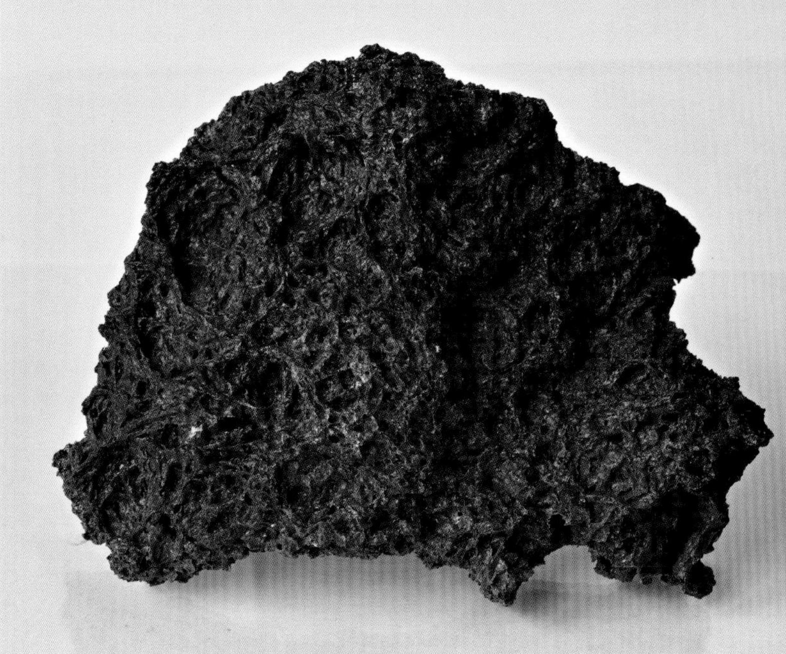

It is these round grains, called chondrules (from the Greek *chondros* = grain), that can be found in certain rock meteorites, the chondrites.

The mineral composition and form of the chondrules (fig. 10.7) are over 4.5 billion years old, frozen in a fine matrix. These are the oldest sand grains ever and also the oldest matter that humans can come into contact with and be able to research. Matter also, of which our earth consists, and with it we ourselves and everything what we take in from this earth into us. After everything we know, the macroscopic universe surrounding us consists of exactly these substances, which are familiar to us from our immediate nearness. In both cases, it's stardust.

The following figures demonstrate this through the example of the important and beautiful mineral olivine, which almost symbolically connects all observed worlds. It is one of the most common in the solar system and has also been detected beyond the solar system in the dust clouds of stars and quantitatively forms the largest part of our earth's mantle. Hidden here, it reaches the surface with some luck and is accessible to us either carried along by volcanoes (fig. 10.9), through the collision of continents (fig. 10.6), or, with good luck, through the impact of meteorites (fig. 10.8). Extraterrestrial, within the earth, and on the earth.

Alexander von Humboldt described the mutual relations among light, matter, earth, stars, and the stones of the air, the aeroliths, in a particularly beautiful and clear-sighted way and targeted them in their temporality and perceptibility by man. We therefore want to let him have his say at the very end—quintessentially, as it were.

Figure 10.4 Piece of the Tarda meteorite, about the size of a sand grain. The meteorite fell over Morocco on August 25, 2020, and belongs to the very rare class of carbonaceous chondrites C2-ungrouped. The bubbly melted crust of the surface is easy to recognize. Without the direct observation of the meteorite falling, this grain would have been transferred unnoticed to the sand of the desert. Width 2 mm.

Figure 10.5 Lava chunks from the volcanic eruption in April 2020 on the peninsula of Reykjanes, Iceland. The tholeiitic, quite thin basaltic source magma originates from a depth of about 12 miles (20 km). The piece cooled directly before being collected with the recognizable bubbly melt crust. Width 95 mm.

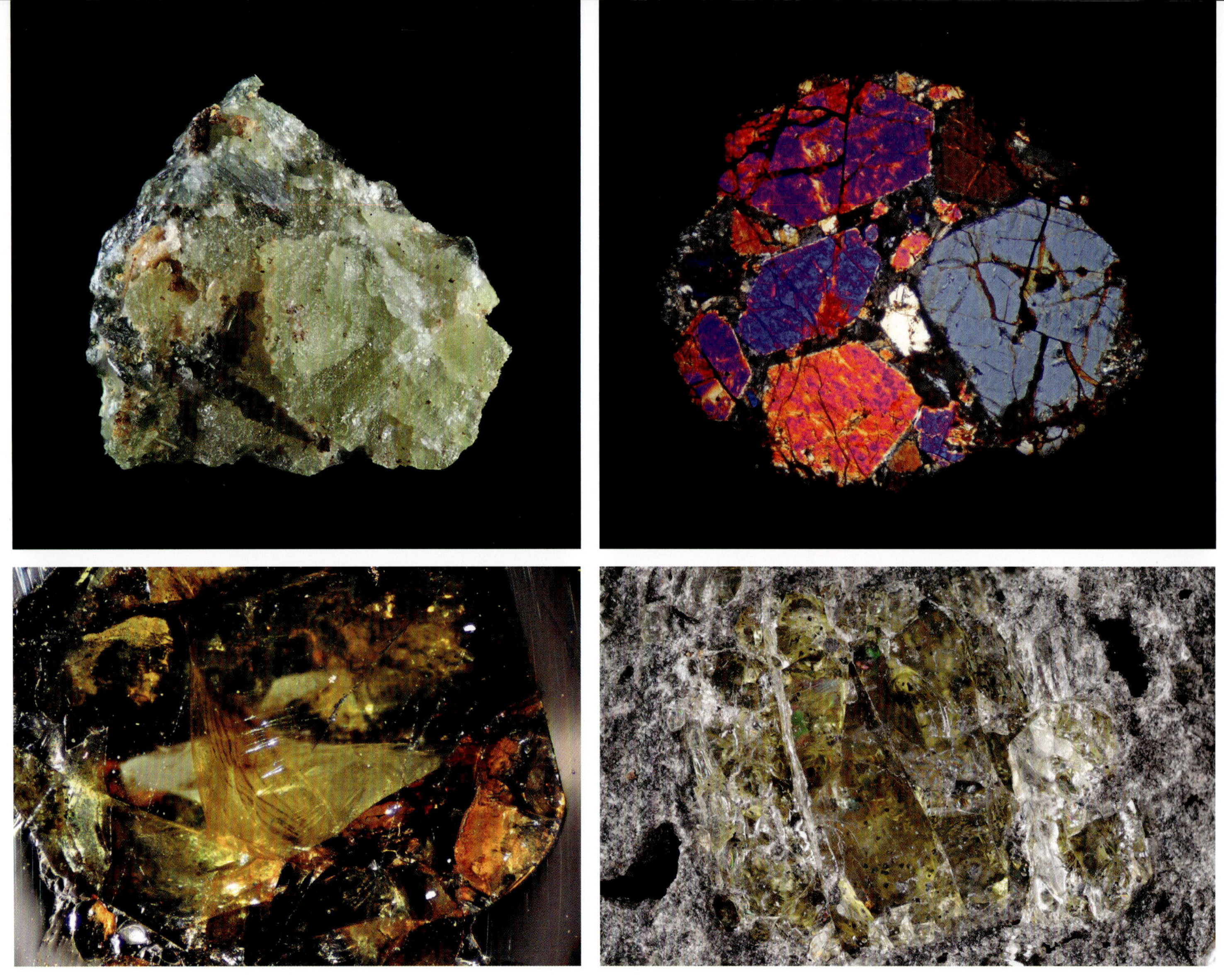

Figure 10.6 *Top left*: Earthly formed olivine in a sand grain of peridotite, a stone containing olivine. Peridotite comes from the interior of the earth, more precisely from the area of the upper mantle. A floe of the mantle is accessible at the site, due to mountain-forming processes. According to estimates, the olivine (modification ringwoodite) contained in the earth's mantle at depths of 249–404 miles (400–650 km) contains more bound water than all the world's oceans together. The green tone of the olivine changes with its iron content. Canobbino streambed near Finero, Ivrea Zone, Italy. Image width 1.4 mm.

Figure 10.7 *Top right*: Extraterrestrial (idiomorphic) olivine crystal with brightly colored interference colors with a chondrule from the Ophara H5 meteorite, which landed in Australia. Thin section, transmitted light with crossed polarizers. The different colors are caused by the differing orientations of the crystal axes. Image width 1.3 mm.

Figure 10.8. *Bottom left*: Extraterrestrial olivine crystal from the Seymchan meteorite (pallasite type). The olivine is bedded in an iron-nickel matrix and shows traces of shock (cracks). The Seymchan meteorite comes from the interior, more exactly the core/mantle border, of an over-four-billion-year-old and later broken asteroid. Olivine is one of the most common minerals of the solar system and can be found in the dust of other stars in our galaxy. Polished plate, reflected and transmitted light. Image width 9 mm.

Figure 10.9 *Bottom right*: Earthly formed olivine crystal from the interior of the earth, brought to the surface by the 1977 eruption of the Fournaise on the volcanic island of Reunion, France. The olivine crystal is embedded in bubbly lava and has numerous mineral inclusions. Image width 7 mm.

"I have already pointed out earlier that we are in contact with the universe and what fills it only by light and heat vibrations; as by the mysterious forces of attraction which distant masses (heavenly bodies) exert according to the quantity of their particles on our globe, its oceans and atmosphere. The light oscillation, which goes out from the smallest telescopic fixed star, from a soluble nebula, and for which our eye is receptive, brings us . . . a testimony of the oldest existence of the matter. An impression of light from the depths of the star-filled celestial spaces leads us back by means of a simple thought connection over a myriad of centuries into the depths of the prehistoric times. Even if the impressions of light coming from shooting[-]star streams, aerolites hurling fireballs, or visible fire meteors may be of quite different nature, even if they first ignite, when they reach the earth atmosphere, nevertheless the falling aerolite offers the only spectacle of a material contact of something that is alien to our planet."[211]

We close at this point, end our journey, and recapitulate that basically the subject of our reflections was quite simply and plainly only one thing: sand.

While you, dear sand lover, have been reading this epilogue, some five hundred billion new grains of sand have been created outside on Earth, none of which resembles the other. Wherever you come into contact with sand in the future, a magnifying glass or microscope will allow you to look into a world whose uniqueness passes us by when we just carelessly shake the sand out of our shoes after a long day at the beach. I hope the book has helped do this at least with a certain respect: pausing.

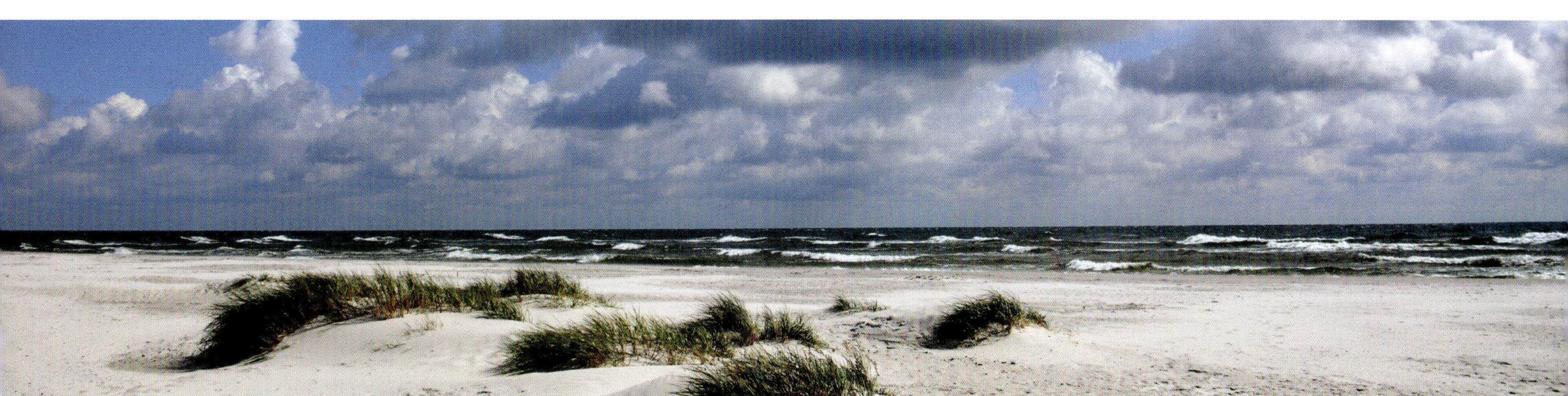

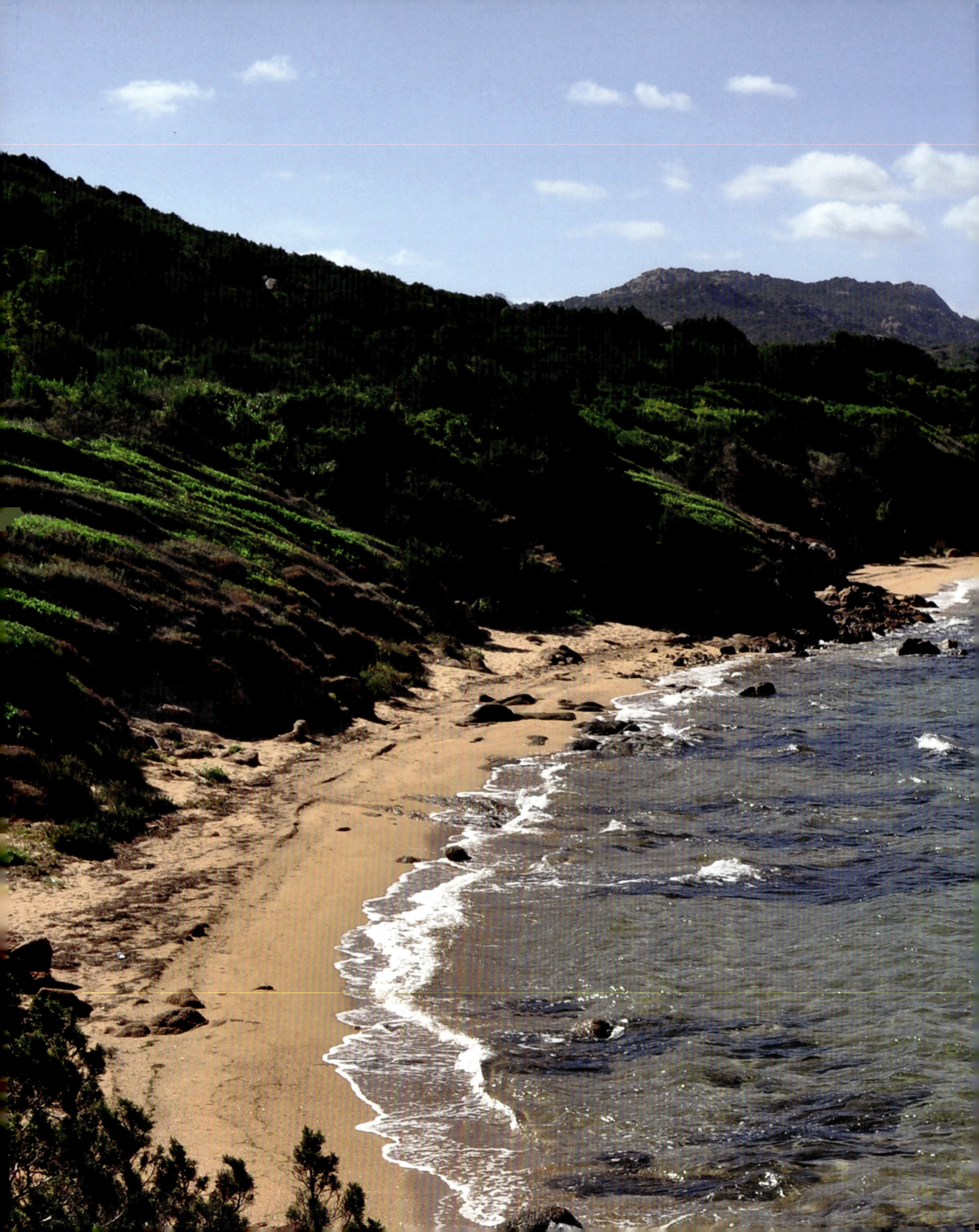

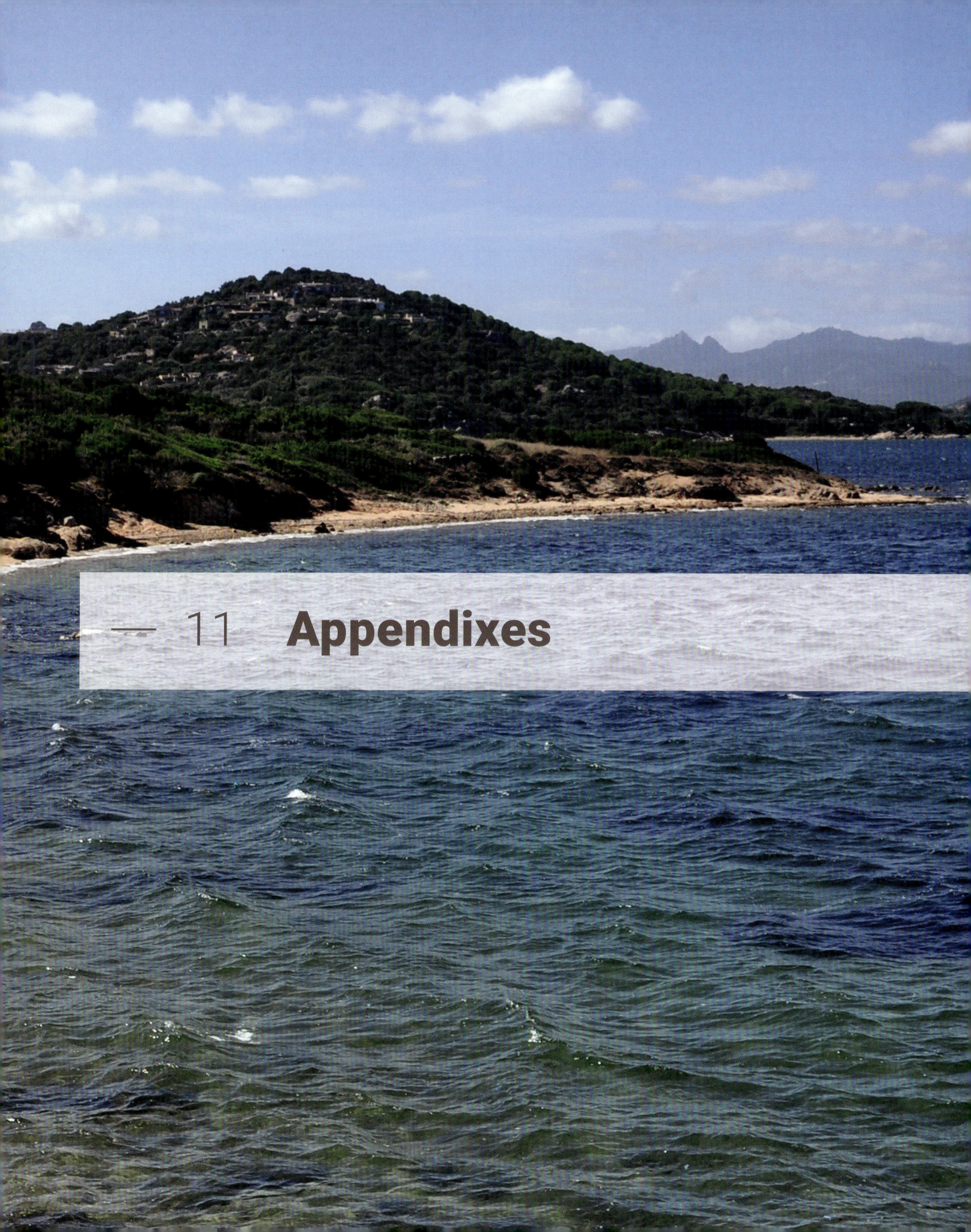

— 11 Appendixes

The source from which existing things derive
their existence is also that to which they return at their destruction,
according to necessity;
for they give justice and make reparation to one another for their injustice,
according to the arrangement of time.

—Anaximander of Miletus, ca. 600 BCE

According to today's knowledge, Anaximander was the first thinker who freed the sciences from mystical ideas. He traced back the things of the world and their origin to a general principle, apeiron, a kind of balancing and unlimited original substance. He was the first to describe the earth as a sphere floating freely in space, and he was also the first to draw a map of the then-known world.

We have tried to clarify the origin and "destruction" of sand, discovered the sand cycle "according to the arrangement of time," and found sand as an almost omnipresent "original substance." We conclude in the following with a supplementing sand selection from places around the earth—our world, which forms an inseparable community of fate as a sphere worth protecting, freely floating in the universe together with us.

11.1 Sands of the World, Additional Selection

So far, we have dealt with sand samples in each case under the aspect of certain criteria or parameters. From grain size to sand color to compositional maturity and the respective mineralogical composition. The origin, depositional environment, and mode of natural transport of the sediment have also been considered. The photographs in the book show either individual grains or a typical section of the sand sample, with image widths of a few millimeters.

Of course, it is neither practical nor possible to achieve even a somewhat complete overview of the sands of the world. Even a reasonably representative selection cannot be presented on a reasonable scale. Nevertheless, at the very end of our discourse we would like to add a number of other sands from all over the world as examples. This shows the diversity, on one hand, but also the similarity of some samples from quite different parts of the world that are impressive and appealing to look at. For simplicity, the samples are shown roughly in alphabetical order of their countries of origin. This creates a random but inspiring comparison.

For reasons of comparability, the overview images were taken with an image width of 5.9 mm. This enlargement allows a good overview of the composition but also makes the individual grains stand out. Some samples with finer sands were photographed at an image width of 3.8 mm. All the images were taken under identical conditions, using natural light without additional lighting, so that they stay true to color.

Previous double page:
Cala Bitta, Sardinia, Italy

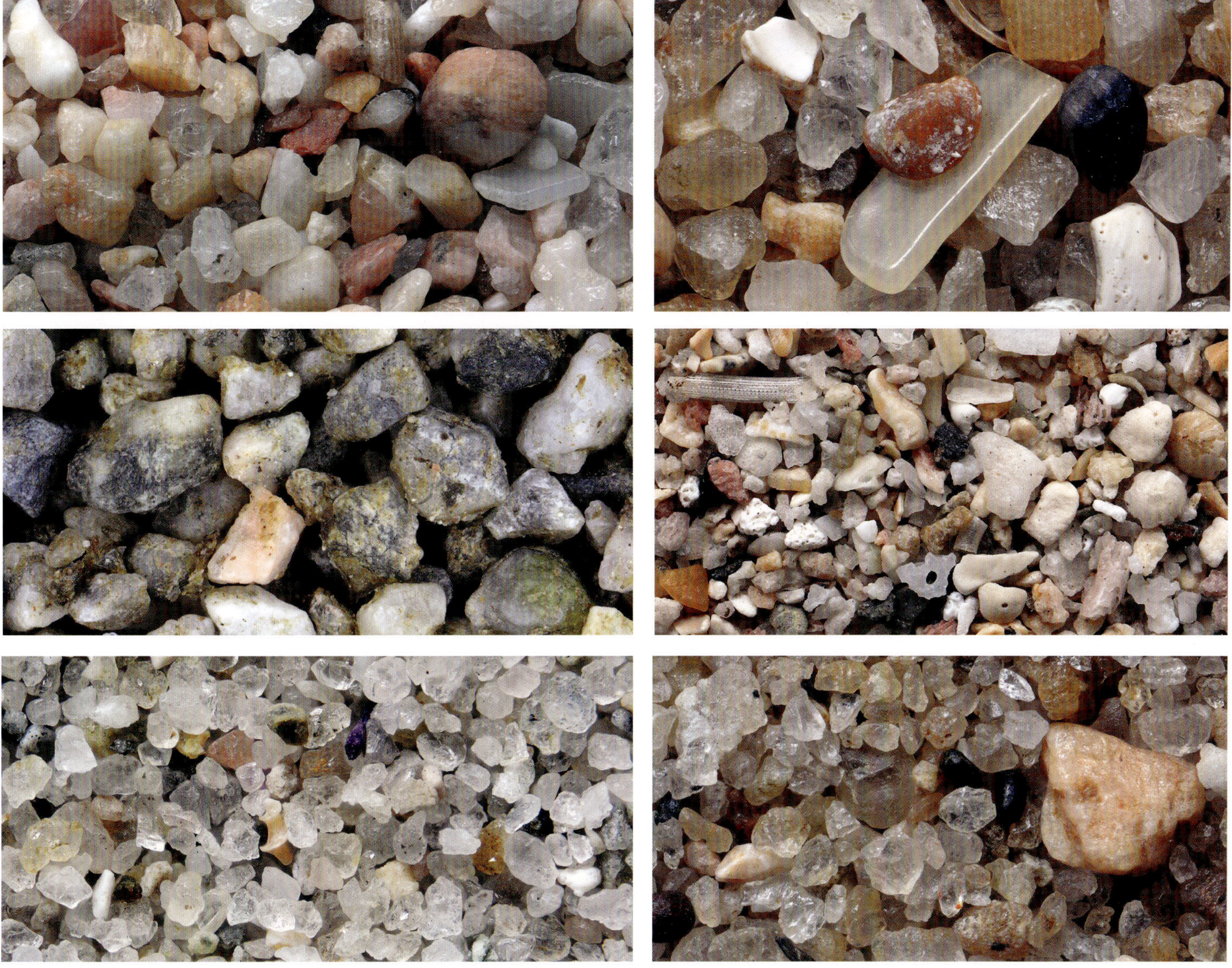

Figure 11.1 Egypt, Red Sea, southern tip of the Sinai Peninsula. The round-edged sand has components of land sediments (quartz, feldspar) as well as many biogenic parts in the form of shell fragments and see urchin spines (*top, right of the middle*). Image width 5.9 mm.

Figure 11.2 Antarctica, Marguerite Bay, Adelaide Island, Rothera Research Station. Even though the sediment sample, taken from the waterline, was washed with water (an exception), algae remains can still be seen. The grains come from the mountains of the backcountry and are angular due to their short transport distances. The sand is very immature and is composed mostly of rock fragments. Image width 5.9 mm.

Figure 11.3 Australia, Tasmania, Nubeena, White Beach. Well-sorted, subangular, almost pure quartz sand with a few biogenic components. Image width 3.8 mm.

Figure 11.4 Bulgaria, Golden Sands. Coarser, relatively unsorted sand with a high proportion of quartz, but also further minerals and shell remains from the animals of the Black Sea. The quartz in the picture is partially clear but also has parts that are covered with a golden iron oxide layer. The long-standing name of the beach probably comes from this, in combination with the color impression of the feldspars. Image width 5.9 mm.

Figure 11.5 Chile, Easter Island, Hanga Roa. Pure, biogenic sand, composed of disc-shaped foraminifera (*right half of the picture*), shell remains, and the spine of a sea urchin (*top left*). Image width 5.9 mm.

Figure 11.6 China, Xiamen, beach on the southern tip. Angular sand from the coast of a densely built urban area. Mostly quartz, barely any biogenic components, moderate to poorly sorted. Image width 5.9 mm.

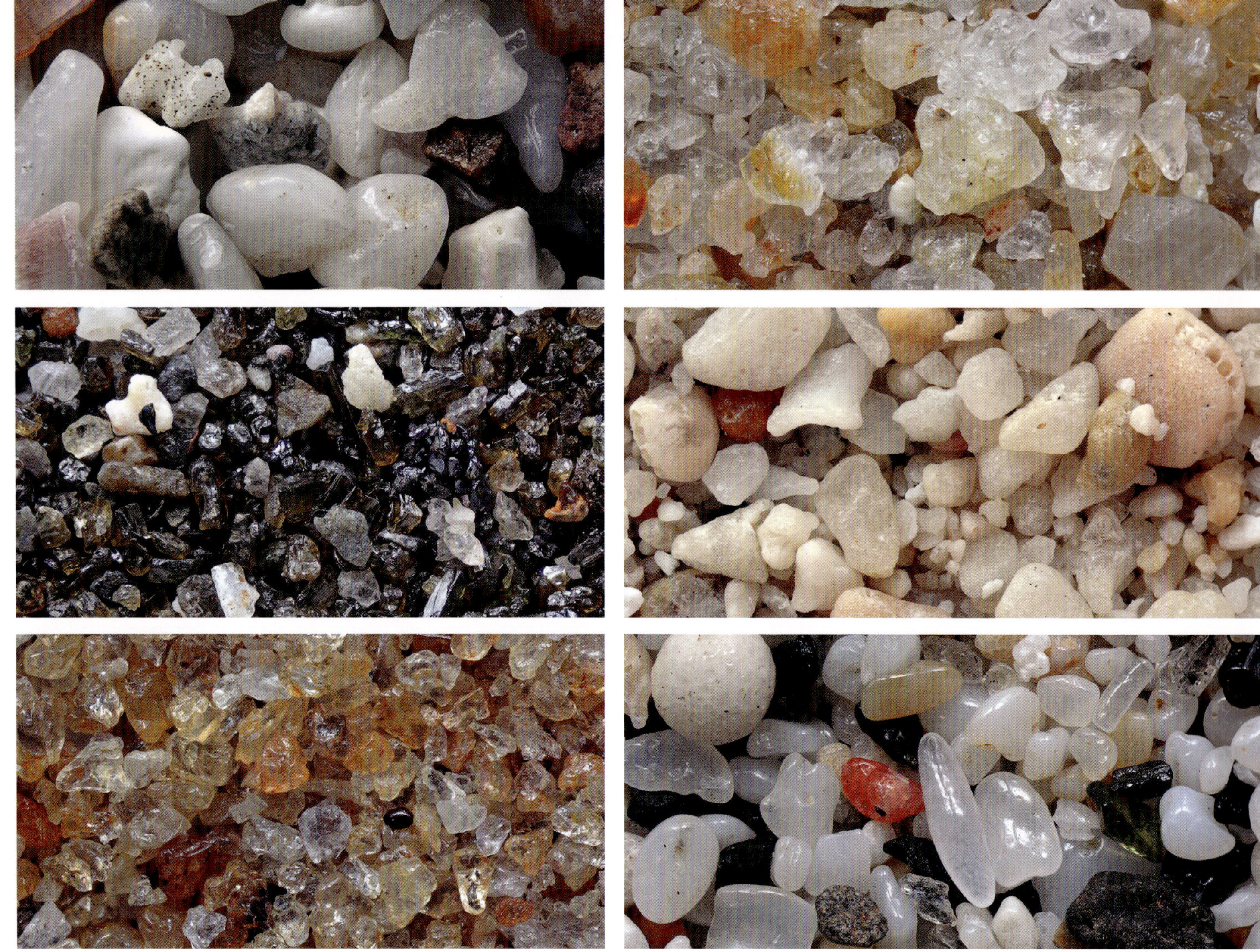

Figure 11.7 Costa Rica, Playa Espadilla Sur. The sand from Costa Rica's Pacific coast is a mix of calcareous fragments of marine animals and minerals of continental origin. Image width 3.8 mm.

Figure 11.8 France, Lesser Antilles, Guadeloupe, Bass-Terre. The volcanic complex of the Lesser Antilles island of Bass-Terre is around 200,000 years old and still active. The black, fine-grained sand is therefore made up of green to dark-green minerals (pyroxene, olivine) and a lot of magnetite, as well as some biogenic fragments. A collection of small, magnetically attached magnet grains can be found (*to the right of the middle*). Image width 3.8 mm.

Figure 11.9 France, French Guiana, Cayenne, Plages de Montjoly. The reddish, almost orange color of the sand comes from the iron oxide–containing coatings on the quartz grains. The iron oxide was probably brought in from the nearby disemboguing rivers that originate in the rainforest. The sand is moderately sorted; the grains are subangular to rounded. A mix with river sediments is probable. Image width 5.9 mm.

Figure 11.10 India, Tamil Nadu, Chennai, Marina Beach. Marina Beach is constantly being filled with sediments from a current running from south to north. Sediment cargo from the disemboguing Adyar River to the south is also being carried. In 2004, the very popular beach was overrun by a devastating tsunami. The sand is moderately sorted, and the grains are angular to slightly rounded. Quartz dominates. Image width 5.9 mm.

Figure 11.11 Indonesia, Sulawesi, Togian Islands, Kadidiri. Kadidiri lies in the middle of the tropical Indonesian group of islands on the northwestern edge of the Togian Islands. The sand is singularly composed of shell fragments of marine lifeforms and is very unsorted. Image width 5.9 mm.

Figure 11.12 Indonesia, Bali, Lovina, Singaraia. Bali is a part of the volcanic island group (Sunda Arc) that was created by the subduction of the Sahul plate below the Sunda plate. The sand is a gorgeous, colorful mix of volcanic rock fragments and minerals such as olivine, clinozoisite, and pyroxene, as well as biogenic grains from sea urchin spines, sclerites, foraminifera (a spherical example, *to the top left*), and diverse shell fragments. All grains are spectacularly polished by the ocean surf. Image width 5.9 mm.

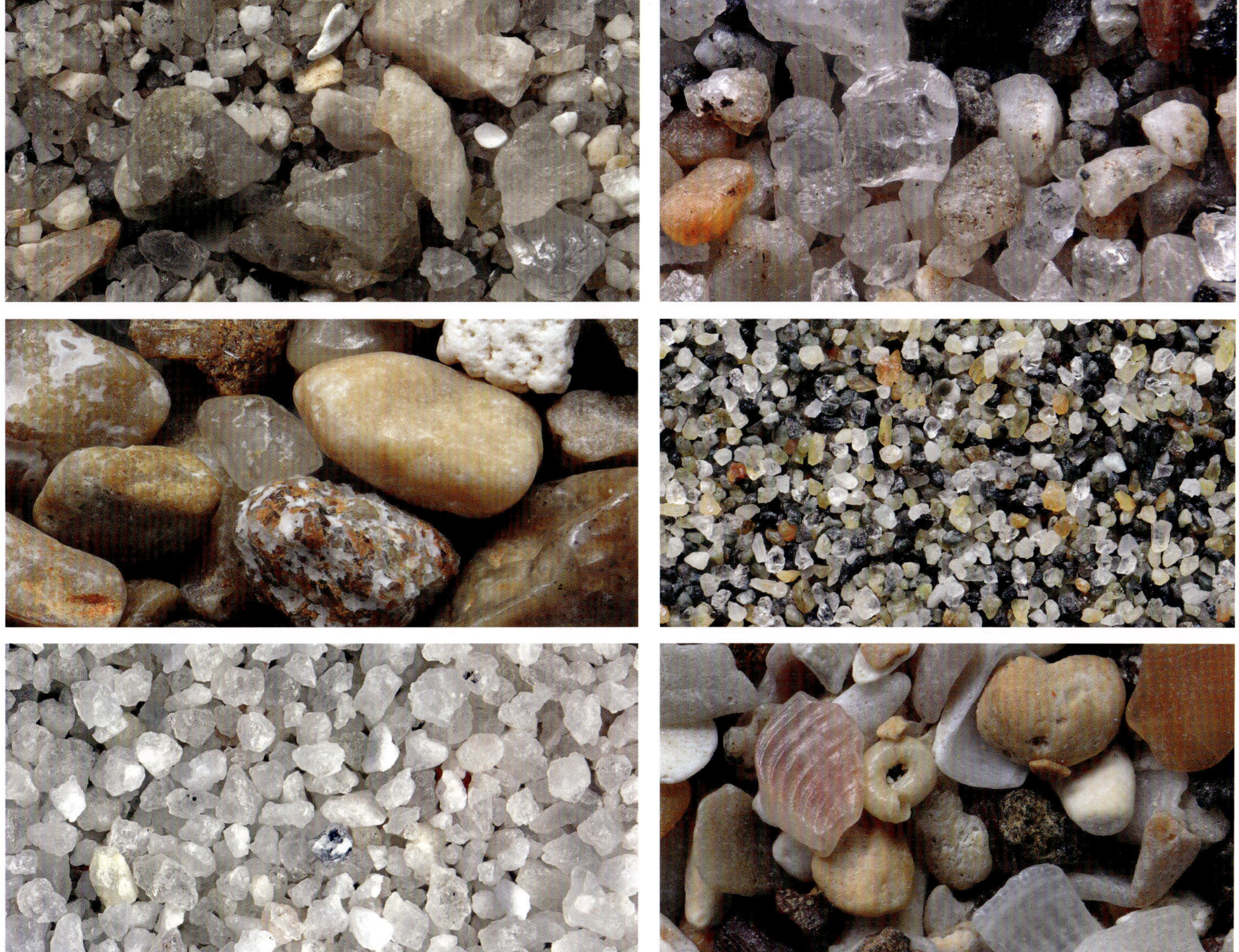

Figure 11.13 Japan, Miyajima. Coarse, unsorted, and angular to very angular sand. Miyajima Island (literally Screaming Island) lies in a bay in front of Hiroshima and is dominated from Mount Misen, which is where the sediment comes from with a short transport distance. Image width 7.9 mm.

Figure 11.14 Jordan, Dead Sea, near Madaba. The sand grains are caked after drying due to the high salt content of the Dead Sea and are well rounded. The coarse sand (on the border to fine gravel) is very immature, with low sorting. Image width 5.69 mm.

Figure 11.15 Cambodia, Saracene Bay, Koh Rong, Sanloem Island. The sand of Sanloem Island has a curious structure. Pure quartz sand, uniform grain size, no biogenic components, angular grains, edges only minimally rounded. For marine sand, this combination of characteristics is very rare to impossible. It is probable that this is a previously sorted beach replenishment with allochthonous material (see chapter 9). Image width 5.9 mm.

Figure 11.16 Canada, Kelowna, Okanagan Lake. Image width 3.8 mm. Okanagan Lake stretches from north to south in the mountain region east of Vancouver. The sand is moderately coarse, immature, and partially angular. Glacially influenced sediment as the original material is suspected. Image width 5.9 mm.

Figure 11.17 Canada, Vancouver Island, Pacific Rim National Park, Long Beach. The very fine sand is well sorted (similar granulation) and has a moderate proportion of quartz. Along with orange-colored feldspars are abundant unspecified green minerals (epidote, pyroxene, amphibole) that give the sand an overall green coloration. The very fine grains are angular and take longer to round than their coarser representatives. Image width 5.9 mm.

Figure 11.18 Korea, Jeju Island. Eastern coast. Jeju Island is south of the Korean Peninsula in the ocean and was formed volcanically. You see this in its typically dark sand, which seems to be sprinkled because of the light shell fragments. When enlarged microscopically, the sand has very colorful elements of violet sea urchin spines, pink shell fragments, beige snail shells and foraminifera, and dark, partially eroded volcanic rock fragments. Image width 5.9 mm.

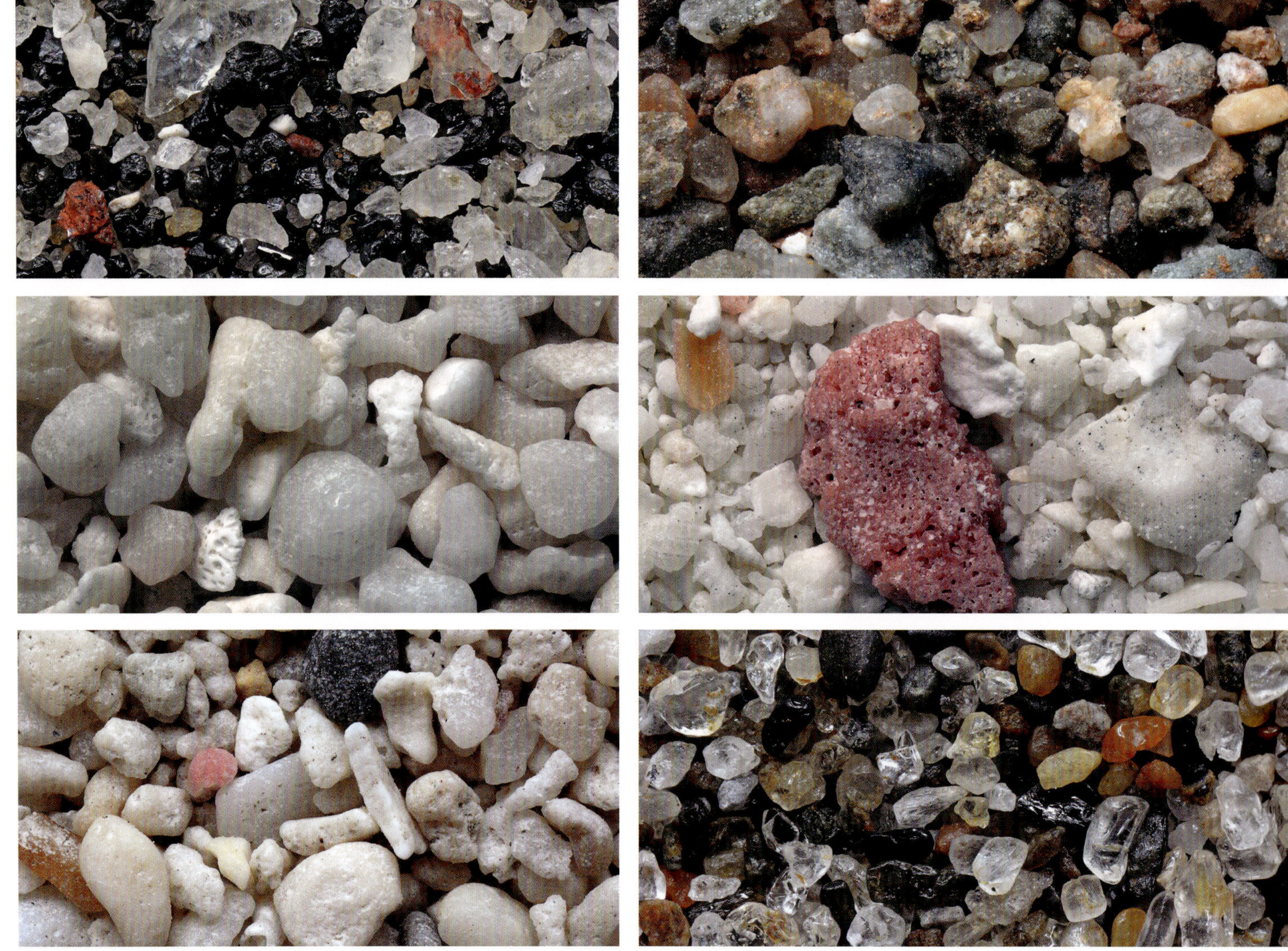

Figure 11.19 Malaysia, Lankawi Island, Black Sand Beach. The Black Sand Beach is found in the northern part of the island. The origins of the partly black, fine-grained sand are not clear—most likely it is a soap that is composed of dark minerals such as ilmenite and tourmaline. Magnetic components are not found. There is no other beach on the island that has this type of coloring. Image width 5.9 mm.

Figure 11.20. Maldives, Lhaviyani Atoll, Madhiriguraidhoo. The sand from the tropical atoll in the Maldives is composed only of lime, and with it the shell remains of marine life form. *To the top right*, a piece of a foraminifer can be seen; *to the bottom left*, feeding traces of gangue-boring algae can been seen in shell remnants. The grain size is moderate to coarse. Image width 5.9 mm.

Figure 11.21 Mexico, eastern coast, Cozumel Island. Cozumel Island lies in the Caribbean Sea on the eastern edge of the Yucatán Peninsula, where, sixty-six million years ago, a meteorite hit the northwestern edge, which led to the extinction of the dinosaurs. The coarse sand is composed mostly of biogenic lime. Image width 5.9 mm.

Figure 11.22 Mongolia, Gobi Desert, Sainshand. Moderately coarse, unsorted, immature, and mostly angular sand. Although Sainshand is in the eastern part of the Gobi Desert, this sample does not have any characteristics of desert sand. The history of this sample is unknown. Image width 5.9 mm.

Figure 11.23 New Calcedonia, Isle of Pines. The impressive white lime sand of the Island of Pines, in the Coral See off the southeastern tip of New Calcedonia, is solely of biogenic nature. Occasionally, coral branches or, as the picture shows, red fragments of the foraminifera *Homotrema rubrum* can be found in the white sand. Image width 5.9 mm.

Figure 11.24 New Zealand, Auckland, Whatipu Beach. The fine, overall black-appearing sand is composed of a variety of distinguished heavy minerals along with a proportion of bipyramidal quartz of probably volcanic origin. Auckland has about fifty small and extinct volcanic points in its famous "Volcanic Field." Image width 5.9 mm.

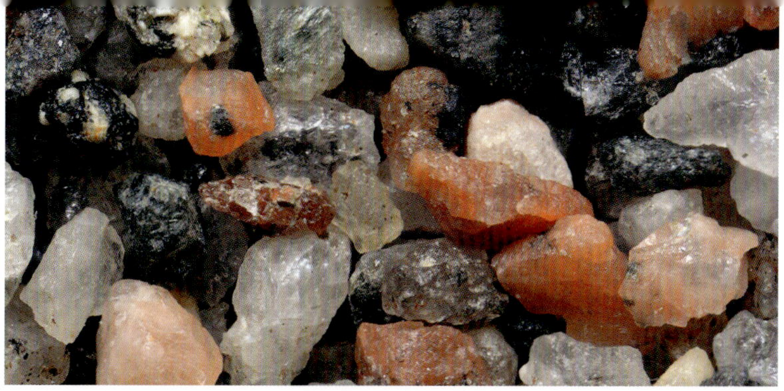

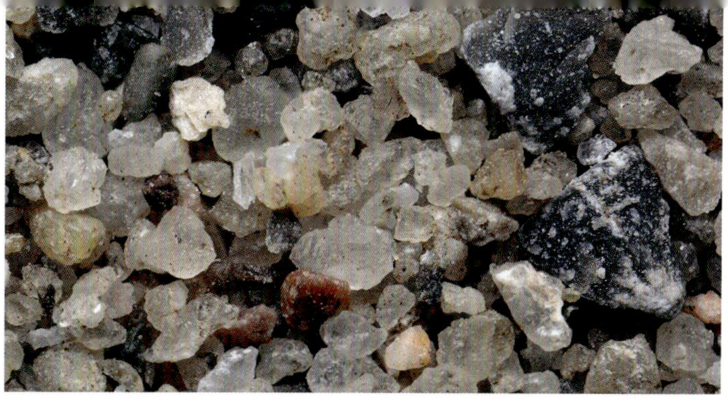

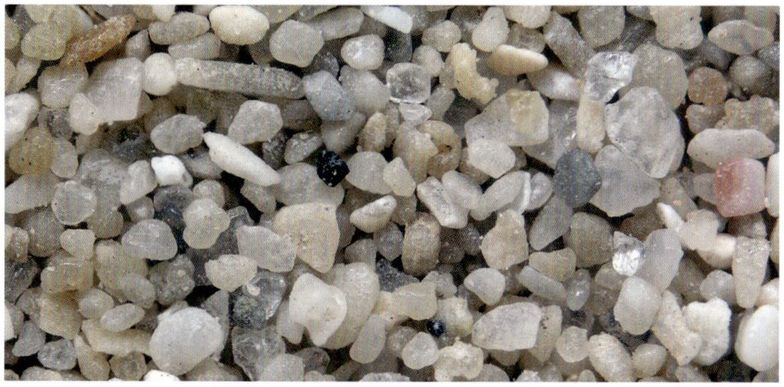

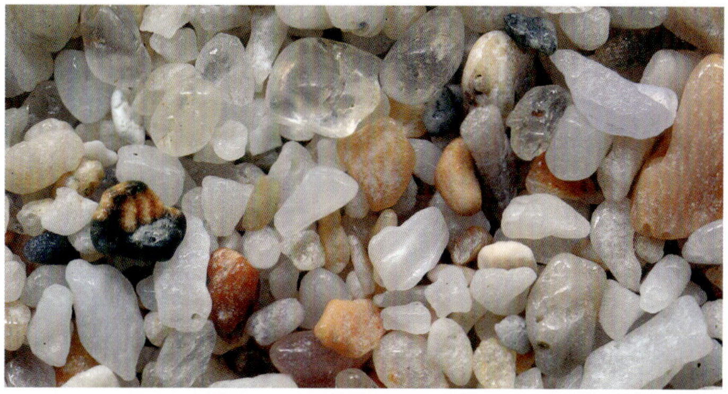

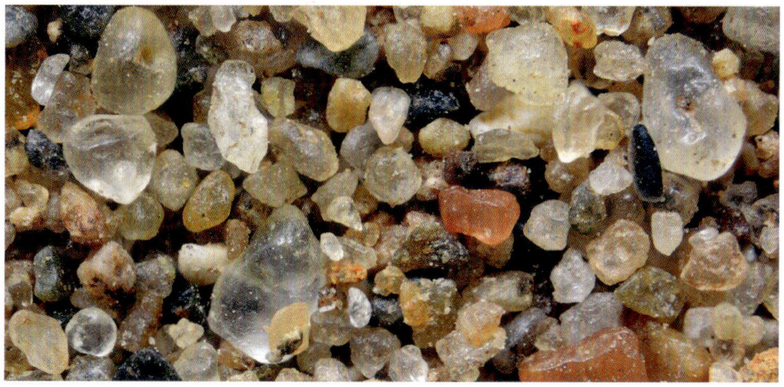

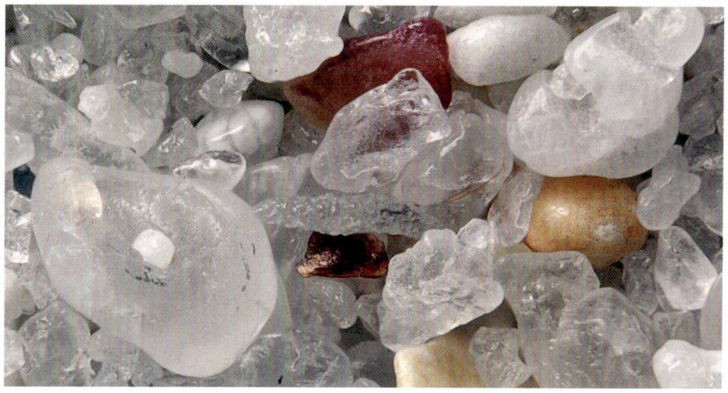

Figure 11.25 Norway, Kirkenes, Beach. The sand shown is from the adjacent Precambrian tonalitic- or granodioritic-composed gneisses and migmatites. Here, as well as on the southern flank of the Varangerfjord to the west, are the oldest bedrock outcrops in Europe, with an age of up to 2.9 billion years. Along with the clear quartz crystals, one can recognize the prominent orange-to-flesh-colored potassium feldspar and a greenish amphibole grain (*left of the center*). Image width 5.9 mm.

Figure 11.26 Oman, Salalah. Fine-grained and well-sorted marine sand, subangular to slightly rounded. Image width 3.8 mm.

Figure 11.27 Russia, Ulyanovsk, Volga River. This sample from the bank of the Volga, originally heavily enriched with suspended solids and fine particles (silt), was rinsed with clear water. The remaining fine and barely sorted sand has partially remarkably well-rounded grains, some even spherically formed. The origin is unclear—rounding caused purely from the river is not probable, meaning that these grains mostly likely have been through multiple erosion cycles. Image width 3.8 mm.

Figure 11.28 Russia, St. Petersburg, Neva River. The sample was taken from the bank of the Neva River in the city of St. Petersburg. The sand is moderately coarse, unsorted, and moderately mature. Allochthonous components cannot be excluded, on the basis of the sampling location. Image width 5.9 mm.

Figure 11.29 Sri Lanka, Unawatuna. Beautiful and impressively polished midsized sand from the Indian Ocean. The sand contains mineral grains as well as a large proportion of biogenic grains of different marine life-forms, as can be seen in the picture. Proportions of heavy minerals cannot be seen in the picture, which can be found in the form of smaller, magnetic, black grains. Image width 5.9 mm.

Figure 11.30. Seychelles, Praslin, Anse Lazio. Over the course of the continental drift, the Indian plate moved from the eastern coast of Africa to its current position two hundred million years ago. From it, continental, plutonic relics were left behind in the Indian Ocean, as well as the Seychelles and Sri Lanka. That is why on the inner Seychelles, continental, mineral grains can be found along with the coral sands: products of erosion from the impressive and granite boulder relicts, which appear to have come out of nowhere. Image width 3.8 mm.3,8 mm.

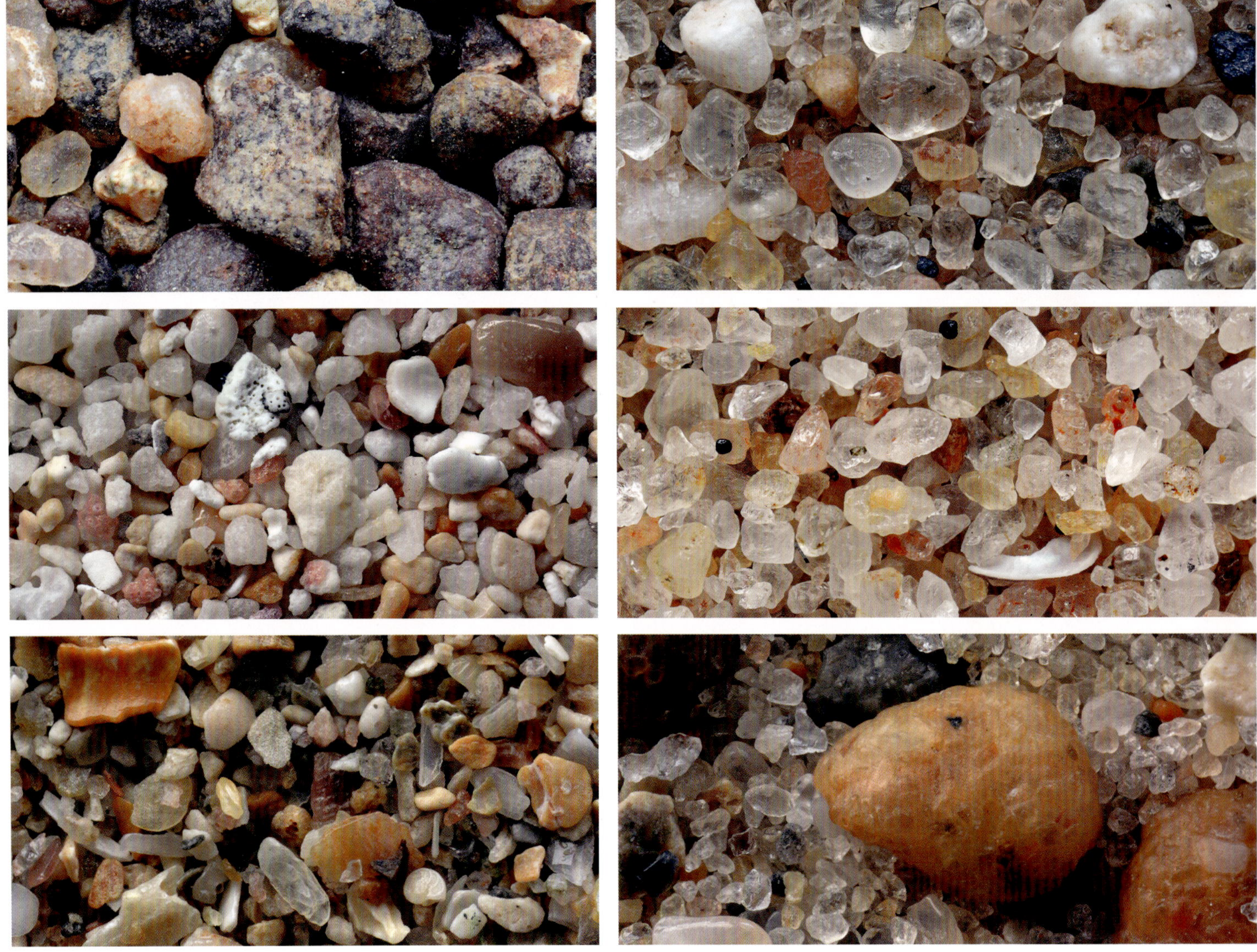

Figure 11.31 Tanzania, Kasulu, street edge. Unsorted coarse sand, where silicate minerals reign. Calciferous coatings are visible by effervescence when hydrochloric acid is added. Many grains show lustrous desert varnish coatings of iron and manganese oxides. The image shows a detailed view of the typical brick-red laterite sediments of Tanzania. Some grains are clearly rounded by edge erosion and wind transport; some even have medium to high sphericity. Image width 5.9 mm.

Figure 11.32 Trinidad and Tobago, Pigeon Point. Fine, biogenic marine sand, formed from shell and housing fragments of coral, mussels, foraminifera (*top of the picture, left of the middle*), snails, and others. Microboreholes of algae are recognizable in the white grain above the middle. Image width 5.9 mm.

Figure 11.33 Morocco, Essaouira. Moderately fine, brown-colored sand from the Moroccan Atlantic coast, which is almost exclusively composed from the shell remains of marine organisms. Mussel shells dominate, and only a few very small quartz grains can be identified: they most likely come from the dunes near the coast. Image width 5.9 mm.

Figure 11.34 USA, Lake Michigan, Sleeping Bar Dunes. Sand from one of the largest freshwater dune regions in the world. The sand is moderately sorted but has very high compositional maturity. It is composed mostly of quartz. The well-rounded grains lead to the inference of aeolian transport, but possibly also of a second erosion cycle. Image width 5.9 mm.

Figure 11.35 Vietnam, MuiNe. Moderately fine, almost pure quartz sand. In the backcountry of Mui Ne there are inland dunes that are stretched out, as well as rocky regions that partially extend to the coast. The sand has low sphericity and is rounded to partially rounded. Image width 5.9 mm.

Figure 11.36 Israel, Tel Aviv—Jaffa. The well-sorted, finely-grained quartz sand is special because it has a second size class of grains that are significantly larger and are composed of reddish-yellow sandstone fragments. Sand grains made of sandstone. The dimensions of their subgrains are similar to the surrounding fine quartz sand, which they clearly erode down to and enter their third cycle (at least). Image width 5.9 mm.

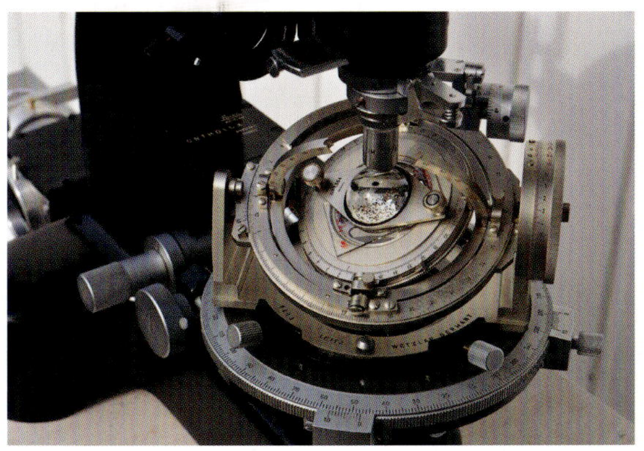

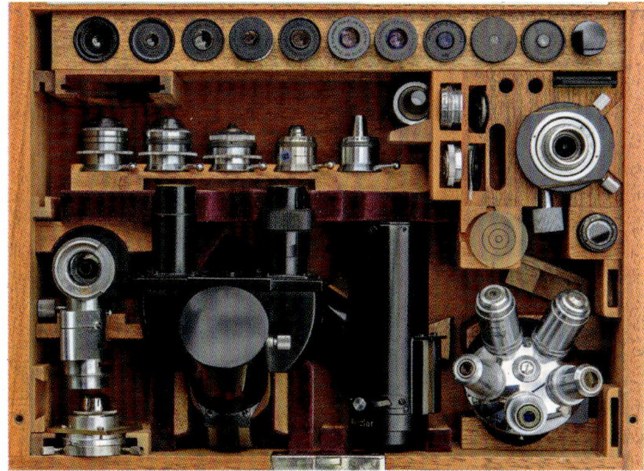

11.2 About the Photographs

Sand grains, although very small, are three-dimensional (i.e., spatially extended objects). The microscope produces planar images of them, each of which represents only a single height layer and has a shallow focal depth. For this reason, images were taken of each object in different layers and then combined with the aid of stacking software to form an overall image with continuous sharpness. Depending on the subject, this required between twenty and one hundred layered images. This procedure was used for images taken both in transmitted light and in reflected light (incident light).

Lighting has a significant influence on the image effect of micro-objects. Specially made miniature soft boxes, diffusers, and filters, such as are also known from portrait photography, together with the light sources used, ensured a natural-looking and true-color illumination of the sand grains in reflected light. Transparent grains, on the other hand, can be examined like diapositives in transmitted light. In most cases, polarizing filters were used, which are also necessary for analysis purposes. If the grain of sand is between two crossed polarizing filters, as described in the text, colorful inference colors are produced against a black background. In addition to their attractive effect, the colors then allow conclusions to be drawn about the crystalline material.

The postprocessing of the image files was limited mainly to a moderate adjustment to the directly experienced visual perception of the objects. In isolated cases, the background was additionally darkened to emphasize the image motif.

The photomicrographs were taken predominantly on a Leitz Ortholux 1 microscope from the 1960s (fig. 1.6 and picture above left with mounted five-axis rotary stage) and occasionally on a Leitz Ortholux 2 Pol BK polarization microscope. Original objectives and ancillary equipment of the time were used to a large extent (*see figures above*). Macroimages were taken with different self-made solutions and different objectives, such as the Micro Nikkor 60 mm 1:2.8 G ED objective and a Laowa Ultramacro 2.5–5:1. The microscopes were equipped with Nikon cameras of the D 5000 and D 7100 series. All images were taken manually; methods with automated feed and such were not used. The original image section is indicated in each case as image width and typically lies between 0.8 mm and 8 mm.

11.3 Notes

1. M. Ritter, "Goethe und Augustin-Pyramus De Candolle," in *Abenteuer der Vernunft: Goethe und die Naturwissenschaften um 1800* (Dresden, Germany: Sandstein Verlag, 2019), 287.
2. Johann Wolfgang von Goethe, "Brief an Charlotte von Stein vom 27. Juni 1785," in *Goethes Briefe*, Hamburger Ausgabe, Band 1, ed. Robert Mandelkow (Munich: Deutscher Taschenbuch Verlag, 1988).
3. Georg Forster, *Reise um die Welt* (Frankfurt: Eichborn, 2004), 183.
4. Friedrich Delitzsch, *Assyrische Lesestücke, mit grammatischen Tabellen und vollständigem Glossar*, 4th ed. (Leipzig: J. C. Hinrichs, 1900), 127.
5. Ibid., 143.
6. Ellermeier, *Sumerisches Glossar* (Nörten-Hardenberg bei Göttingen, Germany: F. Ellermeier, 1979), 258.
7. John A. Halloran, ed., *Sumerian Lexicon: A Dictionary Guide to the Ancient Sumerian Language*, bilingual edition (Los Angeles: Logogram, 2006).
8. Leonard Woolley, *Ur und die Sintflut: Sieben Jahre Ausgrabungen in Chaldäa, der Heimat Abrahams* (Leipzig: F. A. Brockhaus, 1930).
9. Wolf Dieter Blümel, *Wüsten: Entstehung, Kennzeichen, Lebensraum*, UTB Bd. 3882 (Stuttgart: Ulmer, 2013), 115.
10. H. M. Enzensberger, ed., *Alexander von Humboldts Kosmos—Entwurf einer physischen Weltbeschreibung*, book III (Frankfurt: 2004), 9.
11. Ibid., 10.
12. Johann Wolfgang von Goethe, "Brief an Zelter vom 22. Juni 1808," in *Goethes Briefe*, Hamburger Ausgabe, Band 3, ed. Robert Mandelkow (Munich: Deutscher Taschenbuch Verlag, 1988).
13. Johann Wolfgang von Goethe, "Infusions-Tiere; Studien über organische Naturen bis zur Rückkehr aus Italien 1776–1788," in *Schriften zur Morphologie, Sämtliche Werke*, Frankfurter Ausgabe, Serie I, Band 24 (Frankfurt: Deutscher Klassiker Verlag, 1987).
14. Otto Krätz, *Goethe und die Naturwissenschaften*, 2nd corr. ed. (Munich: Callwey, 1998), 173.
15. O. Richard Norton, *Rocks from Space: Meteorites and Meteorite Hunters*. 2nd ed. (Missoula, MT: Mountain, 1998), 13.
16. Jens Soentgen and Knut Völzke, *Staub—Spiegel der Umwelt*, Stoffgeschichten 1 (Munich: Oekom Verlag, 2006), 74.
17. Susanne Klerner, *Materie im frühen Sonnensystem: Die Entstehung von Chondren, Matrix und refraktären Forsteriten* (Universität zu Köln, 2001).
18. Ian S. McKeegan, „Early Solar System Chronology," in Andrew M. Davis, *Meteorites, Comets, and Planets: Treatise on Geochemistry*, vol. 1, 2nd ed. (Amsterdam: Elsevier, 2005).
19. Ludolf Schultz and Jochen Schlüter, *Meteorite* (Darmstadt: Primus, 2012), 52.
20. Wolf von Engelhardt, Christoph Berthold, Thomas Wenzel, and Thomas Dehner, "Chemistry, Small-Scale Inhomogeneity, and Formation of Moldavites as Condensates from Sands Vaporized by the Ries Impact," *Geochimica et Cosmochimica Acta* 69, no. 23 (December 1, 2005): 5611–26.
21. Hoimar von Ditfurth, *Kinder des Weltalls: Der Roman unserer Existenz* (Hamburg, Germany: Hoffmann & Campe, 1970).
22. Johann Wolfgang von Goethe, "Analyse und Synthese," in *Goethes Werke*, Hamburger Ausgabe, Band 13 (Munich: Deutscher Taschenbuch Verlag, 1998), 1.
23. Maurice E. Tucker, *Methoden der Sedimentologie* (Stuttgart: Enke, 1996), 61.
24. Kenneth Pye and Haim Tsoar, *Aeolian Sand and Sand Dunes* (London: Unwin Hyman, 1990), 35.
25. Michael Welland, *Sand* (Berkeley: University of California Press, 2009), 11.
26. Tucker, *Methoden der Sedimentologie*, 74.
27. Robert L. Folk, *Petrology of Sedimentary Rocks* (Austin, TX: Hemphill, 1980), 11.
28. Francis John Pettijohn, Paul Edwin Potter, and Raymond Siever, *Sand and Sandstone* (New York: Springer US, 1973), 83.
29. Dorrik Stow, *Sedimentgesteine im Gelände: Ein illustrierter Leitfaden* (Heidelberg: Spektrum, Akademie-Verlag, 2008), 117.
30. Henri Bortoft, *The Wholeness of Nature: Goethe's Way toward a Science of Conscious Participation in Nature*, Renewal in Science (Hudson, NY: Lindisfarne, 1996).
31. Jürg Meyer, *Gesteine der Schweiz: Der Feldführer* (Bern, Switzerland: Haupt, 2017), 38.
32. Hans Füchtbauer, *Sedimente und Sedimentgesteine*, 4th ed., completely rev., Sediment-Petrologie (Stuttgart: E. Schweizerbart, 1988), 104.
33. Ibid., 512.
34. Martin Okrusch and Siegfried Frimmel, *Mineralogie: Eine Einführung in die spezielle Mineralogie, Petrologie und Lagerstättenkunde*, 8th ed., completely rev., expanded, and updated (Berlin: Springer, 2022), 162.
35. Peter Rothe, *Gesteine: Entstehung–Zerstörung–Umbildung*. 3rd rev. ed. (Darmstadt: Wiss. Buchges, 2010), 23.
36. Füchtbauer, *Sedimente und Sedimentgesteine*, 105.
37. Johann Wolfgang von Goethe, *Dichtung und Wahrheit*, Teil II, Kapitel 8, in *Goethes Werke, Hamburger Ausgabe*, Band 9 (Munich: Deutscher Taschenbuch Verlag, 1998), 343.
38. G. Drosdowski et al., *Duden Bd. 7. Etymologie: Herkunftswörterbuch der deutschen Sprache* (Mannheim, Germany: Bibliographisches Institut, 1963).
39. Olaf Medenbach and Harry Wilk, *Zauberwelt der Mineralien* (Würzburg, Germany; Künzelsau, Germany; Thalwil, Switzerland; Salzburg, Austria: Sigloch, 1977), 76.
40. Per Smed, *Steine aus dem Norden: Geschiebe als Zeugen der Eiszeit in Norddeutschland*, 2nd rev. ed. (Berlin: Borntraeger, 2002), 27.
41. Horst Peter Hann, *Grundlagen und Praxis der Gesteinsbestimmung*, 2nd corr. ed. (Wiebelsheim, Germany: Quelle & Meyer Verlag, 2018), 46.

42. William Scott MacKenzie, *Atlas gesteinsbildender Minerale in Dünnschliffen* (Stuttgart: Enke, 1981), 72.
43. Johann Wolfgang von Goethe, *Sämtliche Werke*, Frankfurter Ausgabe, Serie II, Band 25 (Frankfurt: Deutscher Klassiker Verlag, 1989), 346.
44. Walter Ehrenreich Tröger, *Optische Bestimmung der gesteinsbildenden Minerale*, 2nd ed. (Stuttgart: Schweizerbart'sche Verlagsbuchhandlung, 1969), 522.
45. Walter Maresch, *Gesteine: Systematik, Bestimmung, Entstehung*. 2nd ed., completely rev. (Stuttgart: Schweizerbart, 2014), 26.
46. Füchtbauer, *Sedimente und Sedimentgesteine*, 584.
47. Ibid.
48. Folk, *Petrology of Sedimentary Rocks*, 96.
49. Ibid.
50. Füchtbauer, *Sedimente und Sedimentgesteine*, 125.
51. A. C. Morton, "Stability of Detrital Heavy Minerals in Tertiary Sandstones from the North Sea Basin," *Clay Minerals* 19, no. 3 (June 1984): 287–308.
52. Volker Dietz, "Experiments on the Influence of Transport on Shape and Roundness of Heavy Minerals," in *Contributions to Sedimentology: Stability of Heavy Minerals*, ed. Hans Füchtbauer, J. D. Milliman, A. P. Lisitzyn, and E. Seibold, 1:69–102. Quoted in Füchtbauer, *Sedimente und Sedimentgesteine*, 122.
53. Rothe, *Gesteine*, 37.
54. Donald R. Prothero, "Zirkone—Zeugen der frühen Erdgeschichte," *Spektrum der Wissenschaft* 9 (2018): 60–65.
55. R Boyle, "Auf den Spuren der ältesten Fossilien," *Spektrum der Wissenschaft* 10 (2018): 46–51.
56. Robert Sturm, "Im Mineral eingesperrt eingesperrt—Mikroskopie von Einschlussphasen in magmatischen Kristallen," Mikrokosmos 98 (2009): 283–88.
57. Klaus Binder, *Lukrez—Über die Dinge der Natur* (Berlin: Galiani, 2015).
58. Ibid.
59. Okrusch and Matthes, *Mineralogie*, 86.
60. Pye and Tsoar, *Aeolian Sand and Sand Dunes*, 261.
61. Kenneth Pye, "Post-depositional Reddening of Late Quaternary Coastal Dune Sands, North-Eastern Australia," *Geological Society, London, Special Publications* 11, no. 1 (January 1983): 117–29.
62. Tröger, *Optische Bestimmung der gesteinsbildenden Minerale*, 144.
63. Stow, *Sedimentgesteine im Gelände*, 103.
64. Marcia Bjornerud, *Zeitbewusstheit: Geologisches Denken und wie es helfen könnte, die Welt zu retten* (Berlin: Matthes & Seitz Berlin, 2022), 125.
65. Elisa Bergslien, *An Introduction to Forensic Geoscience* (Chichester, UK: Wiley-Blackwell, 2012), 150.
66. Okrusch and Matthes, *Mineralogie*, 587.
67. Ibid., 588.
68. Rothe, *Gesteine*, 14.
69. Okrusch and Matthes, *Mineralogie*, 190.
70. Rothe, *Gesteine*, 118.
71. Johann Wolfgang von Goethe, *Sämtliche Werke*, Frankfurter Ausgabe, Serie II, Band 39 (Frankfurt: Deutscher Klassiker Verlag, 1999).
72. Goethe, *Sämtliche Werke*, Serie I, Band 25.
73. Ibid., 613.
74. Roland Vinx, *Steine an deutschen Küsten: Finden und Bestimmen* (Wiebelsheim, Germany: Quelle & Meyer Verlag, 2016), 71.
75. Raymond Siever, *Sand, ein Archiv der Erdgeschichte*, Spektrum-Bibliothek 22 (Heidelberg, Germany: Spektrum d. Wiss, 1989), 87.
76. Füchtbauer, *Sedimente und Sedimentgesteine*, 328.
77. Rothe, *Gesteine*, 143.
78. Meyer, *Gesteine der Schweiz*, 320.
79. Quoted in Raoul Schrott, *Erste Erde Epos* (Munich: Carl Hanser Verlag, 2016), 415.
80. Finn Hansen, *Bornholms Natur: 367 Erlebnistouren* (Aakirkeby, Denmark: Verlag Natur Bornholm, 2013), 161.
81. Vinx, *Steine an deutschen Küsten*, 51.
82. Philip Henry Kuenen, *Sand: Its Origin, Transportation, Abrasion, and Accumulation*, Alex. L. du Toit Memorial Lectures 6 (Johannesburg: Geological Society of South Africa, 1959).
83. Ibid.
84. Siever, *Sand*, 108.
85. Kuenen, *Sand*.
86. Siever, *Sand*, 65.
87. Welland, *Sand*, 6.
88. Siever, *Sand*, 68.
89. Ibid., 100.
90. Blümel, *Wüsten*, 114.
91. Antonie van Leeuwenhoek in Dieter Gerlach, *Geschichte der Mikroskopie* (Frankfurt: Deutsch, 2009), 570.
92. Welland, *Sand*, 17.
93. Okrusch and Matthes, *Mineralogie*, 182.
94. Edwin Roedder, *Fluid Inclusions*, Reviews in Mineralogy (Washington, DC: Mineralogical Society of America, 1984), 111.
95. Harry Rosenbusch, *Mikroskopische Physiographie der Mineralien und Gesteine: Ein Hülfsbuch bei mikroskopischen Gesteinsstudien*, 5th rev. ed. (Stuttgart: Schweizerbart, 1924), 779.
96. W. R. Almon, L. B. Fullerton, and D. K. Davies, "Pore Space Reduction in Cretaceous Sandstones through Chemical Precipitation of Clay Minerals," *Journal of Sedimentary Research* 46, no. 1 (March 1, 1976): 89–96.
97. Images from Lenzen, *Sand* (Cologne: Snoeck, 2021), 74–75.
98. Eduard J. Gübelin, *Innenwelt der Edelsteine: Urkunde aus Raum und Zeit* (Zurich: ABC-Verlag, 1973.
99. Otto Leeder, Rainer Thomas, and Werner Klemm, *Einschlüsse in Mineralen* (Stuttgart: Enke, 1987), 16.
100. Roedder, *Fluid Inclusions*, 43.
101. Maria A. Mange and Heinz F. W. Maurer, *Heavy Minerals in Colour* (London: Chapman & Hall, 1992), 41.
102. R. Lindsey, "Ancient Crystals Suggest Earlier Ocean" (NASA Earth Observatory, 2006).
103. Prothero, "Zirkone—Zeugen der frühen Erdgeschichte," 60–65.
104. Gübelin, *Innenwelt der Edelsteine*, 73.
105. Roedder, *Fluid Inclusions*, 181.
106. Jens Götze and Winfried Zimmerle, "Quartz and Silica as Guide to Provenance in Sediments and Sedimentary Rocks," *Contributions to Sedimentary Geology* 21 (January 1, 2000).

107. Folk, *Petrology of Sedimentary Rocks*, 72.
108. Oliver Lenzen, "Darstellung der Oberflächentextur von Sandkörnern aus Quarz in Durchlicht-Hellfeld-Beleuchtung," *Mikrokosmos* 101, no. 1 (2012): 20–26.
109. William Mahaney, *Atlas of Sand Grain Surface Textures and Applications* (Oxford: Oxford University Press, 2002).
110. P. A. Bull and R. M. Morgan, "Sediment Fingerprints: A Forensic Technique Using Quartz Sand Grains," *Science & Justice: Journal of the Forensic Science Society* 46, no. 2 (2006): 107–24.
111. Philip Henry Kuenen, "Experimental Abrasion 4: Eolian Action," *Journal of Geology* 68, no. 4 (July 1960): 206–14.
112. Pye and Tsoar, *Aeolian Sand and Sand Dunes*, 73.
113. David H. Krinsley and John C. Doornkamp, *Atlas of Quartz Sand Surface Textures* (Cambridge, UK: Cambridge University Press, 1973).
114. Pye and Tsoar, *Aeolian Sand and Sand Dunes*, 73.
115. Blümel, *Wüsten*, 157.
116. Soentgen and Völzke, *Staub—Spiegel der Umwelt*, 100.
117. R. A. Bagnold, *The Physics of Blown Sand and Desert Dunes* (New York: Springer, 2012).
118. Blümel, *Wüsten*, 158.
119. Mahaney, *Atlas of Sand Grain Surface Textures and Applications*, 68.
120. Stanley V. Margolis and David H. Krinsley, "Processes of Formation and Environmental Occurrence of Microfeatures on Detrital Quartz Grains," *American Journal of Science* 274 (May 1974): 449–64.
121. Mahaney, *Atlas of Sand Grain Surface Textures and Applications*, 69.
122. Pye and Tsoar, *Aeolian Sand and Sand Dunes*, 77.
123. Jayagopal Madhavaraju, Juan Carlos García y Barragán, Shaik Mohammad Hussain, and Saidapet Pachaiveedu Mohan, "Microtextures on Quartz Grains in the Beach Sediments of Puerto Peñasco and Bahía Kino, Gulf of California, Sonora, Mexico," *Revista Mexicana de Ciencias Geológicas* 26 no. 2 (August 1, 2009), 315–27.
124. Füchtbauer, *Sedimente und Sedimentgesteine*, 866.
125. Kuenen, "Experimental Abrasion 4, " 17.
126. M. Cremer and P. Legigan, *Morphology and Surface Texture of Quartz Grains from ODP Site 645, Baffin Bay*, ed. S. P. Srivastava et al., Proceedings of the Ocean Drilling Program: Scientific Results 105 (College Station: Texas A&M University Press, 1989).
127. David H. Krinsley and Jack Donahue, "Environmental Interpretation of Sand Grain Surface Textures by Electron Microscopy," *Geological Society of America Bulletin* 79, no. 6 (1968): 743–48.
128. Tucker, *Methoden der Sedimentologie* [referencing R. Higgs, "Quartz-Grain Surface Features of Mesozoic-Cenozoic Sands from the Labrador and Western Greenland Continental Margins," *Journal of Sedimentary Petrology* 49, no. 2 (1979)], 238.
129. Krinsley and Donahue, "Environmental Interpretation."
130. Martin Schwarzbach, "Glazigene Sichelmarken als Klimazeugen (Nach Beobachtungen in Finnland)," *E&G Quaternary Science Journal* 28, no. 1 (January 1, 1978), 109–18.
131. Mahaney, *Atlas of Sand Grain Surface Textures and Applications*, 112.
132. Füchtbauer, *Sedimente und Sedimentgesteine*, 877.
133. K. Vos, N. Vandenberghe, and J. Elsen, "Surface Textural Analysis of Quartz Grains by Scanning Electron Microscopy (SEM): From Sample Preparation to Environmental Interpretation," *Earth-Science Reviews* 128 (January 1, 2014), 93–104.
134. Pye and Tsoar, *Aeolian Sand and Sand Dunes*, 78.
135. Johann Wolfgang von Goethe, *Betrachtungen über Morphologie*, in *Goethes Werke*, Hamburger Ausgabe, Band 13 (Munich: Deutscher Taschenbuch Verlag, 1998), 121.
136. Meyer, *Gesteine der Schweiz*, 368.
137. Hans Ulrich Schmincke, *Vulkanismus*, 4th rev. ed. (Darmstadt: Primus, 2013), 65.
138. Bjornerud, *Zeitbewusstheit*, 95.
139. Welland, *Sand*, 53.
140. Maresch, *Gesteine*, 194.
141. Gunnar Gad, "Expedition Mystacocarida," *Mikrokosmos* 99 (2010): 33–45.
142. Gunnar Gad, "Die Loricifera, (Korsetttragerchen)—Winzlinge aus den Sandboden der Ozeane—Teil 2: Die Pliciloricidae." *Mikrokosmos* 94, no. 2 (2005): 104–15.
143. Ibid.
144. Tucker, *Methoden der Sedimentologie*, 252.
145. Schrott, *Erste Erde Epos*, 289.
146. Andrea Rhode, *Auf Fossiliensuche an der Ostsee: Gesteine, Fossilien, Fundmöglichkeiten* (Neumünster, Germany: Wachholtz, 2008), 197.
147. Norbert Welsch, Claus Liebman, and Jürgen Schwab, *Erde und Leben: Die Geschichte einer innigen Wechselbeziehung* (Berlin: Springer, 2017), 249.
148. Schrott, *Erste Erde Epos*, 778.
149. Füchtbauer, *Sedimente und Sedimentgesteine*, 319.
150. Rhode, *Auf Fossiliensuche an der Ostsee*, 255.
151. Füchtbauer, *Sedimente und Sedimentgesteine*, 319.
152. Bergslien, *Forensic Geosciece*, 355.
153. Füchtbauer, *Sedimente und Sedimentgesteine*, 282.
154. Joseph Augustine Cushman, *Foraminifera: Their Classification and Economic Use* (Cambridge, MA: Harvard University Press, 1950).
155. Welsch et al., *Erde und Leben*, 91.
156. Füchtbauer, *Sedimente und Sedimentgesteine*, 284.
15.7 Cushman, *Foraminifera*, 13.
158. Jean-Pierre Debenay, *A Guide to 1,000 Foraminifera from Southwestern Pacific New Caledonia* (Paris: Institut de Recherche pour le Développement, 2013).
159. Rhode, *Auf Fossiliensuche an der Ostsee*, 240.
160. Rudolf Kilias, *Die Weinbergschnecke: Uber Leben und Nutzung von Helix pomatia*, 2nd ed., Die neue Brehm-Bücherei (Heidelberg, Germany: Akademie-Verlag, 1995).
161. Karl Dietrich Adam, *Das Steinheimer Becken: Eine Fundstätte von Weltgeltung*, 2nd ed. (Steinheim am Albuch, Germany: Bürgermeisteramt, 1992).
162. Schrott, *Erste Erde Epos*, 753.

163. Walter Schumann, *Der große BLV Steine- und Mineralienführer: Mineralien, Erze, Edelsteine, Gesteine, Meteorite*. 7th rev. ed. (Munich: blv, 2007), 70.
164. MacKenzie, *Atlas gesteinsbildender Minerale in Dünnschliffen*, 89.
165. Füchtbauer, *Sedimente und Sedimentgesteine*, 284.
166. Welsch et al., *Erde und Leben*, 246.
167. Bjornerud, *Zeitbewusstheit*, 182.
168. James Joyce, *A Portrait of the Artist as a Young Man*, trans. Walter Spiegelberg, Huebers Fremdsprachliche Texte (Munich: Max Hueber, 1964).
169. Archimedes, "Die Sandzahl," in *Ostwalds Klassiker der exakten Wissenschaften*, vol. 201 (Frankfurt am Main: Harri Deutsch, 2003).
170. Johann Wolfgang von Goethe, *"Entwurf einer Farbenlehre, Einleitung,"* in *Goethes Werke*, Hamburger Ausgabe, Band 13 (Munich: Deutscher Taschenbuch Verlag, 1998), 322.
171. Carl Friedrich von Weizsäcker, *Aufbau der Physik* (Munich: C. Hanser, 1985).
172. Johann Wolfgang von Goethe, "Maximen und Reflexionen 591, aus dem Nachlass, Zählung nach Hecker," in *Goethes Werke*, Hamburger Ausgabe, Band 12 (Munich: Deutscher Taschenbuch Verlag, 1998).
173. Johann Wolfgang von Goethe, "Zur Farbenlehre, Historischer Teil, 4. Abteilung, 16. Jahrhundert, Zwischenbetrachtung," in *Sämtliche Werke*, Frankfurter Ausgabe, Erste Abteilung, Band 23/1 (Frankfurt: Deutscher Klassiker Verlag, 1991), 667.
174. Johann Wolfgang von Goethe, *Die Schriften zur Naturwissenschaft*, Erste Abteilung, Band 8 (Weimar, Germany: Böhlau Verlag, 1962), 232.
175. *DIE ZEIT* 23 (June 3, 1988): sec. Feuilleton, 55.
176. Goethe, "Maximen und Reflexionen 591, aus dem Nachlass, Zählung nach Hecker.".
177. Carl Friedrich von Weizsäcker, "Einige Begriffe aus Goethes Naturwissenschaft," in *Goethes Werke: Naturwissenschaftliche Schriften I*, 13:159, Hamburger ed. (Munich: C. H. Beck, 1994), 546.
178. Ernst Jünger, *Das Sanduhrbuch*, 2nd ed. (Frankfurt am Main: Vittorio Klostermann, 1954), 146.
179. Mathematically speaking, there are infinite series that converge to a specific value. In Zeno's time, this was not yet known.
180. Arthur Schopenhauer, *Die Welt als Wille und Vorstellung*, vol. 1, part 2 (Cologne: Könemann, 1997).
181. Marc Aurel, *Selbstbetrachtungen*, reprint, vol. II (Stuttgart: Kröner, 2001).
182. Ludwig Wittgenstein, *Tractatus logico-philosophicus*, proposition 6.4311 (Oxford: Basil Blackwell, 1959).
183. Plato, "Parmenides," in *Phaidros, Parmenides, Theaitetos, Sophistes: Sämtliche Werke: Charmides, Eutyphron, Parmenides (Philosophiehistorische Texte)* Vol. 1. 2nd ed., ed. Walter F. Otto, Ernesto Grassi, and Gert Plamböck (Berlin: Akademie-Verlag, 1989).
184. Zensho W. Kopp, *Der große Zen-Weg: Der Weg zur Erleuchtung mitten im Leben*, Schirner-Taschenbuch (Darmstadt: Schirner, 2004).
185. Dylan Thomas and Hans Bender, *Unter dem Milchwald: Ein Spiel für Stimmen (Under the Milkwood: A Play for Voices)*, trans. Erich Fried (Stuttgart: P. Reclam, 1970), 26.
186. Martin Heidegger, *Sein und Zeit*, 17th ed. (Tübingen, Germany: Max Niemeyer Verlag, 1993), 425.
187. Brian Greene, *Der Stoff, aus dem der Kosmos ist*, trans. Hainer Kober (Munich: Goldmann TB, 2008), 154.
188. Stefan M. Maul, "Im Rückwärtsgang in Die Zukunft," *Spektrum der Wissenschaft* 8 (2010): 72–77.
189. Carlo Rovelli, *Die Ordnung der Zeit* (Reinbek bei Hamburg, Germany: Rowohlt, 2018), 85.
190. Pettijohn et al., *Sand and Sandstone*, 5.
191. Kuenen, "Experimental Abrasion 4," 17.
192. Arie Poldevaart, "Chemistry of the Earth's Crust," in *Crust of the Earth: A Symposium*, ed. A. Poldevaart, 62:119–44. Geological Society of America Special Paper, 1955, 5.
193. Pettijohn, *Sand and Sandstone*, 7.
194. Archimedes, "Die Sandzahl," 349.
195. Johann Wolfgang von Goethe, *Italienische Reise*, in *Goethes Werke*, Hamburger Ausgabe, Band 11: Eintrag vom 25. Februar 1787 (Munich: Deutscher Taschenbuch Verlag, 1998).
196. Vince Beiser, Bernhard Jendricke, Christa Prummer-Lehmair, and Gerlinde Schermer-Rauwolf, *Sand: Wie uns eine wertvolle Ressource durch die Finger rinnt*, Stoffgeschichten (Munich: Oekom Verlag, 2021), 232.
197. "Der Sand wird knapp," *Geo*, June 2016.
198. Bjornerud, *Zeitbewusstheit*, 153.
199. Andreas Sturm, "Sand wird knapp," *Heilbronner Stimme*, February 18, 2019.
200. Beiser et al., *Sand: Wie uns eine wertvolle Ressource durch die Finger rinnt*, 236.
201. Ibid., 131.
202. Gary Greenberg, Carol Kiely, and Kate Clover, *The Secrets of Sand: A Journey into the Amazing Microscopic World of Sand* (Minneapolis: Voyageur, 2015), 95.
203. Beiser et al., *Sand: Wie uns eine wertvolle Ressource durch die Finger rinnt*, 200.
204. Anna Hellge, "Wie Sand am Meer?," *Natur* 7 (2017): 44–48.
205. Beiser et al., *Sand: Wie uns eine wertvolle Ressource durch die Finger rinnt*, 117.
206. Jon Larsen, *In Search of Stardust: Amazing Micrometeorites and Their Terrestrial Imposters*, 2nd ed. (Minneapolis: Voyageur, 2017).
207. Translation: "About this stone, many know much, some know something, and no one knows enough."
208. Bagnold, *The Physics of Blown Sand and Desert Dunes*, 247.
209. H. Joachim Schlichting, "Musikalischer Sand," *Spektrum der Wissenschaft* 12 (2017): 52–55.
210. Johann Wolfgang von Goethe, "Der Versuch als Vermittler von Subjekt und Objekt, " in *Goethes Werke*, Hamburger Ausgabe, Band 13 (Munich: Deutscher Taschenbuch Verlag, 1998).
211. Alexander von Humboldt, *Kosmos: Entwurf einer physischen Weltbeschreibung*, vol. 3 (Berlin: Cotta, 1845), 609.

11.4 Bibliography

Adam, Karl Dietrich. *Das Steinheimer Becken: Eine Fundstätte von Weltgeltung.* 2nd ed. Steinheim am Albuch, Germany: Bürgermeisteramt, 1992.

Almon, W. R., L. B. Fullerton, and D. K. Davies. "Pore Space Reduction in Cretaceous Sandstones through Chemical Precipitation of Clay Minerals." *Journal of Sedimentary Research* 46, no. 1 (March 1, 1976): 89–96. https://doi.org/10.1306/212F6EC9-2B24-11D7-8648000102C1865D.

Archimedes. "Die Sandzahl." In *Ostwalds Klassiker der exakten Wissenschaften*. Vol. 201. Frankfurt am Main: Harri Deutsch, 2003.

Armstrong-Altrin, John S., and Olmedo Natalhy-Pineda. "Microtextures of Detrital Sand Grains from the Tecolutla, Nautla, and Veracruz Beaches, Western Gulf of Mexico, Mexico: Implications for Depositional Environment and Paleoclimate." *Arabian Journal of Geosciences* 7, no. 10 (2013): 4321.

Aurel, Marc. *Selbstbetrachtungen*. Reprint. Vol. II. Kröners Taschenausgabe. Stuttgart: Kröner, 2001.

Bagnold, R. A. *The Physics of Blown Sand and Desert Dunes*. New York: Springer, 2012.

Beiser, Vince, Bernhard Jendricke, Christa Prummer-Lehmair, and Gerlinde Schermer-Rauwolf. *Sand: Wie uns eine wertvolle Ressource durch die Finger rinnt*. Stoffgeschichten 13. Munich: Oekom Verlag, 2021.

Bergslien, Elisa. *An Introduction to Forensic Geoscience*. Chichester, UK: Wiley-Blackwell, 2012.

Binder, Klaus. *Lukrez—Über die Dinge der Natur*. Berlin: Galiani, 2015.

Bjornerud, Marcia. *Zeitbewusstheit: Geologisches Denken und wie es helfen könnte, die Welt zu retten.* Berlin: Matthes & Seitz Berlin, 2020. https://d-nb.info/1201366348.

Blatt, Harvey. Various publications on the origin of quartz grains and their diagnosis.1963–1967.

Blümel, Wolf Dieter. *Wüsten: Entstehung, Kennzeichen, Lebensraum.* UTB Bd. 3882. Stuttgart: Ulmer, 2013.

Boenigk, Wolfgang. *Schwermineralanalyse*. Stuttgart: F. Enke, 1983.

Boothby, Thomas C., Hugo Tapia, Alexandra H. Brozena, Samantha Piszkiewicz, Austin E. Smith, Ilaria Giovannini, Lorena Rebecchi, Gary J. Pielak, Doug Koshland, and Bob Goldstein. "Tardigrades Use Intrinsically Disordered Proteins to Survive Desiccation." *Molecular Cell* 65, no. 6 (March 16, 2017): 975–84.e5. https://doi.org/10.1016/j.molcel.2017.02.018.

Bortoft, Henri. *The Wholeness of Nature: Goethe's Way toward a Science of Conscious Participation in Nature.* Renewal in Science. Hudson, NY: Lindisfarne, 1996.

Boyle, R. "Auf den Spuren der ältesten Fossilien." *Spektrum der Wissenschaft* 10 (2018): 46–51.

Brown, Joan E. "Depositional Histories of Sand Grains from Surface Textures." *Nature* 242, no. 5397 (April 1973): 396–98. https://doi.org/10.1038/242396a0.

Bull, P. A., and R. M. Morgan. "Sediment Fingerprints: A Forensic Technique Using Quartz Sand Grains." *Science & Justice: Journal of the Forensic Science Society* 46, no. 2 (2006): 107–24. https://doi.org/10.1016/S1355-0306(06)71581-7.

Chakroun, A., J.-C. Miskovsky, and D. Zaghbib-Turki. "Quartz Grain Surface Features in Environmental Determination of Aeolian Quaternary Deposits in Northeastern Tunisia." *Mineralogical Magazine* 73, no. 4 (August 1, 2009): 607–14. https://doi.org/10.1180/minmag.2009.073.4.607.

Clark, Jim. "The Myxamoebae of Didymium Iridis and Physarum Polycephalum Are Immortal." *Mycologia* 84, no. 1 (1992): 116–18. https://doi.org/10.2307/3760411.

Cremer, M., and P. Legigan. *Morphology and Surface Texture of Quartz Grains from ODP Site 645, Baffin Bay*. Edited by S. P. Srivastava et al. Proceedings of the Ocean Drilling Program: Scientific Results 105. College Station: Texas A&M University Press, 1989. https://doi.org/10.2973/odp.proc.sr.105.1989.

Cushman, Joseph Augustine. *Foraminifera: Their Classification and Economic Use*. Cambridge, MA: Harvard University Press, 1950.

Davis, Andrew M. *Meteorites, Comets, and Planets: Treatise on Geochemistry*. Vol. 1. 2nd ed. Amsterdam: Elsevier, 2005.

Debenay, Jean-Pierre. *A Guide to 1,000 Foraminifera from Southwestern Pacific New Caledonia.* Paris: Institut de Recherche pour le Développement, 2013.

Delitzsch, Friedrich. *Assyrische Lesestücke, mit grammatischen Tabellen und vollständigem Glossar*. 4th ed. Leipzig: J. C. Hinrichs, 1900.

"Der Sand wird knapp." *Geo*, June 2016.

DIE ZEIT 23 (June 3, 1988): sec. Feuilleton, 55.

Dietz, Volker. "Experiments on the Influence of Transport on Shape and Roundness of Heavy Minerals." In *Contributions to Sedimentology: Stability of Heavy Minerals*, edited by Hans Füchtbauer, J. D. Milliman, A. P. Lisitzyn, and E. Seibold, 1:69–102. Quoted in Füchtbauer, *Sedimente und Sedimentgesteine*, 122.

Dippel, Leopold. *Grundzüge der allgemeinen Mikroskopie*. Braunschweig, Germany: F. Vieweg & Sohn, 1885.

Ditfurth, Hoimar von. *Kinder des Weltalls: Der Roman unserer Existenz*. Hamburg, Germany: Hoffmann & Campe, 1970.

Drosdowski, G., et al. *Duden Bd. 7. Etymologie: Herkunftswörterbuch der deutschen Sprache*. Mannheim, Germany: Bibliographisches Institut, 1963.

Ellermeier, Friedrich. *Sumerisches Glossar: Führer durch die neuere sumerologische Fachliteratur*. Theologische und orientalistische Arbeiten aus Göttingen 4. Nörten-Hardenberg bei Göttingen, Germany: F. Ellermeier, 1979.

Engelhardt, Wolf von, Christoph Berthold, Thomas Wenzel, and Thomas Dehner. "Chemistry, Small-Scale Inhomogeneity, and Formation of Moldavites as Condensates from Sands Vaporized by the Ries Impact." *Geochimica et Cosmochimica Acta* 69, no. 23 (December 1, 2005): 5611–26. https://doi.org/10.1016/j.gca.2005.07.004.

Enzensberger, H. M., ed. *Alexander von Humboldts Kosmos—Entwurf einer physischen Weltbeschreibung*. Books III and V. Frankfurt: Eichborn, 2004.

Folk, Robert L. *Petrology of Sedimentary Rocks*. Austin, TX: Hemphill, 1980. http://hdl.handle.net/2152/22930.

Forster, Georg. *Reise um die Welt*. Frankfurt: Eichborn, 2007.

Füchtbauer, Hans. *Sedimente und Sedimentgesteine*. 4th ed., completely rev. Sediment-Petrologie. Stuttgart: E. Schweizerbart, 1988.

Gad, Gunnar. "Die Loricifera (Korsettträgerchen)—Winzlinge aus den Sandböden der Ozeane—Teil 2: Die Pliciloricidae." *Mikrokosmos* 94, no. 2 (2005): 104–16.

Gad, Gunnar. "Expedition Mystacocarida." *Mikrokosmos* 99 (2010): 33–45.

Georgescu, M. Dan. *Microfossils through Time: An Introduction; First Steps in Micropaleontology*. Stuttgart: Schweizerbart Science Publishers, 2018.

Gerlach, Dieter. *Geschichte der Mikroskopie*. Frankfurt: Deutsch, 2009. https://d-nb.info/980701988/04.

Goethe, Johann Wolfgang von. "Analyse und Synthese." In *Goethes Werke*. Hamburger Ausgabe, Band 13. Munich: Deutscher Taschenbuch Verlag, 1998.

Goethe, Johann Wolfgang von. *Betrachtungen über Morphologie*. In *Goethes Werke*. Hamburger Ausgabe, Band 13. Munich: Deutscher Taschenbuch Verlag, 1998.

Goethe, Johann Wolfgang von. "Brief an Charlotte von Stein vom 27. Juni 1785." In *Goethes Briefe*. Hamburger Ausgabe, Band 1, edited by Robert Mandelkow. Munich: Deutscher Taschenbuch Verlag, 1988.

Goethe, Johann Wolfgang von. "Brief an Zelter vom 22. Juni 1808." In *Goethes Briefe*. Hamburger Ausgabe, Band 3, edited by Robert Mandelkow. Munich: Deutscher Taschenbuch Verlag, 1988.

Goethe, Johann Wolfgang von. "Der Versuch als Vermittler von Subjekt und Objekt." In *Goethes Werke*. Hamburger Ausgabe, Band 13. Munich: Deutscher Taschenbuch Verlag, 1998.

Goethe, Johann Wolfgang von. *Dichtung und Wahrheit*, Teil II, Kapitel 8. In *Goethes Werke*. Hamburger Ausgabe, Band 9. Munich: Deutscher Taschenbuch Verlag, 1998.

Goethe, Johann Wolfgang von. *Die Schriften zur Naturwissenschaft*. Erste Abteilung, Band 8, page 232. Weimar, Germany: Böhlau Verlag, 1962.

Goethe, Johann Wolfgang von. "Entwurf einer Farbenlehre, Einleitung." In *Goethes Werke*. Hamburger Ausgabe, Band 13, page 322. Munich: Deutscher Taschenbuch Verlag, 1998.

Goethe, Johann Wolfgang von. *Goethes Briefe*. Edited by Robert Mandelkow. Hamburger Ausgabe, Band 4. Munich: Deutscher Taschenbuch Verlag, 1988.

Goethe, Johann Wolfgang von. "Infusions-Tiere; Studien über organische Naturen bis zur Rückkehr aus Italien 1776–1788." In *Schriften zur Morphologie, Sämtliche Werke*. Frankfurter Ausgabe, Serie I, Band 24. Frankfurt: Deutscher Klassiker Verlag, 1987.

Goethe, Johann Wolfgang von. *Italienische Reise*. In *Goethes Werke*. Hamburger Ausgabe, Band 11: Eintrag vom 25. Februar 1787. Munich: Deutscher Taschenbuch Verlag, 1998.

Goethe, Johann Wolfgang von. *Italienische Reise*. Weimarer Ausgabe, Serie I, Band 31, page 94. Weimar, Germany: Böhlau Verlag, 1909.

Goethe, Johann Wolfgang von. "Maximen und Reflexionen 512." In *Goethes Werke*. Hamburger Ausgabe, Band 12. Munich: Deutscher Taschenbuch Verlag, 1998.

Goethe, Johann Wolfgang von. "Maximen und Reflexionen 591, aus dem Nachlass, Zählung nach Hecker." In *Goethes Werke*. Weimarer Ausgabe. Weimar, Germany: Böhlau Verlag, 1907.

Goethe, Johann Wolfgang von. *Sämtliche Werke*. Frankfurter Ausgabe, Serie I, Band 25. Frankfurt: Deutscher Klassiker Verlag, 1989.

Goethe, Johann Wolfgang von. *Sämtliche Werke*. Frankfurter Ausgabe, Serie II, Band 25. Frankfurt: Deutscher Klassiker Verlag, 1989.

Goethe, Johann Wolfgang von. *Sämtliche Werke*. Frankfurter Ausgabe, Serie II, Band 39. Frankfurt: Deutscher Klassiker Verlag, 1999.

Goethe, Johann Wolfgang von. "Zur Farbenlehre, Historischer Teil, 4. Abteilung, 16. Jahrhundert, Zwischenbetrachtung." In *Sämtliche Werke*. Frankfurter Ausgabe, Erste Abteilung, Band 23/1, page 667. Frankfurt: Deutscher Klassiker Verlag, 1991.

Göke, Gerhard. *Moderne Methoden der Lichtmikroskopie vom Durchlicht-Hellfeld bis zum Lasermikroskop*. Kosmos-Wissenschaft. Stuttgart: Franckh, 1988. https://d-nb.info/881125792/04.

Golia, Maria. *Meteorite: Nature and Culture*. Earth series. London: Reaktion Books, 2015. https://www.loc.gov/catdir/enhancements/fy1619/2016417226-t.html.

Good, Paul, and Stefan Steiner. *Mein Heraklit*. Cologne: Snoeck, 2021.

Götze, J., and W. Zimmerle. "Quartz and Silica as Guide to Provenance in Sediments and Sedimentary Rocks." *Contributions to Sedimentary Geology* 21 (January 1, 2000): X–91.

Greenberg, Gary. *A Grain of Sand: Nature's Secret Wonder*. St. Paul, MN: Voyageur, 2008. http://catdir.loc.gov/catdir/enhancements/fy0803/2007034731-t.html.

Greenberg, Gary, Carol Kiely, and Kate Clover. *The Secrets of Sand: A Journey into the Amazing Microscopic World of Sand*. Minneapolis: Voyageur, 2015.

Greene, Brian. *Der Stoff, aus dem der Kosmos ist*. Translated by Hainer Kober. Munich: Sidler, 2004.

Greven, Hartmut. *Die Bärtierchen: Tardigrada*. Die Neue Brehm-Bücherei. Wittenberg Lutherstadt, Germany: A. Ziemsen, 1980.

Gübelin, Eduard J. *Innenwelt der Edelsteine: Urkunde aus Raum und Zeit*. Zurich: ABC-Verlag, 1973.

Halloran, John A., ed. *Sumerian Lexicon: A Dictionary Guide to the Ancient Sumerian Language*. Bilingual edition. Los Angeles: Logogram, 2006.

Hann, Horst Peter. *Grundlagen und Praxis der Gesteinsbestimmung*. 2nd corr. ed. Wiebelsheim, Germany: Quelle & Meyer Verlag, 2018. http://deposit.dnb.de/cgi-bin/dokserv?id=dfa2d0e0ec6a41adb-76cf075b55afb10&prov=M&dok_var=1&dok_ext=htm.

Hansen, Finn. *Bornholms Natur: 367 Erlebnistouren*. [Aakirkeby, Denmark]: Verlag Natur Bornholm, 2013.

Heidegger, Martin. *Sein und Zeit*. 17th ed. Tübingen, Germany: Max Niemeyer Verlag, 1993.

Hellge, Anna. "Wie Sand am Meer?" *Natur* 7 (2017): 44–49. https://nb.ieb.kit.edu/index.php/wie-sand-am-meer-2.

Higgs, R. "Quartz-Grain Surface Features of Mesozoic-Cenozoic Sands from the Labrador and Western Greenland Continental Margins." *Journal of Sedimentary Petrology* 49, no. 2 (1979): 559–610. https://doi.org/10.1306/212F779D-2B24-11D7-8648000102C1865D.

Horton, J. W., and G. A. Izett. "Crystalline-Rock Ejecta and Shocked Minerals of the Chesapeake Bay Impact Structure, USGS-NASA Langley Core, Hampton, Virginia, with Supplemental Constraints on the Age of Impact." US Geological Survey Professional Paper, January 1, 2006, E1–24. https://doi.org/10.3133/pp1688E.

Humboldt, Alexander von. *Kosmos: Entwurf einer physischen Weltbeschreibung*. Vols. 3 and 5. Berlin: Cotta, 1845.

Immonen, Ninna. "Surface Microtextures of Ice-Rafted Quartz Grains Revealing Glacial Ice in the Cenozoic Arctic." *Palaeogeography, Palaeoclimatology, Palaeoecology* 374 (March 15, 2013): 293–302. https://doi.org/10.1016/j.palaeo.2013.02.003.

Joyce, James. *A Portrait of the Artist as a Young Man*. Translated by Walter Spiegelberg. Huebers Fremdsprachliche Texte. Munich: Max Hueber, 1964.

Jünger, Ernst. *Das Sanduhrbuch*. 2nd ed. Frankfurt am Main: Vittorio Klostermann, 1954.

Jünger, Ernst. *Subtile Jagden*. Stuttgart: Klett, 1967.

Kilias, Rudolf. *Die Weinbergschnecke: Über Leben und Nutzung von Helix pomatia*. 2nd ed. Die neue Brehm-Bücherei 563. Heidelberg, Germany: Akademie-Verlag, 1995. https://d-nb.info/944424716/04.

Klerner, Susanne. "Materie im frühen Sonnensystem: Die Entstehung von Chondren, Matrix und refraktären Forsteriten." Universität zu Köln, 2001. https://nbn-resolving.org/urn:nbn:de:hbz:38-5357.

Kopp, Zensho W. *Der große Zen-Weg: Der Weg zur Erleuchtung mitten im Leben*. Schirner-Taschenbuch. Darmstadt: Schirner, 2004.

Krätz, Otto. *Goethe und die Naturwissenschaften*. 2nd corr. ed. Munich: Callwey, 1998. http://www.gbv.de/dms/goettingen/245427570.pdf.

Kremer, Bruno P. *Das große Kosmos-Buch der Mikroskopie*. Stuttgart: Kosmos, 2002. https://d-nb.info/96467162X/04.

Krinsley, David H., and Jack Donahue. "Environmental Interpretation of Sand Grain Surface Textures by Electron Microscopy." *Geological Society of America Bulletin* 79, no. 6 (1968): 743–48.

Krinsley, David H., and John C. Doornkamp. *Atlas of Quartz Sand Surface Textures*. Cambridge, UK: Cambridge University Press, 1973.

Krinsley, David H., and Floyd W. McCoy. "Surface Features on Quartz Sand and Silt Grains: Leg 39 Deep Sea Drilling Project." In *Initial Reports of the Deep Sea Drilling Project: 1974 Expedition; Amsterdam, Netherlands to Cape Town, South Africa*. 34:445–52. Washington, DC: US Government Printing Office, 1977. http://www.deepseadrilling.org/39/volume/dsdp39_12.pdf.

Krinsley, David H., and William Wellendorf. "Wind Velocities Determined from the Surface Textures of Sand Grains." *Nature* 283, no. 5745 (January 1980): 372–73. https://doi.org/10.1038/283372a0.

Krot, Alexander N. *Dating the Earliest Solids in Our Solar System*. Hawaii Institute of Geophysics and Planetology 2002. http://www.psrd.hawaii.edu/Sept02/isotopicAges.html

Kuenen, Philip Henry. "Experimental Abrasion 4: Eolian Action." *Journal of Geology* 68, no. 4 (July 1960): 427–49. https://www.jstor.org/stable/30084814.

Kuenen, Philip Henry. *Sand: Its Origin, Transportation, Abrasion, and Accumulation*. Alex. L. du Toit Memorial Lectures 6. Johannesburg: Geological Society of South Africa, 1959.

Langenhorst, Falko. "Shock Metamorphism of Some Minerals: Basic Introduction and Microstructural Observations." *Bulletin of the Czech Geological Survey* 77, no. 4 (2002): 265–82. http://www.geology.cz/bulletin/fulltext/03langenhorstfinal.pdf.

Larsen, Jon. *In Search of Stardust: Amazing Micrometeorites and Their Terrestrial Imposters*. 2nd ed. Minneapolis: Voyageur, 2017.

Leeder, Otto, Rainer Thomas, and Werner Klemm. *Einschlüsse in Mineralen*. Stuttgart: Enke, 1987. https://d-nb.info/870625365/04.

Lenzen, Oliver. "Darstellung der Oberflächentextur von Sandkörnern aus Quarz in Durchlicht-Hellfeld-Beleuchtung." *Mikrokosmos* 101, no. 1 (2012): 20–26. https://www.zobodat.at/pdf/Mikrokosmos_101_1_0001.pdf.

Lenzen, Oliver. *Sand*. Bilingual edition. Cologne: Snoeck, 2020.

Lindsey, R. "Ancient Crystals Suggest Earlier Ocean." NASA Earth Observatory, 2006. https://earthobservatory.nasa.gov/features/Zircon/zircon.php.

MacKenzie, William Scott. *Atlas gesteinsbildender Minerale in Dünnschliffen*. Stuttgart: Enke, 1981. https://d-nb.info/810383233/04.

Madhavaraju, Jayagopal, Juan Carlos García y Barragán, Shaik Mohammad Hussain, and Saidapet Pachaiveedu Mohan. "Microtextures on Quartz Grains in the Beach Sediments of Puerto Peñasco and Bahía Kino, Gulf of California, Sonora, Mexico." *Revista Mexicana de Ciencias Geológicas* 26, no. 2 (August 1, 2009): 367–79.

Mahaney, William. *Atlas of Sand Grain Surface Textures and Applications*. Oxford: Oxford University Press, 2002.

Mange, Maria A., and Heinz F. W. Maurer. *Heavy Minerals in Colour*. London: Chapman & Hall, 1992. http://bvbr.bib-bvb.de:8991/F?func=service&doc_library=BVB01&doc_number=003844030&line_number=0001&func_code=DB_RECORDS&service_type=MEDIA.

Maresch, Walter. *Gesteine: Systematik, Bestimmung, Entstehung*. 2nd ed., completely rev. Stuttgart: Schweizerbart, 2014.

Margolis, Stanley V., and David H. Krinsley. "Processes of Formation and Environmental Occurrence of Microfeatures on Detrital Quartz Grains." *American Journal of Science* 274 (May 1974): 449–64.

Margolis, Stanley V., and David H. Krinsley. "Submicroscopic Frosting on Eolian and Subaqueous Quartz Sand Grains." *Geological Society of America Bulletin* 82 (1971): 3395–3406.

Maul, Stefan M. "Im Rückwärtsgang in die Zukunft." *Spektrum der Wissenschaft* 8 (2010): 72–77.

McKeegan, Ian S. "Early Solar System Chronology." In *Meteorites, Comets, and Planets: Treatise on Geochemistry*, edited by Andrew M. Davis, 2nd ed., vol. 1, Amsterdam: Elsevier, 2005.

Medenbach, Olaf, and Harry Wilk. *Zauberwelt der Mineralien*. Künzelsau, Germany; Würzburg, Germany; Thalwil, Switzerland; and Salzburg, Austria: Sigloch, 1977.

Meyer, Jürg. *Gesteine der Schweiz: Der Feldführer*. Bern, Switzerland: Haupt, 2017. https://d-nb.info/112979959X.

Moral-Cardona, J. P., A. Sanchez Bellón, F. López-Aguayo, and M. A. Caballero. "The Analysis of Quartz Grain Surface Features as a Complementary Method for Studying Their Provenance: The Guadalete River Basin (Cádiz, SW Spain)." *Sedimentary Geology* 106, no. 1 (October 1, 1996): 155–64. https://doi.org/10.1016/0037-0738(95)00155-7.

Moral-Cardona, J. P., J. M. Gutiérrez Mas, A. Sánchez Bellón, F. López-Aguayo, and M. A. Caballero. "Provenance of Multicycle Quartz Arenites of Pliocene Age at Arcos, Southwestern Spain." *Sedimentary Geology* 112, no. 3 (September 1, 1997): 251–61. https://doi.org/10.1016/S0037-0738(97)00040-7.

Morton, A. C. "Stability of Detrital Heavy Minerals in Tertiary Sandstones from the North Sea Basin." *Clay Minerals* 19, no. 3 (June 1984): 287–308. https://doi.org/10.1180/claymin.1984.019.3.04.

Nesse, William. *Introduction to Optical Mineralogy*. 4th ed. New York: Oxford University Press, 2013.

Norton, O. Richard. *Rocks from Space: Meteorites and Meteorite Hunters*. 2nd ed. Missoula, MT: Mountain, 1998.

Okrusch, Martin, and Siegfried Matthes. *Mineralogie: Eine Einführung in die spezielle Mineralogie, Petrologie und Lagerstättenkunde*. 8th ed., completely rev., expanded, and updated. Berlin and Heidelberg, Germany: Spektrum, 2009.

Pettijohn, Francis John, Paul Edwin Potter, and Raymond Siever. *Sand and Sandstone*. New York: Springer US, 1973. https://doi.org/10.1007/978-1-4615-9974-6.

Plato. "Parmenides." In *Phaidros, Parmenides, Theaitetos, Sophistes: Sämtliche Werke*. Vol. 4. Translated by Friedrich Schleiermacher and Hieronymus Müller. Edited by Walter F. Otto, Ernesto Grassi, and Gert Plamböck. Reinbek bei Hamburg, Germany: Rowohlt, 1989.

Poldervaart, Arie. "Chemistry of the Earth's Crust." In *Crust of the Earth: A Symposium*. Edited by A. Poldevaart, 62:119–44. Geological Society of America Special Paper, 1955. https://doi.org/10.1130/SPE62-p119.

Prothero, Donald R. "Zirkone—Zeugen der frühen Erdgeschichte." *Spektrum der Wissenschaft* 9 (2018): 60–65.

Pye, Kenneth. "Post-depositional Reddening of Late Quaternary Coastal Dune Sands, North-Eastern Australia." *Geological Society, London, Special Publications* 11, no. 1 (January 1983): 117–29. https://doi.org/10.1144/GSL.SP.1983.011.01.13.

Pye, Kenneth, and Haim Tsoar. *Aeolian Sand and Sand Dunes*. London: Unwin Hyman, 1990. http://bvbr.bib-bvb.de:8991/F?func=service&doc_library=BVB01&local_base=BVB01&doc_number=002739415&line_number=0001&func_code=DB_RECORDS&service_type=MEDIA.

Ramberg, I. B., I. Bryhni, A. Nottvedt, and K. Rangnes. *The Making of a Land—the Geology of Norway*. Trondheim: Geological Society of Norway, 2008.

Rebecchi, Lorena, Tiziana Altiero, Michele Cesari, Roberto Bertolani, A. M. Rizzo, Paola Corsetto, and Roberto Guidetti. "Resistance of the Anhydrobiotic Eutardigrade Paramacrobiotus Richtersi to Space Flight (LIFE–TARSE Mission on FOTON M3)." *Journal of Zoological Systematics and Evolutionary Research* 49 (April 21, 2011): 98–103. https://doi.org/10.1111/j.1439-0469.2010.00606.x.

Rhode, Andrea. *Auf Fossiliensuche an der Ostsee: Gesteine, Fossilien, Fundmöglichkeiten*. Neumünster, Germany: Wachholtz, 2008.

Ritter, M. "Goethe und Augustin-Pyramus De Candolle." In *Abenteuer der Vernunft. Goethe und die Naturwissenschaften um 1800*. Dresden, Germany: Sandstein Verlag, 2019. http://deposit.dnb.de/cgi-bin/dokserv?id=054aea2096cf4398972574725ecadbe9&prov=M&dok_var=1&dok_ext=htm.

Roedder, Edwin. *Fluid Inclusions*. Reviews in Mineralogy. Washington, DC: Mineralogical Society of America, 1984.

Rosenbusch, Harry. *Mikroskopische Physiographie der Mineralien und Gesteine: Ein Hülfsbuch bei mikroskopischen Gesteinsstudien*. 5th rev. ed. Stuttgart: Schweizerbart, 1924.

Rothe, Peter. *Gesteine: Entstehung–Zerstörung–Umbildung*. 3rd rev. ed. Darmstadt: Wiss. Buchges, 2010. https://d-nb.info/1004978928/04.

Rovelli, Carlo. *Die Ordnung der Zeit*. Reinbek bei Hamburg, Germany: Rowohlt, 2018. https://d-nb.info/1156668018/04.

Rubin, Alan E. "Urtümliche Meteoriten," *Spektrum der Wissenschaft* 3 (2014): 44–49.

Schlichting, H. Joachim. "Musikalischer Sand." *Spektrum der Wissenschaft* 12 (2017).

Schmincke, Hans Ulrich. *Vulkanismus*. 4th rev. ed. Darmstadt: Primus, 2013. https://nbn-resolving.org/urn:nbn:de:101:1-2014102215709.

Schopenhauer, Arthur. *Die Welt als Wille und Vorstellung*. Vol. 1, part 2. Cologne: Könemann, 1997.

Schrott, Raoul. *Erste Erde Epos*. Munich: Carl Hanser Verlag, 2016.

Schultz, Ludolf, and Jochen Schlüter. *Meteorite*. Darmstadt: Primus, 2012.

Schumann, Hilmar, Friedrich Rinne, and Max Berek. *Anleitung zur allgemeinen und Polarisations-Mikroskopie der Festkörper im Durchlicht*. 3rd ed. Stuttgart: E. Schweizerbart'sche Verlagsbuchhandlung, 1973. http://digitale-objekte.hbz-nrw.de/storage/2009/02/01/file_3/2789779.pdf.

Schumann, Walter. *Der große BLV Steine- und Mineralienführer: Mineralien, Erze, Edelsteine, Gesteine, Meteorite*. 7th rev. ed. Munich: blv, 2007. http://media.obvsg.at/AC05810969-1001.

Schwarzbach, Martin. "Glazigene Sichelmarken als Klimazeugen (Nach Beobachtungen in Finnland)." *E&G Quaternary Science Journal* 28, no. 1 (January 1, 1978): 109–18. https://doi.org/10.3285/eg.28.1.09.

Siever, Raymond. *Sand, ein Archiv der Erdgeschichte*. Spektrum-Bibliothek 22. Heidelberg, Germany: Spektrum d. Wiss, 1989. https://d-nb.info/891104275/04.

Smed, Per. *Steine aus dem Norden: Geschiebe als Zeugen der Eiszeit in Norddeutschland*. 2nd rev. ed. Berlin: Borntraeger, 2002. https://d-nb.info/963763652/04.

Soentgen, Jens, and Knut Völzke. *Staub—Spiegel der Umwelt*. Stoffgeschichten 1. Munich: Oekom Verlag, 2006. http://swbplus.bsz-bw.de/bsz254876315inh.pdf.

Stevic, Marijana. "Identification and Environmental Interpretation of Microtextures on Quartz Grains from Aeolian Sediments: Brattforsheden and Vittskövle, Sweden." *Dissertations in Geology at Lund University*, 2015. http://lup.lub.lu.se/student-papers/record/5468853.

Stow, Dorrik. *Sedimentgesteine im Gelände: Ein illustrierter Leitfaden*. Heidelberg, Germany: Spektrum, Akademie-Verlag, 2008. https://d-nb.info/989120783/04.

Sturm, Andreas. "Sand wird knapp." *Heilbronner Stimme*, February 18, 2019.

Sturm, Robert. "Im Mineral eingesperrt—Mikroskopie von Einschlussphasen in magmatischen Kristallen." *Mikrokosmos* 98 (2009): 283–88.

Tetsch, Larissa. "Wiederbelebung bei Bärtierchen." *Spektrum der Wissenschaft* 6 (2017): 22–24.

Thomas, Dylan, and Hans Bender. *Unter dem Milchwald: Ein Spiel für Stimmen* (*Under the Milkwood: A Play for Voices*). Translated by Erich Fried. Stuttgart: P. Reclam, 1971.

Thureau-Dangin, François. *Les Cylindres de Goudéa*. Textes cunéiformes. Paris: Paul Geuthner, 1925. http://pi.lib.uchicago.edu/1001/dig/neh2/32.

Trepmann, C. A., and J. G. Spray. "Post-shock Crystal-Plastic Processes in Quartz from Crystalline Target Rocks of the Charlevoix Impact Structure." Paper presented at the 35th Lunar and Planetary Science Conference, March 15–19, 2004, League City, TX. Abstract 1730, March 1, 2004.

Tröger, Walter Ehrenreich. *Optische Bestimmung der gesteinsbildenden Minerale*. 2nd ed. Stuttgart: Schweizerbart'sche Verlagsbuchhandlung, 1969.

Tucker, Maurice E. *Methoden der Sedimentologie*. Stuttgart: Enke, 1996.

Udayaganesan, P., N. Angusamy, A. R. Gujar, and G. V. Rajamanickam. "Surface Microtextures of Quartz Grains from the Central Coast of Tamil Nadu." *Journal of the Geological Society of India* 77, no. 1 (January 1, 2011): 26–34. https://doi.org/10.1007/s12594-011-0005-8.

Vaas, Rüdiger. "Gehirn und Zeit: Ein Streifzug durch die Dimension unseres Erlebens." *Der Blaue Reiter* 5 (1997): 17–23.

Vinx, Roland. *Steine an deutschen Küsten: Finden und Bestimmen*. Wiebelsheim, Germany: Quelle & Meyer Verlag, 2016. https://d-nb.info/1081093412/04.

Vos, K., N. Vandenberghe, and J. Elsen. "Surface Textural Analysis of Quartz Grains by Scanning Electron Microscopy (SEM): From Sample Preparation to Environmental Interpretation." *Earth-Science Reviews* 128 (January 1, 2014): 93–104. https://doi.org/10.1016/j.earscirev.2013.10.013.

Weizsäcker, Carl Friedrich von. *Aufbau der Physik*. Munich: C. Hanser, 1985.

Weizsäcker, Carl Friedrich von. "Einige Begriffe aus Goethes Naturwissenschaft." In *Goethes Werke: Naturwissenschaftliche Schriften I*, 13:159. Hamburger ed. Munich: C. H. Beck, 1994.

Welland, Michael. *Sand: The Never-Ending Story*. Berkeley: University of California Press, 2009.

Welsch, Norbert, Claus Liebman, and Jürgen Schwab. *Erde und Leben: Die Geschichte einer innigen Wechselbeziehung*. Berlin: Springer, 2017.

Willkomm, Moritz. *Die Wunder des Mikroskops oder die Welt im kleinsten Raume: Für Freunde der Natur und mit Berücksichtigung der studirenden Jugend bearbeitet*. 4th ed., rev. and enlarged. Leipzig: Spamer, 1878. https://digital.deutsches-museum.de/item/BV006313924/.

Wittgenstein, Ludwig. *Tractatus logico-philosophicus*, proposition 6.4311. Oxford: Basil Blackwell, 1959.

Woolley, Leonard. *Ur und die Sintflut: Sieben Jahre Ausgrabungen in Chaldäa, der Heimat Abrahams*. Leipzig: F. A. Brockhaus, 1930.

11.5 Acknowledgments

I'd first like to thank my whole family, especially my wife, who saw and endured my preoccupation with books, cameras, microscopes, ants, stones, stars, birds, meteorites, myxomycetes, tardigrades, and much more, now supplemented by the quite strange and additional preoccupation with the almost invisible, sand, over several years and over hundreds of hours. My very special thanks go to my father, who with inexhaustible patience created adapters and connecting parts for all the individual parts of microscope optics that never fit together, and without whose advice and help I often would not have been able to continue. I'd also like to thank my microscope retailer and expert Mr. Harald Rasche, who left no stone unturned to get hold of the components and optics of years gone by that I coveted, and with the help of which all the photomicrographs in this book were made. I would like to thank Dr. Olaf Medenbach, Ruhruniversität Bochum, twice. Since 1977, his book *The Magic of Minerals* had fascinated me in such a way that I subsequently started to deal more intensively with minerals and rocks, not knowing that forty-five years later I now must thank him for his critical review of my own book. Apart from collecting samples, my old school friend Wolfgang Antes really did not contribute anything to the subject of sand; nevertheless, the Saturday discussions were and are for me formative and completely irreplaceable—may they continue to accompany us in the future. Furthermore, my thanks go to those who have brought me sand samples from their travels in almost all parts of the world. Many of them have been collected, labeled, and documented with loving care, some even under quite adventurous conditions. Without you, this book would not have been possible. I have listed only the initials (first name, surname, title, if applicable), so that those who wish to remain unrecognized remain that way:
Prof. Dr. N.A., Prof. Dr. M.A., W.A. A.A, E.B., R.B., M.B., R.B., M.B, R.B., Prof. Dr. T.B., B.B., Dr. A.B., S.B., Prof. Dr. R.B., F.B., A.B., A.C., Prof. Dr. A.D., H.-W.D., S.D., M.D., M.D., H.E., U.E., Prof. Dr. W.E., B.F., U.F., Prof. Dr. R.F., G.F., T.F., T.G., A.G., N.G., J.G., T.G., G.H., S.H., F.H., B.H., C.H., G.H., B.K., H.K., Dr. G.K., U.K., H.K., J.K., U.K., H.K., Prof. Dr. K.K, L.K., H.K., B.K., D.K., Prof. Dr. H.K., A.K., D.L., Dr. H.L., Prof. Dr. D.L., M.L., H.L., T.M., Dr. P.M., A.M., S.M., R.M., Prof. Dr. A.M., M.M., K.N., H.Ö., M.P., B.P., U.P.-R. W.P., S.R., Prof. Dr. K.R., Prof. Dr. D.R., E.R., Dr J.R., C.R., P.R., K.S., S.S., I.T., E.T., I.V., W.W., Prof. Dr. W.W., Prof. Dr. M.W., K.W., E.W., Prof. Dr. J.W., Prof. Dr. H.W., K.W., B.Z., C.Z.

Finally, I'd like to thank my editor, Dr. Martin Lind, from Haupt Verlag, who has patiently and encouragingly accompanied the project over the course of a decade and has also been able to prevent many a turn into quicksand through gentle encouragement. It was always a great pleasure and enrichment to exchange ideas with him.

11.6 List of Image Sources

All images, unless otherwise noted, were created by the author and made for this book. The figures 1.14, 6.3, 6.6, 6.7, 6.8, 6.15, 6.16, 6.22 (*top left*), 6.40, 6.43, 6.53, 6.63, 7.17, 7.25, 7.26, 7.46, 7.50, 7.64, 7.635, 7.70, 7.82, 7.84, and 7.93 come from the book *Sand* by Oliver Lenzen (Cologne, 2020). The double-page landscape shots before chapters 1, 3, and 7 were taken by Prof. Dr. Wolfgang Elmendorf, and the ones before chapters 2 and 8 were made available by Dr. Martin Lind. Thank you to both.

11.7 Location Index, Sand Samples

The page number refers to pictures of sand grains or overview pictures of sand samples. The latter are highlighted in **bold**. The years written in *italics* refer to the year the sand sample was taken, if known. In the addendum to the book, there is a world map with the marked locations of the sand samples.

11.8 Index

The page numbers in **bold** refer to detailed images and definitions.